POLYMER SCIENCE AND TECHNOLOGY
Volume 14

BIOMEDICAL AND DENTAL APPLICATIONS OF POLYMERS

POLYMER SCIENCE AND TECHNOLOGY

Recent volumes in the series:

Volume 8 POLYMERS IN MEDICINE AND SURGERY
Edited by Richard L. Kronenthal, Zale Oser, and E. Martin

Volume 9 ADHESION SCIENCE AND TECHNOLOGY (Parts A and B)
Edited by Lieng-Huang Lee

Volume 10 POLYMER ALLOYS: Blends, Blocks, Grafts, and Interpenetrating Networks
Edited by Daniel Klempner and Kurt C. Frisch

Volume 11 POLYMER ALLOYS II: Blends, Blocks, Grafts, and Interpenetrating Networks
Edited by Daniel Klempner and Kurt C. Frisch

Volume 12 ADHESION AND ADSORPTION OF POLYMERS (Parts A and B)
Edited by Lieng-Huang Lee

Volume 13 ULTRAFILTRATION MEMBRANES AND APPLICATIONS
Edited by Anthony R. Cooper

Volume 14 BIOMEDICAL AND DENTAL APPLICATIONS OF POLYMERS
Edited by Charles G. Gebelein and Frank F. Koblitz

POLYMER SCIENCE AND TECHNOLOGY
Volume 14

BIOMEDICAL AND DENTAL APPLICATIONS OF POLYMERS

Edited by
Charles G. Gebelein
Youngstown State University
Youngstown, Ohio

and
Frank F. Koblitz
Dentsply International
York, Pennsylvania

PLENUM PRESS • NEW YORK AND LONDON

Library of Congress Cataloging in Publication Data

American Chemical Society Symposium on Biomedical and Dental Applications of Polymers, Houston, Tex., 1980. Biomedical and dental applications of polymers.

(Polymer science and technology; v. 14)

"Based on an American Chemical Society Symposium on Biomedical and Dental Applications of Polymers, held March 23-28, 1980, at the 179th national meeting in Houston, Texas."

Includes index.

1. Polymers in medicine—Congresses. 2. Polymers in dentistry—Congresses. I. Gebelein, Charles G. II. Koblitz Frank F. III. American Chemical Society. IV. Title. I. Series. [DNLM: 1. Biocompatible materials—Congresses. 2. Biomedical engineering —Congresses. 3. Dental materials—Congresses. 4. Polymers—Congresses. QT 34 A111b 1980]

R857.P6A17 1980 610.'.28 80-29429

ISBN 0-306-40632-2

Based on an American Chemical Society Symposium on Biomedical and Dental Applications of Polymers, held March 23–28, 1980, at the 179th National Meeting in Houston, Texas

A Division of Plenum Publishing Corporation
227 West 17th Street, New York, N.Y. 10011

Printed in the United States of America

FOREWORD

The development and use of medical and dental materials are highly interdisciplinary endeavors which require expertise in chemistry, materials science, medicine and/or dentistry, mechanics and design engineering. The Symposium upon which this treatise is based was organized to bring members from these communities together to explore problems of mutual interest.

The biomaterials which are used in medical or dental prostheses must not only exhibit structural stability and provide the desired function, but they must also perform over extended periods of time in the environment of the body. The latter is a very stringent requirement. The oral and other physiological environments are designed by nature to break down many organic substances. Also of importance is the requirement that materials used in the prosthesis not have a deleterious effect on body tissues. Most foreign (to the body) substances are somewhat toxic to human tissues; in fact, few factors are more limiting in the medical prosthesis field than the biocompatibility problem. Some of these problems and the attempts to solve them are discussed in this volume.

In this book particular attention is given to polymers as biomaterials. Most engineering behavior of materials is complex and our understanding of the exact mechanisms and molecular processes which are involved is incomplete. The behavior is particularly complicated in polymers because of their complex morphology and the strong dependence of properties upon this physical structure, as well as the chemical structure. Time, for example, takes on particular significance for polymers--not only are polymeric materials susceptible to aging (physical and chemical changes with time) but, in addition, most exhibit viscoelastic behavior. When these materials are to be used as load-carrying structural elements, these time effects must be understood in order to predict performance. This is a very demanding task, particularly when we consider that the prosthesis will often be expected to function for years in a very unfavorable environment and the consequences of malfunction are, to say the least, very undesirable.

Materials behavior may be approached from several different points of view. Traditionally, designers have treated materials as continua and have observed their behavior from a phenomenological point of view. Other researchers have tried to explain behavior on a molecular basis. Both approaches have had their advantages and disadvantages. The former provides information and parameters that, in a comparatively direct and straightforward manner, facilitate sizing, optimization of shape, etc. In addition, this method is closely akin to methodology used to design components in other areas of technology. This may give the researchers the confidence that comes with using methods that have been tried and tested. On the other hand, it can be argued that in order to reliably predict performance, the laboratory conditions under which the phenomenological criteria are established or the design parameters are determined, must almost exactly duplicate the intended service conditions for which the part is to be designed. If considerable extrapolation between testing and service conditions is desired it is important to have an understanding of the basic molecular mechanisms involved in behavior. Only in this way can we with any confidence predict how behavior might be changed if environment, time, and loading conditions differ from the test conditions.

Texts such as this, as well as the symposium upon which it is based, provide a forum in which members of the various communities might present for discussion their findings and ideas. Hopefully, this will result in more rapid development and use of improved biomaterials.

K. L. DeVries,
Professor and Chairman,
Department of Mechanical and
Industrial Engineering,
University of Utah,
Salt Lake City, Utah

PREFACE

Nearly three thousand years ago, King David stated, "I will praise Thee; for I am fearfully and wonderfully made: marvelous are Thy works; and that my soul knoweth right well" (Psalm 139:14). Anyone who has considered the design and functions of the human body would readily agree with this observation. Unfortunately, this body is subject to various diseases and/or genetic problems that often produce defects or damage to the body. In such cases, some form of medication or medical treatment is necessary in order to enable the person to pursue a relatively normal life. Through the centuries many approaches have been developed in the attempts to restore health to an ailing human body. In this book we consider the role that polymeric materials are playing in this important line of service.

The present volume is the fruit of seeds planted at the American Chemical Society 179th National Meeting in Houston, Texas, March 23-28, 1980. The symposium from which this present volume is harvested was sponsored by the Division of Organic Coatings and Plastics Chemistry and short versions of most of these papers appeared in their Preprint Volume Number 42. This symposium was unique in that it spanned an unusually wide range of biomedical applications of polymers including artificial organs, cardiovascular uses, dental applications, implants and medication applications. To our knowledge, this was the first American Chemical Society symposium to cover this broad a range of topics and was certainly the first such symposium to dwell at any length on dental applications.

The arrangement of the revised and expanded Chapters in this volume differs considerably from the original order in the symposium in an attempt to make the book more consistent in its development of this broad topic. Basically these thirty five papers have been grouped into four categories: (1) General Biomaterial Applications of Polymers, (2) Cardiovascular Applications of Polymers, (3) Applications of Polymers in Medication and (4) Dental Applications of Polymers. Frequently there is some overlap of the information in one section with that in another section. This is unavoidable and even desirable to an extent. Very often a material used in one application could also have utility in a totally different area.

Likewise, the solution of a problem in one area could aid in overcoming difficulties in another area. One major purpose for this present book, and the original symposium, is to spread information amongst a broader body of scientists then normally would occur at a more specialized meeting which might concentrate nearly exclusively on a more restricted area, such as cardiovascular uses or dental applications. Although not every area of the biomedical applications of polymers is discussed in this book, we do feel that this goal has been accomplished. This book does cover most of the major applications of polymers in medicine, and the papers have been written by leading scientists in these fields.

These chapters present overviews, historical background, state of the art and current results. They represent work done in academic, industrial, government, medical or dental school and research institute laboratories in the United States, Canada, England and Italy and refer to work done in other countries as well. Approximately half the papers are from academic laboratories.

We wish to thank Drs. Robert Lalk, Larry Thompson and the other officers of the Organic Coatings and Plastics Division for their support. We also wish to acknowledge the contributions of the session chairmen: Drs. J. M. Anderson, K. L. DeVries, R. I. Leininger and J. M. Whiteley. We wish to thank the authors and their supporting institutions for their excellent contributions and these are acknowledged in the Table of Contents and in each Chapter heading. Several members of the Central Research Laboratory of Dentsply International, Inc. assisted in presenting the symposium and in preparing this book. Mrs. Rita Loveland was especially helpful. Last, and certainly not least, we wish to thank our families for bearing with us while we were going through 'labor pains' to give birth to this book. Their patience and warm smiles helped far more than any salve we might have used. We hope sincerely that this book will help to advance biomedical science and help to alleviate human suffering.

Charles G. Gebelein,
Department of Chemistry,
Youngstown State University,
Youngstown, Ohio, 44555
and
Department of Pharmacology,
Northeastern Ohio Universities,
College of Medicine,
Rootstown, Ohio

and

Frank F. Koblitz,
Central Research Laboratory
Dentsply International, Inc.
York, Pennsylvania, 17405

CONTENTS

SECTION II

SECTION III

SECTION I

GENERAL BIOMATERIAL APPLICATIONS OF POLYMERS

Although this entire book deals with the biomaterial applications of polymers, the cardiovascular, medication and dental applications are grouped together in special sections. This section is concerned with all other biomedical applications of polymers. Basically the plan of this section is to survey the entire biomedical polymer field (Gebelein) and then consider some specific problems with implants (Anderson). Following this, the areas of artificial lungs (Gray), temporary skin substitutes (Wang/Samji), poly(amido-amine) applications (Ferruti/Marchisio) and covalent bonding of collagen and acrylates (Lloyd/Burns) are examined. This section then concludes with a paper on the clinical analysis applications of polymer coated oxygen sensing devices (Hahn et al). Thus this section considers artificial organs (lungs, skin and general considerations), new polymer applications in microclinical analysis (oxygen detection in blood, etc.) and the pathological problems with implants. Some aspects of cardiovascular applications (Anderson, Ferruti) and medication applications (Gebelein) are considered also in this section as minor themes.

BIOMEDICAL POLYMERS IN THEORY AND PRACTICE

Charles G. Gebelein

Department of Chemistry
Youngstown State University
Youngstown, Ohio 44555

The field of biomedical polymers continues to show steady growth in both the basic research and applications areas. The only limit to the suggested uses of polymers in medicine seems to be a limit in our imagination. Some recent popular articles suggest that nearly all parts of the human body could be replaced by some type of a plastic and/or metal device (1,2). Many major problems remain to be solved before this vision could become fact, however. A large number of recent surveys and books attest to the great interest in this important field of science (3-31). These books but scratch the surface and hundreds of articles are published each year in a variety of journals including such specialized ones as the Journal of Biomedical Materials Research and the Transactions of the American Society of Artificial Internal Organs. Courses in the field of biomedical polymers and biomaterials are now offered at many different universities (32).

Obviously it would not be possible to cover every detail of biomedical polymers in this paper or even in a single book. The basic purpose of this paper is to overview the various uses of polymers in medically related applications and to note some of the advantages and limitations in these uses. Table I shows some typical biomedical polymer applications. The forty five applications listed in Table I do not comprise every possible application, nor do they even cover all the areas under current research, but they merely indicate the wide range of applications being studied. According to statistics from the University Hospitals of Case-Western Reserve University, 9-15% of the autopsies reveal some type of an implant in the patient (33). If these statistics apply to the USA as a whole, this would mean that 20-34 million people would have some type of implant, exclusive of dental fillings, dentures and contact lenses. Each year there are in excess of 400 million dental fillings made with 33-50% of these utilizing polymers. In addition, over eleven million dentures are made annually and most of these materials are acrylic polymers with some polyurethanes and fluoro-polymers. At the present time this dental polymer market is about $50 million per year (34). The contact lense market was over six million lenses in 1974 (35) and many of these involve hydrogel polymers as 'soft contact lenses' (28,36).

The materials requirements for these biomedical applications vary markedly according to the application being considered and it is improbable that any single material would be useful in all these areas. Possibly the most difficult, and the most spectacular, area of application is in artificial organs. Ideally an artificial organ would be capable of implantation into the body and be able to replace totally the function of a diseased or otherwise disabled organ. In actual fact this is seldom possible with our present state of technological development. More frequently, an artificial organ actually consists of an extra-corporal device, such as a hemodialyzer, which can be attached to a patient. Even where devices can be made small enough to permit implantation, such as an artificial heart,

Table 1

TYPICAL BIOMEDICAL POLYMER APPLICATIONS

Artificial Blood	Heart Assist Devices
Artificial Heart	Heart Valves
Artificial Kidney	Hydrocephalus Shunts
Artificial Limbs	Immobilized Enzymes
Artificial Liver	Implantable Pumps
Artificial Pancreas	Inner Ear Repairs
Artificial Penis	Joint Replacement
Biomedical Polypeptides	Pacemakers
Bladder Replacement	Plasma Extenders
Bone Cements	Plastic Surgery
Bone Replacement	Polymeric Drugs
Casts	Polymeric Food Additives
Catheters	Pseudoenzymes
Contact Lenses	Reinforcing Mesh
Controlled Release Drugs	Replacement Blood Vessels
Cornea Replacement	Replacement Skin
Dental Fillings	Soft Tissue Replacement
Dentures	Surgical Adhesives
Drainage Tubes	Surgical Tape
Drug Administration Devices	Sutures
External Ear Repairs	Testicle Replacement
Eye Lense Replacement	Visual Prothesis
	Wound Dressings

external power sources are necessary. One of the major problems encountered with artificial organs is blood compatibility since many of these organs either handle blood directly (heart, blood vessels) or come into contact with the blood in a membrane exchange reaction (kidneys, lungs, etc.). The question of blood compatibility has been reviewed in several places (37-40) and will be considered further in several articles in this book. Numerous polymeric systems have been explored in regards to blood compatibility in the cardiovascular system but at the present time the most promising materials appear to be some of the polyether urethane ureas (PEUU)(38). These polymers are being studied in several laboratories for use in total artificial hearts, heart assist devices and replacement blood vessels. Most intra-aortic balloon assist devices are made of this polymer. Most large blood vessel replacements, however, are done using Dacron mesh tubing or, to a lesser extent, Teflon tubing. (The use of saphenous veins is usually preferred but often these blood vessels have become too deteriorated or clogged to be of use and an artificial substitute must be used.) These Dacron tubes cannot be used for small blood vessel replacement since the neointima ingrowth would completely fill the tube space and another type of material is needed with the PEUU being the strongest candidate at this time.

In moving organs, such as a heart, the durability of the polymer also becomes critical. An artificial heart will undergo about 368 million flexes in a decade but few, if any, polymeric materials can survive this severe a test. (This same problem also exists for heart valve material in addition to the blood compatibility problems.) While much progress has been made on artificial hearts, the fact remains that these devices have not functioned continuously for ten years and they always require an external energy source for the pumping action (usually compressed air). Obviously any cardiovascular device must also exhibit compatibility with the other body fluids and tissues. Several other articles in this book will deal with this area extensively.

Much research is in progress on other artificial organs such as kidneys and lungs but most of these are a long way from being an implantable replacement organ even if this is the ultimate goal of some of this research. The problems here are mainly to achieve adequate blood compatibility along with good membrane characteristics. Generally the desired surgical route for kidney replacement is a transplant but this does require careful donor-recipient matching and this is not always possible. The net result is that the patient must be on a hemodialyzer for many years even if this is considered to be a temporary measure. This problem has been alleviated somewhat by the development of small, wearable devices (41) but a completely implantable, permanent device would be more desirable. Unfortunately, this does not appear likely to be developed in the immediate future.

The situation with artificial lungs is even poorer and only fixed external devices are likely in the near future. The lung is a much more difficult organ to replace for several reasons. While one could survive for a fairly long period of time without functioning kidneys, death would occur in only a few minutes if the functions of the lung were to cease. In addition, the blood-membrane interface area is at least fifteen times greater with the lung than with the kidney. As a consequence, blood compatibility becomes even more critical here. Mechanical devices, such as bubble oxygenators, tend to damage the blood and, in addition, would be an unlikely disign for implantation. Many styles of membrane devices have been made utilizing several different types of polymeric materials for the oxygen exchanging membrane (42,43). These will be discussed in more detain in a later chapter of this book section. In any event, these present 'artificial lungs', both the current heart-lung machines and the experimental laboratory devices, are extra-corporal and implanted replacement lungs are still future. We might note, however, that even transplanting 'natural lungs' is much more difficult than kidney transplants.

Many of the other biomedical applications of polymers involving implants and related functions do not have direct blood contact as occurs with the artificial organs and the problem of long term blood compatibility is not nearly as improtant. Nevertheless, these applications also have specific requirements and problems. Polymers for wound dressings and/or artificial skin must have the flexibility and permeability properties of natural skin and also be able to maintain these features for long time periods. Obviously, this time requirement is greater with an artificial skin which must also be capable of matching the color of the remaining skin and not undergoing discoloration on exposure to light, heat or the normal everyday environment. A wound dressing or a temporary skin replacement, on the other hand, must permit proper healing of the subdermal tissue, which had been damaged by a burn or other source of injury, and permit ready skin grafting at a future time. These requirements have been summarized in some recent papers (18,44,45) and wound dressings are discussed further in another paper in this book.

A major use of plastics in surgery is to replace soft tissue such as a prosthetic breast, testicles, etc. The major polymer used here is poly(dimethylsiloxane) and these uses have been summarized in some recent articles (46,47). The major requirements for soft tissue replacements are (a) a consistency similar to the natural tissue being replaced, (b) no change in this consistency on aging in the body, (c) no fibrous ingrowth and (d) no adverse reactions with the body tissues. Porous and/or woven materials are usually unsuitable for this use since fibrous ingrowth normally leads to a pronounced loss in flexibility and a hard feel to the implant. Soft tissue implants usually do not come into direct contact with the blood and this problem is not particularly critical in this use. In a similar manner, soft tissue replacements normally do not require the material to be permeable. (On the contrary, this would usually be a disadvantage for these implants.)

Much progress has been made in recent years on artificial limbs and joint replacement. Generally the larger prosthetic joints (such as a hip) consist of a plastic socket, often made from Teflon or polyethylene, and a metal ball which is affixed to the other part of the joint (10,11,23). In many cases these joints work very well but the plastic often wears away permitting play in the joint and a subsequent looseness in the fit of the joint. Smaller joints (as in a finger) are usually made from specially designed strips of a polysilicone or a poly(1,4-hexadiene) material which sometimes are reinforced with metal. These can undergo many thousands of flexing operations and these replacements can alleviate a severe arthritic condition (48-50). It appears that reasonable satisfactory materials exist for these smaller joint replacements but that better materials are still needed for the larger joint prostheses. While part of this difficulty may be a design problem, a more durable plastic is needed here in order to achieve long term joint stability. We might note in passing that some major advances are being made in motor powered artificial limbs and some of this work, especially that involving myoelectric interactions, borders on the fantastic. This area is, however, primarily an engineering problem and is beyond the scope of this present article.

The application of polymers in delivering medications is a relatively new development. There are several ways this can be accomplished including (a) plastic drug administration devices, (b) biomedical polypeptides, (c) controlled release systems, (d) synthetic enzyme-like polymers, (e) bound enzymes and (f) polymeric drugs. The primary object in these areas is to treat a diseased organ or tissue more directly or selectively without producing an adverse interaction with other parts of the body which would result in undesired toxic side effects. In principle, it would be possible to develop long term medication systems using these polymeric approaches. For example, a birth control device has been developed which can deliver the antifertility drug directly to the uterus in a very small dosage level for more than a year. The use of these smaller dose levels reduces the risk of undesired side effects, such as heart attacks, while placing this drug directly at the target site organ allows this lower concentration to be effective (51).

A number of laboratories are attempting to synthesize new polypeptides that often are variations of natural hormonal polypeptides in the hope of achieving a material that will have a specific medical function. Many derivatives and variations have been made for such hormones as oxytocin, kallidin, corticotrophin and others with this goal in mind. Much recent research has centered on the endorphins and enkephalins as possible new pain relievers.

Some of these synthetic variations have over 850 times the analgesic activity of morphine and over 28,000 times the activity of the natural methionine enkephalin (52).

Much research has been done on placing a drug into a polymeric matrix in order to control the release of this drug in a physiological environment. These controlled release systems have been studied for the release of many different types of drugs including birth control agents, hypertension agents, steroids, arrhythmia agents and fluoride ions for treating a wide variety of illnesses. The basic concept here is that the drug will diffuse through the polymeric matrix and be released into the diseased area at a relatively constant concentration for a prolonged period of time. A detailed discussion of this area is beyond the scope of this article but these systems do show promise as a new application of polymers in medicine (27-29).

Many researchers have studied synthetic analogs of enzymes in hope of developing a biologically active enzyme-like system. Many of these systems developed do show a high degree of catalytic activity and this area could lead to medical applications in the future. Several recent reviews have appeared in this area (53-55). Much work has also been done on bound enzymes and related species for possible medical applications but a review of this is beyond our present scope and several reviews have appeared recently in this area (56-58).

Polymeric drugs comprise a relatively new approach to the use of polymers in medication. In this case, the drug unit is attached to or contained within the polymer backbone. These polymeric drugs could act as a source of controlled release of the drug or could be biologically active as the polymer. Examples of each case can be found in the literature. Several papers in this book will consider this area in more detail and several reviews are available (30, 59-63) for this promising approach to medication.

In conclusion we note that there are a wide variety of current applications of polymers in medicine and that much research is in progress. While essentially all of these applications still need improvement, these biomedical applications of polymers are helping to alleviate human suffering and future advances will aid in achieving this goal more fully.

REFERENCES

1. J. A. Miller, Science News, 112, 154 (1977)
2. M. Pines, Smithsonian, 9(8), 50 (1978)
3. H. J. Sanders, C&EN, April 5, 1971, p. 31; April 12, 1971, p. 68
4. S. N. Levine, Ann. N.Y. Acad. Sci., 146, 3 (1968)
5. B. D. Halpern, Ann. N.Y. Acad. Sci., 146, 193 (1968)
6. J. Autian, Ann. N.Y. Acad. Sci., 146, 251 (1968)
7. Y. Nose, P. Phillips & W. J. Kolff, Ann. N.Y. Acad. Sci., 146,

271 (1968)
8. H. F. Mark, Pure & Appl. Chem., 16, 201 (1968)
9. G. W. Hastings, Chemistry in Britain, 7, 119 (1971)
10. D. A. Sonstegard, L. S. Matthews & H. Kaufer, Sci. Am., Jan. 1978, p. 44
11. J. Charnley, Plastics & Rubber, 1(2), 59 (1976)
12. W. A. Mackey, J. A. MacFarlane & M. S. Christian, editors, "Arterial Surgery", Pergamon Press, New York (1964)
13. L. Stark & G. Agarwal, editors, "Biomaterials", Plenum Press, New York, (1969)
14. Y. Nose & S. N. Levine, editors, "Cardiac Engineering", Interscience, New York, (1970)
15. M. Gutcho, "Artificial Kidney Systems", Noyes Data Corp., Park Ridge, N.J. (1970)
16. S. N. Levine, editor, "Advances in Biomedical Engineering and Medical Physics", Vol. 4, Interscience, New York, (1971)
17. T. D. Sterling, E. A. Bering, Jr., S. V. Pollack & H. G. Baughan, Jr., editors, "Visual Prosthesis", Academic Press, New York (1971)
18. A. L. Bement, Jr., editor, "Biomaterials", U. Washington Press, Seattle (1971)
19. H. Lee & K. Neville, "Handbook of Biomedical Plastics", Pasadena Press, Pasadena, CA (1971)
20. A. Rembaum & M. Shen, editors, "Biomedical Polymers", M. Dekker, New York (1971)
21. B. Block & G. W. Hastings, "Plastic Materials in Surgery", C. C. Thomas, Springfield, IL (1972)
22. H. E. Stanley, editor, "Biomedical Physics and Biomaterials Science", M.I.T. Press, Cambridge, MA (1972)
23. S. F. Hulbert, S. N. Levine & D. D. Moyle, editors, "Materials and Design Consideration for the Attachment of Prostheses to the Musculo-Skeletal System", Interscience, New York (1973)
24. J. W. Boretos, "Concise Guide to Biomedical Polymers", C. C. Thomas, Springfield, IL (1973)
25. J. C. Deaton, "New Parts for Old", Franklin Publ., Palisade, NJ, (1974)
26. H. P. Gregor, editor, "Biomedical Applications of Polymers", Plenum Press, New York (1975)
27. R. L. Kronenthal, Z. Oser & E. Martin, editors, "Polymers in Medicine and Surgery", Plenum Press, New York (1975)
28. J. D. Andrade, editor, "Hydrogels for Medical and Related Applications", American Chemical Society, Washington, DC (1976)
29. D. R. Paul & F. W. Harris, editors, "Controlled Release Polymeric Formulations", American Chemical Society, Washington, DC (1976)
30. L. G. Donaruma & O. Vogl, editors, "Polymer Drugs", Academic Press, New York (1978)
31. J. B. Park, "Biomaterials, an Introduction", Plenum Press, New York (1979)
32. C. G. Gebelein, Org. Coatings & Plastics Chem., 41, 243 (1979)
33. J. M. Anderson, Org. Coatings & Plastics Chem., 42, 613 (1980)

34. F. F. Koblitz, Private Communication.
35. J. J. Falcetta, Polymer News, 1, (11/12), 3 (1974)
36. B. J. Tighe, British Polymer J., 1976, 71
37. R. I. Leininger, R. D. Falb & G. A. Grode, Ann. N. Y. Acad. Sci., 146, 11 (1968)
38. D. J. Lyman, Angew. Chem. Internat. Edit., 13, 108 (1974)
39. S. D. Bruck, "Blood Compatible Synthetic Polymers", C. C. Thomas, Springfield, IL (1974)
40. S. D. Bruck, Polymer, 16, 409 (1975)
41. W. J. Kolff in Ref. 27, p. .
42. P. M. Galletti in "Advances in Biomedical Engineering and Medical Physics", Vol. 2, S. N. Levine, editor, Interscience, New York (1968), p. 121
43. D. N. Gray, Org. Coatings & Plastics Chem., 42, 616 (1980)
44. I. V. Yannis & J. F. Burke, J. Biomed. Mat. Res., 14, 65 (1980)
45. P. Y. Wang & N. Samji, Org. Coatings & Plastics Chem., 42, 628 (1980)
46. S. Bradley in Ref. 18, p. 277
47. S. Bradley in Ref. 13, p. 67
48. J. L. Goldner & J. R. Urbaniak in Ref. 23, p. 137
49. Trade Literature, Dow Corning Corp., Midland, MI
50. Trade Literature, Lord Corp., Erie, PA
51. A. Zaffaroni, Chemtech, 6, 756 (1976)
52. S. H. Snyder, C&EN, Nov. 28, 1977, p. 26
53. A. S. Lindsey in "Reviews in Macromolecular Chemistry", G. B. Butler & K. F. O'Driscoll, editors, M. Dekker, New York (1970), p. 1
54. T. Shimidzu in "Advances in Polymer Science", Vol. 23, H. J. Cantow, editor, Springer-Verlag, New York, (1977), p. 55
55. J. A. Pavlisko & C. C. Overberger, Org. Coatings & Plastics Chem., 42, 537 (1980)
56. H. H. Weetal, editor, "Immobilized Enzymes, Antigens, Antibodies and Peptides", M. Dekker, New York (1975)
57. T. M. S. Chang, Chemtech, 5, 80 (1975)
58. G. Manecke & J. Schlusen in Ref. 30, p. 39
59. K. Takemoto, J. Polymer Sci., Symp. 55, 105 (1976)
60. C. G. Gebelein, Polymer News, 4, 163 (1978)
61. S. M. Samour, Chemtech, 8, 494 (1978)
62. H. Ringsdorf, J. Polymer Sci., Symp. 51, 135 (1975)
63. H. G. Batz in "Advances in Polymer Science", Vol. 23, H. J. Cantow, editor, Springer-Verlag, New York (1977), p. 25

SELECTED EXAMPLES OF PATHOLOGIC PROCESSES ASSOCIATED WITH HUMAN POLYMERIC IMPLANTS

James M. Anderson, Ph.D., M.D.

Departments of Pathology and Macromolecular Science
Case Western Reserve University
Cleveland, Ohio 44106

Continuing advances in the design of prostheses and techniques for their implantation have produced impressive results in the length and quality of survival in patients who receive these polymeric prostheses or devices. With the ever increasing number of implants over the past two decades, basic and clinical scientists have used a wide and varied number of techniques to appreciate or determine the suitability of a given polymer for a given implant application. These techniques, aimed at determining the implant/host interaction, have generally been included under the term "biocompatibility". Investigators commonly deal with biocompatibility in terms of whether the implant material or its degradation products, if any, initiate adverse tissue responses in the host or conversely whether deleterious changes in the chemical, physical and/or mechanical properties of the implant material are caused by the host environment. The vast majority of fundamental studies of biocompatibility involve animal models. The ultimate test for biocompatibility of a polymer, device or prosthesis, is human implantation. With this in mind, we have sought to examine implants and tissue obtained by surgical removal from humans or at autopsy. The simultaneous examination of polymeric implants and tissues removed from humans offers an opportunity to directly appreciate the various aspects of the implant/host interaction.

Toward this goal, a formal program of Implant Retrieval and Evaluation has been developed in the Department of Pathology at Case Western Reserve University. The objective of this program has been to analyze, whenever possible, the tissue response and document any material changes of all implants recovered by the Department's Autopsy and Surgical Pathology Services. Assisting in the evaluation of these implants are members of the Department of Macromolecular Science and Department of Biomedical Engineering.

Over the past six years, 2,819 autopsies performed in the Department of Pathology, University Hospitals, Case Western Reserve University, have yielded 323 implants. The overall percentage for this six-year period is 11%. Table shows the statistics for each individual year and it can be seen that from 9 to 15% of the autopsies per year yielded implants. Table 2 lists the common types of implants found at autopsy, and it should be noted that various types of catheters have been included as implants. The main reason for this inclusion is their predilection toward infection, that is, their ability to serve as a pathway for bacteremia or septicemia. Implants retrieved at autopsy come from both children and adults and rarely has an implant been associated with patient morbidity or mortality. Implants retrieved from the Surgical Pathology Service, however, are usually associated with patient morbidity. It is not the purpose of this paper to present or identify the various modes or mechanisms by which polymeric implants may fail when implanted in humans. Rather, the goal of this paper is to identify selected factors related to the polymeric implant which may lead to a compromise in the biocompatibility of the implant and an inappropriate implant/host interaction. Numerous monographs and reviews are now available which deal in much greater depth with the biocompatibility of specific types of prostheses (1-6). Several cases will be presented which illustrate problems associated with various aspects of polymeric implant biocompatibility.

Table 1. Autopsy Implant Retrieval and Evaluation

Year	Autopsies	Implants	Percentage
1978	480	65	14
1977	424	43	10
1976	431	39	9
1975	474	72	15
1974	497	60	12
1973	513	44	9

Table 2. Types of Implants Retrieved at Autopsy

Total Hips	Intubation Tubes
Total Knees	Tracheotomy Tubes
Total Ankles	
Finger Joints	Umbilical Artery Catheters
Orthopaedic Nails	Umbilical Vein Catheters
Orthopaedic Pins	Hyperalimentation Catheters
	Venous Catheters
Cardiac Valves	
Pacemakers	CSF Shunts
Vascular Grafts	Soft Tissue Implants
Septal Defect Patches	Breast Augmentation Implants
Arterial-Venous Shunts	IUDs
	Artificial Testicles

FIBROUS CAPSULE FORMATION

The extent and duration of tissue reactions to polymeric implant materials is important in assessing biocompatibility. In general, the ideal foreign body reaction in soft tissue to a polymeric implant is one in which the acute inflammatory response, marked by polymorphonuclear leukocytes, exudation and hyperemia, resolves quickly with little or no chronic inflammatory phase, marked by lymphocytes, plasma cells or histiocytes, and increasing fibrosis and eventual encapsulation of the implant by fibrous tissue occurs (7-9). Interaction of the tissue with the polymeric implant may result in a continuation of the chronic inflammatory phase or lead to increased fibrosis. This increased fibrosis or over-development of the fibrous capsule may involve adjacent organs or tissues and lead to adverse reactions (10, 11).

Two examples of this phenomenon, a vascular graft and a total hip replacement, show how the fibrous capsule which develops around the implant may involve and extrinsically compress other organs. In these two cases, the fibrous encapsulation of an aorto-femoral vascular graft and polymethylmethacrylate bone cement from the acetabular component of the total hip prosthesis, respectively, have also encapsulated a ureter leading to obstruction of the urine flow from the kidney to the bladder. In each of these cases, the fibrous capsule formation is normal, that is, it is no different histologically from fibrous capsules which form around biocompatible polymeric prostheses.

Fibrous capsule formation with subsequent contraction of the fibrous capsule around the implant is the most frequent complication of mammary augmentation (12). The etiology of this phenomenon remains unknown although investigators have sought to show the dependency of fibrous capsule contraction on migration of silicone from the implant and the presence of myofibroblasts in the capsule (13-15). Gayou has recently shown that capsule thickness or vascularity, collagen alignment, presence of foreign bodies or the presence of an inner cell layer of macrophages did not correlate with the clinical finding of firmness in mammary augmentations (16).

Recent research efforts are being directed toward developing a better understanding of the tissue/implant interface and the role that leukocytes, macrophages, myofibroblasts and fibroblasts play in determining the interactions at this interface and the end result, <u>i.e.</u> fibrous capsule formation. Several groups are using random copolymers of varying composition, charge and hydrophobicity as probes for determining these interactions (17,18). Of particular interest along these lines is the variation in chemotaxis and cell differentiation that may result from variations in the tissue/implant interaction.

INTERFACE INTERACTIONS IN TOTAL JOINT PROSTHESES

In describing the components of total joint prostheses and their interaction with adjacent tissues, many authors identify the articulating interface (metal/polymer), the metallic prosthesis - polymethylmethacrylate bone cement interface and the bone cement-bone interface. Unfortunately, the bone cement-bone interface may be transient with a fibrous capsule eventually interposing itself between the bone cement and the bone to create the bone cement-fibrous capsule-bone interface (19,20). As described earlier, the formation of a fibrous capsule around implants is a common occurrence. In the case of total joint prostheses where in situ polymerizing methyl methacrylate is used as bone cement, this fibrous capsule formation may be accelerated by the heat of polymerization or the toxicity of the monomer, both of which may lead to localized tissue destruction and cell death (21,22). These changes in the bone and tissue can lead to an inflammatory response with the result being fibrous capsule formation and eventually a loosening of the prosthesis.

Aseptic loosening is the most frequent postoperative complication of total joint replacement in the weight-bearing extremities. In general, the peak incidence occurs early, zero to two years postoperatively, or late, beyond four years postoperatively. The purpose of this study was to differentiate microscopic findings at the cement-bone interface in early loosening, usually less than one year, and late loosening, greater than four years. In addition, relationship of the developing fibrous capsule which was interposed between the bone cement and trabecular bone with developing changes in the underlying trabecular bone was evaluated.

Twenty-three loose total joint prostheses (10 hips, 12 knees and 1 ankle) were evaluated. The time postoperatively prior to symptoms was greater than four years in 12 patients, and less than one year in seven patients. The radiographic findings included a radiolucent line around the cement-bone interface and in several cases fracture of the cement and/or metallic component. Microscopically, five criteria were utilized for evaluation of the cement-bone interface.

Early loosening was predominantly characterized by osteoid formation, thickened fibrous capsule at the interface and the trabeculae were randomly oriented to the cement interface. Late loosening was characterized by reorientation of the trabeculae to being parallel to the cement interface with marked trabecular hypertrophy. There was also an increased incidence of cement particulate in the fibrous capsule. Examination of the interface of the non-loose component in late loosening revealed the same trabecular hypertrophy and reorientation, but minimal cement particulate and a thin fibrous capsule. The microscopic findings in early and late loosening are summarized in Table 3.

Microscopic Findings in Early and Late Loosening of Total Joint Replacements

Criteria	Early	Late
Fibrous capsule thickening	++++	+
Osteoid	++	+
Cement Particulate	0	+++
Trabecular Reorientation	0	++++
Trabecular Hypertrophy	0	++++

Early loosening can result from excessive loading of the cement-bone interface with possible trabecular fractures combined with undesired features in the preparation and insertion of the cement. Micromotion leads to a thickened fibrous capsule. During the early postoperative course, the randomly oriented trabecular bone may act as a sufficient energy absorbing medium between the cement and cortical bone unless unduly stressed. With time, the trabeculae respond to stress by reorientation and hypertrophy of the trabeculae. Frequently, the diameters of the trabeculae adjacent to the interface were two to six times that of more distant trabeculae. The cement particulate at the interface ranged from 300 to 600 microns in thickness.

Aseptic loosening which occurs after four years postoperatively may be related to the reorientation and hypertrophy of the trabeculae at the cement-bone interface. During remodelling, unequal trabecular resorption or new bone formation may result in stress concentrations creating fractures of the trabeculae and/or cement. In both early and late loosening, the fibrous capsule (radiolucent line) which is interposed between the bone and cement is part of the normal healing response and may be accelerated in its growth by poor preparation of the implant bed or by microfracture of the bone cement.

The implants reported in these series are representative of how the normal events of wound healing and the end result of the inflammatory response, i.e. fibrosis, may lead to implant failure. It is obvious that new methods and techniques of artificial joint fixation or stabilization are necessary if these joints are to be used in more active individuals for extended periods. It is hoped that investigators working in this area will take advantage of the healing response and incorporate the healing response into the stabilization technique.

ATHEROMA FORMATION IN VASCULAR GRAFTS

A major need in the area of vascular grafts is a biomaterial which will exhibit high patency when used as a replacement for medium or small vessels. Such a material may be expanded polytetrafluoroethylene (PTFE) which in experimental and clinical studies has shown great promise as an arterial and venous substitute (23,24). Unlike Dacron or Teflon grafts, expanded PTFE grafts develop a smooth, thin neointimal lining and this may be the reason for the success of these grafts. Graft failure with PTFE has primarily resulted from progression of atherosclerotic disease at the vessel/graft anastomoses resulting in partial to complete occlusion of the vessel, i.e. junctional stenosis (25).

We have recently found, in several expanded PTFE grafts retrieved from humans, atherosclerotic changes in the neointima which are independent of the anastomoses. Unlike the thin, smooth neointima which usually develops in these grafts and appears to be fibrin, the focal areas of atherosclerosis show cellular proliferation with microfoci of hypocellularity, cellular degeneration, focal calcification and lipid-laden macrophages. These histologic changes are consistent with the early atherosclerotic lesion.

While the occurrence of atheromatous plaques within expanded PTFE grafts may be rare, their presence indicates that the disease for which the patient received the vascular graft, atherosclerosis, may involve the graft material. Thus, the expected life of the graft may be decreased and the graft truly represents a temporary measure in controlling the progression of atherosclerosis and its systemic effects.

WEAR AND PARTICULATE EMBOLIZATION OF PROSTHETIC HEART VALVES

The ability of polymeric materials in functioning implants such as heart valves and total joint prostheses to withstand wear is inadequate at best. Since prosthetic heart valves were first introduced, wear of the component parts has ranked with nonthrombogenic behavior as major criteria for the acceptability of candidate materials for prosthetic heart valves.

The erosion and subsequent embolization of ball occluders from cage-ball cardiac valve prostheses has been reported (26,27). This phenomenon has been noted in two cases from our Implant Retrieval and Evaluation Program and associated with this finding has been the presence of characteristic perivascular foreign body giant cells containing refractile, irregular, nonbirefrigent material in the liver, spleen and kidneys. The material in the foreign body giant cells could not be stained with hematoxylin and eosin, PAS or trichrome stains. Secondary X-ray dispersion analysis with scanning electron microscopy showed the presence of the $K\alpha$ and $K\beta$ peaks for silicon. While it is unclear whether the silicon is from polydimethylsiloxane or silicon dioxide, the identification of silicon within these perivascular foreign body giant cells coupled with the marked variance of the silicon occluder balls is strong evidence for the embolization of wear particles from the occluder ball.

The inability of polymers to withstand wear remains as a major problem in the application of these materials to implantable prostheses. The Teflon sewing rings of cardiac valve prostheses have also been noted to undergo degradation and embolization to the liver and spleen (28). While Teflon is generally believed to be inert in the body, the mechanism of degradation is unknown.

INFECTIONS AND POLYMERIC IMPLANTS

The infectious susceptibility of implants remains as a major unsolved problem in the area of implants (29,30). Implants may serve as a route for the entry of an infectious agent or may serve as a preferential site for the deposition of the agent with subsequent growth, embolization if in the cardiovascular system or infiltration into adjacent tissue. In many of these cases, removal of the prosthesis is necessary to achieve a cure.

Nearly every type of prosthesis used in humans has been associated with serious infections. These include orthopedic protheses (joint space or bone infections), prosthetic heart valves (endocarditis), vascular prostheses (graft infections and bacteremia), intraocular lens prostheses (endophthalmitis), hemodialysis (hepatitis, bacteremia and shunt site infections), cerebrospinal fluid shunts (ventriculitis), and all types of catheters. In many cases, fulminant infectious processes may be present in patients who present clinically with minimal signs and symptoms.

While we have seen numerous cases of implant related infections, two cases point out the common problems associated with the diagnosis and therapy of these infections. The first case involved the growth of a large candida fungus ball on a pacemaker electrode lead in the right ventricle. The fungus had invaded the endocardium of the right ventricle, the pulmonic valve and the tricuspid valve. This patient had a chronic indwelling urinary catheter and at autopsy the bladder was infected with candida. It is proposed that the urinary catheter provided access for the candida, the patient developed candida septicemia or fungemia, which led to involvement of the pacemaker lead and infiltration of the cardiac tissues. This is a common pathway by which infectious agents may gain access to the blood and organs. It should be noted that at no time was the extent of this patient's problems appreciated and there was no clinical evidence of the cardiac infection. This is a common finding where the involvement of blood and organs by infectious agents associated with implants may be marked but the clinical presentation is minimal. For this reason, implant associated infections have a significant morbidity and mortality.

The other case involves the infection of a Beall prosthetic heart valve. Prosthetic valve endocarditis is a serious complication of cardiac valve replacement and may lead to high rates of mortality because of the occult nature of this disease process and the ability of these infections to serve as nidi for the dissemination of the infectious agents.

It is clear that the involvement of implants by infections is a major problem in the clinical application of these devices and prostheses. While numerous clinical methods and techniques will be used to reduce the morbidity and mortality of these infections, there exists an opportunity for the polymer scientist to develop infection resistant materials or sustained release devices which can be used to combat these infections (31).

CONCLUSION

While these cases are uncommon to rare in their occurrence, they represent factors which must be addressed in determining the biocompatibility of polymeric implants in humans. From another perspective, the altered implant/tissue interactions presented in these cases offer an opportunity to the polymer scientist who wishes to examine these complex problems. Included in these problems is an elucidation of those tissue and material variables which lead to variations in fibrous capsule formation. Little research has been done in this area and yet, since almost all human implants are partially or totally encapsulated, this area would appear to be of prime importance in determining the biocompatibility of human implants. The other major area which appears to be common to the majority of implants is the infectious susceptibility of implants. Does the implant/tissue interaction predispose the tissue to an infectious process through a biochemical or physiologic mechanism? What is the influence of the surface properties of the implant on interaction with bacteria and other organisms?

As better materials and more sophisticated devices are implanted in humans, new problems in the biocompatibility of these materials and devices will appear as a result of the improved functional longevity of the materials and devices. These so-called second generation problems will be recognized and solved by a better understanding of the tissue/implant interaction.

REFERENCES

1. P. N. Sawyer and M. J. Kaplitt, editors, "Vascular Grafts", Appleton-Century-Crofts, New York, (1978).
2. H. Dardik, editor, "Graft Materials in Vascular Surgery", Year Book Medical Publishers, Chicago, Illinois, (1978).
3. B. C. Syrett and A. Acharya, editors, "Corrosion and Degradation of Implant Materials", American Society for Testing and Materials, Philadelphia, Pa., (1979).
4. G. Chapchal, editor, "Arthroplasty of the Hip", G. Thieme, Stuttgart, West Germany, (1973).

5. Total Joint Arthroplasty Symposium in Mayo Clinic Proceedings, 54, 557-601, (1979).
6. E. A. Lefrak and A. Starr, "Cardiac Valve Prostheses", Appleton-Century-Crofts, New York, (1979).
7. R. H. Rigdon, "CRC Critical Reviews in Food Science and Nutrition", August, 1975, p. 435.
8. D. L. Coleman, R. N. King, and J. D. Andrade, J. Biomed. Mater. Res., 8, 199, (1974).
9. D. E. Ocumpaugh, and H. L. Lee, J. Macromol. Sci.-Chem., A4(3), 595, (1970).
10. S. Hirsch, H. Robertson, and M. Gornowsky, Arch. Surg., III, February (1976).
11. P. Casagrande, and P. Darahy, J. Bone and Joint Surg., 53, 167, (1971).
12. C. A. Vinnick, Plastic and Reconstructive Surgery, 58, 555 (1976).
13. H. Wagner, F. K. Beller, and M. Pfautsch, Plastic and Reconstructive Surgery, 60, 49, (1977).
14. R. Rudolph et al. Plastic and Reconstructive Surgery, 62, 185, (1978).
15. D. E. Barker, M. I, Retsky, and S. Schultz, Plastic and Reconstructive Surgery, 61, 836, (1978).
16. R. M. Gayou, Plastic and Reconstructive Surgery, 63, 700, (1979).
17. D. F. Gibbons, G. Picha, and S. R. Taylor, Unpublished results.
18. D. K. Gilding, Unpublished results.
19. H.-G. Willert, J. Ludwig, and M. Semlitsch, J. Bone and Joint Surgery, 56-A, 1368 (1974).
20. F. W. Reckling, M. A. Asher, and W. L. Dillon, J. Bone and Joint Surgery, 59-A, 355, (1977).
21. F. W. Reckling and W. L. Dillon, J. Bone and Joint Surgery, 59-A, 80, (1977).
22. C. D. Jefferiss, A.J.C. Lee, and R.S.M. Ling, J. Bone and Joint Surgery, 57-B, 511, (1975).
23. A. Florian, L. Cohn, G. Dammin, and J. Collins, Arch. Surg., 111, 267, (1976).
24. C. D. Campbell, D. H. Brooks, M. W. Webster et al. Surgery, 85, 177, (1979).
25. M. P. Morgan, G. J. Dammin, and J. M. Lazarus, Trans. Am. Soc. Artif. Intern. Organs, 24, 44, (1978).
26. K. Hameed, S. Ashjaq, and D. O. W. Waugh, Arch. Path., 86, 520, (1968).
27. J. C. Hylen, F. E. Kloster, A. Starr, and H. E. Griswold, Annals of Internal Medicine, 72, 1, (1970).
28. N. R. Niles, J. Thoracic and Cardiovascular Surgery, 59, 794, (1970).
29. W. E. Stamm, Annals of Internal Medicine, 89, 764, (1978).
30. R. J. Duma, editor, "Infections of Prosthetic Heart Valves and Vascular Grafts", University Park Press, Baltimore, (1977).
31. L. S. Olanoff, J. M. Anderson, and R. D. Jones, Trans. Am. Soc. Artif. Intern. Organs, 25, 334, (1979).

THE STATUS OF OLEFIN-SO_2 COPOLYMERS AS BIOMATERIALS

D. N. Gray

Owens-Illinois, Inc.
Corporate Technology
Toledo, Ohio 43666

INTRODUCTION

In the late sixties a new series of low cost monomers were commercially introduced as an industry response to the ban on non-biodegradable (hard) detergents. These linear alpha-olefins were available as individual monomers in the even-number C_{10} to C_{18} series. When sulfonated these olefins yielded biodegradable surface active materials. Later these heretofore rare olefins were used as a base for the manufacture of synthetic lubricants with a projected industry capacity of twelve million gallons in 1985 (1).

Polyalkylsulfones are copolymers of alpha-olefins and sulfur dioxide and were prepared as early as 1898 (2) and again in 1915 (3) but were not well characterized as polymers until 1935 (4). The polymer series described here were reported in 1971 (5). Excellent reviews of polyolefin sulfones have been published by Fettes and Davis (6), and Ivin and Rose (7).

The polymerization is catalyzed by the usual radical initiators such as peroxides, azo compounds, light, redox systems and hard radiation. Polymerization systems which have been used include bulk, solution, emulsion and suspension. A unique feature of the solution system is that one of the monomers, sulfur dioxide, may be used in excess as an inexpensive and easily removed solvent. A threshold temperature for polymerization exists and it is different for each olefin. This characteristic, termed the ceiling temperature (T_c) has been extensively studies and the T_c for a large number of monomers has been determined. In contrast to many addition polymer systems, olefin-sulfur dioxide comonomer systems generally have low ceiling temperatures (many at room temperature or below).

We have studied the series of olefin-SO_2 copolymers prepared from the C_8-C_{18} even numbered carbon series of alpha-olefins and sulfur dioxide. The properties of this polymer series vary greatly depending upon the chain length of the alpha-olefin. The polymer prepared from 1-hexene (PAS-6) is a hard, brittle material while PAS-16 (prepared from hexadecene) is elastomeric. Physical properties such as density, tensile strength, elongation and especially gas permeability vary regularly up to PAS-16 but reverse themselves with PAS-18 as shown in Figure 1. As the length of the side chain of the alpha-olefin increases the polymers become more flexible, have lower specific gravities, have a lower tensile modulus, become more preme-able to gases and exhibit higher percent elongation at yield up to PAS-16. An explanation of this reversal of property trend at the PAS-18 based polymer has been proposed (5,8).

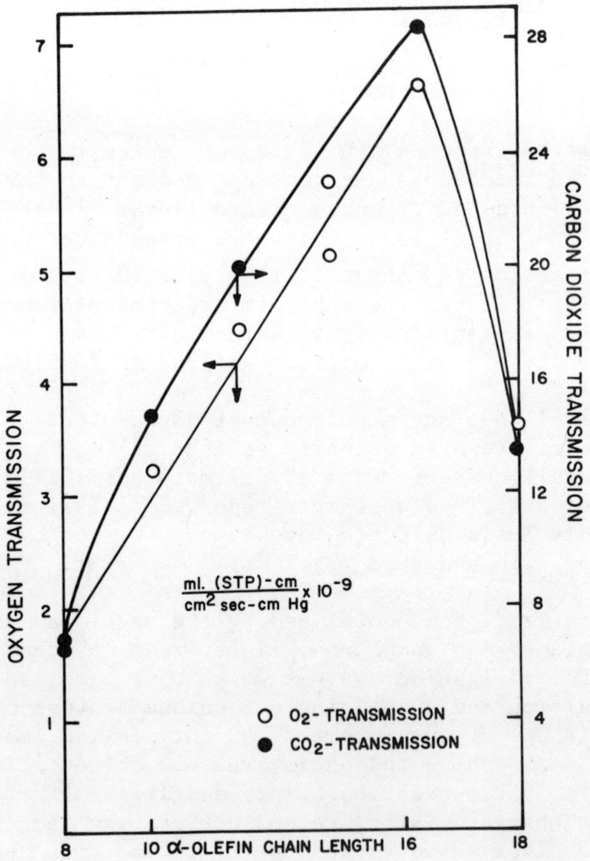
OXYGEN TRANSMISSION
CARBON DIOXIDE TRANSMISSION
ml. (STP)-cm / cm² sec-cm Hg x 10⁻⁹
O₂-TRANSMISSION
CO₂-TRANSMISSION
α-OLEFIN CHAIN LENGTH
1 2 3 4 5 6 7
4 8 12 16 20 24 28
8 10 16 18

Figure 1. Polymer permeability to oxygen and carbon dioxide as a function of co-monomer side chain length.

BLOOD OXYGENATORS

The first extracorporeal blood oxygenators were designed to give optimum direct contact between oxygen and blood using bubbles passing thru the blood or discs to increase the surface area between the blood or discs to increase the surface area between the blood and the gas phase. The development of these devices permitted the exchange of oxygen and carbon dioxide in extracorporeal blood either, partially, when lung function is not adequate to allow the required perfusion or, totally, when used in procedures such as open heart surgery or complete, but temporary, natural lung failure. While these devices represented a major development, the blood-gas interface causes a certain level of red blood cell damage and denaturation of essential proteins (9).

In the 1950's it was shown that perfusing blood via contact with a gas permeable membrane caused less hematological damage to blood. Such membrane materials as Teflon, silicones, silicone-copolymers were investigated using a number of configurations to reduce the surface area of the blood-polymer contact. Silicones, having the highest permeability (gas-to-gas), have been the polymer system most used as the membrane material and a comparison of various designs using silicone membranes has been given by Ketteringham and Zapol (9).

In addition to homogenous, permeable polymer films, microporous polymeric films have been also used as membrane materials. Difficulties with their use are the possible production of microemboli due to inadvertant sparging, high water vapor flux and blood damage due to the nature of the microporous material (10).

When the permeability versus polymer composition relationship (as shown in Figure 1) were discovered the polymer with the highest permeability (gas-to-gas) was compared to other polymers used for artificial lung membranes as shown in Table 1.

Table 1

PERMEABILITY OF MEMBRANE MATERIALS

Material	Permeability[a]	
	O_2	CO_2
Silicone (polydimethylsiloxane)	50	270
Silicone/polycarbonate copolymer (MEM 213®)	16	97
PAS 16[b]	6	25
Teflon	1.1	2.5

a. $\frac{\text{ml(STP)-cm}}{\text{cm}^2\ \text{sec-cm Hg}} \times 10^9$

b. Prepared from hexadecene and sulfur dioxide

®MEM 213 - Trademark, General Electric

Obviously the method of increasing the total gaseous flux passing from the gaseous phase into the blood for a given surface area for any permeable membrane material is to decrease the effective thickness of the membrane. Here a compromise must be made between the desired thickness (preferably low) and the requisite strength of the membrane required for device manufacture and to withstand the fluid pressures encountered in use.

ULTRATHIN MEMBRANES

In order to eliminate the potential problems with microporous membranes but still utilize their intrinsically high gas fluxes, permeable, biocompatible polymers were cast as ultra-thin coatings on microporous supports. Ketteringham and coworkers (10) at A. D. Little first prepared ultrathin membranes of PAS-16 cast on microporous polypropylene (Celgard®). These membrane structures, in a variety of artificial lung geometries, have been intensively studied at A. D. Little and at the Artificial Lung Laboratory, Department of Anesthesia, Harvard Medical School at the Massachusetts General Hospital by Dr. Warren Zapol. PAS-16 was chosen because of its relatively high gas permeability, good blood compatability, potential low cost and, expecially, the ease of fabrication on the microporous membranes by solvent casting from dilute solution. The gas-to-gas permeability comparison data is given in Table 2. Note that the dual layer polymeric membrane of PAS-16 on porous

Table 2

COMPARATIVE MEMBRANE PERMEATION RATES OF AVAILABLE MEMBRANE MATERIALS (25°)

		THICKNESS-MICRONS		PERMEATION RATE*	
POLYMER	FABRICATION	TOTAL	MEMBRANE	O_2	CO_2
Polydimethyl siloxane	Reinforced with polyester knit	190	160	140	770
Polydimethyl siloxane/ polycarbonate copolymer	Homogeneous film	50	50	146	885
Polyalkylsulfone	On porous polypropylene	25	2.5	1100	4600

*cc(STP)/min-meter^2.atm.

® registered trademark, Celanese Corporation.

polypropylene gives very high gas flux but still retains the excellent strength of the microporous support.

Blood flowing past a membrane, at least at the fluid velocities permitted in membrane lungs, forms a laminar boundary layer adjacent to the membrane. This phenomena limits the gas transfer capabilities of the device. To exploit the true high potential gas transfer capabilities of membrane lungs, Bellhouse and coworkers (15) have investigated the concept of vortex shedding, secondary flow techniques to increase membrane to blood gas transfer. This is accomplished by impressing a secondary pulsitile flow on the main flow in order to disturb the laminar layer. The frequency and intensity of this secondary flow must be chosen to minimize blood trauma. Using these techniques the A. L. Little group under Ketteringham have developed vortex shedding devices, using high performance ultra-thin layer PAS-16 membranes, with one-fifth the membrane area and one-tenth the average pressure drop of conventional devices as tested under _in vitro_ conditions (16).

STERILIZABILITY & BLOOD COMPATIBILITY

The polymer melting temperatures of the series of _alpha_-olefins polymers described here, including PAS-16, are lower than the temperatures used for either normal steam (autoclave) sterilization or dry heat sterilization. Therefore any articles, devices etc. fabricated from the PAS-16 polymer must be sterilized by low temperature methods.

Plastic articles are commonly treated with ethylene oxide to impart sterility, a common gas composition consists of 600 mg. of a EtO-Freon 12 mixture per liter of air. Exposing test strips of PAS-16 to this gas composition for four hours rendered them sterile without changing the physical properties (11). Zapol and Ketteringham report that PAS coated Celgard® devices can be sterilized by ethylene oxide, radiation and alcohol (8).

Good blood compatibility is of the utmost importance in tissue contact prosthetic devices in general and membrane lungs in particular. The combination of high membrane permeability, efficient device design and good blood compatibility are desirable. Such a system should have low priming volume, low surface area contact and a benign surface/blood contact reaction - all contributing to the safety and utility of the membrane lung.

An initial test of biocompatibility is to screen a candidate material by comparing whole blood clotting times versus both glass and siliconized glass surfaces (the modified Lee-White test). Clotting times in excess of those obtained with siliconized glass (20 minutes) are considered to offer promise for further testing. In one such study of PAS-16, average whole blood clotting times of about 26 minutes were obtained (11).

Local cytotoxicity of PAS-16 was shown to be nil (11) using protocols involving intramuscular implantation as described by Stetson and Guess (13). In addition standard U. S. Pharmacopeia tests (14) were performed showing no systemic toxicity or intracutaneous tissue reactions with saline or cottonseed oil extracts (11,12).

The most useful and practical test of blood compatibility involves fabricating a prototype membrane lung and using it to perfuse experimental animals under conditions closely approximating anticipated clinical use. Zapol and Ketteringham found awake, alert lambs to be an ideal model to study changes induced in blood by membrane oxygenators. The details of subject choice, blood train ancillary equipment and methodology has been documented (9). Given that PAS does not cause hemolysis and is as nonthrombogenic as silicone the longterm monitoring of perfused blood has been concerned with platelet damage and platelet loss (9).

In addition to the platelet damage studies described above a number of other tests have been carried out or are in progress to determine the degree of biocompatibility of PAS-16. These tests include Gott ring tests, Kusserow renal infarct tests, protein adsorption tests, tests for renal emboli from surgically implanted PAS coated rings, microscopic examinations of protein and platelet adhesion and critical contact angle measurements. The initial results have been summarized by Ketteringham (16) and indicate that smooth, pure adherent coatings of PAS-16 show a high degree of biocompatibility.

REFERENCES

1. C&EN, March 31, 12-13 (1980)
2. W. Solonia, Zh. russk. fiz.-khim. Obshch., 30, 836 (1898)
3. F. E. Matthews and H. M. Elder, British Patent 11,635 (1914)
4. D. S. Frederick, H. D. Cogan and C. S. Marvel, J. Am. Chem. Soc., 56, 1815 (1934)
5. J. E. Crawford and D. N. Gray, J. Appl. Polym. Sci., 15, 1881, (1971).
6. F. M. Fettes and F. O. Davis, in High Polymers, Volume XIII - Part III, Interscience, NY, (1962)
7. K. J. Ivin and J. B. Rose, in Advances in Macromolecular Chemistry, Vol. 1, Academic Press, NY, (1968)
8. J. M. Crawford and D. N. Gray, U. S. Patent 3,928,294; December 23, 1975.
9. W. M. Zapol and J. M. Ketteringham, Polymers in Medicine and Surgery, Polymer Science and Technology, Volume 8, Plenum Press, NY, (1975)
10. J. M. Ketteringham, R. deFillippi and J. D. Birkett, Ultra-Thin Membranes For Membrane Lungs, in Artificial Lungs for Acute Respiratory Failure, Academic Press, NY, (1976)
11. D. N. Gray, Polym. Eng. Sci., 17, 719 (1977)
12. J. M. Ketteringham, W. M. Zapol, D. N. Gray, K. K. Stevenson,

A. A. Massucco, L. L. Nelsen and D. P. Cullen, Trans. Amer. Soc. Artif. Int. Organs, XIX (1973)
13. J. B. Stetson and W. L. Guess, Anesthesiology, 33, 635 (1970)
14. U. S. Pharmacopeia XVIII, p. 926 (1970)
15. B. J. Bellhouse, F. M. Bellhouse, C. M. Curl, T. I. MacMillan, A. J. Gunning, E. M. Spratt, S. B. MacMurray, and J. M. Nelems, Trans. ASAIO, 19, 72, (1973)
16. J. M. Ketteringham, T. J. Driscoll, A. A. Massucco and L. Wu, "Development and Evaluation of Polyalkylsulfone as A Biomaterial and its Application in The Membrane Lung", Third Annual Progress Report, NIH-NO1-HV-3-2916, April, 1979.

TEMPORARY SKIN SUBSTITUTE FROM NON-ANTIGENIC DEXTRAN HYDROGEL

Paul Y. Wang and Nimet A. Samji

Institute of Biomedical Engineering
Faculty of Medicine
University of Toronto
Ontario, Canada M5S 1A4

INTRODUCTION

Accidental injury with loss of skin is considered to be a "neglected disease", partly because of the superficial nature of damage to the protective layer of the body. Surveys conducted in the U.S. show that such injury is among the principal causes of death in North America between ages 1 and 44 [1]. In the Canadian province of Ontario, there were over 50,000 people suffered loss of skin in 1977 due to trauma (including surgical autograft), disease (e.g., ulceration) or burns from electrical, chemical or thermal agents. Patients often must change their occupations or alter their normal activities, and in addition many working or school days are lost annually through lengthy stays in hospital as well as re-admission for revision of scars. Besides pain and the high incidence of bacterial infection, loss of skin is followed by excessive evaporation of vital body fluids which initiates a series of physiological responses as illustrated schematically in the following diagram:

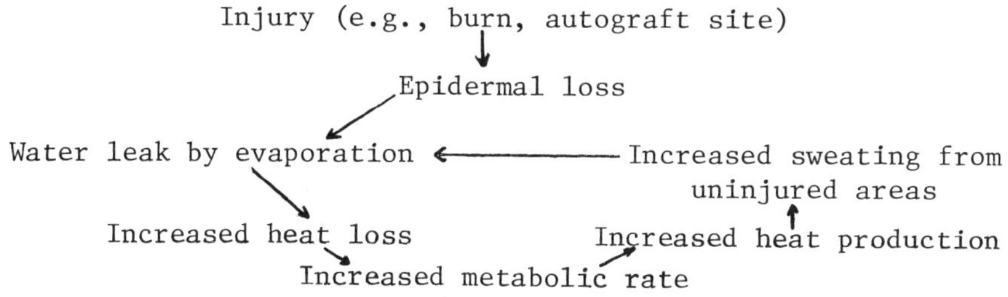
Injury (e.g., burn, autograft site)
Epidermal loss
Water leak by evaporation
Increased sweating from uninjured areas
Increased heat loss
Increased heat production
Increased metabolic rate

(Modified from S. Cohen, Plast. Reconstr. Surg., 37, 475 (1966)

Currently, there are a number of methods devised to meet this problem. The most commonly used, and yet probably the least satisfactory by comparison, are the pre-formed plastic sheet dressings [2], because these dressings do not seal the wound edges to prevent evaporative loss and bacteria ingress. Often they are difficult to apply on the hands, neck, etc., especially on children. The porcine xenograft overcomes some of these shortcomings, but the cost is high, quality varies, and the risk of the immunoresponse of the host cannot be totally overlooked [3]. Recent innovations on an experimental basis use primarily two approaches and, if successful, may greatly facilitate the therapy of injury to skin. The first is to apply an adhesive polymer material in situ that will conform and adhere to the injured area [4,5,6]. The second is to use a hydrogel composed of a hydrophilic polymer mass swelled in water. The polymer mass is formed by cross-linking water soluble polymers. The hydrogels have many interesting biomedical applications such as in sustained release medication, soft contact lenses, etc. However, recent observations in our laboratory indicate that a number of water soluble polymers can induce antibody production in mice [7]. This observation imposes some restriction on the choice of polymers for the preparation of hydrogel to be used as implants or other biomedical purposes. This paper describes the preparation of a hydrogel membrane from a non-antigenic natural polymer and the preliminary evaluation of this membrane as a temporary skin substitute for surface injuries.

EXPERIMENTAL

Dextran with a molecular range of 60,000 to 90,000 was purchased from BDH Chemicals, Toronto. The polysaccharide (10 g) was dissolved in distilled water and filtered to separate small amount of insoluble matters. Sodium borohydride (0.5 g) was dissolved in distilled water (10 ml) and added slowly to the dextran solution with stirring at 10°C over a period of 3 h. Unreacted borohydride was decomposed by dilute acetic acid, and inorganic salts were removed by dialysis (24 h); the dextran was recovered by lyophilization. The reduction eliminated the reducing end-groups on the polysaccharide and prevented degradation by "peeling" from the terminal glucosidic units during the cross-linking reaction to prepare the gel. Epichlorohydrin was used as the cross-linking agent to prevent the dispersion of dextran in water according to the procedure of Goldstein et al [8]. After reaction, the cross-linked dextran was placed in water to remove any by-products of the reaction. After 36 to 48 h, the gel became fully hydrated and its dimension was recorded to obtain the volume by calculation. The weight of the cross-linked dextran before and after dehydration in air (7 to 9 d for a 2 cm by 3.5 cm cylinder) was also recorded, and the water regain value [9] was calculated in ml water/g dry dextran.

To prepare a membrane, the required amount of dextran and epichlorohydrin was added to a polycarbonate mould containing a fine, flat sheet of surgical cotton gauze to serve as a reinforcement

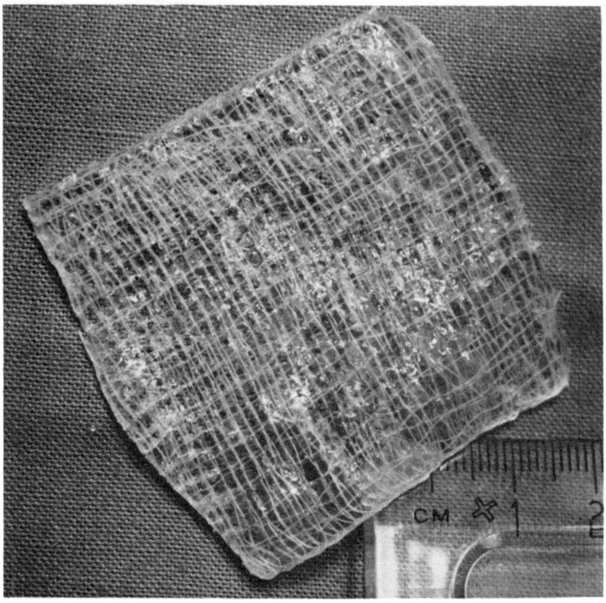
CM
1
2

Figure 1. Reinforced dextran hydrogel membrane.

matrix for the dextran gel after cross-linking; when not reinforced, the gel was fragile to manipulate after full hydration. One piece of reinforced dextran hydrogel membrane (Fig. 1) approximately 5 cm by 5 cm in size and 0.3 cm in thickness was sterilized by autoclaving and implanted subcutaneously into each of the 5 Wistar rats (body weight: ∿400 g). On days 3, 7, 14, 21, and 30, the membranes were exposed for gross visual inspection. Another 5 animals were used to evaluate the reinforced dextran hydrogel as a temporary skin substitute. A circular piece (diameter: 2 cm) of skin in full thickness was excised about 2 cm below the occiput. An incision was then made near the sacrum and a piece of sterilized dextran hydrogel membrane, equilibrated in a saline solution containing penicillin G (1 mg/ml), was inserted subcutaneously until it was positioned evenly over the dorsal musculature just exposed by removal of a disc of skin in full thickness (Fig. 2). A thin layer of petroleum jelly was applied over the exposed membrane, and the lower dorsal incision was closed by suturing. The membrane was changed every 3 or 4 days and after the 14th day it was removed and the wound was covered by autograft. During the *in vivo* experiments, regulations made under the Animals for Research Act of Ontario were observed.

To evaluate the release rate of penicillin G *in vitro*, hydrogel membranes (2 x 2 x 0.3 cm^3, when fully hydrated) were partially dehydrated until they contained one-half the original amount of

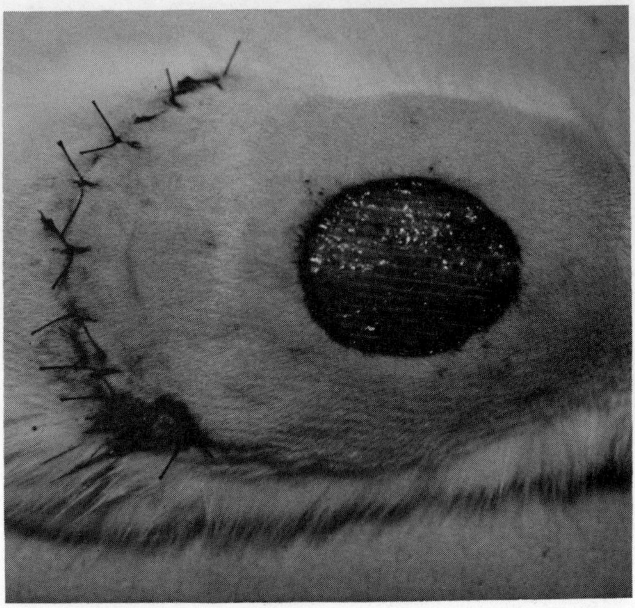

Figure 2. Reinforced dextran hydrogel membrane used as a temporary skin substitute on Wistar rat.

water. Each partially dehydrated membrane was transferred to a solution (5 ml) of penicillin G (1 mg/ml). The antibiotic solution quickly entered the interior during rehydration to the full water regain value. The fully hydrated membrane was then quickly rinsed and immersed in 5 ml saline. At various intervals, the amount of penicillin G released from the hydrogel was determined spectrophotometrically [10] at 263 nm.

The *in vivo* release rate of penicillin from the hydrogel was evaluated using ^{14}C-labelled penicillin G. In this experiment, each of the 3 Wistar rats received 4 pieces of the impregnated membrane implanted subcutaneously about 4 cm apart from each other on both sides of the dorsal spine. On different days, the hydrogel implants were retrieved and the amount of radioactivity remaining in the hydrogel was counted by a standard procedure [11].

RESULTS AND DISCUSSION

The yield from the treatment of dextran with borohydride was almost quantitative. The reduction made the dextran more chemically stable during the cross-linking reaction. A gel made from the unreduced dextran had a slight yellow coloration, but otherwise exhibited no apparent effect on its property. The coherence of the hydrogel is partly a function of the water regain value which can be regulated by the amount of epichlorohydrin used to cross-link the dextran polymer chains. At a constant amount of dextran when the percentage ratio of dextran to epichlorohydrin was 6, the resulting gel had a high water regain value and the coherence was poor (Table 1). When the ratio was reduced to 1.3 or less, the cross-linking was extensive but it appeared to be less uniform in the gel matrix. As a result, the cross-linked product has a tendency to disintegrate and hydrate unevenly when placed in water. If the concentration of epichlorohydrin was held constant, increasing amount of dextran yielded insufficiently cross-linked products when the percentage ratio was 4 or beyond (Table 2). An amount of dextran below a ratio of 1.3 resulted in no coherent gel at all. The best combination was at a ratio of 3.0. The gel made at this ratio and reinforced with a fine cotton gauze provided a membrane with good handling characteristics as well as adequate retention of penicillin G.

Table 1

EFFECT OF EPICHLOROHYDRIN CONCENTRATION ON WATER REGAIN VALUE (FIXED AMOUNT OF DEXTRAN)

% Ratio of Dextran to Epichlorohydrin	H_2O Regain Value
6.0	∞*
3.0	20
2.2	24
1.8	21
1.3	22
0.8	13

*Incoherent hydrogel

Table 2

EFFECT OF DEXTRAN CONCENTRATION ON WATER REGAIN VALUE (FIXED AMOUNT OF EPICHLOROHYDRIN)

% Ratio of Dextran to Epichlorohydrin	H_2O Regain Value
4	∞*
3	20
2	∞
1	∞

*Incoherent hydrogel

Visual inspection of the dextran hydrogel implants between days 3 to 30 at approximately weekly intervals showed that they remained optically clear without discoloration and no apparent change in size or integrity due to biodegradation or body movements of the test animal. There were no signs of encapsulation of the implant, fluid accumulation, abscess, hemorrhage or adhesion due to cellular or fibrous tissue ingrowth.

Attempts were first made to create a surface wound by microtoming or by controlled thermally induced trauma [4], and then cover the injured area with the dextran hydrogel membrane. However, due to the small size of Wistar rat, its acute dorsal curvature, and activity in confinement, it was very difficult to create a physical situation resembling totally the use of a surface wound covering in conventional clinical applications. Instead, the insertion of a reinforced dextran hydrogel membrane subcutaneously beneath the circular full thickness skin wound to protect the exposed dorsal musculature appeared to result in less objection from the test animal. A total of 5 Wistar rats out of 8 was cooperative in such a procedure. With occasional changes the antibiotic impregnated hydrogel membrane was removed after 14 days and replaced by autograft. The appearance of the wound site upon removal of the hydrogel membrane was generally healthy with no apparent signs of discolored necrotic tissues. In this series of preliminary evaluation, all the autografts were taken without complication (Fig. 3).

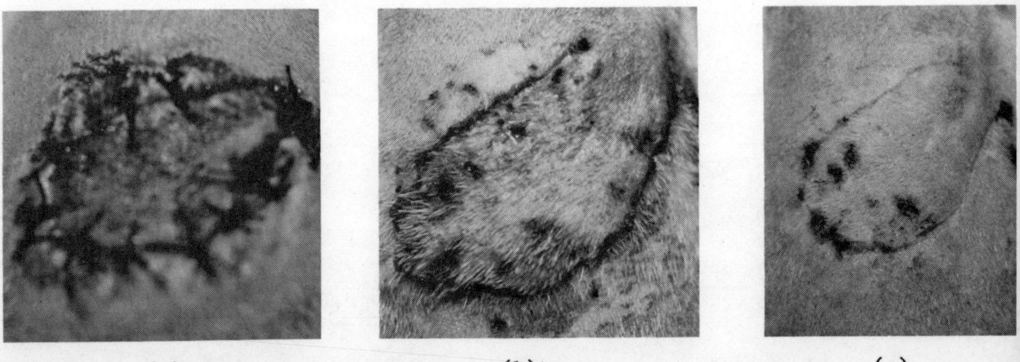

(a) (b) (c)

Figure 3. Pedicle autograft after removal of the dextran hydrogel membrane: (a) day 1, (b) day 4, (c) day 9.

Due to the amount of water within the matrix, the dextran hydrogel could not be secured like a piece of synthetic or natural membrane in a dual compartment diffusion chamber so that its permeability to drugs or other macromolecules could be evaluated. An alternative approach to gain some insights to the permeability characteristics was to dehydrate partially the hydrogel membrane, and then allow it to rehydrate in a solution containing a solute whose transport through the hydrogel matrix was to be estimated. When immersed at 50% partial dehydration, the dextran hydrogel membrane with a thickness of 0.3 cm when fully hydrated could retain 0.41 mg penicillin G/cm^2. About one-half of this amount was released from the hydrogel membrane within 20 min when immersed in saline (Table 3), and the remainder took 6 times as long to migrate from the hydrogel interior. The results from implanting pieces of the hydrogel membrane containing ^{14}C-labelled penicillin G into Wistar rats were also characterized by an initial surge [12], followed by a period of sustained release exceeding the duration of the experiment (Table 3). To minimize the initial surge of medication from the dextran hydrogel, it might be possible to formulate a restraining phase for inclusion in the hydrogel matrix.

Table 3

RELEASE OF PENICILLIN G FROM DEXTRAN HYDROGEL MEMBRANE

Time (h)	In Vitro[a] (mg/ml)	Time (d)	In Vivo[b] (cpm)
0.00	0.000	0	8.5×10^5
0.25	0.105	1	2.8×10^3
0.45	0.150	4	3.0×10^3
0.75	0.165	7	8.0×10^2
1.00	0.185	11	2.3×10^2
1.25	0.195	13	2.4×10^2
1.50	0.205		
1.75	0.205		
2.00	0.240		
2.50	0.240		

[a]Amount released upon immersion in saline and measured spectrophotometrically at 263 nm.

[b]Amount of ^{14}C retained in hydrogel membrane counted by liquid scintillation.

Hydrogels have been prepared from many water swellable polymers or water soluble ones after cross-linking. However, it is often very difficult to predict the permanency of a hydrogel *in vivo*. Any change may contribute to the dissolution of the original polymer which might be antigenic as we have already observed for several water soluble synthetic macromolecules [13]. Our choice of the low molecular weight dextran for the preparation of the hydrogel membrane is mainly intended to avoid this problem. At present, the same dextran is used as a plasma expander clinically, but higher molecular weight dextran is known to be antigenic in man [14]. The excellent biocompatibility of the hydrogel membrane observed in our experiments is probably due to the presence of the commonly occurring α-D-glucose backbone in the polysaccharide structure of dextran, and the relatively simple 1,3-glycerol ether cross-linking units introduced. An important factor contributing to the successful autograft after the removal of the dextran hydrogel membrane may be its capacity to absorb and transmit excess fluids which, if present at wound site, will impede healing and facilitate bacterial growth. In addition, the release *in vivo* of the impregnated penicillin G, and a slight tackiness of the dextran hydrogel membrane which helped to seal the edges of the circular wound, probably have limited any possible further contamination of the unexposed neighboring tissue. To be equally successful in future clinical trials of the hydrogel membrane, modifications in technique and experimental designs may be required to meet the relatively more complex clinical situations.

The potentials of a hydrogel made from the non-antigenic dextran as a biomedical material have not been fully explored. The cost of the hydrogel membrane is only about $10/ft^2. If results from future clinical trials corroborate with our observations in the animal experiment, it will be interesting to evaluate other possible biomedical applications for this material.

ABSTRACT

Injury with loss of skin is considered to be a "neglected disease". The evaporation of vital body fluids follows the damage to the protective layer and initiates a series of physiological responses. Together with bacterial infection, they constitute the main cause of death in North America between ages 1 and 44. Various plastic films have been used satisfactorily to some extent as dressings; an alternative is the more expensive porcine xenograft. Recent innovations employ synthetic polymer hydrogels which permit gas exchange, fluid regulation at wound site, and antibiotic diffusion without bacterial ingress. However, current studies in our laboratory indicate that some synthetic polymer components of the hydrogel are antigenic. To avoid the antigenic problem, dextran of m. wt. $<$ 100,000 clinically used as a plasma expander has been cross-linked by epichlorohydrin to form a hydrogel with water regain values of 10 to 20. Membranes of any size and shape can be made by reinforcing this hydrogel in a thickness of 0.3 cm with a fine cotton gauze matrix. Subcutaneous implantation in Wistar rats showed that the dextran dressing caused no fibrous tissue encapsulation, hemorrhage, fluid accumulation, or tissue discoloration. Other advantages are its low cost, adherence to moist tissue surface, permeability to penicillin, and non-antigenicity.

ACKNOWLEDGMENT

Supported by Medical Research Council of Canada grant number MA-5556.

REFERENCES

1. The U. S. National Institute of Health Guide for Grants and Contracts, 6 (11), (1977)
2. G. B. Park, J. M. Courtney, A. McNair and J. D. S. Gaylor, Engineering in Medicine, 7, 11 (1978)
3. H. M. Bruch, in "Basic Problems in Burns", R. Vrabec, Z. Konickova and J. Moserova, eds., Springer-Verlag, New York, 17 (1975)
4. P. Y. Wang, D. W. Evans, N. Samji and L. Thomas, J. Surg. Res., 28, 182 (1980)
5. M. M. Zeigler, H. Maguire and C. F. Barker, J. Surg. Res., 22, 643 (1977)
6. P. Nathan, E. J. Law, B. G. MacMillan, D. F. Murphy, S. H. Ronal, M. J. D'Andrea and R. A. Abrahams, Trans. Amer. Soc. Artif. Int. Organs, 22, 30 (1976)
7. P. Y. Wang, N. Samji and L. Sun, Paper presented at annual meeting, Can. Biomaterials Soc., Toronto, May 1979
8. J. Lonngren, I. J. Goldstein and K. Bywater, FEBS Letters, 68, 31 (1976)
9. "Sephadex-Gel Filtration in Theory and Practice", Pharmacia, Canada, Dorval, Quebec, 4 (1974)
10. D. C. Grove and W. A. Randall, "Assay Methods of Antiobiotics", Med. Encyclop., Inc., New York, 200 (1955)
11. P. Y. Wang and D. W. Evans, Biomat., Med. Dev., Artif. Org., 5 (3), 277 (1977)
12. P. J. Blackshear, Sci. Amer., 241 (6), 52 (1979)
13. P. Y. Wang and N. Samji, Proceedings of First World Biomaterials Congress, Vienna (in press)
14. E. A. Kabat, "Structural Concepts in Immunology and Immunochemistry", second ed., Holt, Rinehart and Winston, New York, 21 (1976)

BIOMEDICAL APPLICATIONS OF POLY(AMIDO-AMINES)

Paolo Ferruti, Istituto Chimico dell'Università,
Via Mezzocannone 4, 80134 Naples (Italy)
and Maria A. Marchisio, Intituto di Chimica Industriale
del Politechnico, Milan, Italy

SYNTHESIS OF POLY(AMIDO-AMINES)

Poly(amido-amines) are a class of polymers characterized by the presence of amido-and tertiary amino-groups regularly arranged along the macromolecular chain. Additional functional groups may be introduced as side substituents.

Linear poly(amido-amines) are obtained by polyaddition of primary monoamines, or bis-(secondary amines), to bis-acrylamides (1-8) (Scheme 1).

Scheme 1

a) $\quad x \ CH_2=CH-CO-N(R^1)-R^2-N(R^3)-CO-CH=CH_2 \ - \ x \ H_2N-R^4 \longrightarrow$

$$\longrightarrow \left[CH_2-CH_2-CO-N(R^1)-R^2-N(R^3)-CO-CH_2-CH_2-N(R^4) \right]_x$$

b) $\quad x \ CH_2=CH-CO-N(R^1)-R^2-N(R^3)-CO-CH=CH_2 \ + \ x \ HN(R^4)-R^5-NH(R^6) \longrightarrow$

$$\longrightarrow \left[CH_2-CH_2-CO-N(R^1)-R^2-N(R^3)-CO-CH_2-CH_2-N(R^4)-R^5-N(R^6) \right]_x$$

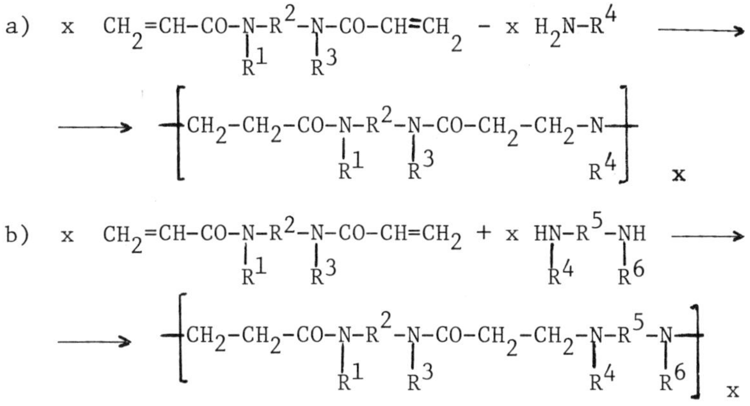

The reaction takes place readily in water or alcohols, at room temperature, and without added catalysts. Aprotic solvents are not recommended, if high molecular weight products have to be obtained (1,2). The above method is a general one, as far as aliphatic or cycloaliphatic amines are concerned. Under the same conditions, aromatic amines do not give high polymers.

All other conditions being equal, the reaction time necessary to obtain fairly high molecular weight polymers (as indicated by intrinsic viscosity values), is strongly dependent on the steric hindrance of the N-substituents of the amine, ranging from a few hours in the case of piperazine and methylamine, to several weeks in the case of N,N'-di-isopropylethylene diamine (2-4). Some examples of poly(amido-amines) are listed in Table 1, together with some physico-chemical characterizations. It may be observed that, owing to their regular structure, many poly(amido-amines) can crystallize. Some crystallinity is even observed in some cases, where a certain degree of irregularity has been introduced (see polymer II, Table 1).

As pointed above, poly(amido-amines) carrying additional functions as side substituents can be easily obtained, starting with the appropriate monomers (1,7,8). In fact, hydroxyl groups, tertiary amino groups, allylic groups, etc., if present in the monomers, do not interfere with the polymerization process. Some functionalized poly-(amido-amines) are listed in Table 2.

It may be added that polymers of related structures can be obtained by substituting either bis-(acrylic esters) or divinylsulfone for bis-acrylamides (9), or hydrazines (10), or phosphines (11), for amines. Some of these polymers are listed in Table 3.

PROPERTIES OF POLY(AMIDO-AMINES)

Many linear poly(amido-amines) are water soluble. They are also soluble in several organic solvents, such as chloroform, dichloromethane, and lower alcohols.

Poly(amido-amines) are basic polymers of medium strength (5-8). Their protonation behavior has been thoroughly investigated, and found to be quite unusual in the polyelectrolytes domain. "Real" constants, in fact, have been determined in the case of poly(amido-amines), while as a rule only 'apparent' constants can be determined in the case of polymeric acids and bases. This means that in poly-(amido-amines) the basicity of the aminic nitrogens of a given unit do not depend on the degree of protonation of the whole macromolecule (5-8). The number of basicity constants is always equal to that of the aminic nitrogens present in the repeating unit.

This particular behavior has been confirmed by thermodynamic studies (12), and is even observed in the case of poly(amido-amines) having additional tertiary amino groups as side substituents, such as, for instance, polymers XVIII and XIX (Table 2) (6). The

TABLE 1

SOME EXAMPLES OF POLY(AMIDO-AMINES)

Polymer No.	Structure of the repeating unit	[η] (a) (dl/g)	Crystallinity	M.p. (^{o}C)	Ref.
I	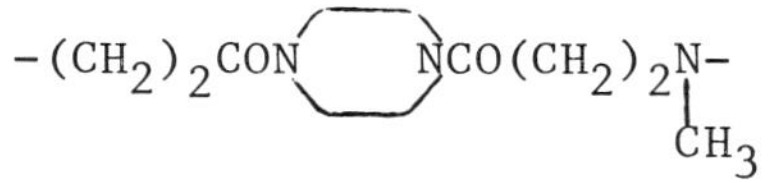	0.40	high	n.d.	1,4,5
II	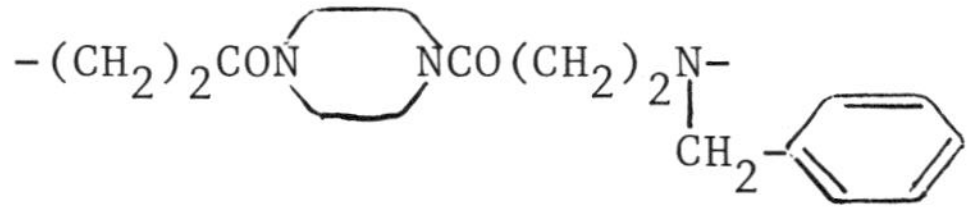	0.15	low	n.d.	1,4
III	$-(CH_2)_2CON(C_2H_5)(CH_2)_2N(C_2H_5)CO(CH_2)_2N(CH_3)-$	0.20	amorph.	--	1,4
IV	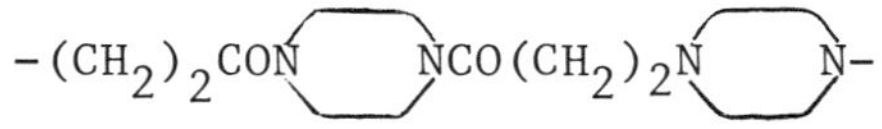	0.46[b]	high	270 (dec.)	1,2,5
V	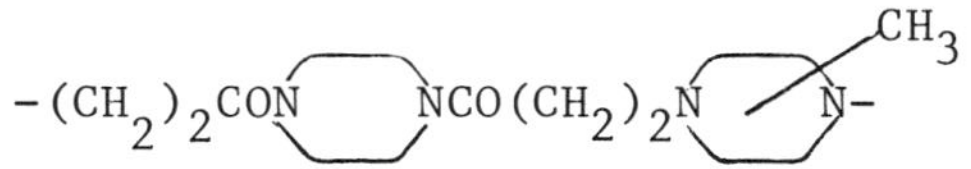	0.81	low	218	1,2
VI	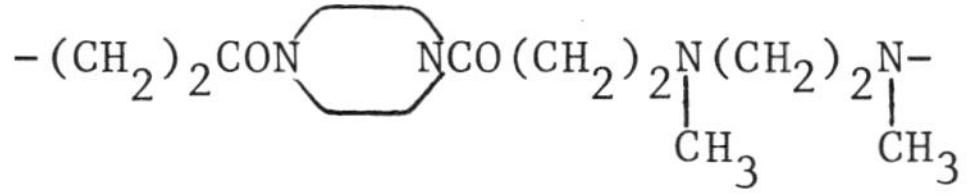	0.41	high	113	1,3,5
VII	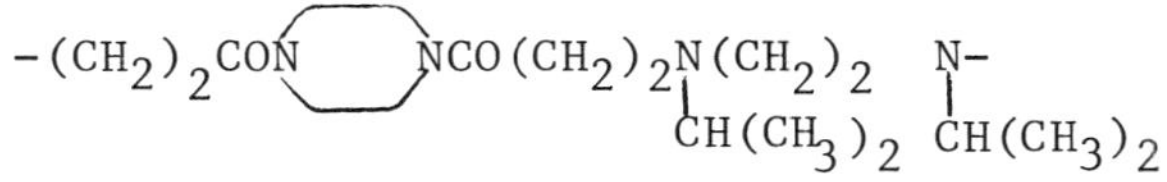	0.12	high	97	1,3

TABLE 1 (CONTINUED)

Polymer No.	Structure of the repeating unit	η (a) (dl/g)	Crystallinity	M.p. (°C)	Ref.
VIII	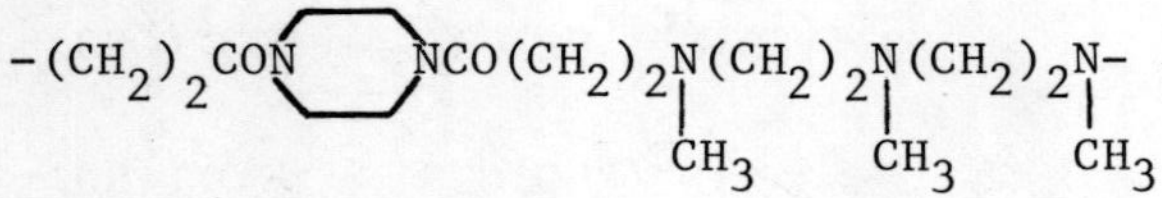	0.27	amorph.	--	5,6
IX	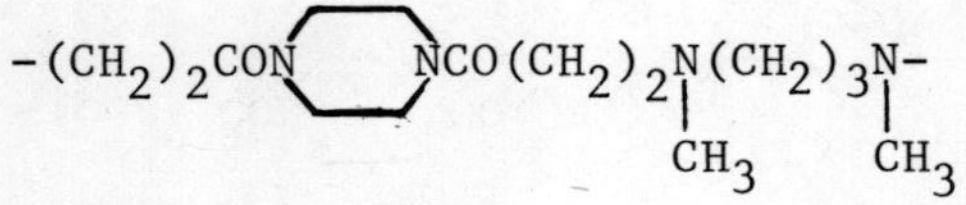	0.43	high	n.d.	7
X	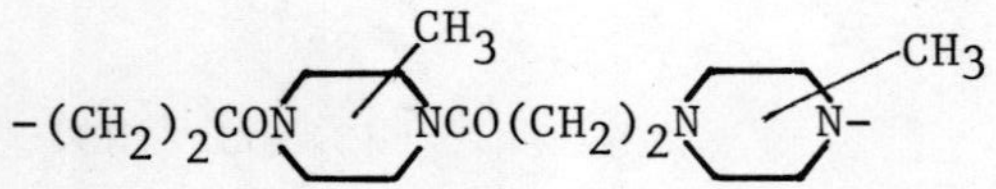	0.31	amorph.	--	1,3
XI	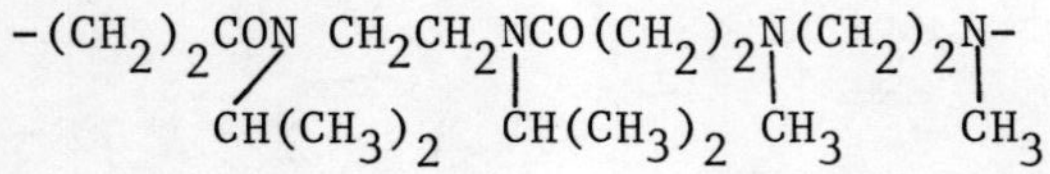	0.17	medium	102	1,3
XII	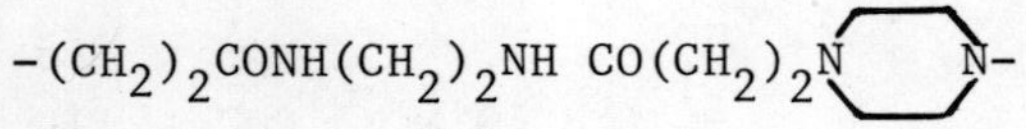	0.12 (c)	high	231 (dec.)	1,3

(a) in chloroform at 30°

(b) in aq. 0.1 M HCl/1 M NaCl

(c) in aq. 0.1 M CH_3COOH/1 M CH_3COONa

TABLE 2

SOME EXAMPLES OF POLY(AMIDO-AMINES) CARRYING FUNCTIONAL GROUPS AS SIDE SUBSTITUENTS

Polymer No.	Structure of the repeating unit	[η] (dl/g)	Ref.
XIII	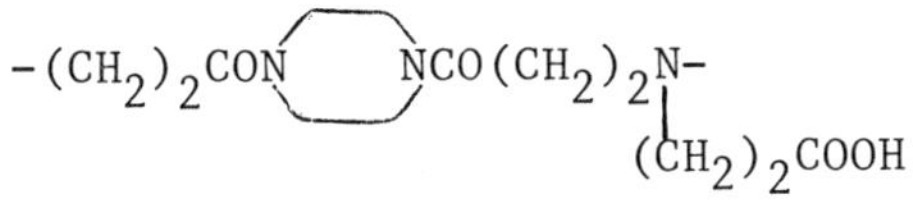	0.27 [a]	1
XIV	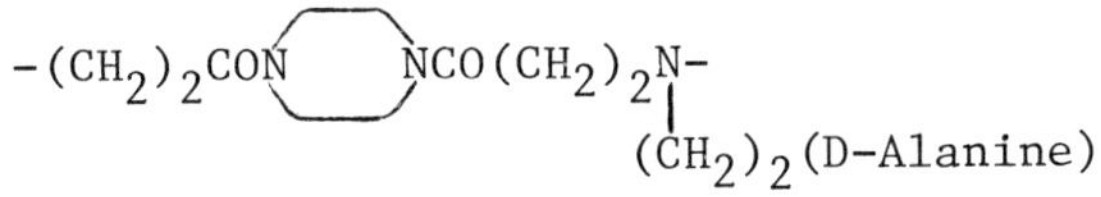	0.25 [a]	1 [d]
XV	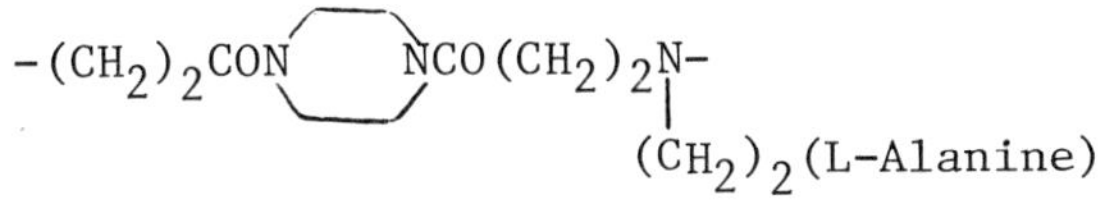	0.10 [a]	1 [e]
XVI	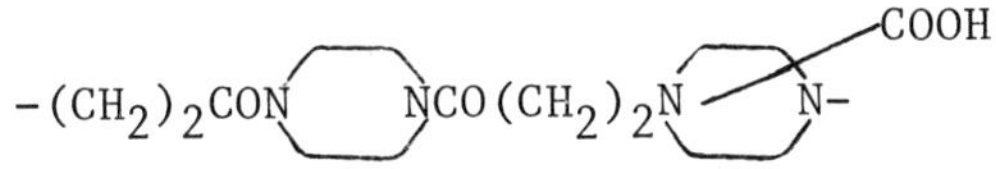	0.28 [b]	1
XVII	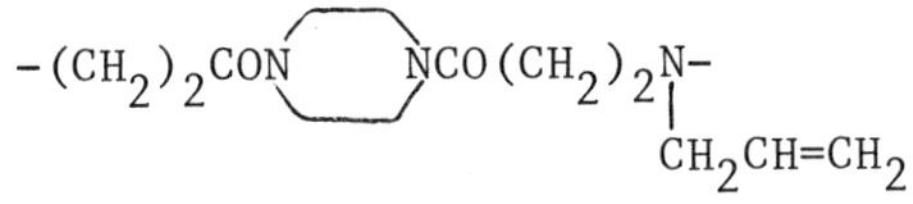	0.33 [c]	(f)
XVIII	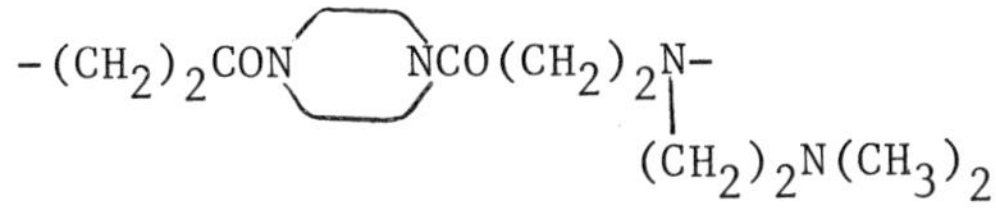	0.19 [c]	8
XIX	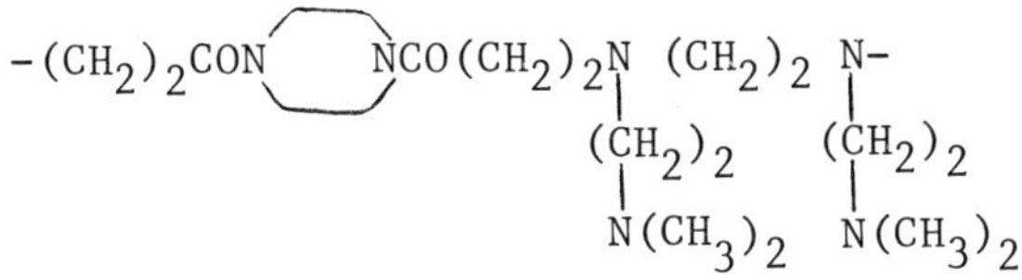	0.25	8
XX	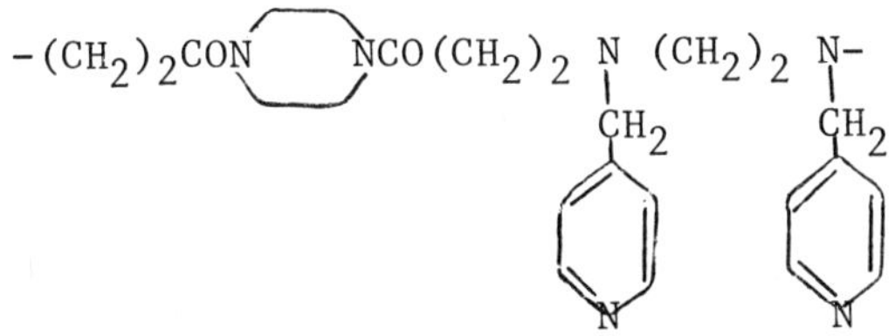	0.18 [c]	7
XXI	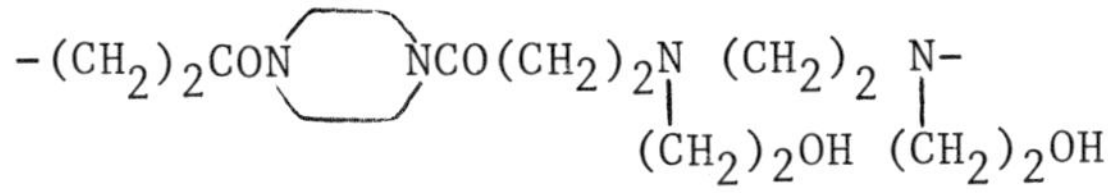	0.37 [c]	7

(a) η inh (C = 0.5) in 90% methanol/10% water at 30°.

TABLE 2 (CONTINUED)

(b) as above, in 80% methanol.

(c) in chloroform at 30°.

(d) $[\alpha]_D^{25} = + 10,4^o$

(e) $[\alpha]_D^{25} = - 9,8^o$

(f) P. Ferruti and M. A. Marchisio, unpublished.

TABLE 3

SOME EXAMPLES OF POLY(ESTERE-AMINES), POLY(SULFONE-AMINES), POLY(AMIDO-HYDRAZINES), AND POLY(AMIDO-PHOSPHINES)

Polymer No.	Structure of the repeating unit	[η] (a) (dl/g)	Ref.
XXII	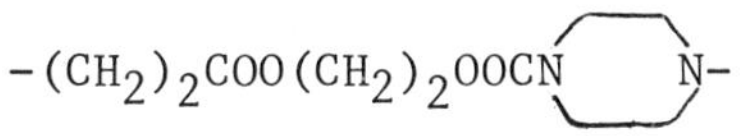	0.41	1,9
XXIII	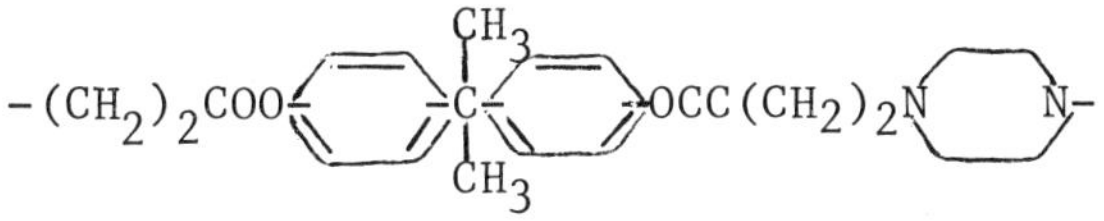	0.23	1,9
XXIV	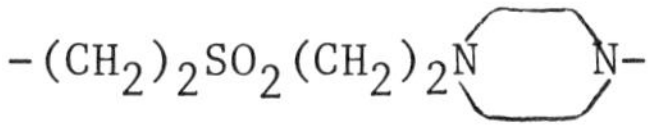	0.60[b]	1,9
XXV	$-(CH_2)_2SO_2(CH_2)_2N(CH_3)(CH_2)_2N(CH_3)-$	0.19	7
XXVI		0.18	1,10
XXVII	$-(CH_2)_2CON(CH(CH_3)_2)(CH_2)_2N(CH(CH_3)_2)CO(CH_2)_2NH\ NH-$	0.16	1,10
XXVIII	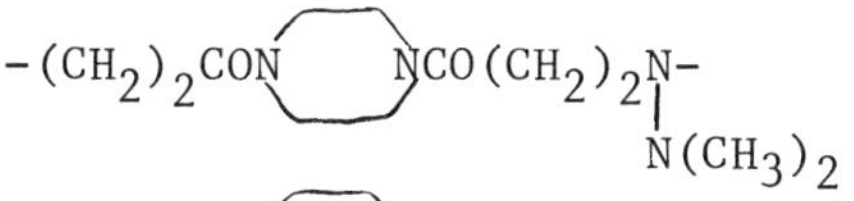	0.10	1,10
XXIX	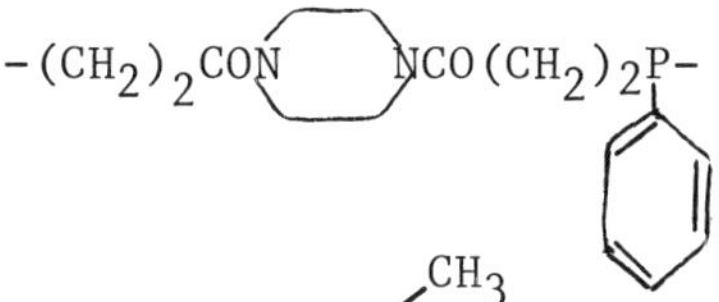	0.25	1,11
XXX	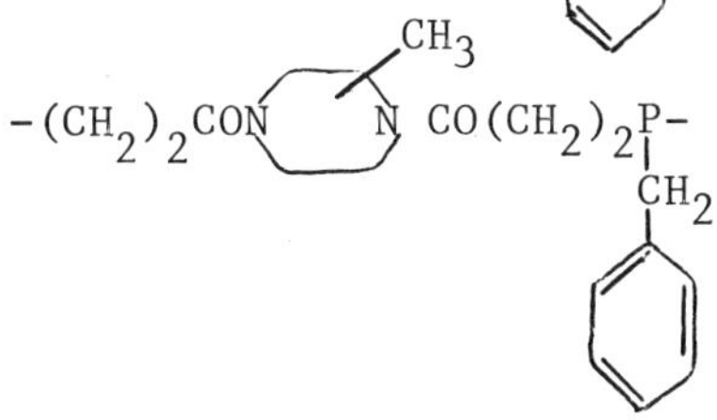	0.23	1,11

(a) in chloroform at 30°

(b) in dimethylsulfoxide at 100°

basicity constants of some poly(amido-amines) are reported in Table 4.

TABLE 4

BASICITY CONSTANTS OF SOME POLY(AMIDO-AMINES)*

Poly(amido-amine)	log K_1	log K_2	log K_3	log K_4	Ref.
I (Tab. 1)	7.79	--	--	--	5
V (Tab. 1)	7.01	2.98	--	--	12
VI (Tab. 1)	8.09	4.54	--	--	5
VIII (Tab. 1)	8.08	6.94	1.8	--	5
XVIII (Tab. 2)	8.87	4.10	--	--	8
XIX (Tab. 2)	9.02	7.91	4.47	2.15	8

*in 0.1 M NaCl at 20°.

Poly(amido-amines) having more than one basic nitrogen in their repeating unit are also capable of forming in aqueous solution stable complexes with heavy metal ions, e.g., with the Cu^{2+} ion, with the exception of those in which the aminic part has a piperazinic structure (13,14). When complexes are formed, sharp stability constants may be determined. Furthermore, solid Cu^{2+} complexes have been isolated in several instances, having a well defined composition, namely one Cu^{2+} ion per unit. Thus, in complex formation with metal ions poly(amido-amines) also exhibit a quite unusual behavior among polyelectrolytes.

HEPARIN ADSORBING RESINS BASED ON POLY(AMIDO-AMINES)

An interesting feature of several poly(amido-amines) is their ability to form stable complexes with heparin in aqueous solution at physiological pH. Heparin is a well known anticoagulant agent, widely used in the clinical practice. It is a mucopolysaccharide containing carboxyl- and sulfonic groups. In aqueous solution heparin behaves as a polyanion with a high density of negative charges (15).

In many cases, the anticoagulant activity of heparin must be inhibited when no longer needed. This is usually obtained by administering protamine sulfate. Protamine is a highly basic natural polymer of polypeptide structure. It is itself a powerful anticoagulant when administered alone, or in excess over heparin. Moreover, it has several untoward side effects, and the neutralization of

BIOMEDICAL APPLICATIONS OF POLY(AMIDO-AMINES)

heparin by protamine may be followed the so-called heparin rebound (15-19).

We have found that several poly(amido-amines) are able, in a linear form, to neutralize the anticoagulant ability of heparin, much as protamine sulfate does (20). This is apparently due to their heparin-complexing ability, which has been studied by H^1 and C^{13} NMR spectroscopy, and low-angle X-ray scattering. Some heparin-complexing poly(amido-amines) are listed in Table 5.

TABLE 5

SOME POLY(AMIDO-AMINES) ABLE TO NEUTRALIZE THE ANTICOAGULANT ACTIVITY OF HEPARIN

Polymer No.	Structure of the repeating unit(s)
I	See Table 1
VI	" "
XVIII	See Table 2
XXXI	$\left[-(CH_2)_2CON\bigcirc NCO(CH_2)_2-\right]_x - \left[\begin{array}{c}-N-\\|\\CH_2\\|\\CH_2\\|\\N(CH_3)_2\end{array}\right]_{0.5x} - \left[\begin{array}{c}-N-\\|\\CH_2\\|\\COOH\end{array}\right]_{0.5x}$ (a)
	$\left[-(CH_2)_2CON\bigcirc N CO(CH_2)_2-\right]_x - \left[\begin{array}{c}-N-\\|\\CH_2\\|\\CH_2\\|\\N(CH_3)_2\end{array}\right]_{0.5x} - \left[-N\bigcirc N-\atop COOH\quad COOH\right]_{0.5x}$ (a)

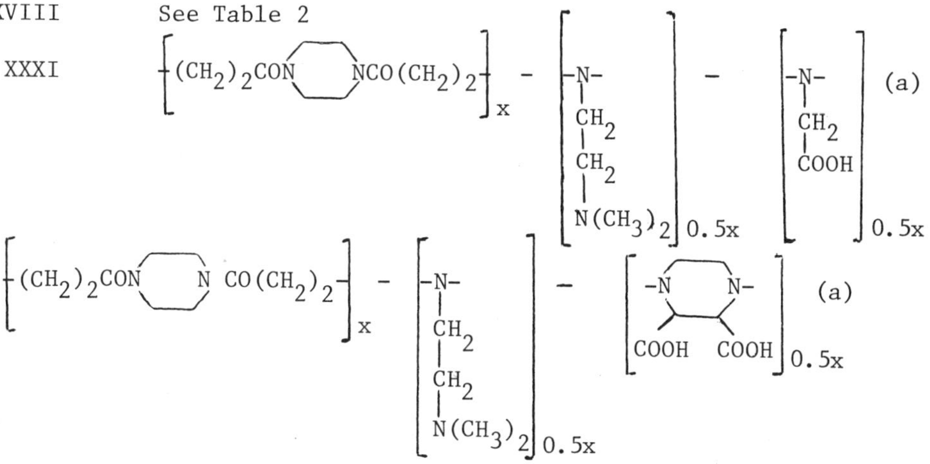

(a) Each aminic unit regularly alternates with an amidic unit, but the two aminic units are not expected to be regularly distributed, with respect to each other, along the macromolecular chain.

The heparin-complexing ability of poly(amido-amines) in aqueous solution lead us to assume that if the polyamido-amines could be obtained in a crosslinked, water-insoluble form, they would act as heparin-adsorbing resins.

It is apparent from the general scheme of poly(amido-amines) synthesis, that bis-primary amines act as tetrafunctional monomers. Consequently, crosslinked poly(amido-amines) can be obtained by substituting in the polymerization process a part of the aminic monomers for ethylenediamine, or other α, ω-bisamines (Scheme 2).

Scheme 2

x (Bisacrylamide) + x-a ("difunctional" amine) + a $H_2N(CH_2)_nNH_2 \longrightarrow$

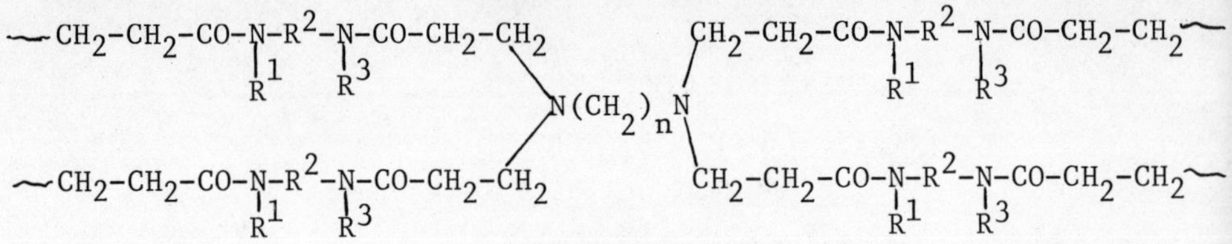

Crosslinked poly(amido-amines) obtained by this way are obviously insoluble in water. In aqueous media, however, they swell to a large extent, giving rather brittle gels. This was disadvantageous when these materials were used as de-heparinizing filters (see below). Consequently, we endeavoured to find another way to obtain cross-linked poly(amido-amines), giving products of better mechanical properties when swollen in aqueous media.

By substituting in the polymerization process a part of aminic monomers for allylamine, linear poly(amido-amines) having pendant allylic groups were obtained. Subsequently, these poly(amido-amines) were treated with a hydrophilic monomer (vinylpyrrolidone), in the presence of radical initiators. Crosslinked products in which the poly(amido-amine) chains were connected by polyvinylpyrrolidone chains, were finally obtained (Scheme 3). The reaction condition could be adjusted to have about 50% by weight of poly(amido-amine) chains in the final products (21).

The hydrophilic resins prepared by both methods proved to be able to remove heparin from plasma or blood, without producing unfavorable side effects on either the plasma proteins or the blood cells, including platelets. They can adsorb from 50 to 100% by weight of heparin.

Crosslinked poly(amido-amines) are by themselves inert towards the coagulation parameters of blood. We have demonstrated that they do not affect recalcification time, prothrombin time, thrombin time, or fibrinogen content of unheparinized blood, at any speed of filtration from 2 to 200 ml/min. Their biological effects and physical and chemical properties lead us to visualize their being used to make molecular filters capable of de-heparinizing blood. Application of the filters in clinical apparatus can be envisaged (21-23).

Scheme 3

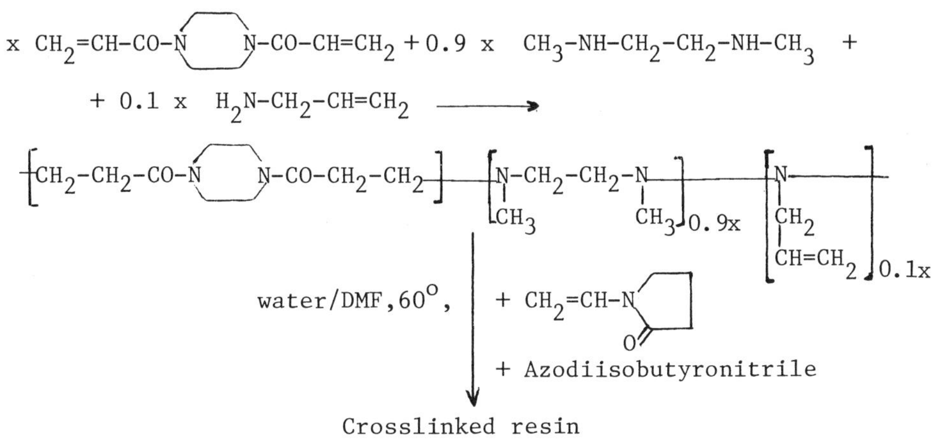
x CH2=CH-CO-N N-CO-CH=CH2 + 0.9 x CH3-NH-CH2-CH2-NH-CH3 +
+ 0.1 x H2N-CH2-CH=CH2
CH2-CH2-CO-N N-CO-CH2-CH2
N-CH2-CH2-N
CH3 CH3 0.9x
N
CH2
CH=CH2 0.1x
water/DMF,60°,
+ CH2=CH-N
O
+ Azodiisobutyronitrile
Crosslinked resin

BLOCK AND GRAFT COPOLYMERS OF POLY(AMIDO-AMINES) AS POTENTIALLY NON-THROMBOGENIC POLYMERIC MATERIALS

It is well known that the thrombogenic properties of most (if not all) non-physiological materials constitutes one of the most serious problems usually encountered in the field of artificial prostheses designed for long-term implantation in contact with blood.

An interesting approach to the preparation of non-thrombogenic materials is the adsorption of heparin on the surface of synthetic polymers, by using quaternary ammonium salts (24-37). In most cases, the main disadvantage of this method is a slow release of the quaternary ammonium salts into the bloodstream. Furthermore, it is known that most compounds containing ammonium groups have haemolytic properties and may exert adverse effects on platelets (31), and this probably applies also to quaternary ammonium group-bearing surfaces.

Crosslinked poly(amido-amines), besides being able to adsorb heparin stably, show on the whole an excellent haemocompatibility (see above). Consequently, they were good candidates for a use in the biomedical field as heparinizable materials. However, they do not possess satisfactory mechanical properties for constructing artificial prostheses. Consequently, we decided to modify their structure by preparing block copolymers containing poly(amido-amine) blocks and blocks of conventional polymers, in order to obtain products possessing good mechanical properties and retaining the ability to form stable complexes with heparin.

The synthesis of poly(amido-amine) block and graft copolymers was performed starting from poly(amido-amines) having purposely tailored end groups. Most of our studies have been performed starting from poly(amido-amine) VI (see Table 1), deriving from the polyaddition of N,N'-dimethylethylenediamine to 1,4-bis acryloyl piperazine. In the poly(amido-amine) synthesis (see Scheme 1), by using an excess of amine or of bisacrylamide in the monomeric mixture, poly(amido-amines) can be obtained predominantly end-capped with secondary-amino groups or acrylamido groups, respectively. The average molecular weight of the products can be roughly predetermined by using the stoichiometric ratio between the two types of the reacting functions, according to the general rules of the two-monomer stepwise polyadditions. That the poly(amido-amines) systems follow these rules has been previously demonstrated (2).

We have found that poly(amido-amines) having terminal acrylamido groups give block copolymers when treated with conventional vinyl monomers in the presence of radical initiators. This is a convenient way to obtain block copolymers, similar to that described by Smets and Schmets, who obtained graft copolymers from macromolecular matrices containing pendant polymerizable vinyl groups (38).

Polystyrene- poly(amido-amine) block copolymers were prepared by this way (39-42). In accord with the above considerations, their synthesis was carried out in two steps. First, a poly(amido-amine) of the desired $\bar{M}n$, and with terminal vinyl groups, was prepared. This poly(amido-amine) was then dissolved in a suitable solvent, and treated with styrene in the presence of radical initiators, thus obtaining the final product (Scheme 4). Any unreacted poly(amido-amine) which might have been present at the end of the reaction could be extracted with methanol.

The overall composition of the reaction products could be varied by using different amounts of styrene in the second step. The average $\bar{M}n$ of the poly(amido-amine) blocks ranged from 10,000 to 16,000. The polystyrene blocks has a $\bar{M}v$ ranging from 40,000 to 100,000, as determined in benzene solution after hydrolysis of the poly(amido-amine) blocks (42).

Poly(amido-amines) having terminal sec-amino groups may be prepared by using an excess of amine in the polymerization process. These poly(amido-amines) can be used to prepare other types of segmented polymers. For instance, coupling with isocyanate end-capped polyurethanes leads to poly(amido-amine) - polyurethane block copolymers (40). Furthermore, polyethylene - poly(amido-amine) graft copolymers have been prepared (43) by reacting partially chlorosulfonated polyethylene with an excess of poly(amido-amine) in the presence of triethylamine, and extracting with methanol any unreacted poly(amido-amine) at the end of the reaction (Scheme 5).

The physical properties of both polystyrene-poly(amido-amine) and polyethylene-poly(amido-amine) copolymers with a poly(amido-amine) content of 8-16% by weight have been evaluated, and found to be similar to that of slightly plasticized polystyrene and polyethylene

Scheme 4

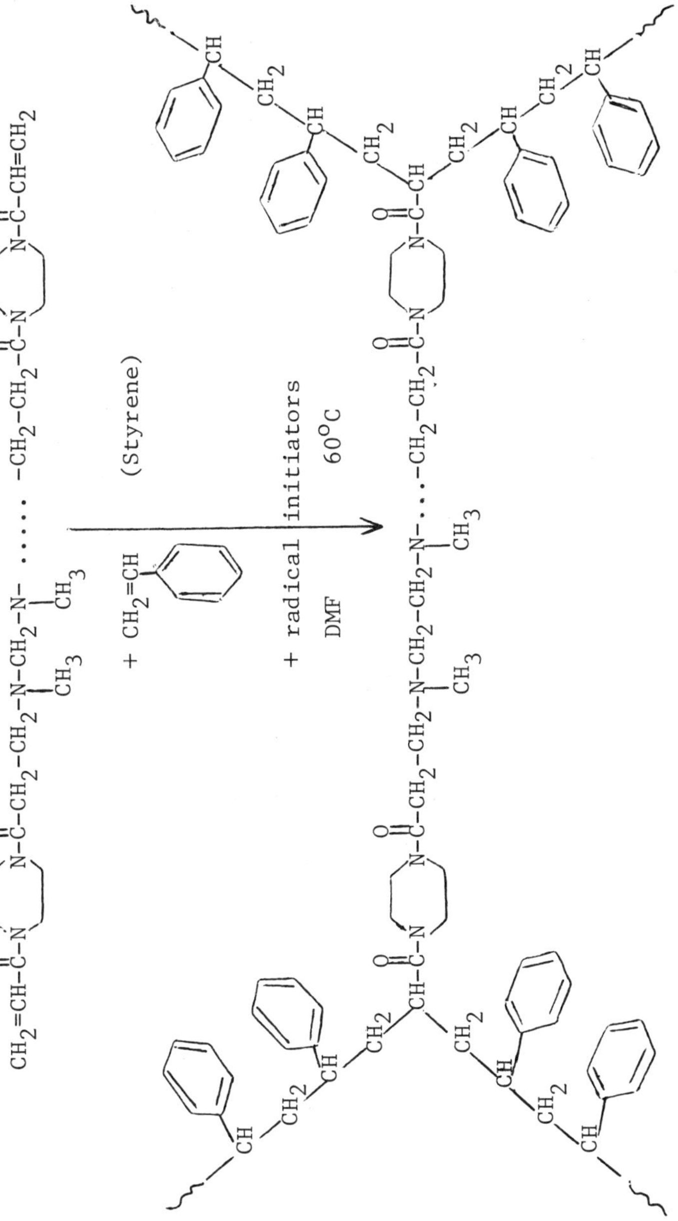
(Styrene)
+ radical initiators
DMF
60°C

respectively (42-44).

From both PS-PAA and PE-PAA copolymers, probes have been prepared for in vivo biocompatibility tests. These probes were in the form of short tubes, and after heparinization were inserted in the inferior vena cava of test animals (dogs) according to a well known antithrombogenicity test (45). The results obtained so far are very encouraging. After 2-weeks implantation, no thrombus formation was observed at the interior of heparinized probes of both PS-PAA and PE-PAA, having in both cases a PAA content of 10 to 16% by weight. Some animals were sacrificed after longer periods, up to 6 months. Also in this case, no thrombus formation took place on the probes. These results compare favorably with the literature data on the nonthrombogenic materials described so far.

Scheme 5

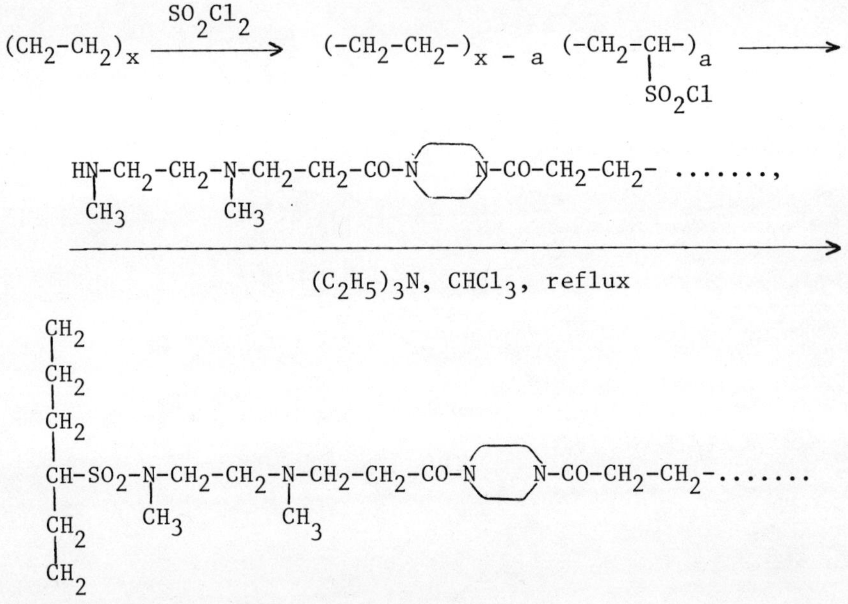

SO_2Cl_2
$(CH_2-CH_2)_x$
$(-CH_2-CH_2-)_{x-a}$ $(-CH_2-CH-)_a$
SO_2Cl
$HN-CH_2-CH_2-N-CH_2-CH_2-CO-N$ $N-CO-CH_2-CH_2-$ ......,
CH_3 CH_3
$(C_2H_5)_3N$, $CHCl_3$, reflux
CH_2
CH_2
CH_2
$CH-SO_2-N-CH_2-CH_2-N-CH_2-CH_2-CO-N$ $N-CO-CH_2-CH_2-$.......
CH_3 CH_3
CH_2
CH_2

GRAFTING OF POLY(AMIDO-AMINES) ON THE SURFACE OF MATERIALS

A further development of the above research, which is still in progress, is to graft poly(amido-amines) on the surface of several materials. In fact, the heparin-adsorbing ability is a surface property. It follows that grafting poly(amido-amine) chains on a given surface would impart to it the same properties found in the above graft copolymers.

Very promising results have been obtained in the case of glass. We have been able to graft poly(amido-amines) on glass by several methods, starting from a sec- amino end capped poly(amido-amine). This was reacted with a glass surface on which acylchloride groups had been introduced, by treating glass either with thionyl chloride, followed by glycerine and isophthaloylchloride (Scheme 6) or with γ-aminopropyltriethoxysilane followed by isophthaloylchloride (Scheme 7).

Scheme 6

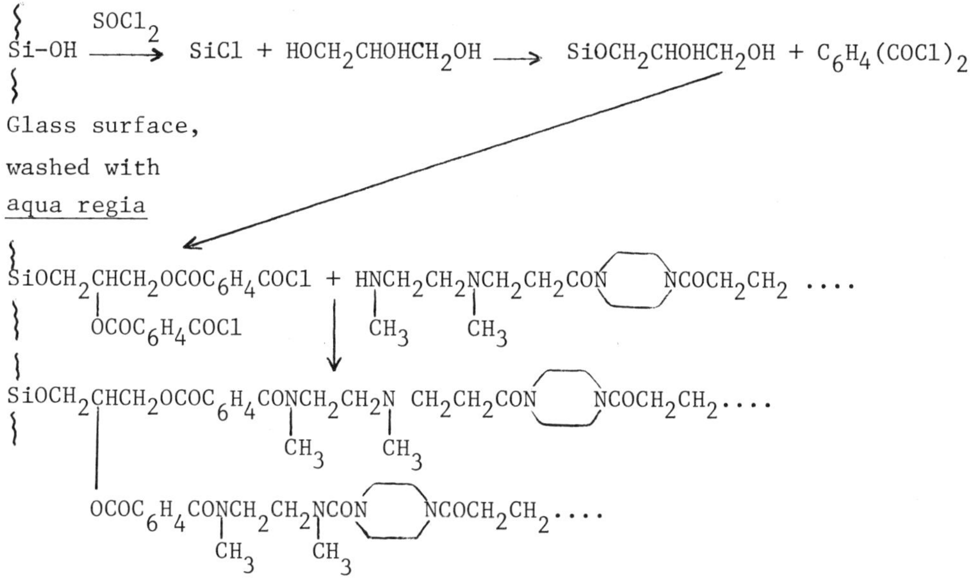

$SOCl_2$
Si-OH
SiCl + $HOCH_2CHOHCH_2OH$
$SiOCH_2CHOHCH_2OH$ + $C_6H_4(COCl)_2$
Glass surface,
washed with
aqua regia
$SiOCH_2CHCH_2OCOC_6H_4COCl$
$OCOC_6H_4COCl$
$HNCH_2CH_2NCH_2CH_2CON$
$NCOCH_2CH_2$ ....
CH_3
CH_3
$SiOCH_2CHCH_2OCOC_6H_4CONCH_2CH_2N$ CH_2CH_2CON
$NCOCH_2CH_2$....
CH_3
CH_3
$OCOC_6H_4CONCH_2CH_2NCON$
$NCOCH_2CH_2$....
CH_3
CH_3

A third method involves the preparation of a poly(amido-amine) containing γ-aminopropyltriethoxysilane groups as side substituents. This poly(amido-amine) can be directly attached to glass surfaces (Scheme 8). However, the amount of poly(amido-amine) grafted by this method appears to be much smaller than in the previous cases.

All reaction steps were monitored by ESCA. The final products are able to adsorb considerable amounts of heparin on their surface.

The biological evaluation of the heparinizable glass surfaces so obtained is still in progress. However, the first results are encouraging. This may have a considerable impact on the field of glass containers for blood storage, where the necessity to add relatively high quantities of anticoagulant agents gives often a number of problems in the medical and surgical practice. It may be noted, in fact, that the shape of the glass object to be treated is irrelevant, as far as the grafting reaction is concerned, as it is, to our present knowledge, the type of glass involved.

We are now endeavoring to find reliable methods to graft poly-

(amido-amines) on the surface of several polymeric materials which already find application in the biomedical field, such as, for instance, poly(vinylchloride) and silicone polymers. Thus we reasonably expect to be able in a future to prepare several new materials covering a wide range of physical and mechanical requirements, while maintaining the ability to become, after heparinization, permanently non-thrombogenic.

Scheme 7

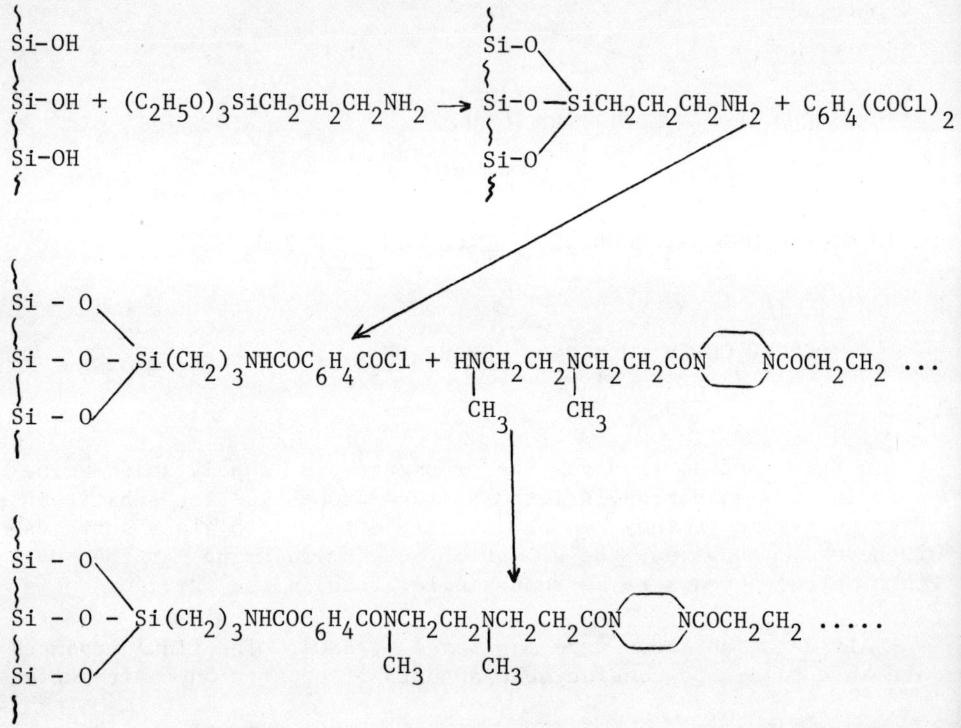
Si-OH
Si-OH + (C2H5O)3SiCH2CH2CH2NH2 → Si-O —SiCH2CH2CH2NH2 + C6H4(COCl)2
Si-OH
Si-O
Si-O
Si - O
Si - O - Si(CH2)3NHCOC6H4COCl + HNCH2CH2NCH2CH2CON NCOCH2CH2 ...
Si - O
CH3 CH3
Si - O
Si - O - Si(CH2)3NHCOC6H4CONCH2CH2NCH2CH2CON NCOCH2CH2 .....
Si - O
CH3 CH3

Scheme 8

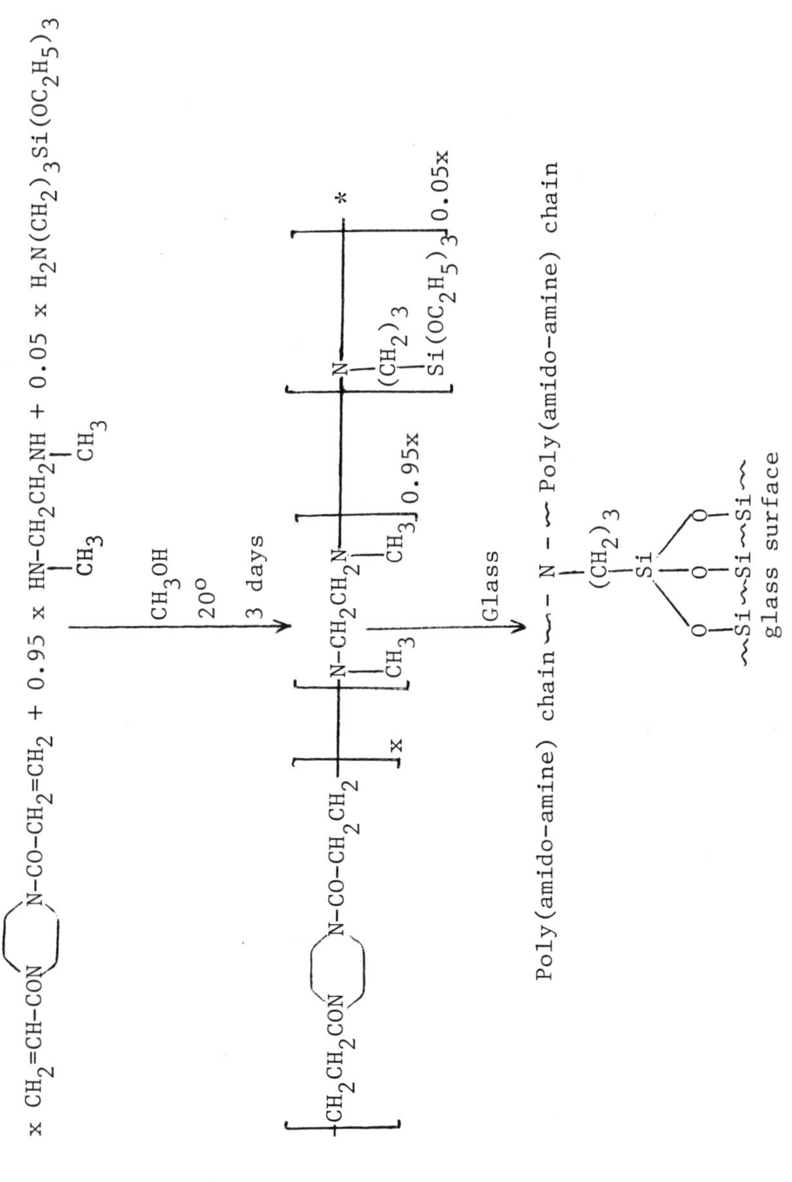

*See footnote to Table 5.

REFERENCES

1. F. Danusso, and P. Ferruti, Polymer, 11, 88 (1970)
2. F. Danusso, P. Ferruti, and G. Ferroni, Chimica e Industria (Milan),49, 271 (1967)
3. F. Danusso, P. Ferruti, and G. Ferroni, Chimica e Industria (Milan),49, 453 (1967)
4. F. Danusso, P. Ferruti, and G. Ferroni, Chimica e Industria (Milan),49, 587 (1967)
5. R. Barbucci, P. Ferruti, C. Improta, M. Delfini, A. L. Segre, and F. Conti, Polymer, 19, 1329 (1978)
6. R. Barbucci, P. Ferruti, C. Improta, M. La Torraca, L. Oliva, and M. C. Tanzi, Polymer, 20, 1298 (1979)
7. P. Ferruti, and R. Barbucci, in preparation
8. R. Barbucci, and P. Ferruti, Polymer, 20, 1061 (1979)
9. F. Danusso, P. Ferruti, and G. Ferroni, Chimica e Industria (Milan), 49, 826 (1967)
10. P. Ferruti, and Z. Brzozowski, Chimica e Industria (Milan), 50, 441 (1968)
11. P. Ferruti, and R. Alimardanov, Chimica e Industria (Milan), 49, 831 (1967)
12. R. Barbucci, P. Ferruti, M. Micheloni, M. Delfini, A. L. Segre, and F. Conti, Polymer, 21, 81 (1980)
13. R. Barbucci, V. Barone, and P. Ferruti, Atti Acc. Naz. Lincei (Rome), 64, 481 (1978)
14. P. Ferruti, L. Oliva, R. Barbucci, and M. C. Tanzi, Inorganica Chim. Acta, in press.
15. I. B. Jaques, in progress in Medical Chemistry, G. P. Ellis and G. B. West, Eds., Butterworths, London, vol. 5, p. 139 (1967)
16. E. T. Kimura, P. R. Young, R. J. Stein, and R. K. Richards, Toxicology and Appl. Pharmacology, 1, 185 (1959)
17. H. A. Perkins, G. Harkins, F. Gerbade, M. R. Roles,and D. I. Acra, J. Clin. Invest., 40, 1421 (1961)
18. J. N. Shamberge, H. Fukin, J. Lab. Clin. Med., 69, 927 (1967)
19. P. G. Frick, Surgery, 59, 721 (1966)
20. M. A. Marchisio, T. Longo, and P. Ferruti, Europ. Surg. Research, 4, 312 (1972)
21. M. A. Marchisio, P. Ferruti, T. Longo, and F. Danusso, U. S. Pat. 3,865,723 (1975)
22. M. A. Marchisio, T. Longo, P. Ferruti, and F. Danusso, Europ. Surg. Research, 3, 240 (1971)
23. M. A. Marchisio, T. Longo, and P. Ferruti, Experientia, 28, 93 (1973)
24. D. J. Lyman, Rev. Macromol. Chem., 1, 355 (1966)
25. T. Akutsu, "Artificial Heart", Igaku Shein Ltd., Tokyo (1975)
26. V. L. Gott, J. D. Whiffen, and R. C. Dutton, Science, 142, 1297 (1963)
27. V. L. Gott, J. D. Whiffen, D. E. Koepke, R. L. Dagget, W. C. Boake, and W. P. Young, Trans. Amer. Soc. Artif. Int. Organs, 10, 213 (1964)
28. J. D. Whiffen, and V. L. Gott, J. Thorac. Cardiovasc. Surg., 50,

31 (1965)
29. R. I. Leininger, H. H. Epstein, R. D. Falb, and G. A. Grode, Trans. Amer. Soc. Artif. Int. Organs, 12, 151 (1966)
30. J. G. Eriksson, G. Gilbert, and H. Lagergren, J. Biomed. Mater. Res., 1, 301 (1967)
31. R. I. Leininger, R. D. Falb, and G. A. Crode, Ann. N.Y. Acad. Sci., 146, 11 (1968)
32. G. A. Grode, S. J. Anderson, H. M. Grotta, and R. D. Falb, Trans. Amer. Soc. Artif. Int. Organs, 15, 1 (1969)
33. H. Lagergren, and J. G. Eriksson, Trans. Amer. Soc. Artif. Int. Organs, 17, 10 (1971)
34. L. S. Hersh, H. H. Weetall, and I. W. Brown, J. Biomed. Mat. Res. Symp., 1, 99 (1971)
35. R. I. Leininger, J. P. Crowley, R. D. Falb, and G. A. Grode, Trans. Amer. Soc. Artif. Int. Organs, 18, 312 (1972)
36. H. S. Hersh, V. L. Gott, and F. Najjar, J. Biomed. Mat. Res. Symp., 3, 85 (1972)
37. R. D. Falb in "Polymers in medicine and surgery", R. L. Kronenthal, Z. Ozer, and E. Martin Eds., Polymer Science and Technology, Vol. 8, Plenum Press, New York, (1975)
38. G. Smets, and J. Schmets, Bull. Soc. Chim. Belg., 62, 358 (1953)
39. P. Ferruti, and L. Provenzale, Transplant. Proceed., 8, 103 (1976)
40. P. Ferruti, E. Martuscelli, F. Riva, and L. Provenzale, U. S. Patent, 4,093,677 (1978)
41. P. Ferruti, L. Provenzale, E. Martuscelli, and F. Riva, Chimica e Industria (Milan), 58, 539 (1976)
42. P. Ferruti, D. Arnoldi, M. A. Marchisio, E. Martuscelli, M. Palma, and F. Riva, J. Polymer Sci., Polym. Chem. Ed., 15 2151 (1977)
43. E. Martuscelli, L. Nicolais, F. Riva, P. Ferruti, and L. Provenzale, Polymer, 19, 1063 (1978)
44. P. Ferruti, E. Martuscelli, L. Nicolais, M. Palma, and F. Riva, Polymer, 18, 387 (1977)
45. V. L. Gott, and A. Furuse, Fed. Proc., 30, 1679 (1971)

COVALENT BONDING OF COLLAGEN AND ACRYLIC POLYMERS

Douglas R. Lloyd and Charles M. Burns

Department of Chemical Engineering
Virginia Polytechnic Institute & State University
Blacksburg, Virginia 24061

University of Waterloo
Waterloo, Ontario
Canada, N2L 3G1

INTRODUCTION

The ability to establish chemical bonding between biological substrates and polymeric prosthetic devices has long been a goal of biomedical researchers. Although significant advancements have been made towards improved biocompatibility of synthetic polymeric materials and biomedical adhesives, the ability to establish permanent bonding under physiological conditions is a goal yet to be fully realized. Attempts to provide the desired adhesion through mechanical bonding have proven to be neither convenient nor permanent (1). For this reason attention has turned to adhesion by chemical bonds. One approach to this problem has been to establish secondary chemical bonds (2-6). These bonds, however, usually do not provide as strong or as permanent a bond as does the establishment of primary chemical bonds (5). We report here the initial stages of work designed to develop an adhesive agent capable of establishing primary chemical bonding between biological substrates and synthetic polymers of biomedical interest.

The literature reveals the demonstrated biomedical acceptability of many acrylic-based polymers (7-12). The work reported here, therefore, is concerned with acrylic-based synthetic polymers. A great deal of work has been done in the area of adhesion of synthetic polymers to both soft and hard tissue (1-6). In light of the multitude of biological substrates of possible interest, and in an attempt to make the research as broad as possible, we concentrate on adhesion or coupling to collagen, the most abundant protein in the mammalian body (13). Due to the variety of conditions encountered in the physiological environment and in an attempt to establish as strong a bond as possible between these two polymeric materials, the goal of covalent primary bonding was selected.

A survey of the literature reveals a number of areas in which the coupling of collagenlike polypeptides and synthetic polymers has been attempted, namely in enzyme immobilization (14-17), leather tanning (18-20), and biomedical adhesives (6). Considering the polymeric systems of interest here, the type of bonding desired, and the imposed restrictions for coupling reactions in a physiological environment (Table I), the material selected to serve as a coupling agent was a water-soluble carbodiimide (WSC). The particular carbodiimide employed in this study was 1-cyclohexyl-3-[2-morpholinyl-(4)-ethyl]carbodiimide metho-p-toluene sulfonate:

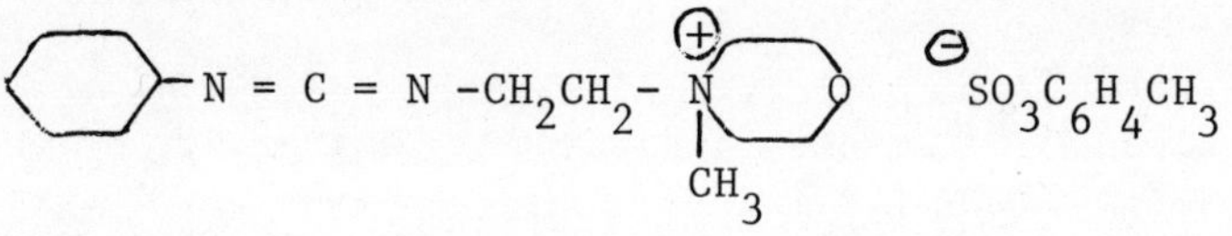
N = C = N -CH2CH2- N
O
CH3
SO3C6H4CH3

The chemistry of carbodiimides has been reviewed by Khorana (21) and more recently by Kurzer and Douraghi-Zadeh (22). They point out that under proper conditions carbodiimides have been made to react with, among other functional groups, carboxylic groups (23), amino groups (24), guanidines (25), phenolic groups (26), amino alcohols (27), and sulfhydryls (28), all of which are found in collagen. The reaction which is most relevant to the present research, because of the nature of the synthetic polymer and the reaction conditions employed, is that of carbodiimides reacting with carboxylic groups. Taking place in an aqueous environment of slightly acidic pH and at room temperature, it has been shown possible to couple carboxylic-containing macromolecules (M_1—COOH) and macromolecules containing nucleophilic functional groups (M_2—H) through the use of carbodiimides (21,23,29,30). The overall reaction sequence leading to coupling has been postulated by Khorana (21) and may be summarized as

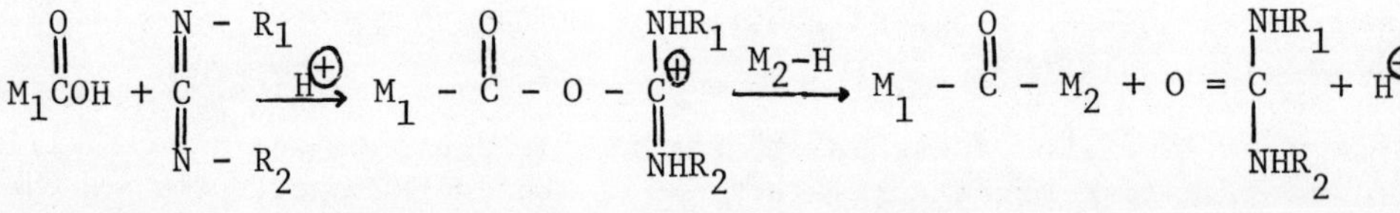
M1COH
N - R1
N - R2
H
M1 - C - O - C
NHR1
NHR2
M2-H
M1 - C - M2 + O = C
NHR1
NHR2
+ H

The coupling mechanism originally postulated by Khorana (21) entails the modification of a pendant carboxylic group by the protonated carbodiimide. The modified carboxyl subsequently undergoes nucleophilic attack. In the case under study it is assumed that because of the greater availability of carboxyl groups on the acrylic polymer relative to the protein the majority of the modified carboxyl groups are located on the acrylic polymer. One purpose of the present study was to investigate the role played by several of the amino acid functional groups in the completion of the coupling reaction. To carry out this investigation of the effect of functional groups two sets of experiments were conducted. In the first a series of homopolymers of α-amino acids was used. In the second a blocking technique was used to modify some of the functional groups of the collagen.

To investigate the effect of the conformation of the protein, the coupling of collagen and its denatured counterpart gelatin were studied under identical conditions. Because of the heterogeneity of the reaction and the macromolecular nature of the collagen triple helix, accessibility of functional groups on the protein should play a role in determining the extent of coupling.

EXPERIMENTAL

Materials

Acid-soluble collagen and the homopolymers of a number of α-amino acids were obtained from the Sigma Chemical Co., St. Louis, Missouri, and used without further purification.

The water-soluble carbodiimide 1-cyclohexyl-3-[2-morpholinyl-(4)-ethyl] carbodiimide metho-p-toluene sulfonate was supplied by the Aldrich Chemical Co., Milwaukee, Wisconsin, and used as received.

Acrylic acid (AAc), from Eastman Chemical Co., Rochester, New York, was distilled before use. Acrylamide (AAm), from J. T. Baker Chemical Co., Toronto, Ontario, was recrystallized from chloroform before use. N,N'-methylene-bisacrylamide (Bis), from Eastman Chemical Co., was used as received.

Succinic anhydride (highest purity grade) and ninhydrin (certified grade) were obtained from Fisher Scientific Co., Toronto, Ontario, and used without further purification.

All other materials were reagent grade and used as received.

Table I

RESTRICTIONS IMPOSED ON PHYSIOLOGICAL COUPLING REACTIONS

(1) Should take place at temperatures between 20 and 37°C.
(2) Must not be highly exothermic.
(3) Must be relatively rapid under restriction (1).
(4) Must not be negated by the presence of oxygen or water.
(5) Must take place at or near normal pH values.
(6) The established bond must not break down under adverse conditions of pH, temperature, etc.
(7) The reactants and products must not be toxic.

Table II

COMPOSITION OF CROSSLINKED COPOLYMERS, co(AAc/AAm)

Copolymer designation	Monomer feed composition[a]: Weight AAc / Weight (AAc + AAm) (%)	Weight AAm / Weight (AAc + AAm) (%)	Reaction conditions: pH	Temperature (°C)	Time (min)	Copolymer composition: Weight AAc / Weight (AAc + AAm + Bis) (% ± 2%)	Weight (AAm + Bis) / Weight (AAc + AAm + Bis) (% ± 2%)
co(AAc/AAm)I	0	100	6.5	0	30	0	100
co(AAc/AAm)II	50	50	6.5	0	30	19	81
co(AAc/AAm)III	95	5	6.5	0	30	46	54
co(AAc/AAm)IV	97.5	2.5	4.0	25	30	79	21
co(AAc/AAm)V[b]	...	...	...	..	..	95	5

[a]The reactive feed, expressed as wt %, was 93.73% monomer mixture, 4.93% Bis, 1.23% ammonium persulfate, 0.11 N,N,N',N'-tetramethylethylenediamine.

[b]co(AAc/AAm)V was purchased from BioRad Laboratories of Richmond, California; their reference for the material is Bio-Gel CM2, "a fully carboxylated acrylamide gel".

Preparation of Acrylic Copolymers

A series of copolymers of acrylic acid (AAc) and acrylamide (AAm), cross-linked with Bis, were either prepared or, as in the case of co(AAc/AAm)V, purchased. Those synthesized in the laboratory were produced from freshly distilled AAc and AAm recrystallized from chloroform. The solution copolymerization and crosslinking, when carried to approximately 5% conversion, yielded an insoluble gel that was subsequently broken in a blender and forced through a 20-mesh screen, washed in 0.5N HCl, washed in distilled water, freeze-dried, and stored in a desiccator under vacuum. By controlling both the pH of the reaction medium and the feed ratio of the two monomers, the series of copolymers listed in Table II was prepared, the analysis being by the titration, with 0.1N NaOH, of a suspension of the copolymer in 0.1N HCl to determine the AAc content of the copolymer.

Coupling Method and Analysis

The coupling experiments were carried out in an aqueous, heterogeneous mixture with insoluble copolymer and soluble protein and carbodiimide. The solvent employed was 0.05M KH_2PO_4 adjusted to a pH of 3.7. Stock solutions of the protein were prepared approximately 20 hr before use and stored at 5°C. Similarly, a stock solution of WSC in 0.05M KH_2PO_4 was prepared. The reactions were carried out in a stirred 30 ml beaker by adding an aliquot of the protein solution containing the desired amount of protein. The appropriate weight of copolymer was added next, and finally, the predissolved WSC. In all cases, the total volume of solvent plus solvated material added was 7.0 ml. The reactions were carried out at 20 $\pm$ 1°C in air.

The reaction was halted at the desired time by swamping the vessel with an additional four times the volume of solvent, thereby limiting the possibility of further intermolecular reactions. The contents of the reaction vessel were then immediately transferred to a fritted glass filter, where they were washed and filtered. The complete time lapse between the initial addition of excess solvent and the finishing of the filtering was less than 30 sec. The gel was then transferred back to the beaker, washed with fresh 0.05M KH_2PO_4 for 30 min, and filtered. This procedure was repeated with distilled water, 0.1M $NaHCO_3$, 0.001M HCl, 0.5M NaCl, and finally distilled water.

The filtrate collected above was sampled for analysis, and the volume of filtrate was recorded. The protein content of the filtrate was determined via the method of Lowry et al (31), and by difference, the protein attached to the copolymer was determined. The gel was lyophilized, weighed, and hydrolyzed in 6N HCl for 36 hr at 105°C. The contents were then cooled and filtered to separate the hydrolyzed amino acids from the copolymer gel. From this solution, samples were prepared and analyzed on a Bechman 105B auto amino acid analyzer.

From the amount of amino acid in the analyzed sample, it was possible to calculate the amount of protein coupled to the copolymer.

RESULTS AND DISCUSSION

Preliminary Experiments

Since there is little or no buffering capacity in the 0.05M KH_2PO_4, it was thought that the pH may drop to levels that could cause the denaturation of the collagen. Therefore, the pH was monitored during a number of preliminary experiments, typical results being presented in Table III. It can be seen that the pH varied over a narrow range of values. A survey of the literature reveals that for temperatures as high as 37°C, calf-skin collagen is stable to denaturation in solvents of pH as low as 1.2 (32,33). It can therefore be assumed that for the pH values being studied here, the collagen is stable.

In order to verify the validity and accuracy of the methods of coupling and analysis, preliminary experiments were carried out. In the first experiment, the collagen in solution and co(AAc/AAm)II were stirred together without the WSC for 17 hr and then washed and filtered in the usual fashion. Within the experimental error of ± 1%, 100% of the collagen was found to be present in the filtrate, indicating that without the WSC, no collagen was being picked up by the acrylic gel. In a second experiment, a solution of collagen without the acrylic gel was charged with an appropriate amount of WSC and the reaction allowed to proceed for 17 hr. The mixture was then transferred to the filter, washed, and the filtrate analyzed. Again, within experimental error, 100% of the collagen was recovered in the filtrate, indicating that the WSC and collagen by themselves were not forming an insoluble product. A third experiment was to test the accuracy of the two analytical techniques relative to each other. The two methods produced the same answer to within 0.1 mg in a total of 5 mg. The method of conducting the experiments and the techniques of analysis have been shown to be valid and accurate.

Investigation of Coupling Reaction Variables

The influence of a number of possible variables affecting the coupling reaction was studied through a series of experiments. The results for each set of experiments are presented in Figures 1-6. In each set of experiments, the total volume of liquid plus solvated reactants was 7.0 ml, and the temperature was 20 ± 1°C. The other parameters were varied according to the variable under investigation. "Coupling efficiency" is used as a measure of success and is defined as the weight percentage of the initial protein present that is successfully coupled to the copolymer.

Table III

VARIATION OF pH DURING COUPLING REACTION

		pH
(a)	Solvent alone	3.74
(b)	Solvent plus collagen at 3.5 mg/ml	4.30
(c)	(b) plus 140 mg co(AAc/AAm)II	3.21
(d)	WSC/solvent solution at 40 mg/ml	3.77
(e)	(c) upon adding (d) (Note: (e) is initial pH of reaction mixture)	3.27
(f)	Reaction mixture after 1 min	3.19
(g)	Reaction mixture after 30 min	3.18

Influence of Copolymer Composition. The major course of the coupling reaction had been assumed to be via the carbodiimide modification of the carboxyl groups (21,23,29,30). In order to verify this assumption, a series of tests was conducted. In these tests, all parameters were held constant at the values given in the heading for Figure 1, the composition of the copolymer being the only variable. A series of copolymers of approximately 0, 19, 46, 79, and 95% by weight acrylic acid was employed as the synthetic copolymer. The results shown in Figure 1 indicate a trend of increasing coupling efficiency with increasing acrylic acid content of the copolymer.

The relationship is essentially linear in the region of acrylic acid (AAc) content greater than the 0.20 weight fraction of the copolymer. In the region of AAc content between 0.0 and 0.2, the relationship is sharply curved. The coupling of protein to the copolymer of approximately zero percent AAc ($\pm$ 2%) is attributed to a small amount of the AAm functional groups having been hydrolyzed to the carboxyl form during the highly exothermic polymerization.

The results shown in Figure 1 confirm that the carboxyl group is a main ingredient in the coupling reaction. The conclusion may be drawn that the greater the carboxyl content of the copolymer, the greater will be the chances of coupling to the collagen. Since the most suitable substrate, co(AAc/AAm)V, was purchased and its preparation and characteristics therefore were not readily available, the copolymer chosen to be used in the remaining experiments was the copolymer of 79% by weight acrylic acid, co(AAc/AAm)IV. The coupling efficiency yielded by this copolymer in these experiments was 18%.

Influence of Reaction Time. The rapid drop and leveling off in pH evidenced in Table III indicated that the reaction was essentially completed in a very short time. To investigate this aspect further,

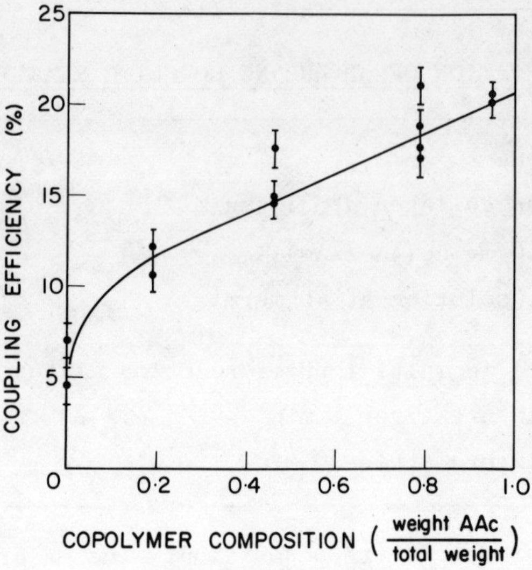
25
20
15
10
5
0
0·2
0·4
0·6
0·8
1·0
COUPLING EFFICIENCY (%)
COPOLYMER COMPOSITION (weight AAc / total weight)

Figure 1. Increase of coupling efficiency with proportion of acrylic acid in copolymer. Conditions: time, 17 hr; collagen, 2.5 mg/ml; copolymer, 20.0 mg/ml; coupling agent, 11.4 mg/ml.

a complete series of runs was carried out. The results of these tests are shown in Figure 2. It is evident that the reaction does take place in a very short time span. If we take 18% as maximum coupling efficiency under these conditions, it can be seen that within 5 min, 45% of this level is reached. In 15 min, 71% of the maximum is reached, and in 30 min, 90% of total coupling has been achieved. Although precise kinetic studies of the reaction have not been made, the results of these experiments have been beneficial. They have indicated that maximum coupling is attained within 60 min. This result allowed the reaction time for all further experiments to be shortened from 17 to 1 hr.

In a somewhat similar work, Hoare and Koshland investigated the kinetics of the reaction between glycine methyl ester, N-benzyl-N'-3-di-methylamino-propyl carbodiimide, and various enzymes in 1.0M acetate buffer at pH 4.75 (29). A comparison of the present results with their data is given in Table IV. It can be seen that their results tend to confirm the present findings.

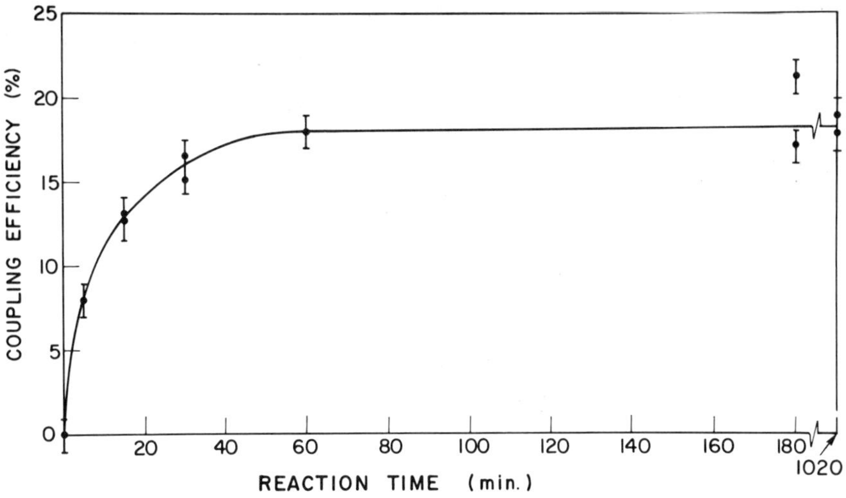
COUPLING EFFICIENCY (%)
25
20
15
10
5
0
20
40
60
80
100
120
140
160
180
1020
REACTION TIME (min.)

Figure 2. Dependence of coupling efficiency on time of reaction. Conditions: co(AAc/AAm)IV, 20 mg/ml; collagen, 2.5 mg/ml; coupling agent, 11.4 mg/ml.

Table IV

COMPARISON OF REACTION RATE WITH DATA FROM LITERATURE (29)

Time of reaction (min)	Present results (% of maximum coupling)	Data of ref. 29 (% of maximum conversion)
0	0	0
5	45	42
15	71	77
30	90	95

Influence of Copolymer co(AAc/AAm)IV Concentration. A set of experiments was conducted, in which the composition of the copolymer was fixed and the availability of carboxyl groups was varied by varying the concentration of the copolymer. The results in Figure 3 show an increase in coupling efficiency with increasing availability of carboxyl groups (i.e., copolymer concentration). These results again confirm that the carboxyl functional groups are a major participant in the coupling reaction.

The leveling off of the coupling efficiency can be explained because all, or at least the majority, of the carbodiimide added has been consumed by the leveling-off point, and that additional acrylic acid groups added beyond this point remain inactive. A second factor contributing to the leveling off of the coupling efficiency versus copolymer concentration curve may be viscosity. Since at the upper concentrations stirring begins to get difficult, this limits the accessibility and freedom of movement of the collagen chains. It is assumed that this contributes to a lesser degree than the factor discussed above.

The result of this set of experiments may be used as an indication as to what to expect when solid surfaces are tried. The coupling efficiency will obviously be a function of the accessibility of functional groups--both carboxyl on the polymer and nucleophile on the collagen.

Based on the results of this set of experiments and on the experiments that have gone before, the choice for concentration was 20 mg/ml for the remaining experiments.

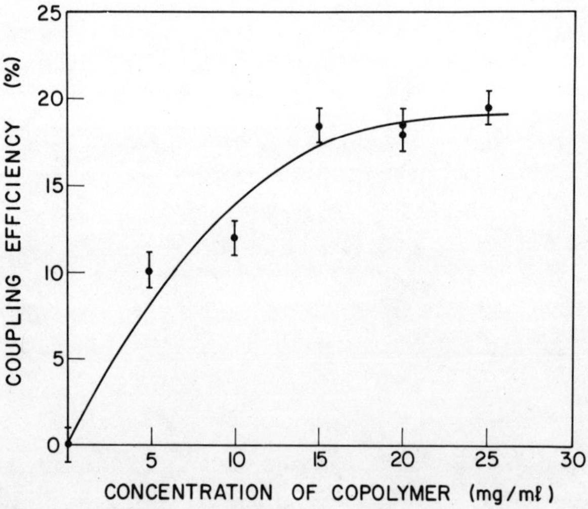
25
20
15
10
5
0
COUPLING EFFICIENCY (%)
5
10
15
20
25
30
CONCENTRATION OF COPOLYMER (mg/ml)

Figure 3. Increase of coupling efficiency with concentration of co(AAc/AAm)IV. Conditions: time, 60 min; collagen, 2.5 mg/ml; coupling agent, 11.4 mg/ml.

Influence of Concentration of Coupling Agent. The fourth variable to be investigated was the concentration of the coupling agent, 1-cyclohyxyl-3-[2-morpholinyl-(4)-ethyl]carbodiimide metho-p-toluene sulfonate. The results, presented in graphical form in Figure 4, indicate that as the concentration of coupling agent increases, so does the coupling efficiency. This trend holds until a concentration of approximately 10 mg/ml is attained. At this point, the coupling efficiency levels off, the addition of more coupling agent being superfluous. For consistency with previous experiments and experimental convenience, a concentration of 11.4 mg/ml WSC was selected for subsequent experiments.

The leveling off of the curve in Figure 4 indicates that the carbodiimide is not the limiting reactant. Since in each experiment a portion of the collagen remains uncoupled, collagen is not the limiting reactant. These two factors lend support to the postulate that the carboxyl group seems to be the limiting reactant. These findings will be discussed further in a later section.

Influence of Concentration of Acid-Soluble Collagen. A series of experiments of variable collagen concentration was conducted. When coupling efficiency versus concentration of collagen is plotted (Fig. 5), the results indicate that at the lower limit of relative

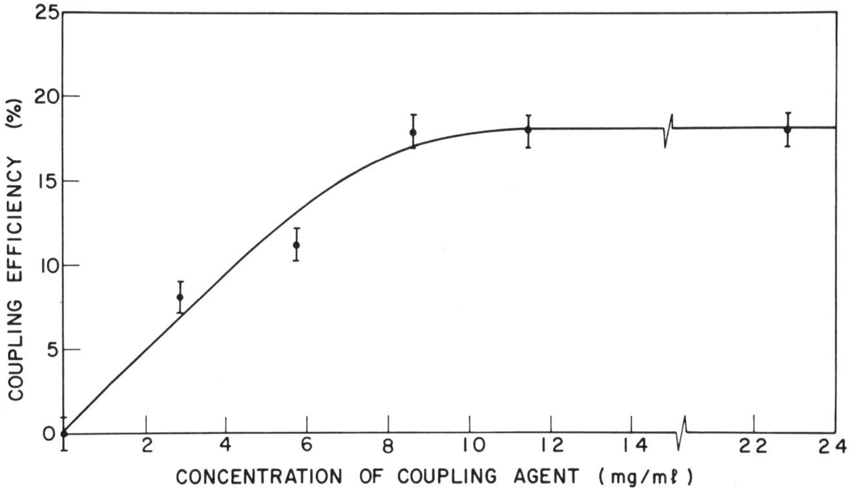
COUPLING EFFICIENCY (%)
25
20
15
10
5
0
2
4
6
8
10
12
14
22
24
CONCENTRATION OF COUPLING AGENT (mg/mℓ)

Figure 4. Variation of coupling efficiency with the concentration of coupling agent 1-cyclohexyl-3-[2-morpholinyl-(4)-ethyl] carbodiimide metho-p-toluene sulfonate. Conditions: time, 60 min; collagen, 2.5 mg/ml; co(AAc/AAm)IV, 20 mg/ml.

protein concentration, total coupling is possible. The variation of amount of collagen coupled with concentration of collagen initially present is also plotted in Figure 5 and serves as a more meaningful basis for discussion. As the concentration of collagen present is increased and the molecules couple to the available modified carboxyl groups on the copolymer, the number and accessibility of reaction sites on the copolymer decreases. The result is that not all of the collagen is capable of reacting with the limited number of modified carboxyl groups available. Thus, the coupling efficiency decreases, although the absolute amount of collagen coupled increases. Eventually the limit is reached where all of the potentially reactive

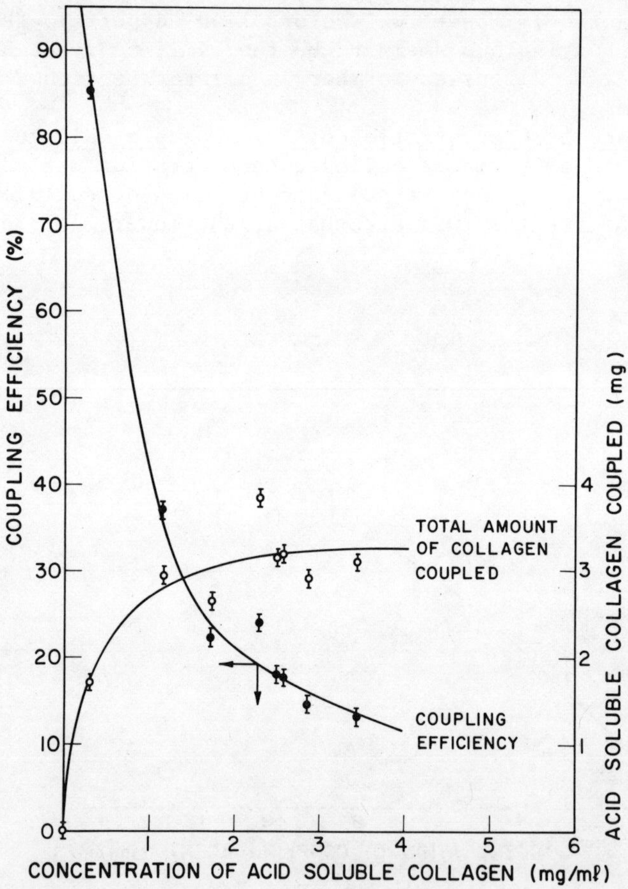
COUPLING EFFICIENCY (%)
ACID SOLUBLE COLLAGEN COUPLED (mg)
CONCENTRATION OF ACID SOLUBLE COLLAGEN (mg/mℓ)
TOTAL AMOUNT OF COLLAGEN COUPLED
COUPLING EFFICIENCY
0
10
20
30
40
50
60
70
80
90
1
2
3
4
5
6

Figure 5. Dependence of coupling and coupling efficiency on concentration of acid-soluble collagen. Conditions: time, 60 min; co(AAc/AAm)IV, 20 mg/ml; coupling agent, 11.4 mg/ml.

carboxyl groups have been reacted or masked, and the addition of more collagen is redundant. Therefore, the amount of collagen coupled has reached a plateau, and the plot in Figure 5 levels off. The coupling efficiency curve does not level off, however, since the addition of more collagen just decreases the efficiency (recalling that coupling efficiency is defined as the weight of collagen coupled divided by the total weight of collagen added).

These results again confirm the assertion that the limiting factor is the accessibility of carboxyl groups.

For the remainder of the experiments, it was decided to use a concentration of collagen that was safely in the plateau region of Figure 5; specifically, a concentration of 2.5 mg of acid-soluble collagen per millilitre of reaction volume.

Influence of pH of the Reaction Medium. In these experiments, the solvent was a 0.05M KH_2PO_4 solution, the pH of which, before dissolving the collagen, was adjusted by the addition of either concentrated HCl or 10M NaOH, depending on whether a pH below or above the normal pH of 0.05M KH_2PO_4 (pH 4.4) was required. The results of pH 3.7, 4.4, and 5.4 were collected, using the optimal conditions of concentration, time, etc. Attempts to determine the coupling efficiency at pH 6.4 proved unsuccessful, since the collagen was not soluble at the desired concentrations, apparently due to the proximity of the isoelectric point (34).

The result shown in Figure 6 indicate that the coupling efficiency increases with increasing pH, or conversely, decreases with increasing concentration of hydrogen ions. Theoretically, a certain

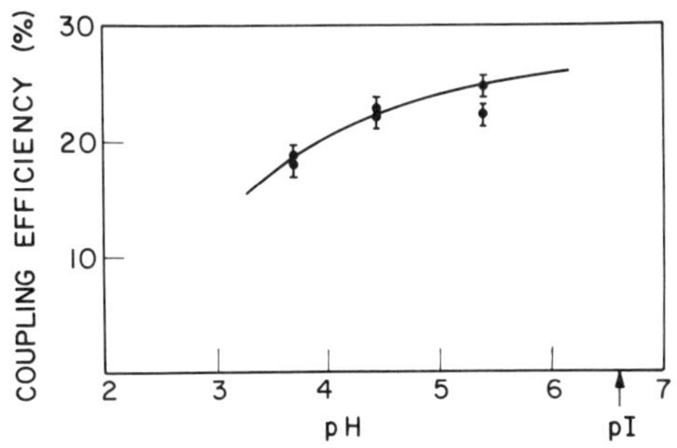
COUPLING EFFICIENCY (%)
30
20
10
2
3
4
5
6
7
pH
pI

Figure 6. Dependence of coupling efficiency on initial pH of the reaction medium. Conditions: time, 60 min; collagen, 2.5 mg/ml; co(AAc/AAm)IV, 20 mg/ml; coupling agent, 11.4 mg/ml.

concentration of hydrogen ions is required to cause the protonation of all of the carbodiimide, thereby initiating the proposed coupling action. Any excess hydrogen ions therefore would cause a shift in the equilibrium between the ionized and unionized forms of the nucleophilic groups, the ionized (cation) form resulting in a reduction of the number of available nucleophiles and a reduced potential for the coupling reaction. The results presented in Figure 6 are consistent with this interpretation.

Swelling of the copolymer was observed to increase with increasing pH, which was expected, since the pK_a for carboxyl groups is in the range of 3.0--4.5. This swelling will also contribute to the increased coupling with increased pH by increasing the number of accessible carboxylic groups on the copolymer.

Coupling of Homopolymers of α-Amino Acids

To study the roles played by the various functional groups of collagen in completing the coupling reaction, a number of homopolymers of α-amino acids were obtained. If polymers of the α-amino acids are used (as opposed to the monomeric amino acid) the end-group influence is minimized, although not completely eliminated. In addition, the conditions are representative of those found in the experiments with collagen, although conformation of the different macromolecules in solution are not likely to be the same. Those polymers that were soluble in 0.05M KH_2PO_4 at pH 4.4 were used in coupling reactions under the conditions determined as optimal in earlier studies (i.e., time = 60 min, temperature = 20° ± 1°C, copolymer of 79% by weight AAc and 21% by weight AAm plus Bis, concentrations: copolymer = 20 mg/ml, protein = 2.5 mg/ml, WSC = 11.4 mg/ml). The results of coupling these poly(amino acid) macromolecules are presented in Table V.

To present the results for the poly(amino acid)s on a comparative basis two adjustments to the raw data were made. First, to make direct comparisons between amino acids one particular functional group was selected as a basis. Franzblau, Gallop, and Seifter (35), among others, have described the ability of amino functional groups to complete the reaction with the modified carboxyl group; therefore all results were reported relative to the coupling capabilities of poly-L-lysine. The second adjustment to the raw data accounts for the different number of repeat units [i.e., degree of polymerization (DP)] and thus for the different number of functional groups found in the various homopolymers. This adjustment is necessary to put the coupling capabilities of the various functional groups on an identical basis of number of functional groups per terminal group (because terminal amino groups may enter the reaction). By doing so the influence of the terminal groups is further minimized. Therefore all results were normalized to a degree of polymerization of 385 which corresponds to that reported by the suppliers of poly-L-lysine.

Table V

COUPLING OF HOMOPOLYMERS OF α-AMINO ACIDS TO ACRYLIC COPOLYMER

Protein	Molecular weight	Degree of polymerization	Coupling efficiency %	Coupling as a percent of that for poly-L-lysine (normalized to DP = 385)
Poly-D, L-Ala Type I	5,000	67	1.15, 1.01	8.6, 7.5
Poly-L-Lys, Type I-B	85,000	385	2.4	100
Poly-L-Pro, Type II	5,700	52	1.32	7.5
Poly-L-Asp, Type II-A	7,700	52	4.1, 4.4	22.6, 24.7
Poly-L-Glu, Type I	14,700	91	5.1	50.5
Poly-L-Asn, Type I-B	8,000	70	1.3	9.7

The results presented in Table V clearly indicate that the amino functional groups of the lysine repeat units possess a greater ability to complete the postulated coupling reaction than any of the other functional groups found on the repeat units tested. The results also suggest that for the polymers of Ala, Pro, and Asn the coupling (less than 10% of that for poly-L-lysine) may be attributed to coupling by the terminal groups. The same may not be said, however, for the polymers of Asp and Glu. The fact that their functional groups are carboxylic suggests that the functional groups located on the amino acid are equally susceptible to activation as the groups on the acrylic polymer. This concept introduces the possibility of reaction between terminal groups of individual polypeptides. The result of this crosslinking could lead to the prevention of passage through the filter during the separation procedure and the inflation of the apparent coupling to acrylic polymer. The possibility that WSC will lead to the crosslinking of poly-L-Asp or poly-L-Glu, and the extent of such a reaction, is presently under investigation. The alternative explanation for the apparently high coupling of the poly-L-Asp and poly-L-Glu is that the modified carboxyls of these polymers are coupled to the copolymer via the copolymer's acrylamide groups. Considering the inert character of the amide group under these conditions, the possibility of such a coupling is remote. The fact that the results for poly-L-Asp and poly-L-Glu do not fit comfortably into the trend predicted for coupling due solely to terminal group coupling suggests that further investigations into these systems of excess carboxyl groups is warranted.

Coupling of Modified Collagen

To investigate further the role played by the amino functional groups of the lysine repeat unit studies were done with modified collagen. A 88 mg sample of acid-soluble collagen was reacted with 34.4 mg of succinic anhydride at a constant pH of 8.0, a procedure used to block the free amino groups of proteins (36). The reaction mixture was then dialyzed extensively and freeze-dried. The analysis of the modified collagen materials was done by the ninhydrin analysis (37). The succinic anhydride-modified collagen (soluble in 0.05M KH_2PO_4 adjusted to pH values of 6.4 or greater) contained only 50% of the amino acid groups found on unmodified acid-soluble collagen, which indicated that 50% of the amino acid functional groups had been blocked.

The results of coupling experiments with modified collagen are given in Table VI. Partly modified collagen was not soluble in the desired concentration at pH 4.4 but it did become soluble at pH 6.4. Although this does not represent the optimal conditions determined earlier, it at least provided an opportunity to investigate the coupling of a partly modified collagen. The results of these experiments are that the coupling efficiency has an average value of 16%. An attempt to couple unmodified collagen at this pH for comparison purposes failed because the collagen is not soluble in the desired concentration at this pH. [Note: pI for acid-soluble collagen is 6.6 (34)]. Even though a direct comparison is not possible, it can be seen by comparison with Figure 6, that a coupling efficiency of 16% is considerably reduced from any expected coupling efficiency for unmodified collagen. It should be noted that the amino content of the collagen was reduced by 50%. If the amino groups were solely responsible for the coupling, we would expect a 50% reduction in coupling efficiency. Because the comparison with Figure 6 does not indicate a drastic reduction in the coupling efficiency, we must leave open the possibility that functional groups other than amino groups may contribute to the coupling.

Table VI

COUPLING OF MODIFIED COLLAGEN TO co(AAc/AAm)IV[a]

Succinic anhydride modification	Amino groups available (%)	Amount coupled (mg)	Coupling efficiency (%)
Run No. 1	50	3.0	17.1
Run No. 2	50	2.7	14.9

[a]Conditions: temperature, 20 $\pm$ 1°C; time, 1 hr; solvent, 0.05M KH_2PO_4; pH 6.4 for succinic anhydride modified collagen; volume, 7.0 ml; concentrations: collagen, 2.5 mg/ml; coupling agent, 11.4 mg/ml; co(AAc/AAm)IV, 20 mg/ml.

These results support the conclusion drawn from the coupling studies with model polypeptides that the free amino groups of collagen play a determining role in the coupling reaction.

Influence of Protein Conformation

To investigate the influence of the conformation of the protein on the coupling reaction the acid-soluble collagen and gelatin were coupled to the acrylic copolymer under the previously determined optimal conditions. The results indicate that by unraveling the triple-stranded helix of collagen, and thereby increasing the accessibility of amino functional groups, we can increase the coupling efficiency from 22.6% for collagen to 32.6% for gelatin. In "intact" collagen approximately 30% of the "free" amino groups are tied up in intramolecular, noncovalent bonding, thus rendering them less accessible to the coupling reaction. The results of these studies then suggest possible pretreatments of tissue material before application of the adhesive (for example, acid denaturation of surface collagen) to increase accessibility and therefore bonding.

INTERPRETATION OF THE RESULTS

The interpretation of the results presented here requires that a number of factors be considered. In this section we discuss the stoichiometric, kinetic, and steric considerations individually and then use these discussions to tie together the experimental results.

Stoichiometric Considerations

Let us consider the complete coupling mechanism proposed by Khorana (21). The initial phase in the reaction of the carboxylic acid with a WSC has been postulated as the protonation of the WSC,

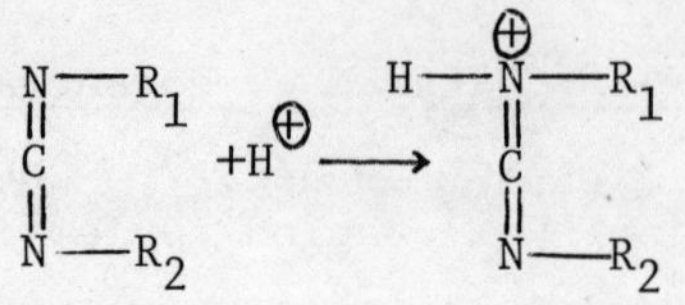

$$\begin{array}{c} N-R_1 \\ \| \\ C \\ \| \\ N-R_2 \end{array} + H^{\oplus} \longrightarrow \begin{array}{c} H-\overset{\oplus}{N}-R_1 \\ \| \\ C \\ \| \\ N-R_2 \end{array} \tag{1}$$

the cation being attacked by the acid anion to form the O-acylisourea:

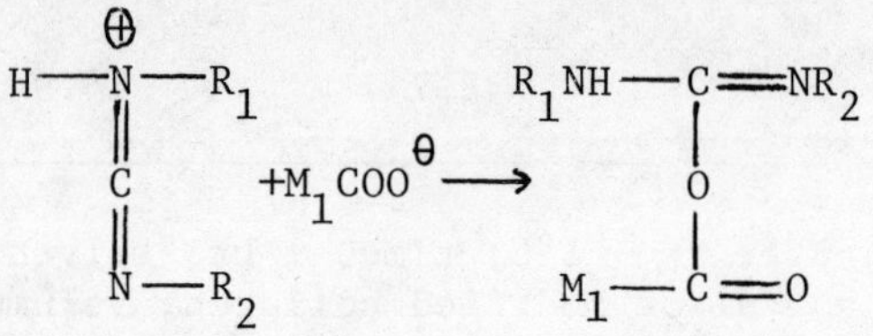

$$\begin{array}{c} H-\overset{\oplus}{N}-R_1 \\ \| \\ C \\ \| \\ N-R_2 \end{array} + M_1 COO^{\ominus} \longrightarrow \begin{array}{c} R_1 NH-C=NR_2 \\ | \\ O \\ | \\ M_1-C=O \end{array} \tag{2}$$

The latter may (i) rearrange by way of cyclic electronic displacement to the stable N-acylurea:

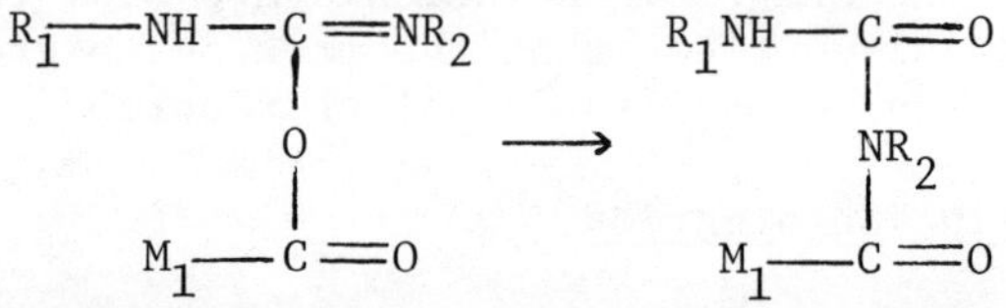

$$\begin{array}{c} R_1-NH-C=NR_2 \\ | \\ O \\ | \\ M_1-C=O \end{array} \longrightarrow \begin{array}{c} R_1 NH-C=O \\ | \\ NR_2 \\ | \\ M_1-C=O \end{array} \tag{3}$$

or (ii) be protonated to the cation:

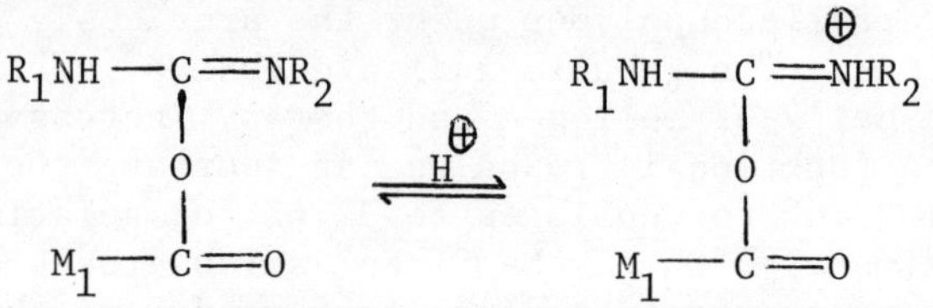

$$\begin{array}{c} R_1 NH-C=NR_2 \\ | \\ O \\ | \\ M_1-C=O \end{array} \overset{H^{\oplus}}{\rightleftharpoons} \begin{array}{c} R_1 NH-C=\overset{\oplus}{N}HR_2 \\ | \\ O \\ | \\ M_1-C=O \end{array} \tag{4}$$

which may (iia) react with water to regenerate the carboxylic acid while forming the N, N'-disubstituted urea:

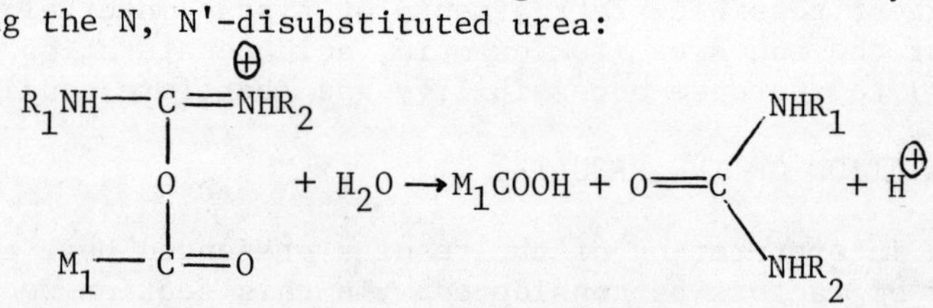

$$\begin{array}{c} R_1 NH-\overset{\oplus}{C}=NHR_2 \\ | \\ O \\ | \\ M_1-C=O \end{array} + H_2O \longrightarrow M_1 COOH + O=C\begin{array}{c} NHR_1 \\ \\ NHR_2 \end{array} + H^{\oplus} \tag{5}$$

or (iib) be converted by the attack of a second anion into the N,N'-disubstituted urea and an acid anhydride:

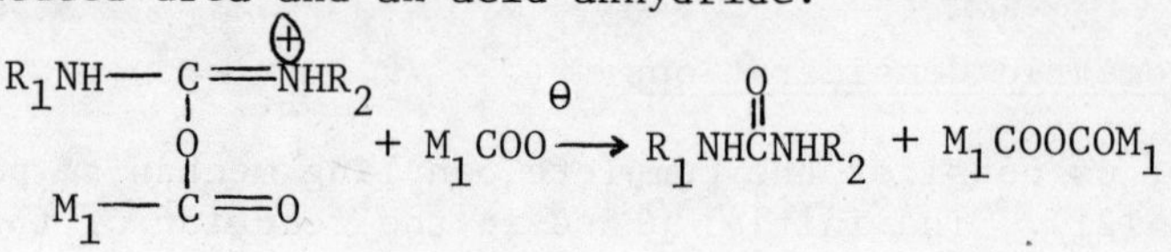

$$\begin{array}{c} R_1 NH-\overset{\oplus}{C}=NHR_2 \\ | \\ O \\ | \\ M_1-C=O \end{array} + M_1 COO^{\ominus} \longrightarrow R_1 NH\overset{O}{\overset{\|}{C}}NHR_2 + M_1 COOCOM_1 \tag{6}$$

Route (ii) is highly favored over route (i) (23). In the presence of an added nucleophile, however, (iia) and (iib) may become insignificant and the result is the modification of the carboxylic group by the nucleophile:

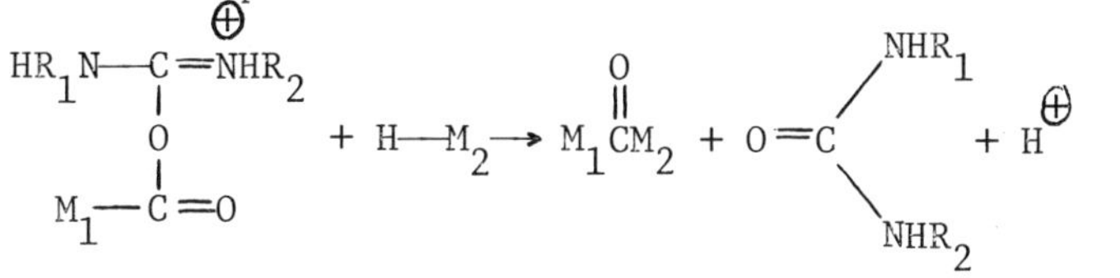
(7)

On the basis of the results let us consider the major participating nucleophile to be the amino group, therefore the polymers are joined by a peptide bond (M_1CONHM_2). The coupling reaction required equimolar amounts of $H^{\oplus}$ ions, carboxyl ions, carbodiimide molecules, and nucleophilic functional groups to establish 1 mole of covalent bonds between macromolecules. A calculation of the concentrations of the reactive components at the optimal conditions, however, reveals that the components are required in concentrations other than the stoichiometric amounts. At a pH of 4.4 in 0.05M KH_2PO_4 it is reasonable to assume that all the carbodiimide present (188.9 μmole) is readily activated. The number of carboxyl groups present is definitely in excess: 1534.8 μmole AAc in the copolymer plus 7.4 μmole Asp and 11.9 μmole Glu in the collagen. Assuming a molecular weight for collagen of 300,000 and using the amino acid analysis given in the literature for calf skin collagen (38), we can calculate the number of μmoles of nucleophiles (4.8 μmole Lys, 1.1 μmole Lys-OH; the total, including terminal amino groups, equals 6.0 μmole). The limiting reactant then is apparently the nucleophilic group-- a mere 6.0 μmole in 7.0 ml compared with 188.9 μmole carbodiimide and 1534.8 μmole of available AAc. This number would be further limited in "intact" collagen where it has been shown that, due to intramolecular crosslinking, only approximately 70% of the amino groups are available for reaction (39). This situation suggests that increasing the concentration of collagen beyond the 2.5 mg/ml level should increase the coupling efficiency. As shown in Figure 5, however, this increase in coupling efficiency with increasing collagen concentration is not possible. It is therefore evident that factors other than stoichiometry play a dominant role in determining the ultimate coupling efficiency.

Kinetic Consideration

It has been stated that the proposed coupling mechanism is represented by the sequence of eqs. (1), (2), (4), and (7). Although this is a sequence that ultimately leads to the coupling of the protein and the synthetic copolymer, it is not the only possible reaction sequence. Others are eqs. (3), (5), and (6).

Some of the last three reactions result in the consumption of carbodiimide without leading to coupling. Because stoichiometric considerations show that nucleophilic groups are in very low concentrations in relation to other possible reactants such as water and anions, it is possible that the coupling reaction sequence is dominated by the other noncoupling reactions.

As a means of investigating the extent of the participation of these side reactions an experiment was designed to determine the number of AAc carboxyl groups that are converted to these inactive forms. To carry out this experiment three samples of 140.0 mg of co(AAc/AAm)IV were weighed into individual 200 ml beakers. The first sample was suspended in 7.0 ml of 0.05M HCl and used as a control; the second was reacted with 7.0 ml of a 11.4 mg/ml carbodiimide--0.05M HCl solution; and the third was reacted with 7.0 ml of a 22.8 mg/ml carbodiimide--0.05M HCl solution. After the standard 60 min reaction time the samples were diluted with 20 ml 0.1N HCl. Then, using the usual titration technique, the carboxyl content was determined. If the carboxyls were unreacted or if they had been reacted but reverted back to the carboxyl form, they would show up in the titration as carboxyl groups. If the carboxyl had undergone one of the other possible reactions that would lead to a nonreactive carboxyl, the group would not have shown up in the titration. The results of the titrations, using the first sample as a control, showed that when carbodiimide is added at 11.4 mg/ml the carboxyl content is reduced approximately 12%. Because the concentration of carbodiimide at the determined optimal reaction conditions is 12.3% of the carboxyl concentration, it can be assumed that when the carbodiimide is added almost all of it reacts with carboxyl groups. In fact, under these conditions, when no collagen is added, almost all the carbodiimide is consumed in the production of the modified carboxyls that are not capable of coupling to collagen.

The results of this section are verified by the findings of Hoare and Koshland (23), who have concluded that in the presence of no added nucleophile the reaction with water is much more rapid than rearrangement to N-acylurea. The regenerated carboxyl is again attacked by carbodiimide. Because of the cyclic nature of this set of reactions, they also conclude that N-acylurea eventually accumulates. They find that this trend can be eliminated only if the added nucleophile is present in high concentrations. Because this is not the case in the present experiments, it can be assumed that the formation of N-acylurea is still competing with the desired nucleophilic attack. The added nucleophile is present in sufficient quantity, however, to cause the reaction sequence to produce some intermolecular bonds. Although the production of urea was not monitored in these studies, these experiments would prove informative and are presently being conducted.

The results of the third experiment, in which the concentration of the carbodiimide was doubled, also proved informative. It was expected, if the carbodiimide were the limiting reactant in this experiment, then doubling the carbodiimide would result in the blocking of twice as many carboxyl groups. This was not found, however. The doubling of the carbodiimide had no effect on the percentage of carboxyls modified. The results of the titration again showed that approximately 12% of the carboxyl groups were reacted, an indication that the additional carbodiimide, with a molecular weight of 423.6, was not capable of penetrating into the swollen copolymer and that steric considerations also play a role in determining the extent of coupling.

The possibility that WSC is being consumed in further cross-linking of the acrylic copolymer is presently under investigation. This consumption would, of course, limit the extent to which intermolecular coupling can take place.

Steric Considerations

The facts outlined in the preceding discussion show that the addition of carbodiimide beyond the 188.9 μmole level is superfluous. Evidently only carboxyl groups at or near the surface of the swollen granule are capable of being activated. It is also evident that the number of accessible carboxyl groups is approximately equal to 188.9 μmole. This result confirms the findings in Figure 4 that additional coupling agent beyond 11.4 mg/ml does not improve coupling efficiency.

In conducting these experiments we observed qualitatively that the degree to which the copolymer swells is dependent on a number of factors: AAc content of the polymer, salt content of the solvent, and pH of the solvent. Under the optimal conditions in effect the copolymer swells to approximately five times its original size. The increased coupling with increased pH (Fig. 6) can be attributed in part to increased copolymer swelling and therefore increased accessibility with increased pH. This fact lends support to the postulate of steric control.

It should be pointed out that if accessibility is a controlling factor in the activation of the carboxyl groups by the carbodiimide it will be of even greater importance in the actual coupling reaction of the collagen and the copolymer. In activation a molecule of molecular weight 423.6 and some bulk must be brought into contact with the carboxyl groups at or near the surface of the copolymer. If, as we have shown, accessibility problems are encountered in the activation stage, then surely the larger collagen molecule will be even more sterically hindered in its approach. In the coupling of gelatin the accessibility of amino groups is improved with the result that coupling is improved.

Bearing in mind the above discussions it is possible to tie together the experimental results.

It has been shown that an increase in the concentration of coupling agent beyond a certain level had no significant influence on the coupling of gelatin or collagen, all other factors being equal. It has also been shown that it is the ability of the carbodiimide to come into contact with suitable reacting partners which dictates this limitation.

It was apparent that the addition of extra carboxyl groups, if accessible, would increase the coupling efficiency. Therefore experiments were carried out. In one method of increasing accessible carboxyls the carboxyl content of the copolymer was increased. In a second method the concentration of the copolymer was raised. It is postulated that because the carbodiimide is present only in slight excess when dealing with the copolymer of 79% AAc at a concentration of 20 mg/ml an increase in the carboxyl group content of the reaction mixture beyond this level serves no purpose as shown in Figure 2. Further increases in the coupling efficiency are made possibly by raising the number of accessible carboxyls by increasing the AAc content of the copolymer (Fig. 1).

The influence of concentration of collagen on coupling efficiency is a problem of the accessibility of the carboxyl groups and the reactions subsequent to activation. At acidic pH values, moreover, the protonated form of the nucleophilic amino groups may be favored and is less reactive. These factors lead to limiting the possible maximum of coupling efficiency. The higher the pH (and therefore the lower $H^{\oplus}$ concentration), the less protonated the nucleophilic groups situated on the collagen molecule and the more swollen the copolymer. Thus it is easier to activate AAc carboxyl groups, and the probability of this activated group coming in contact with a nucleophilic group is increased. Therefore coupling efficiency increases with increasing pH.

Another way in which the pH may influence the coupling of collagen to a copolymer of acrylic acid and acrylamide should also be mentioned. Because the operating pH (4.4) is below the isoelectric point of the collagen (pI = 6.6—6.8), it is assumed that the protonated forms of the basic groups (Arg, Lys, His, etc.) dominate, with the result that an overall positive charge occurs on the molecule (33). Similarly, it can be assumed that at pH 4.4 the copolymer bears an overall negative charge (40). Coupling may therefore be facilitated by the absorption of the positively charged collagen solute onto the negatively charged copolymer substrate, followed by covalent bonding.

In studies of the coupling of homopolymers of α-amino acids and modified collagen the situation is somewhat different. In the homopolymers of Asp and Glu the carboxyls on the protein are in competition with the AAc of the copolymer for the carbodiimide. The result is the complete consumption of the coupling agent, but not so many of the AAc carboxyls are activated as usual. Therefore the chances of the few terminal amino groups coupling to the active sites on the copolymer are reduced. There is limited coupling, however, which attests to the mobility of the terminal amino groups on these more flexible chains. The possibility that crosslinking reactions involving these homopolymers alone must be considered. The polymer of Pro contains only terminal carboxyl and nucleophilic amino groups, which severely limits coupling, although some is still possible. On the other hand, the randomly coiled homopolymer of Lys has an abundance of nucleophilic groups. This fact should influence the degree to which the side reactions occur. The limiting factor here still remains the accessibility of groups on the copolymer. The case of poly-D, L-alanine is similar to that of poly-L-Asp and poly-L-Glu in that any coupling attained can be attributed to terminal amino groups. The considerations in the succinic anhydride-modified collagen are similar to those for copolymer with untreated collagen except that the number of nucleophilic groups on the collagen is decreased. This fact lessens the chances of coupling and results in decreased coupling efficiency.

CONCLUSIONS

The coupling together of acrylic-based copolymers and acid-soluble collagen via 1-cyclohexyl-3-[2-morpholinyl-(4)-ethyl] carbodiimide metho-p-toluene sulfonate has proven successful. From the results presented here it is evident that, for the heterogeneous reaction carried out in air, the following conclusions may be drawn;

(1) The coupling reaction occurs in reasonably short time. The time required to assure that maximum coupling has been accomplished was found to be 60 min. Substantial coupling can be obtained within 5 min.

(2) The coupling reaction occurs at moderate temperatures. A convenient and safe temperature to conduct the experiments was found to be 20 ± 1°C.

(3) For maximum coupling efficiency, the carboxyl content of the copolymer should be high. With material synthesized in our laboratory, maximum coupling was obtained with a copolymer containing 79% by weight acrylic acid.

(4) For maximum coupling efficiency, the concentration of the acid-containing copolymer must be optimized at a high concentration without interfering with agitation; 20 mg/ml was selected.

(5) The carbodiimide coupling agent must be present in slight excess, e.g. 11.4 mg/ml, for maximum coupling efficiency.

(6) The acid-soluble collagen must be present in slight excess,

e.g., 2.5 mg/ml, for maximum coupling.

(7) The pH must be optimized at a value that provides a concentration of $H^{\oplus}$ ions high enough to initiate completely the carbodiimide, but low enough to swell the copolymer, and yet not so low as to favor too strongly the ionized form of the amino groups. A pH of 4.4 was selected.

(8) Substantial coupling efficiency--up to 22.6%--can be obtained for the collagen under these optimal conditions.

The role of carboxylic groups of the synthetic copolymer is significant. The postulated mechanism of carbodiimide modification of this group appears to be substantiated in the present studies. For this reason, both the availability of these groups (number of groups) and the accessibility of these groups (ability to participate in the coupling reaction) is important.

The coupling mechanism involved in chemically bonding collagen and acrylic polymers by a water-soluble carbodiimide has been studied in a heterogeneous solution reaction. The conclusions drawn are that the carbodiimide, in protonated form, attacks and modifies the carboxyl groups located on the insoluble acrylic polymer. The reaction is then completed by the nucleophilic attack of the protein. It has been shown that the initial step in this mechanism is sterically controlled and that only the accessible carboxyl groups (i.e., on the particle surface) are activated. In a series of tests of homopolymers of α-amino acids and modified collagen molecules it has been demonstrated that one significant nucleophile in the coupling reaction is the amino group ($—NH_2$) located on the protein. It has been shown that the overall coupling reaction is controlled by steric, kinetic, and stoichiometric factors, among which the steric factors are the most prominent. This role of accessibility is evident in the activation of the carboxyl groups on the synthetic, insoluble polymer and in the completion of the reaction by the soluble protein. Gelatin couples to a greater extent than collagen.

ABSTRACT

A means of coupling synthetic polymers of biomedical interest with collagen under mild conditions was investigated as a part of preliminary studies in the development of a biomedical adhesive. The coupling agent, 1-cyclohexyl-3[2-morpholinyl-(4)-ethyl]carbodiimide, provides a covalent bond between the acrylic polymer and collagen. The variables affecting the establishment of this bond in solution are concentration of the carboxylic group on the acrylic polymer, concentration of the acid-soluble collagen, concentration of the coupling agent, pH, and time of reaction. Optimum values of these variables have been obtained experimentally. The coupling mechanism involved in the coupling of acrylic polymers and collagen by a water-soluble carbodiimide was investigated. By the use of a series of homopolymers of α-amino acids, support was given to the postulate that a major proportion of the coupling takes place via the amino functional group. In light of the results presented the reaction is sterically controlled and the steric, stoichiometric, and kinetic aspects of the reaction are discussed.

REFERENCES

1. M. Buonocore, in "Adhesion in Biological Systems", R. S. Manly, ed., Academic, New York, 1970, Chap. 15
2. R. L. Bowen, J. Dent. Res., 44, 895 (1965)
3. D. C. Smith, Br. Dent. J., 125, 381 (1968)
4. T. Tosa, T. Mori, N. Fuse, and I. Chibata, Biotech. Bioeng., 9, 603 (1967)
5. G. Baum, F. B. Ward, and H. H. Weetall, Biochim. Biophys. Acta, 268, 411 (1972)
6. R. S. Manly, ed., "Adhesion in Biological Systems", Academic, New York, 1970, Chaps. 15-17
7. D. C. Smith, J. Can. Dent. Assoc., 37, 22 (1971)
8. E. Masuhara, M. Ilido, E. Furuya, S. Kawachi, and J. Tarumi, Clin. Orthoped. Rel. Res., 100, 279 (1974)
9. R. L. Bowen, J. Dent. Res., 58, 1493 (1979)
10. A. D. Wilson, Organic Coatings and Plastic Chemistry, 42, 215 (1980)
11. J. F. Glenn, Organic Coatings and Plastic Chemistry, 42, 186 (1980)
12. B. D. Halpern and W. Karo, Organic Coatings and Plastic Chemistry, 42, 315 (1980)
13. J. Gross, in "Comparative Biochemistry", Vol. V., M. Florkin and H. S. Mason, eds., Academic, New York, 1963
14. R. G. Carbonell and M. D. Kostin, AIChE J., 18, 1 (1972)
15. K. L. Smiley and G. W. Standberg, Adv. Appl. Microbiol., 15, 13 (1972)
16. O. R. Zaborsky, "Immobilized Enzymes", C.R.C. Press, Cleveland, 1973
17. R. Goodman, L. Goldstein, and E. Katchalski, in "Biochemical Aspects of Reactions on Solid Supports", G. R. Stark, ed., Academic, New York, 1971, Chap. 1
18. R. C. Page and E. P. Benditt, FEBS Lett., 9, 49 (1970)
19. Y. Nayudamma, R. Hemalatha, and K. T. Joseph, J. Am. Leather Chem. Assoc., 64, 444 (1969)
20. T. J. Munton and A. D. Russell, Br. Med. J., 3, 372 (1971)
21. H. G. Khorana, Chem. Rev., 53, 145 (1953)
22. F. Kurzer and K. Douraghi-Zadeh, Chem Rev., 67, 107 (1967)
23. D. G. Hoare and D. E. Koshland, J. Biol. Chem., 242, 2447 (1967)
24. C. P. Joshua, J. Ind. Chem. Soc., 37, 621 (1960)
25. L. E. A. Godfrey and F. Kurzer, J. Chem. Soc., 3561 (1962)
26. K. L. Carraway and D. E. Koshland, Biochim. Biophys. Acta, 160, 272 (1968)
27. B. Adcock, A. Lawson, and D. Miles, J. Chem. Soc., 5120 (1961)

28. K. L. Carraway and R. B. Triplett, Biochim. Biophys. Acta, 200, 564 (1970)
29. D. G. Hoare and D. E. Koshland, J. Am. Chem. Soc., 88, 2057 (1966)
30. A. Previero, J. Derancourt, M. A. Coletti-Previero, and R. A. Laursen, FEBS Lett. , 33, 135 (1973)
31. O. H. Lowry, N. J. Rosenbrough, A. L. Farr, and R. J. Randall, J. Biol. Chem., 193, 265 (1951)
32. R. E. Burge and R. D. Hynes, J. Molec. Biol., 1, 155 (1959)
33. P. H. von Hippel and K. Y. Wong, Biochemistry, 2, 1387 (1963)
34. H. R. Mahler and E. H. Cordes, "Biological Chemistry, 2nd ed.", Harper and Row, New York, 1971, Chaps. 3, 4
35. C. Franzblau, P. M. Gallop, and S. Seifter, Biopolymers, 1, 79 (1963
36. K. D. Hapner and P. E. Wilcox, Biochemistry, 9, 4470 (1970
37. E. G. Frame, in "Standard Methods of Clinical Chemistry", D. Seligson, ed., Academic, New York, 1963, Vol 4, p. 1
38. M. Schubert and D. Humerman, "A Primer on Connective Tissue Biochemistry", Lee and Febiger, Philadelphia, 1968, Chap. 2
39. A. Veis, in "Treatise on Collagen", Vol. 1, G. N. Ramachandran, ed., Academic, New York, 1968
40. K. Martensson and K. Mosback, Biotechnol. Bioeng., 14, 715 (1972)

GLOW DISCHARGE POLYMER COATED OXYGEN SENSORS

Allen W. Hahn and Michael F. Nichols, Dalton Research Center, University of Missouri-Columbia, Columbia, Missouri 65211
Ashok K. Sharma and Eckhard W. Hellmuth, Department of Chemistry, University of Missouri-Kansas City, Kansas City, Missouri 64110

INTRODUCTION

The need to measure reliably and accurately oxygen concentration in biological media is often crucial. Presently used commercially available and laboratory fabricated systems usually utilize polarographic sensors to measure oxygen partial pressure in aqueous media. These sensors use a noble metal (platinum or gold) cathode connected to a source of electrons, (commonly a battery) and are referenced to a suitable anode such as Ag/AgCl to complete the circuit. Many methods have been used to improve sensor drift, sensitivity, and response time by permuting cathode geometry, construction details, membrane type and thickness, and electronic correction techniques. For a review of these methods the interested reader is referred to (1,2).

Although conventional sensor configurations (with a membrane) are able to prohibit poisoning of the cathodic surface by macromolecules, even these are not stable enough for quantitative results with continuous long term use without intermittent calibrations (i.e., every 6-12 hours (3) and often more frequently). Previous studies (4,5) have shown that poisoning and aging of the catalytic surface of the cathode are the major problems to be addressed if stability is to be achieved. Poisoning of the cathode surface, a rapid phenomena, can be prevented by membranes which exclude proteins, macromolecules, etc. provided that they also have good oxygen solubility characteristics. A slower poisoning phenomena may occur when the permeability of the membrane to oxygen is altered by the absorption or adsorption of contaminants. However, this effect is usually observed in conjunction with aging of the catalytic surface. Aging refers to the comparatively slow changes in response characteristics which occur even in the "purest" media. This slower "aging" process, taking place on the surface of the cathode, alters its sensitivity and hence is perceived as a drift in the sensor output.

In order to understand this process of aging and drift, and to be able to predict sensor current output, one must first look at what constitutes the measured output as a result of the reduction process occurring at the cathode. For practical considerations, two principal pathways for oxygen reduction at noble metal cathodes are generally accepted (6). These pathways are:

(Equation 1)

$$O_2^{\,b} \xrightarrow{\text{diffusion}} O_2^{\,s} \xrightarrow[k_1]{2e^-} H_2O_2^{\,a} \xrightarrow[k_2]{2e^-} H_2O \quad \text{(A)}$$

$$H_2O_2^{\,a} \xrightarrow[\text{diffusion}]{} H_2O_2^{\,b} \quad \text{(B)}$$

The superscripts b, s and a refer to bulk, surface and adsorbed, respectively, k_1 and k_2 are reaction rate coefficients and e^- represents the electronic charge; the total current then being the total flow of electrons used in the above reaction. Based on the above, a model was developed to determine the maximum current one should expect for a given cathode size so that a starting point for investigating the aging process could be identified (7,8). This equation

$$i = 8\,FD_{O_2}\,RC_{O_2}^{\,b}\,[1 + \{1 + 4\,RD_{H_2O_2}\,[AK_2]^{-1}\}^{-1}] \quad \text{(Equation 2)}$$

is an expression of the oxygen electrode current observed during oxygen diffusion limited conditions. In equation 2, F is Faraday's constant, D_{O_2} and $D_{H_2O_2}$ are the diffusivities of O_2 and H_2O_2 respectively, R the electrode radius, C_{O_2} the oxygen concentration and A the sensor electrode surface area. To predict the maximum current (I_{max}) for a given sensor size, one need only consider the case where all the H_2O_2 is reduced to water (equation 1-part (A)). This corresponds, in equation 2, to the case where $AK_2 >> 4\,RD_{H_2O_2}$ which then simplifies to:

$$i = I_{max} = 16\,FD_{O_2}\,RC_{O_2}^{\,b} \quad \text{(Equation 3)}$$

From the above, we could then predict that a freshly polished and anodized cathode of 50 μm sensing surface diameter would have a

I_{max} value between 19-21 nA in air saturated saline at room temperature. This could then be used as an initial point in our current monitoring studies.

Realizing that no single technique may be capable of alleviating all of the problems associated with polarographic oxygen sensors, we chose an approach to the sensor drift problem which would allow us to prevent poisoning at the cathode, maintain the maximum sensitivity possible for the electrode size, and prohibit or retard the aging process. Our approach was to place an ultrathin (∿0.1 μm) polymer coating over cathodic surface by the process of "glow discharge" polymerization. Polymer membranes formed by conventional synthesis (Polyethylene or Teflon) are permeable to oxygen but not to ions unless pinholes are present. Therefore one must design the sensor with both the cathode and anode behind the membrane and an ionic medium present for charge transfer. This is the basic design of the Clark electrode (9). However, size and construction limitations of the Clark-type electrode make it impractical for use as a microsensor where only fine wires may be used for oxygen tension measurements. Coating these microsensors with materials such as colloidon (permeable to oxygen and ions) have proven to be less than satisfactory as they tend to be fragile, and conventional application processes produce non-uniform thicknesses so that sensitivity is neither stable nor predictable. For these reasons we chose this unique form of polymer synthesis which would allow us to coat the active cathodic surface uniformly with a coating of appropriate characteristics, regardless of the eventual construction details.

The "glow discharge" or plasma polymerization technique, although not new, has more recently shown to be an active field of research in the area of surface modification (10) and biocompatibility studies (11). In the past, the glow discharge polymerization procedure has produced products with a wide range of applications. Some of these include the formation of reverse osmosis membranes, barrier coating, dielectrics, and aiding the adhesion of noncompatible materials. General references to these applications and many others, as well as the fundamentals of glow-discharge polymerization may be found in several locations (12,13,14,15).

MATERIALS AND METHODS

1) Electrode Construction

The cathodes were made from 1 cm. lengths of platinum wire (99.999% Pt) with a diameter of 50 μm. The Pt was bonded to a copper lead wire and this assembly was placed inside a glass capillary tube. The glass was drawn over the platinum and after sealing, the end was ground flat with a graded series of grits and cloths (Figure 1). In all instances following polishing, the electrodes were cleaned in glass distilled water, alcohol, distilled

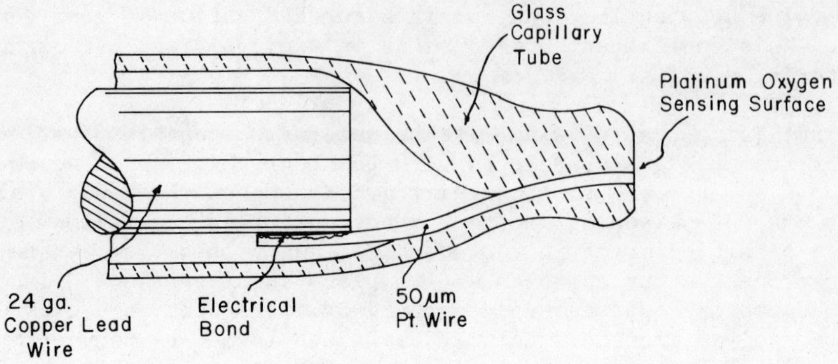
Glass
Capillary
Tube
Platinum Oxygen
Sensing Surface
24 ga.
Copper Lead
Wire
Electrical
Bond
50 µm
Pt. Wire

Figure 1. A sketch of the oxygen sensing electrode. The ultrathin polymer film is applied to the exposed platinum end.

water again, all in an ultrasonic bath. The cathodes were anodically polarized with respect to another platinum electrode for 5 minutes at 2 volts in 1.0 N H_2SO_4. The cathodes were again rinsed in distilled water in an ultrasonic bath. This procedure produced a uniform Pt disk at the tip of the cathode, the characteristics of which have been described elsewhere (8). A series of polarograms (at different O_2 tensions) using a saturated calomel electrode (SCE) as the anode were then done on each electrode to establish that the precoating characteristics were in agreement with calculated I_{max} values.

2) Glow Discharge Coating

The deposition of the thin "glow discharge" or plasma polymerized propylene (GDPP) coatings was carried out in a reactor as shown in Figure 2. The RF energy (27.12 MHz) was inductively coupled to the reactor with power supplied by a Tomac Diathermy unit, Model 1565. An input power level of 100 W (current density setting of 125 ± 5 mA) for all polymerizations and approximately 140 W (current density setting of 175 ± 5 mA) for the etching operations, was delivered to the coils. All polymerizations were conducted at pressures of approximately 20 Pa maintained by a combination of a liquid nitrogen cold trap, an oil diffusion pump, and a roughing pump. Monomeric propylene was obtained from Matheson Gas Products and was used as supplied.

The average thickness of the polymeric films was initially calculated from the mass of the polymer deposited per unit area on an aluminum foil of the substrate, and from the average density of the polymer. The mass gain due to the deposited polymer was determined by weighing with an electrobalance, and its density

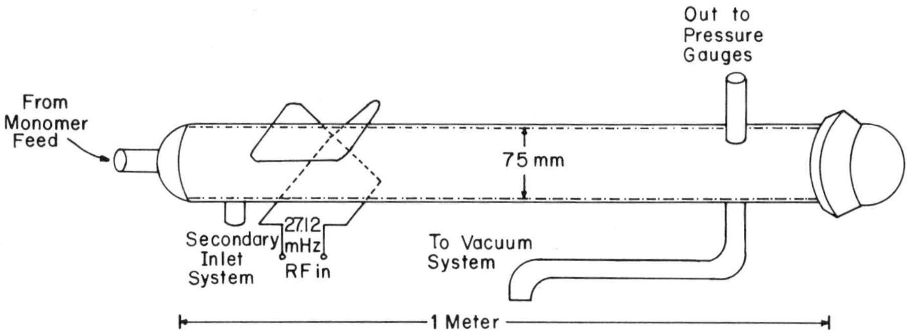
Out to
Pressure
Gauges
From
Monomer
Feed
75 mm
27.12
mHz
Secondary
Inlet
System
RF in
To Vacuum
System
1 Meter

Figure 2. The "glow discharge" reactor used for synthesizing the polymer films.

determined by the density gradient column method (16) using carbon tetrachloride, and n-heptane as the density medium. However, we found, by scanning electron microscopy of the polymer coated sensor tip (scratched with a scapal blade), that the actual thickness of the coating on Pt-surface was at least three times more than the calculated thickness, i.e. the polymer coating appears to be about 1 μm thick for an expected value of 0.28 μm. In all probability, this difference is due to different geometries.

The entire electrode assembly, except the tip, was masked with aluminum foil and mounted, three at a time, in an electrode holder which fit in the sleeves of the reactor and was provided with loops at the center for holding electrodes along the axis of the reactor. The electrodes were argon-etched for 20 minutes before the deposition of polymer coatings. The deposition of propylene glow discharge polymer was done at a position 140 mm from the center of the discharge. A monomer flow rate of 340 $\pm$ 10 μmol min^{-1} and a constant discharge power of 100 W were used. Typical film thickness were in the range of 0.1 μm to 0.3 μm as finally measured by the SEM technique (supra vide). The details of instrument design and polymerization procedure are described elsewhere (17,18,19).

3) Sensing Electrode Testing

Typical sensitivity (O_2 partial pressure vs. current output), stability (current output vs. time) and polarograms (current output vs. cathodization voltage) were made in phosphate buffered isotonic saline (0.9%) solutions (pH 7.4). After coating, each sensor was subjected to an initial series of polarograms at room temperature and various oxygen tensions (0-21%). Then, under a constant cathodization voltage of 700 mV, the output currents were monitored for up to 36 hours. After this initial trial, a polarogram (at 21% O_2)

was repeated for each sensor and the long term trials again run. For some runs, bovine serum albumin (BSA-5% w/v) was placed in the medium to determine the protein sensitivity of the coated sensors. Currents were monitored with a Keithley model 416 picoammeter and recorded on either an Esterline Angus (Model E1124-FM) Multipoint Recorder or on the system which includes a PDP 11/03 microprocessor (Digital Equipment Corp.), floppy disk drives (Charles Rivers Data Systems Corporation), A/D and D/A converters (ADAC Corp.), and appropriate software for data acquisition.

RESULTS: (as applied to O_2 sensing)

Figure 3 shows polarograms of oxygen sensing electrodes before and after coating with GDPP films of 0.15 μm thickness. For all coated electrodes, the plateau regions are extended over a larger voltage range. (Note that polarographic sensors are used in the plateau region-usually between -500 to -700 mV). The effect is more clearly demonstrated in Figure 4 where the polarogram of 0.1 μm thick polymer coated electrode at different oxygen concentrations, may be compared with an uncoated electrode (dashed line) up to a cathodization voltage of -1.5 V.

Figure 5 shows the calibration response of a 0.1 μm thick GDPP coated electrode to varying concentrations of dissolved oxygen in 0.9% saline solution, measured over 72 days. The electrodes were kept soaked in the saline solution during this time. Similar trials

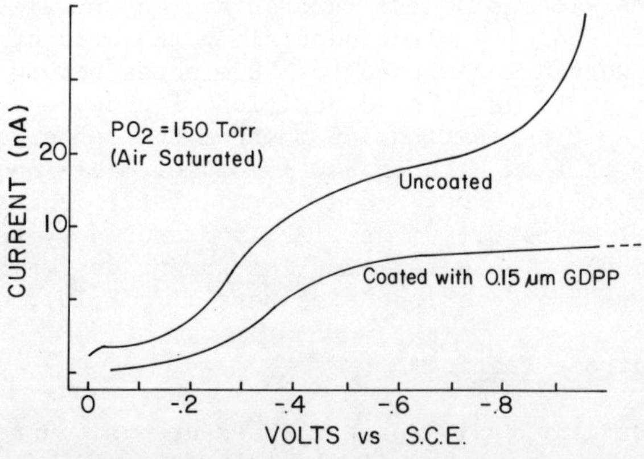
CURRENT (nA)
20
10
PO_2 = 150 Torr
(Air Saturated)
Uncoated
Coated with 0.15 μm GDPP
0
-.2
-.4
-.6
-.8
VOLTS vs S.C.E.

Figure 3. Current-voltage plots (polarograms) of platinum oxygen sensing microelectrodes in an air saturated 0.9% NaCl solution. A Saturated Calomel Electrode was used as reference. Note that the plateau region is extended on the coated electrodes although the magnitude of the current has dropped.

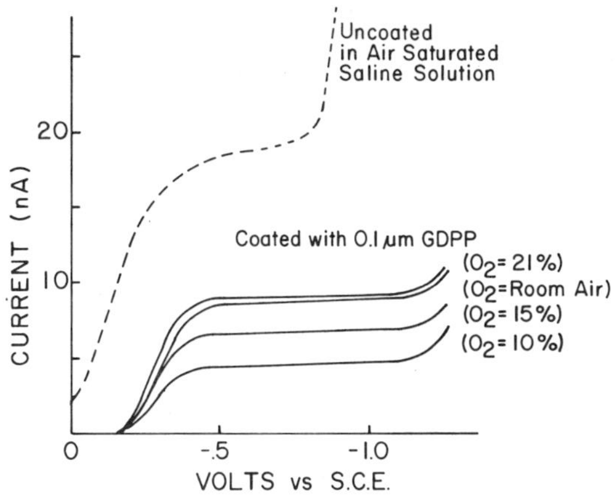
Uncoated
in Air Saturated
Saline Solution
Coated with 0.1 μm GDPP
(O2= 21%)
(O2=Room Air)
(O2= 15%)
(O2= 10%)
CURRENT (nA)
20
10
0
-.5
-1.0
VOLTS vs S.C.E.

Figure 4. Polarograms at platinum oxygen sensing microelectrodes in 0.9% saline solution and different oxygen concentrations. Response of an uncoated sensor is shown as a dashed line. Note the linear change in response to varying oxygen concentrations.

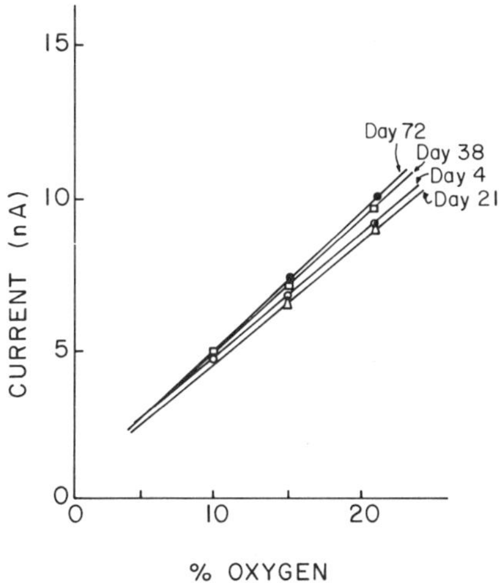
Day 72
Day 38
Day 4
Day 21
CURRENT (nA)
15
10
5
0
0
10
20
% OXYGEN

Figure 5. Calibration responses of a 0.1 μm thick GDPP coated oxygen sensor in 0.9% NaCl solution. Calibration was repeated over 72 days during storage in 0.9% saline. Note the small changes observed.

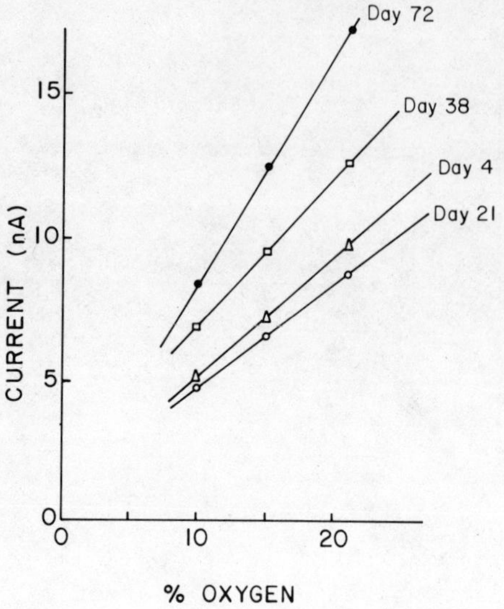
Day 72
Day 38
Day 4
Day 21
CURRENT (nA)
% OXYGEN
0
5
10
15
0
10
20

Figure 6. Calibration responses of a 0.3 μm thick GDPP coated oxygen sensor in 0.9% saline solution. These were also done over a 72 day storage period. Note the marked change toward an uncoated state during this time.

were conducted on 0.05 and 0.3 μm thick GDPP coated sensing electrodes. The thicker coatings (0.3 μm) showed marked changes in slope after 35 days although the linearity of response to oxygen concentration was still maintained (Figure 6).

The response of the uncoated and GDPP coated electrodes to longer periods of time and to proteins is depicted in Figures 7 and 8 respectively. Electrodes with 0.1 μm thick polymer coatings were tested. The results in Figure 7 show that an uncoated platinum oxygen electrode, when placed in 0.9% NaCl solution at ambient oxygen saturation (about 150 torr) suffers an initial large drop (50 percent I_{max}) in output current and when BSA is added an additional decrement of about 15% of the previous stable current, occurs. Figure 8 shows the current output of four GDPP coated oxygen electrodes plotted on the same scale and run under the same conditions as the uncoated in Figure 7 for periods of 24 to 35 hours. One of these electrodes was run at 37.7°C and the others at 25°C. The addition of proteins in BSA had minimal effects (1-3% change) on the current output of coated electrodes.

DISCUSSION

The results of these studies clearly show that current output

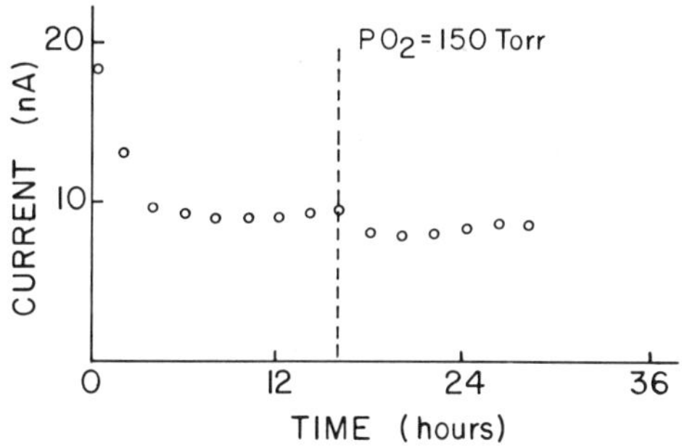

PO_2= 150 Torr
CURRENT (nA)
20
10
0
12
24
36
TIME (hours)

Figure 7. The long term response of an uncoated platinum oxygen sensor in air saturated 0.9% saline solution. Note the initial drift from I. Bovine serum albumin (5% w/v) added at line.

of polarographic oxygen sensors can be stabilized by the application of an ultrathin (~0.1μm) glow discharge polymer coating. This is of signal importance to the long term (days) measurement of oxygen concentration, especially in physiological preparations. We were impressed by the ability of these coatings to not only protect against the addition of protein to the measuring medium but also the immediate (minutes) stabilization of the sensor to a two-electron reduction reaction (path B in Equation 1).

Just as important has been the ability of the 0.1 μm coated sensors to maintain their calibration over a 72 day time interval.

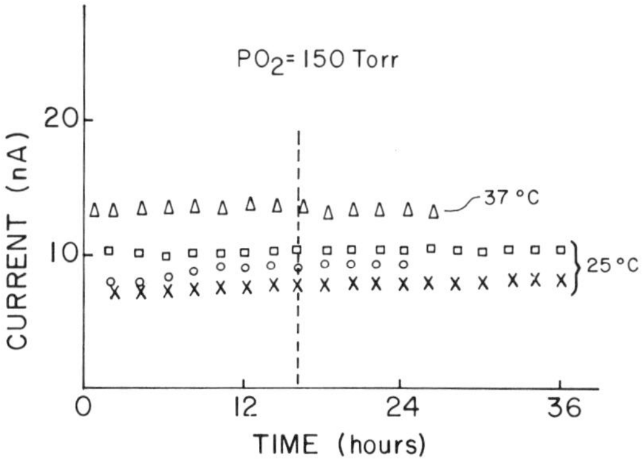

PO_2= 150 Torr
CURRENT (nA)
20
10
37 °C
25 °C
0
12
24
36
TIME (hours)

Figure 8. The long term response of coated oxygen sensors in air saturated 0.9% saline solution. Sensor was run at -37°C and the others at 25°C. Note that there is no initial change from I_{start} and that the addition of BSA has only a minimal effect.

Thicker coatings (~0.3μm) still allowed sensors to maintain their linearity but the output currents moved toward those expected from an uncoated electrode. This phenomenon made us suspicious that the coating was peeling from sensor surface. Subsequent scanning electron micrographs showed that, indeed, that was happening. Thinner coatings (~0.03μm) were very unpredictable in altering the response characteristics of oxygen sensors and we have since abandoned using them.

The exact mechanism of O_2 reduction at the GDPP coated Pt-cathode is still open for discussion. In the diffusion-limited process only half the number of electrons may be involved to yield half the reduction current compared to bare platinum sensing electrodes. Equation 1 shows that oxygen can be reduced to H_2O_2 (2 electrons) or to H_2O (4 electrons). This reduction could take place on the surface of the coating, at any place inside the coating, or on the metal surface itself. Evidence for redox reactions on glow discharge polymerized acrylonitrile surfaces has recently been reported and charge transfer through the membranes could be explained by assuming tunneling of electrons between active sites having specific chemical structures (20).

GDPP coatings usually have hydrophobic properties. The effect of hydrophilicity on the electrochemical properties was studied. Oxygen plasma etching of the polymer applied to the electrodes produced more hydrophilic surfaces. We found that the electrode response did not depend on the surface properties although it is quite possible that the response in biological media could depend on these properties.

Glow discharge polymers are different than those formed by conventional means. Their physical or chemical properties depend on reaction parameters, such as R.F. power, flow rate, and pressure. They have little in common with their conventionally synthesized counterparts. The monomer gas is the most likely source of elements forming the films (12). The propylene polymers used for the application to oxygen sensors were characterized by a density of 1.11-g/cm^3, refractive index 1.540, and chemical composition $(C_1H_{1.59}O_{0.006})_X$, compared to conventionally polymerized polypropylene which has a density of 0.90 g/cm^3, refractive index of 1.49, and chemical composition of $(C\,H_2)_X$.

We are presently investigating "glow discharge" formed polymers of other organic monomers such as methane, ethylene, tetrafluoroethylene, the siloxanes, etc. It is possible that unique properties of these might offer additional advantages. Regardless of the type of polymer film applied, much additional work needs to be done when the sensors are used in a chronic physiological experiment.

In general, the results on glow discharge polymerized propylene coatings on oxygen electrodes are very promising. The tightly adherent coating of the polymer on oxygen sensing surfaces stabilizes the reduction surface and protects it from poisoning effects. Protected and stabilized in this fashion, suitable and reliable results during *in vivo* experiments with oxygen electrodes could be a reality.

ACKNOWLEDGMENTS

Supported in part by the John M. Dalton Research Center, UMC, in part by NINCDS Contract No. NO1-NS-82393, the U. S. Army Surgical Research Directorate, and a U.M.K.C. Faculty Research Grant. Discussions and contributions by Drs. R. E. Barr, F. Millich, V. G. Murphy and T. E. Tang are gratefully acknowledged. Our thanks also to Ms. Twilla Day, Ms. Victoria Watts and Mr. Richard Lee for their superb technical assistance.

REFERENCES

1. I. Fatt, "Polarographic Oxygen Sensors", CRC Press, Cleveland (1976)
2. M. L. Hitchman, "Measurement of Dissolved Oxygen", John Wiley & Sons, Switzerland (1978)
3. A. V. Berman, G. Y. Shigezawa, D. A. Whiteside, H. N. Yeung, and R. F. Huxtable, J. Appl. Physiol., 44(6), 969 (1978)
4. D. W. Davies, "Physical Techniques in Biological Research", W. L. Nastak, Ed., New York, (1962), p. 162
5. I. A. Silver, "Chemical Engineering in Medicine", ACS, Washington, D.C., (1973), p. 345
6. M. Forbes, and S. Lynn, AIChE J., 21, 763 (1975)
7. T. E. Tang, R. E. Barr, V. G. Murphy, and A. W. Hahn, "Oxygen Transport to Tissue-III", I.A. Silver, M. Erecinsky, and H. I. Bicher, eds., Plenum Press, New York (1978), p. 9
8. T. E. Tang, V. G. Murphy, A. W. Hahn, and R. E. Barr, J. of Bioengineering, 2, 381 (1978)
9. L. C. Clark, Jr., Trans. Am. Soc. Intern. Organs, 2, 41 (1956)
10. J. E. Evans and T. Kuwana, Anal. Chem., 51, 358 (1979)
11. M. F. Nichols, A. W. Hahn, J. R. Easley, and K. Mayhan, J. Biomed. Mater. Res., 13, 299 (1979)
12. H. K. Yasuda, "Thin Film Processes", J. L. Vossen and W. Kern Ed., Academic Press, New York, (1978), p. 361
13. T. Wydeven and J. R. Hollahan, "Techniques and Applications of Plasma Chemistry", J. R. Holland and A. T. Bell, eds., John Wiley and Sons, New York (1974)
14. P. J. Dynes and D. H. Kaelble, "Plasma Chemistry of Polymers", M. Shen, ed., Marcel Decker, New York, (1976), p. 167
15. D. T. Clark, A. Dilks, and D. Shuttleworth, "Polymer Surfaces", D. T. Clark and W. J. Feast, eds., John Wiley and Sons, London, (1978), p. 185
16. G. Oster and M. Yamamoto, Chem. Rev., 63, 257, (1963)
17. A. K. Sharma, Ph.D. Thesis, Univ. Missouri-Kansas City (1979)
18. A. K. Sharma, F. Millich and E. W. Hellmuth, J. Appl. Phys.,

49, 5055 (1978)
19. A. K. Sharma, F. Millich and E. W. Hellmuth in "Plasma Polymerization", ACS Symposium Series No. 108, M. Shen and A. T. Bell, editors, Amer. Chem. Soc., Washington, D.C., (1979), p. 65
20. K. Doblhofer, and D. Nolte, Phys. Chem., 82, 703 (1978)

SECTION II

CARDIOVASCULAR APPLICATIONS OF POLYMERS

This section is concerned with the application of polymers in heart assist devices, total artificial hearts and blood vessel replacement. The overriding consideration in all this work is the compatibility of these systems with the blood. The first paper in this section considers this problem carefully (Leininger). The next two papers deal with heart assist devices and total artificial hearts (Murabayashi/Nose and Akutsu, et al.) and consider the concept of biolized surfaces and the specific types of polymers and designs that have been tried in the past work and present work. The final three papers are concerned primarily with materials for blood vessel applications although the same biomaterials are frequently used in artificial hearts, etc. The first of these three papers considers coating the plastic with cell cultures to avert blood clotting (Eskin, et al.) while the last two papers (Lyman, et al and Knutson/ Lyman) are primarily concerned with the block copolyurethane polymer structure and its use in blood vessel replacement. While the papers in this section clearly show that the currently most widely used biomaterials in cardiovascular work are the segmented polyurethanes, they also show that better materials are still needed for long term applications. These papers give a good balance between the current research trends and the historical background in this biomaterial area.

PROGRESS AND PROBLEMS IN BLOOD-COMPATIBLE POLYMERS

R. I. Leininger

Battelle, Columbus Laboratories
505 King Avenue
Columbus, Ohio 43201

Compatibility with blood means, in broadest terms, no adverse effect on blood or any of its components. The complex nature of blood with its formed elements, coagulation system, and multitude of proteins puts not only a variety of restrictions, largely undefined, on the blood-contacting surface, but also makes the determination of blood compatibility an added significant problem. Further complicating the matter is the reverse requirement--no adverse effects by the blood on the polymer. Of the possible adverse effects of a foreign surface on blood, thrombogenesis--promotion of clotting--being most obvious and of immediate consequence, has received, the most attention. Fortunately, this very serious and disabling aspect of incompatibility can be overcome sufficiently by the administration of systemic anticoagulants such as heparin or coumadin to permit the use of the lifesaving devices such as heart-lung and kidney machines and artificial heart valves. Early postulates as to factors increasing thromboresistance included increased hydrophobicity, increased negative surface charge, and increased surface smoothness. The first real indication that thromboresistance was within grasp was the finding by a group at the University of Wisconsin (1) that heparin, ionically bonded to a surface, did provide a very significant degree of thromboresistance.

Very shortly after this finding, a concerted effort was begun in 1964 to find or develop thromboresistant polymers. This effort was the biomaterial portion of the Artificial Heart Program of the National Heart Institute. Descriptions of the approaches taken and progress made have been published in some detail. The early work on individual polymers is not described or referenced here because of space limitation. Consult references (2) and (3) for more specific information. These early efforts were in directions such as:

(1) Anionic polymers - as by the inclusion of acid groups to form surfaces having negative charges as does the blood vessel intima. An extension of this was the use of electrets.

(2) Low surface energy polymers - in accord with the early finding that paraffined surfaces delayed the clotting of blood compared to glass surfaces.

(3) Hydrogels - both because there were early indications that polyhydroxyethylmethacrylate (HEMA) possessed some degree of thromboresistance and because of the hypothesis that a water gel surface would be less recognizable as a foreign surface to the blood. Polymers and copolymers of this type, sometimes including anionic and/or cationic groups, have been extensively investigated.

(4) Polyurethanes - have received most attention because of the wide variety of compositions made possible by variations in the choice of components -- diisocyanate, polyether or polyester, and chain extender. In addition, a commercial spandex-type polyurethane, Lycra®, was found to have appreciable thromboresistance, as was a polyurethane-polydimethyl siloxane block copolymer, Avcothane 51®. (Trade names-- Lycra, Dupont; and Avcothane 51, Avco-Everett Research Lab., Inc.)

(5) Biologically modified surfaces - were a different approach in that if surface treatments could be found that were compatible, or at least thromboresistant, substrates could be chosen on the basis of suitability for the application. Because of the promise shown by the ionically-bonded heparinized surfaces, considerable study was devoted to methods of heparinization, by both ionic and covalent bonding.

Parallel to the development of thromboresistant and hopefully completely compatible polymers or surface treatments, research proceeded on (a) the development of methods to evaluate blood compatibility, and (b) the understanding of blood-material interactions that define compatibility. The latter would not only contribute to the development of badly needed improved evaluation methods but also form the basis for the development of truly blood-compatible surfaces. The most recent comprehensive review of blood-materials interactions is found in Reference No. 4.

It was known early that the first reaction when blood contacted a foreign surface was the adsorption of proteins (5) and it was recognized that it was the nature of this protein adsorbate that governed the further reactions that, *in toto*, governed the compat-

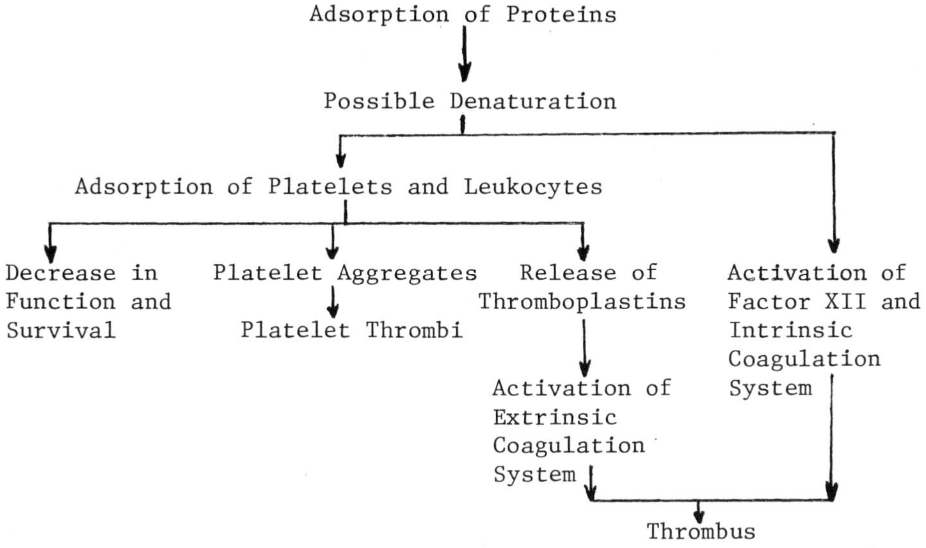
Adsorption of Proteins
Possible Denaturation
Adsorption of Platelets and Leukocytes
Decrease in Function and Survival
Platelet Aggregates
Platelet Thrombi
Release of Thromboplastins
Activation of Extrinsic Coagulation System
Activation of Factor XII and Intrinsic Coagulation System
Thrombus

FIGURE 1

ibility of the surface. Figure 1 shows the possible course of reactions that may occur.

To add to the complexity of the above, still other reactions are possible. Components, usually impurities, may leach into the blood and, in turn, blood components may be absorbed by the polymer. Immune reactions may occur, but as of now they have not been seen as a problem. Red blood cells have been shown to undergo surface-induced changes, but mechanical injury from fluid shear stresses appears to be more significant and can be kept adequately low by proper design of the blood-handling device.

Receiving the most attention has been the adsorption of proteins and the platelet-surface reactions leading to platelet contact and adsorption which in turn lead to platelet release reactions, aggregation, and decrease in platelet survival. Less attention has been paid to effects on leukocytes, even though it has been found that surfaces can cause release of thromboplastins (6) and impair the phagocytic activity of the leukocytes. Somewhat surprisingly, little attention has been given to thrombus formation by the direct activation of Factor XII and consequent initiation of the intrinsic clotting system.

The emphasis in the study of protein adsorption has been the composition of the adsorbed layer, rate of adsorption, and possible changes, such as denaturation of the proteins effected by the adsorption process. Such studies, in relation to surface character, surface thrombogenicity, and platelet adsorption, have primarily been done on solutions of single proteins or simple mixtures, using the primary blood proteins, albumin, gamma-globulin, and fibrinogen.

Rates of adsorption and equilibrium adsorption have been measured on a large number of surfaces using methods such as tagging of the proteins radioactively or with fluors, analysis of eluted protein, with antisera to the individual proteins, and ellipsometry. Difficulties and uncertainties arise in applying such measurements to the behavior of blood itself. The adsorption of proteins is known to be a competitive process so that adsorption from nonphysiological solutions of only a few proteins is far from the adsorption of dozens of proteins from blood itself. It should be noted that other proteins are beginning to be considered as important in blood-materials interactions. For example, fibronectin (cold-insoluble globulin), is being studied (7) as to its role in platelet adhesion.

Other difficulties in in vitro studies arise from the finding that adsorption is changed by flow conditions and by the presence of blood formed elements. In addition, with blood, a stable protein adsorbate appears to be formed only after exposure of a day or more. Despite these difficulties, some adsorption characteristics appear to correlate with thrombogenicity. Preferential adsorption of albumin and low fibrinogen adsorption appear to denote thromboresistant surfaces and this is used by many as an indicator of thromboresistance.

The effects of foreign surfaces on platelets has been intensively studied because of the place of platelets in thrombotic processes and because the visible deposition of platelets is evidence of some surface-platelet interaction. This latter is not strictly true because protein adsorption precedes platelet adsorption so that in reality it is a platelet-protein adsorbate reaction in which the nature of the adsorbate reflects the nature of the underlying surface. In vitro studies, primarily using platelet-rich plasma, have involved measurement of the adsorption rate, number adsorbed, morphological changes, and more recently, the release reaction. Again, translating such results to in vivo reactions is made difficult by the finding that platelet adsorption is modified by the presence of red blood cells (8), the effects of flow rate, the uncertain effects of anticoagulants, the differences between animal and human platelets, and the fact that platelet deposition in vivo rises to a maximum and then decreases to a low value in periods of a few hours to a day or so. (This is further evidence that stabilization of the protein adsorbate required appreciable time.) A further, and perhaps more important, surface effect on platelets is that of decreasing the survival time of the platelets. This decrease has been shown by Harker (9) in a baboon model to vary with different surfaces. This work may have uncovered a larger problem in that polymers showing least effects on platelet survival were those considered thrombogenic which may mean that those two aspects of compatibility are not related. The relationship between platelet reactions, especially release reactions, and thrombogenicity is in doubt. It has been shown that platelet release in plasma or blood can occur without activation of the clotting sequence proceeding to the formation of fibrin (10). Other work has shown that platelet release products (by ADP) do not activate Factor XII (11). Further evidence casting doubt on a direct relationship between platelet adsorption and thrombogenicity is shown below by measurements of platelet adsorption and FPA production from blood contacting unmodified poly(ethylene), sulfated poly(ethylene), and heparinized poly-(ethylene) (12).

	Platelets/cm^3 x 10^6	FPA mg/ml Plasma
Poly(ethylene)	4.3 $\pm$ 3.4	403 $\pm$ 454
Sulfated poly(ethylene)	0.4 $\pm$ 0.2	1008 $\pm$ 666
Heparinized poly(ethylene)	0.8 $\pm$ 0.3	19 $\pm$ 12

Thus reduction in platelet adsorption in itself did not reduce thrombogenesis as measured by FPA production. Leukocyte release, however, has been shown to include thromboplastins.

As shown above, detailed knowledge of the interactions of blood with foreign surfaces and the significance of these reactions to overall blood compatibility is, as yet, incomplete. As a result, research directed toward improved materials has been hindered. Despite this, especially in the last few years, research in this area has increased, based either on approaches previously found promising or on structural features now felt to be capable of reducing adverse reactions. Broadly classified, these recent and current approaches are in the areas of bulk polymers and copolymers, surface modifications, and modifications of mechanical properties or the use (or nonuse) of fillers. Other approaches to blood compatible materials such as treated umbilical veins and porcine heart valves or the use of polymers as reservoirs for the release of anticoagulants or platelet protective agents are not included in this discussion. Also, the review of these approaches is not intended to be exhaustive, but to summarize representative ongoing research in this field.

Bulk Polymers or Copolymers

Early in the search for blood-compatible polymers (2,3), polymers were evaluated on the assumed value of properties such as surface negativity, ionic or zeta potential, low surface energy, or presumed "inertness". Interest has revived in anionic surfaces. For example, complexes between anionic and cationic polyelectrolytes varying as to net negative or positive charge are being investigated as to the effects of charge and chemical structure on platelets (13). In a similar approach, copolymerization of carboxylic monomers with olefins has shown promise in in vivo studies of catheters. Newer approaches to improved polymers have been based on structural features such as side chain mobility (14,15,16) which are varied, for example, by the length of a pendant alkyl group. The influence of the microstructure of domain-forming block copolymers such as hydroxyethylmethacrylate-butadiene or poly(ether)-urethanes has been of interest to several groups (17,18,19). Here the differential protein adsorption of the domains and the size of the domains offer further means of minimizing blood reactions. It must be noted that the domain structure is affected not only by the copolymer composition but also by the processing conditions (20). Domain structure can be characterized by appropriate differential staining and by transmission electron microscopy. Closely related to this surface heterogeneity is the extension of the use of critical surface tension, an averaging value, to measurement of the polar and dispersive force components of the critical surface tension. The ratio of these two components has been shown to affect platelet deposition (21).

Surface Modification

Surface modification has long been seen as offering the advantage of very wide variety in the chemical nature of the blood-presenting surface while allowing the choice of the substrate to be on the bases of the mechanical properties needed. The surface modification can also be the grafting of polymers, such as hydrogels, which are inherently too weak mechanically to be useful as unsupported polymers except in limited applications. This grafting of hydrogels was begun early and is continuing (22,23) with the emphasis on varying the hydrophillicity and ionic character of the hydrogel by inclusion of nonhydroxy monomers, e.g., ethylmethacrylate, or by carboxyl or basic monomers.

The use of biologically active compounds as surface modifiers began in 1963 (1) with the electrostatic bonding of heparin. Interest in heparinization has continued and devices heparinized by the one-step tridodecylmethylammonium chloride-heparin complex (24) or by the gluteraldehyde-treated two-step hexadecylamine-heparin process (25) are in clinical use. Electrostatic bonding of heparin has also been extended to polyamides (26) by reaction of the polyamides with ethylenimine oligomers or polymers followed by quaternization and complexing with heparin. Somewhat similar binding has been accomplished (27) by first treatment of a surface with a palladium or rhodium salt which then will complex heparin.

Methods of covalently bonding heparin are being pursued again through reaction with surface hydroxyl groups via gluteraldehyde. In the recent work, hydroxyl groups are generated by glow discharge treatment of a styrene-butadiene-styrene block copolymer (28). In another instance (29), hydroxyl groups are obtained by hydrolysis of vinyl acetate copolymers. Heparin has also been directly grafted to DEAE cellulose acetate membranes by γ-radiation which may have resulted in both ionic and covalent bonding (30). Heparin has also been copolymerized with acrylic acid initiated by heparin free radicals created by Ce^{+4} salts in aqueous solution (31). The copolymer had heparin activity and could be coupled to aminated surfaces by carbodiimide with preservation of activity.

The very promising results seen with heparinized surfaces have led to the direct production of heparin-like structures on polymer chains. In one approach (32), isoprene polymers or copolymers were reacted with chlorosulfonyl isocyanate to form N-chlorosulfonyl β-lactan groups which upon reaction with NaOH yields the following heparin-like group:

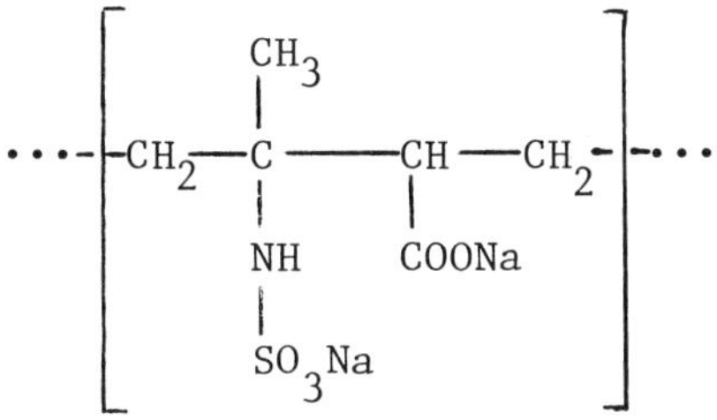
CH3
CH2
C
CH
CH2
NH
COONa
SO3Na

Toward a similar objective, various α-amino acids were reacted with chlorosulfonated cross-linked polystyrene (33) to give the following structure:

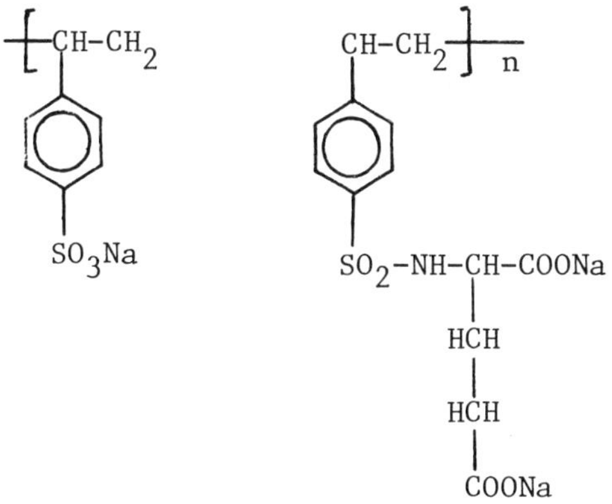
CH-CH2
CH-CH2
n
SO3Na
SO2-NH-CH-COONa
HCH
HCH
COONa

These materials suspended in plasma were shown to have heparin-like activity, with glutamic acid derivative having the greatest activity. To date, however, this reaction has been applied only to insoluble polystyrene spheres.

Although heparin has received most attention as a biologically active surface modifier, two other types of biologic agents have been studied. Lysing agents such as urokinase and streptokinase were surface bonded (34) under the rationale that any thrombi formed would be immediately lysed. This approach has been continued using brinolase (35) bonded to an amine-containing surface through isothiocyano groups formed by reaction with thiophosphene. Proteolytic enzymes such as trypsin, α-chymotrypsin, and actinase were also immobilized on the surface or within a polymer after first acylating the enzyme with acrylic or methacrylic anhydride or acid chloride to provide active groups for covalent bonding (36,37).

Platelet protective agents have also been bonded in an effort to reduce platelet adsorption. Covalent bonding of PGE_1 has been shown effective (38) and more recently Ditazole and dipyridamole have been bonded to polymers with a consequent reduction of platelet adsorption (39).

A different approach to surface modification by a biological material is the use of a protein, usually gelatin, cross-linked by glutaraldehyde (40). Anchorage is via a textured surface and *in vivo* results in heart assist devices or total artificial hearts have been promising.

Physical Modification of Polymers

In the use of solid wall conduits as opposed to fabric prostheses, poor performance has been ascribed to a mismatch in compliance of the tube and receiving artery. One approach (41) has been to match moduli by proper choice of a poly(urethane). An alternate approach has been by the construction of a composite, the modulus of which can be varied by drawing one or more layers. Variation in fillers has also been found useful. The addition of conductive carbon black to a silicone (42) and to a poly(urethane) (43) has been reported to decrease platelet interactions and thrombogenicity. The possibility of changing the domain structure by changes in processing has already been mentioned. Finally, the effect of purity must be considered. Leachables have long been known as the prime cause of adverse tissue reactions, but there is also evidence that thrombogenicity is affected (44) as in the significantly improved thromboresistance of highly purified poly(methylmethacrylate) and the improved performance of washed or extracted fabric grafts (45).

Summary

In summary, development of truly blood-compatible polymers as yet is hindered by both an incomplete understanding of blood-materials interactions and the lack of definitive methods of evaluating candidate materials. This latter includes also the choice of an animal model to serve as a surrogate for the human, which is a major problem in itself. Despite this, several bulk materials and surface-modified polymers have shown significant thromboresistance and are in at least limited clinical use. Fortunately a number of other polymers are usable either because the duration of use is limited or because systemic anticoagulation mitigates to a large degree their thrombogenicity.

In the opinion of the author, the current approaches offering the most promise are:

1. Control of the size and chemical nature of microphase polymers
2. Surface treatments utilizing biologics
3. Incorporation in polymers of synthetic structures having heparin-like activity.

It must be noted that effort in one area of blood-compatible polymers is largely neglected. This is the area of the effects of blood on the polymer. Isolated studies (46,47,48) have shown that degradation takes place, and this aspect of blood compatibility will become increasingly important as long-term, load-bearing polymer materials are needed.

REFERENCES

1. V. L. Gott, J. D. Whiffen, and R. C. Dutoon, Science, 142, 1297 (1963)
2. Artificial Heart Program Conference Proceedings, National Institutes of Health (1969)
3. R. I. Leininger, Critical Reviews in Bioengineering, 1 (3), 233 (1972)
4. Behavior of Blood and Its Components at Interfaces, Ann. N. Y. Acad. Sci., 283 (1977)
5. R. I. Leininger, in "Biophysical Mechanisms in Vascular Homeostasis and Intravascular Thrombosis", P. N. Sawyer, ed., Appleton-Century-Crofts, New York, (1965) p. 288
6. J. Niemetz, T. Muhlfelder, et al., Ann. N.Y. Acad. Sci., 283, 208 (1977)
7. J. H. Ihlanfeld, T. R. Mathis, et al., Trans. Amer. Soc. Art. Intern. Organs, XXIV, 727 (1978)
8. J. L. Brash, J. M. Brophy, and I. A. Feverstein, J. Biomed. Mat. Res., 10, 429 (1976)
9. S. R. Hanson, L. A. Harker, B. D. Ratner, and A. S. Hoffman, J. Lab. and Clin. Med., 95 (2), 289 (1980)
10. K. L. Kaplan, H. L. Nossel, et al., Brit. J. Haematology, 39, 129 (1978)
11. W. J. Vicic, O. D. Ratnoff, H. Saito, and G. H. Goldsmith, Brit. J. Haematology, 43 (1), 91 (1979)
12. R. Larsson, J-C. Eriksson, H. Lagergren, and P. Olsson, Thromb. Res., 15, (1/2), 157 (1979)
13. K. Kataoka, T. Akaike, Y. Sakurai, and T. Tsuruta, Makromol. Chem., 179, 1121 (1978)

14. E. W. Merrill, Ann. N.Y. Acad. Sci., 283, 6 (1977)
15. D. N. Gray, Org. Coatings and Plastics Chem., 42, 616 (1980)
16. S. A. Barenberg, J. M. Anderson, and K. A. Mauritz, Abst. First World Biomat. Cong., Europ. Soc. Biomat., Vienna, P2.5 (1980)
17. T. Okano, S. Nishiyama, I. Shinohara, R. Akaike, and Y. Sakurai, Polymer J., 10 (2), 223 (1978)
18. K. Knutson and D. J. Lyman, Org. Coatings and Plastics Chem., 42, 621 (1980)
19. K. Knutson and D. J. Lyman, Abst. First World Biomat. Cong., Europ. Soc. Biomat., Vienna, 4.3.3 (1980)
20. G. L. Wilkes, T. S. Dziemianowca, and Z. H. Ophir, J. Biomed. Mat. Res., 13 (2), 189 (1979)
21. E. Nyilas, W. A. Norton, R. D. Cumming, D. M. Lederman, and T.-H. Chiu, J. Biomed. Mat. Res. Symp., 8 (1), 51 (1977)
22. R. S. Schaffnit, R. A. Horbett, B. D. Ratner, and A. S. Hoffman, Abst. First World Biomat. Cong., Europ. Soc. Biomat., Vienna, 2.2.1 (1980)
23. D. L. Coleman and J. D. Andrade, op. cit., P2.3
24. R. I. Leininger, J. P. Crowley, R. D. Falb, and G. A. Grode, Trans. Amer. Soc. Art. Intern. Organs, 17, 312 (1972)
25. R. Larsson, M-B. Hjelte, J-C. Eriksson, H. Lagergren, and P. Olsson, Thrombos. Haemostas., 37, 262 (1977)
26. A. Sugitachi, K. Takagi, and Y. Yabushita, Ger. Offen. 2,748,858, 3 May 1979
27. A. E. Bergstroem and B. V. Soedervall, PCT Int. Appl. 79 00,638, 6 Sep 1979
28. M. F. A. Goosen and M. V. Sefton, J. Biomed. Mat. Res., 13 (3), 347 (1979)
29. N. A. Peppas and T. W. B. Gehr, Trans. Amer. Soc. Art. Intern. Organs, XXIV, 404 (1978)
30. A. S. Chawla and T. M. S. Chang, Biomat., Med. Dev., Art. Org., 2 (2), 157 (1974)
31. D. Labarre, A. Dincer, J. Lindon, D. Brier-Russell, M. Jozefowicz, E. Merrill, and E. Salzman, Abst. First World Biomat. Cong., Europ. Soc. Biomat., Vienna, 1.3.6 (1980)
32. L. Van der Does, T. Beugeling, P. E. Froehling, and A. Bantjes, J. Polymer Sci: Polymer Symp. 66, 337 (1979)
33. C. Gougnot, J. Jozefowicz, and M. Jozefowicz, Abst. First World Biomat. Cong., Europ. Soc. Biomat., Vienna, P2.1 (1980)
34. B. K. Kusserow, R. Larrow, and J. Nichols, Trans. Amer. Soc. Art. Intern. Organs, XVII, 1 (1971)
35. A-L Nguyen and G. L. Wilkes, J. Biomed. Mat. Res., 8, 261 (1974)
36. [Patents] U.S.S.R. 666,815, 5 Dec 1979 and U.S.S.R. 665,639, 5 Dec 1979
37. L. I. Valuev, M. A. Al-Nuri, N. S. Egorov, N. A. Plate, and V. V. Navrotskaya, Otkrytiya, Izobret., Prom. Obraztsy, Tovarnye Znaki, (45) 253 (1979)
38. G. A. Grode, J. Pitman, J. P. Crowley, R. I. Leininger, and R. D. Falb, Trans. Amer. Soc. Art. Intern. Organs, XX, 38 (1974)

39. W. Marconi, F. Bartoli, E. Mantovani, F. Pittalis, L. Settembri, C. Cordova, A. Musca, and C. Alessandri, Trans. Amer. Soc. Art. Intern. Organs, XXV, 280 (1979)
40. R. J. Kiraly, G. B. Jacobs, and Y. Nosé, Proc. Int'l Soc. Artif. Organs, 2 (Suppl.), 234 (1978)
41. D. J. Lyman, F. J. Fazzio, H. Voorhees, G. Robinson, and D. Albo, Jr., J. Biomed. Mat. Res., 12, 337 (1978)
42. T. Kolobow, E. W. Stool, P. K. Weathersby, J. Pierce, F. Hayano, and J. Suaudeau, Trans. Amer. Soc. Art. Intern. Organs, XXA, 269 (1974)
43. B. G. Miller, K. A. Dyer, B. C. Taylor, J. I. Wright, and W. V. Sharp, op. cit., p. 91
44. E. Nyilas, Private communication.
45. P. N. Sawyer, B. Stanczewski, R. Turner, and H. Hoffman, Trans. Amer. Soc. Art. Intern. Organs, XXIV, 215 (1978)
46. F. Zartnack, W. Dunkel, K. Affeld, and E. S. Bucherl, op. cit., p. 600
47. E. Roggendorf, J. Biomed. Mat. Res., 10, 123 (1976)
48. A. Hiltner, H. Dickenson, and J. M. Anderson, Abst. First World Biomat. Cong., Europ. Soc. Biomat., Vienna, 3.4.5 (1980)

BIOLIZED MATERIAL FOR CARDIAC PROSTHESIS

Shun Murabayashi and Yukihiko Nose

Department of Artificial Organs
Cleveland Clinic
Cleveland, Ohio 44106

HYPOTHETICAL PROCESS TO GENERATE BIOCOMPATIBLE MATERIAL - "BIOLIZATION"

It has been well documented that most artificial materials, whether hydrophilic, hydrophobic, made electroconductive by the dispersion of conductive particles in it, smooth, rough or heparinized, were covered with protein immediately following implantation into the cardiovascular system (1,2). Regardless of this protein coating, some of the materials had higher thromboresistance than others. Many theories have been proposed to explain these thromboresistant properties, but a definitive answer to these questions would be premature at this time. We have also found that not every protein surface shows thromboresistance; however, when the protein surface was treated with an aldehyde or heat, it resulted in a reproducibly improved thromboresistant surface.

This protein can be mixed in the synthetic material such as in aldehyde treated albumin natural rubber, coated over plastic material, or it can be in natural tissue. Not only protein but probably other biological components such as polysaccharides (heparin), have a similar effect.

It is obvious that protein or biological components alone on a surface will not always enhance its thromboresistance, however when this protein is treated by aldehyde, heat, or some other methods, which are generally applied for sterilization or antiseptic procedures, it becomes quite thromboresistant. This treatment probably makes the protein insoluble, cross-linked, or denatured. What kind of process is actually taking the main role is not completely understood at that time so it was difficult to select an appropriate term to describe it. Thus, the term "biolization" was made to describe this process. Together with this term, hypothetical processes for the production of biocompatible and particularly blood compatible materials were proposed (3).

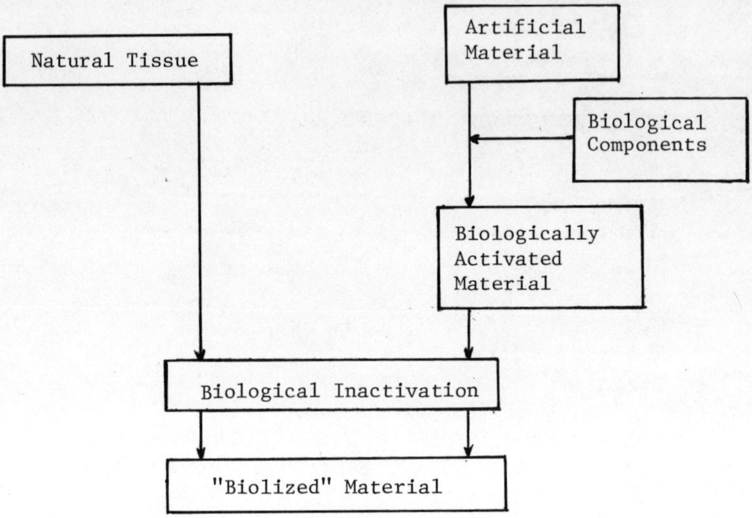
Natural Tissue
Artificial Material
Biological Components
Biologically Activated Material
Biological Inactivation
"Biolized" Material

(Thromboresistant) (Biocompatible)

"BIOLIZATION", HYPOTHETICAL PROCESS TO GENERATE BIOCOMPATIBLE MATERIA

Material to be used for this process will be either of natural or artificial origin, but it should contain protein, polysaccharides, or other biological components. Artificial synthetic materials can be subjected to this condition either by the mixture of biological components or coating with biological components (biologically activated). When the surface is treated with aldehyde or heat it becomes "biologically inactivated". Biological inactivation probably means insolubilization, denaturation, or cross-linkage of the biological components or others.

Natural tissue of a tissue typing group different or remote can be treated and inactivated in the same way so that this tissue can be used for implantation out of the target for intensive immuno-rejection process. Heterologous or homologous preserved aortic valves have already been successfully used clinically (4,5). A low incidence of thromboembolic complications was unanimously reported.

BACKGROUND FOR THE HYPOTHESIS

Method For Testing Thromboresistant Surface Properties - Kinetic Clotting Test

A simple in vitro screening test was developed in our laborator and named the "kinetic clotting test"(6). The kinetic method is applicable to any type of film material and yields four important facts on thrombosis: 1) initiation of blood coagulation, 2) rate of blood coagulation on the surface; 3) amount of thrombus generated by this material; and 4) end point (time) of blood coagulation. These can be interpreted from the curve obtained by the kinetic method. Silastic is generally used for the control.

Synthetic Material With Biological Components

Three different biological materials, namely, albumin (3-27%), gelatin 3-14%), and heparin (3-18%) were mixed singly or in combination in natural rubber. After adding the biological components, these surfaces were treated in one of four ways: 1) 4-10% formaldehyde solution; 2) heated at 90°C; 3) 0.5-2% glutaraldehyde solution; and 4) combination of the above methods.

A total of 63 different samples were prepared and tested (7). In these materials the hydrophilic properties of latex rubber were more enchanced; water content of the rubber was close to three times that of ordinary rubber. The following conclusions were made from these studies.

1. Aldehyde treatment was necessary to enchance the thromboresistance of the natural rubber mixed with albumin. Material made from natural rubber mixed with albumin did not show any enchanced thromboresistance without the aldehyde treatment.

2. The effect of formaldehyde concentration was not evident, however, a concentration of about 4% seems to be favorable.

3. Formaldehyde treatment of ordinary natural rubber, which did not have additional protein components, did not improve its thromboresistance.

4. The dehydration of the aldehyde treated albumin rubber decreased its thromboresistance. Subsequent one-hour soaking in water did not result in any improvement of thromboresistance.

5. Albumin content between 10-20% gave almost the same results. A higher amount of albumin (27%) decreased the thromboresistant property.

6. Aldehyde treated gelatinized natural rubber gave good reproducibility of its thromboresistant property over albuminized natural rubber. In all the samples containing between 6-14% gelatin, good thromboresistant tendency was experienced.

7. The thromboresistant property achieved with glutaraldehyde

treatment was as effective as that obtained with formaldehyde treatment or heat treatment. The lower concentrations of glutaraldehyde give better results. Concentrations of 0.5% are used to preserve clinical homografts.

8. The pH of the aldehyde is quite important in the treating process.

9. The heparin and gelatin mixed material showed excellent thromboresistance even when samples were soaked in distilled water for one week. The best treatment was combination of heat and glutaraldehyde treatment.

10. Heating gelatinized natural rubber for 2.5 hours resulted in thromboresistance comparable to the aldehyde treated material.

Surface Treatment of Plastic Materials

It has been our experience that an artificial heart implanted the second time is more thromboresistant than the first time (7). Any material implanted in the bloodstream will be coated with protein almost instantaneously. Typically, in preparation for a second implantation, formaldehyde solution is used for disinfectant purposes after removing the unit from the first animal. The formaldehyde treatment apparently plays an important part in the improved results achieved during second and subsequent implants. To confirm this experience, a similar type of experiment has been performed with our in vitro model. Silastic samples were exposed to dog blood for 3 hours at room temperature and then afterward one surface was treated with 4% formaldehyde and soaked in distilled water. The other sample was untreated. A new surface and the untreated used surface did not show significant differences in the subsequent blood clotting test, but the surface treated with formaldehyde showed improved thromboresistance.

It is known by clinicians, patients, and technicians, that the second use of a hemodialysis membrane in the dialyzer results in an easier blood collection after the procedure is finished. It appears that platelet and blood cell adhesion on dialysis membranes is less. After the first use (where the membrane was exposed to proteins) it is sterilized using 4% formaldehyde.

Biological Tissue Materials and Their Fixation

Natural tissue was treated in the same way as we treated the natural rubber material. Glutaraldehyde (0.5%) preserved or formaldehyde (4%) preserved natural tissue showed improved thromboresistance (7). The bovine aorta and pericardium were treated in the aldehyde solution for at least one week at room temperature. Afterward all the materials to be tested were soaked in physiologic saline solution for at least one week at 4°C before the test. Treated aortic wall and pericardium showed superior thromboresistance over Silastic or Hydron.

Biolized Materials for Cardiac Prosthesis

The primary requirement for a permanent assist pump is long-term reliability ensured by materials selected for durability, stability and blood compatibility. Rather than search for a single ideal material, it is possible to develop composite materials that lead to the optimization of a number of these properties.

Biolized materials have been used in our cardiac prostheses since 1969. The first application utilized glutaraldehyde treated bovine aortic valves in a Dacron fabric pump termed a "partially biolized heart". Since thick pseudoneointima (PNI) formation and calcification was observed, the Dacron covered surface was replaced with natural tissue material (9). This original totally biolized heart was a sac-type with a flexing element of natural rubber lined on the blood side with aldehyde treated bovine pericardium. The outside case of the device was made from polyurethane. Early in 1973, a calf implanted with this artificial heart lived for a then-remarkable seventeen days (10). Termination of the experiment was caused by a crack in the flexing sac. A passive implant of this device in the aorta did not show any thrombus formation during 5.5 years implantation.

In 1974, a major design change resulted in a diaphragm-type pump, utilizing a high flex life polyolefin rubber called Hesxyn® developed by the Goodyear Tire and Rubber Company. The blood side of the polyolefin rubber diaphragm was bonded to a Dacron® velour fabric coated with glutaraldehyde treated gelatin. Using this redesigned artificial heart we could extend the experimental calf's survival to approximately one month. One of the major problems with the new configuration was an increase in diaphragm stiffness due to the fabric bonded to the diaphragm. In vivo experiments have shown that a relatively thick PNI formed on the diaphragm due to the difficulty of completely covering the fabric with gelatin (11).

To overcome this drawback, a technique to texturize polyolefin rubber was developed. The 100 μ thick porous surface is then impregnated and covered with gelatin. Results obtained using the biolized textured diaphragm were excellent. No diaphragm failure has ever been experienced and in vivo experiments have demonstrated the excellent blood compatibility of glutaraldehyde treated gelatin (12). In 1976, a total artificial heart of this configuration maintained a calf for 145 days (12). In 1977, a biolized left ventricular assist device was implanted in a calf for 10 months. Anticoagulants were not used nor has it been necessary to use anticoagulants with any of the biolized cardiac prostheses.

These experiments demonstrate that surface impregnation with a thin layer of glutaraldehyde treated gelatin results in excellent long-term blood compatibility, while retaining the physical and mechanical properties of the substrate material. This gelatin coating technique is being applied to our pusher-plate-type LVAD currently being developed for chronic human use. All blood contacting surfaces are biolized with glutaraldehyde treated gelatin coated on the textured substrate (14).

Gelatin as Biomaterial

Gelatins are denatured proteins produced by thermal or chemical treatment of collagen (15). The reversible gel-sol conversion of the gelatin solution makes it easy to produce smooth surfaces. Glutaraldehyde treatment of the gelatin in the gel state makes it insoluble in water due to cross-linking reactions (16).

After in vivo implantation, the glutaraldehyde-treated gelatin surfaces were generally clean macroscopically. Under scanning electron microscopy, the surface of the gelatin did not elicit a fibrin network formation, but was coated with a very thin proteinaceous layer. This smooth surface is highly blood compatible as suggested by the finding that platelets attaching to it were both the "contact" and "spread" type, but without aggregation. There were leukocytes observed on the surface, some of these being very similar to the precursors of the endothelial cells observed previously on glutaraldehyde treated pericardium lining of blood pumps (17). Leukocyte adhesion on the foreign materials was explained as being elicited by platelet aggregation (18). It was also suggested that gamma globulin adsorbed on the material surface plays an important role in leukocyte adhesion (19). Although the proteinaceious material adsorbed on the gelatin surface is still to be analyzed, it is not conceivable that the leukocytes on the gelatin surface were mediated by platelet aggregation, since there were no such aggregations adjacent to the leukocytes.

Proteins can be bonded onto substrates by chemical treatments or ionization methods such as γ-ray irradiation. However, with such treatments, special attention should be paid to residues from the chemical reagents and property changes in the substrate. On the other hand, texturization by the salt casting method needs only sodium chloride, which, even if not completely removed, is not harmful biologically, and does not change the chemical or mechanical properties of the substrate. We believe that glutaraldehyde-treated gelatin on textured substrates is one of the best materials for use in blood pumps and is useful for other biomaterial applications as well.

ABSTRACT

In the composition of any prosthetic device which is in contact with blood, blood compatible material is essential. In 1971, the hypothesis of "Biolization" was proposed to provide blood compatible material. Biolized materials have been employed in our cardiac prostheses since then. The biolized surface chosen to apply for cardiac prosthesis consists of either protein coating (gelatin) or natural tissue which was crosslinked with aldehyde treatment. Excellent blood compatibility of biolized materials was demonstrated after long term implantation in calves up to 10 months without the use of anticoagulants. Glutaraldehyde-treated gelatin coating was utilized for the intrathoracic left ventricular assist device for chronic human use now under development. The hypothesis of "Biolization" and its background were described. Biolized cardiac prostheses developed in our laboratory were also reviewed.

REFERENCES

1. R. E. Baier, ed., "Applied Chemistry at Protein Interfaces", Advances in Chemistry Series No. 145, ACS, Washington, CC. (1975)
2. L. Vroman, E. F. Leonard, ed., "The Behaviour of Blood and its Components at Interfaces", Ann. N.Y. Acad. Sci., 283, New York Academy of Science, New York (1977)
3. Y. Nose, K. Tajima, Y. Imai, M. Klain, G. Mrava, K. Schriber, K. Urbanek, and H. Ogawa, Amer. Soc. Artif. Int. Organs., 17, 482 (1971)
4. A. Carpentier, G. Lemaigre, L. Robert, S. Carpentier, and C. Dubust, J. Thorac. Cardiovasc. Surg., 58, 469 (1969)
5. J. W. Yarbrough, W. C. Roberts, and R. L. Reis, Ibid, 64, 364 (1973)
6. Y. Imai, and Y. Nose, J. Biomed. Mater. Res., 6, 165 (1972)
7. Y. Imai, K. Tajima, and Y. Nose, Trans. Amer. Soc. Artif. Int. Organs, 17, 6, 1971
8. Y. Imai, K. von Bally, and Y. Nose, Ibid, 16, 16 (1970)
9. Y. Nose, Y. Imai, K. Tajima, H. Ogawa, M. Klain, K. von Bally and D. B. Effler, J. Thor. Cardiov. Surg., 62, 714 (1971)
10. M. Nakazono, K. Koiso, T. Komai, T. Agishi, J. Uruzu, R. Kiraly, H. Kambic, R. Surovy, C. Carse, and Y. Nose, Jpn. Heart J., 15, 485 (1974)
11. Y. Nose, R. Kiraly, G. Jacobs, C. Arancibia, K. Nakiri, G. Picha, H. Kambic, N. Morinaga, Y. Mitamura, and T. Washizu, "Development and Evaluation of Cardiac Prostheses", Report N01-HV-4-2960-1 (1975)
12. H. Harasaki, R. Kiraly, S. Murabayashi, M. Pepoy, A. Fields, H. Kambic, D. Hillegass, and Y. Nose, "Cross-linked Gelatin as a Blood Contacting Surface", Proc. 2nd Int. Soc. Artif. Organs Meeting (in press).
13. N. Tsushima, S. Kasai, I. Koshino, G. Jacobs, N. Morinaga, T. Washizu, R. Kiraly, and Y. Nose, Trans. Amer. Soc. Artif. Organs, 23, 526 (1977)
14. K. Ozawa, J. Snow, R. Sukalac, S. Takatani, M. Kitagawa, F. Valdes, H. Harasaki, D. Hillegass, C. Castle, G. Jacobs, R. Kiraly, and Y. Nose, Trans. Amer. Soc. Artif. Int. Organs, (1980) in press.

15. A. Veiss, "The Macromolecular Chemistry of Gelatin", Academic Press, New York (1964)
16. H. Kambic, S. Murabayashi, H. Harasaki, S. Suwa, M. Pepoy, K. Hayashi, D. Hillegass, R. Kiraly, and Y. Nose, "Compositive Polymeric Materials: Evaluation of Crosslinked Gel Protein Surfaces", 2nd Int. Soc. Artif. Organs Meeting (in Press)
17. H. Harasaki, R. Kiraly, and Y. Nose, Trans. Amer. Soc. Artif. Organs, 24, 415 (1978)
18. R. E. Baier, Ann. N.Y. Acad. Sci., 283, 17 New York Academy of Science, New York (1977)
19. L. Vroman, A. L. Adams, M. Klings, G. S. Fischer, P. C. Munoz, and R. P. Solensky, Ibid, 283, 65 (1977)

PLASTIC MATERIALS USED FOR FABRICATION OF BLOOD PUMPS

Tetsuzo Akutsu, Noboru Yamamoto, Miguel A. Serrato, John Denning, and Michael A. Drummond

Texas Heart Institute
Houston, Texas 77025

The various primary plastic materials which have been used for the fabrication of blood pumps since 1957 number less than fifteen. The surface property of these materials can be divided into three general categories: 1) smooth, nonpermeable, 2) porous or rough, and 3) biolized. Among the various factors to be considered for candidate materials, the two most important are mechanical durability and blood compatibility, particularly antithrombogenicity. This paper will present an historical review of plastic materials used for the fabrication of blood pumps, with the emphasis on antithrombogenicity not only in relation to the material itself, but also with respect to its design and fabrication.

The material used for the first ventricular assist device developed by Kusserow was natural rubber (1). He used natural rubber to fabricate the diaphragm and polyethylene to fabricate the housing of a right ventricular assist device which was implanted in the abdominal cavity of dogs. A later improved model (2) incorporated a more durable housing material (methylmethacrylate). Atsumi (3) extended the use of natural rubber to the total artificial heart (TAH). Natural rubber's utility as a blood pump material however was short-lived because of its short fatigue life and its tendency to promote blood clotting and hemolysis. Polyvinyl chloride was used in total artificial hearts (4,5), but this also was soon abandoned because it was cumbersome to handle and the surface was very thrombogenic.

In 1958 Schollenberger (6) introduced a new prospective biocompatible polymer, Estane V.C. (polyester urethane). This plastic appeared to be an ideal material because it exhibited very high tensile strength and flex life and was resistant to radiation, elevated temperatures, and strong acids and bases. Kolff (7) was the first to incorporate this polymer in a TAH. Although its use significantly reduced foreign body reaction, the ester linkage was susceptible to hydrolysis. Hydrolysis related degeneration of the polymer chain occurred after implantation which translated into decreased tensile strength and ultimately a brittle, useless material (8).

In 1960 Silastic (medical grade silicone rubber) became available and with it the first significant improvement in blood compatibility. Seidel (9) and Pierce (10) incorporated a Silastic rolling diaphragm made by the Bellofram Corporation in their blood pumps. Although Silastic is a demonstrably better blood contact material, its antithrombogenic properties were substantially masked by other complicating factors. In the former case pump design inadequacies inadvertantly created stagnation regions between the diaphragm and the housing. Any regions of flow stagnation whether in a blood pump or in a natural vessel, invariably lead to thrombus formation. An additional exacerbating factor was the high stresses at the diaphragm-housing (D-H) junction generated by the large bending radius and diaphragm displacement necessitated by the design. When coupled with Silastic's inherently poor tensile and tear strength, this resulted in an extremely limited flex life. In the latter case (Pierce) the complication arose from the fact that methylmethacrylate, a thrombogenic material, was used for the housing (blood contact).

To augment the poor mechanical properties of Silastic, Akutsu (11) fabricated a sac-type artificial heart from calendered Silastic sheets reinforced with Dacron mesh. Because it was easily workable, they were able to construct a heart pump in one piece, but unfortunately not without seams. So even though the flex life of the material was extended, thrombus formation (at the seams) remained a problem.

Liotta (12) constructed his artificial heart from a combination of different materials - Lucite, teflon, polyester urethane, and silk. It is difficult to evaluate the thrombogenic potential of this combination of materials because the longest survival was 13 hours in dogs. The principle cause of death in these animals was low cardiac output secondary to inadequate venous return. The following year in a different series of experiments Liotta (13) tried a different combination of materials. Here, although the experiments were acute in scope, thrombus formation at the blood-plastic interface was a major problem. The left ventricular assist device (LVAD) was a tube-type with the housing and valves constructed of Estane. The internal elastic tube was made of either natural rubber, Silastic, or natural rubber covered externally with Silastic.

These early attempts to find suitable materials for use in blood pumps were fraught with difficulties. First, most of the plastics were industrial grade materials that were never intended for medical use. Second, the poor mechanical strength of these materials (natura rubber and Silastic) when complicated by poor design resulted in approximately one third of the experiments terminating because of material failure. Third, the relatively poor compatibility was not always the fault of the material but of poor pump design and/or fabrication. Such was the case with Silastic.

Having been discouraged by the surface of smooth materials, investigators began to turn their attention to porous or rough materials. This interest was generated by the healing process of pseudointima formed inside Dacron vascular grafts. They reasoned that if the pseudointima completely covered the inside of the blood pump, thrombus would not form on it. Since the vascular graft is porous, the formed neointima can firmly anchor onto the surface where it is nourished by both perivascular tissue and circulating blood. Hence, true neointima can develop from both sides of the anastomosis. Most blood pumps, however, were driven by compressed air, so the materials composing the housing and moving diaphragm are, out of necessity, nonpermeable. In such a situation, we cannot expect either vital tissue growth or cellular ingress through the housing wall from the periprosthetic tissue, or even firmly adhered coagulum formation on the surface. Despite such conjecture, several attempts to apply porous materials such as knitted cloth, velour, and flock were made.

Liotta (14) used Silastic lined with Dacron fabric to make a tube-type LVAD. This devise was designed to have the feasibility of long-term support of the left ventricle without heparin, and this led to the first clinical application of an implantable LVAD in a human subject (15). At approximately the same time, they began using a loose-knitted Dacron fabric liner backed with Silastic (16,17). A fibrin network was laid down on these materials, but it often broke loose, forming thromboemboli. Later they substituted the knitted Dacron in favot of velour fabrics (18) because they thought that the velour surface would form a tenacious mechanical bond with the fibrin deposit and therefore the chance of dissection and thromboembolization would be infinitesimal.

An alternative approach to the problem of anchoring fibrin to nonpermeable surfaces was first advanced by Sharp (19) in arterial prostheses. Moderate success was attained by spraying either Dacron or Teflon flock onto the inner surface (blood contacting) or arterial substitutes. In artificial organs Bernhard's group (20) sprayed Dacron fibrils onto Silastic to make LVADs. The shortcoming of this technique was poor bonding of the fibrils to the Silastic. Silastic then gave way to spraying Dacron fibrils onto uncured polyurethane. This latter technique is still employed in Thermo Electron Corporation's LVAD.

A study of nonpermeable prostheses showed that the fibrinous neointima healed by ingrowth of modified smooth muscle cells from both suture lines (21). The non-porous wall blocked the ingress of cells from the surrounding fibrous connective tissue and spontaneous healing did not occur in some locations. Ghidoni's experimental approach called for the application of tissue culture technique in conjunction with a fabric-lined nonpermeable vascular prosthesis and pump diaphragm. He employed a suspension of connective tissue fragments prepared from autologous skeletal muscle (22,23). The rate of healing and endothelilization clearly increased and the cultured cellular lining appeared to have antithrombogenic properties similar to those of control artificial vascular grafts. Unfortunately the healing was of insular fashion and in some areas cellular proliferation was totally lacking. In addition, the coagulum was twice the thickness of the control In order to regulate the thickness of the coagulum, Adachi (24,25) developed a technique to make a fibrin coagulum membrane preoperatively. This process promoted rapid, even cellular growth which produced a smooth even surface. This approach, however, was never applied to the fabrication of blood pumps. Bernhard (26) also applied tissue culture techniques but in a different way. Bovine, fetal fibroblasts were seeded on the surface of Dacron fibril flocked polyurethane. This technique was used for lining LVADs with the expectation of minimizing the immunological response of the recipient animal.

Although acceleration of pseudoendothelial formation, organization of protein layers, and cellular proliferation were confirmed, these efforts with tissue culture techniques did not yield significant improvement in compatibility. Not only were the results similar for flock treatment versus flock treatment plus tissue culture, but also tissue culturing was very time consuming, costly, and difficult. It is understandable then that application of tissue culture technique became virtually extinct around 1972.

Kolff's group tried a myriad of materials such as smooth surface Silastic, heparin-treated Silastic, albumin-treated Silastic, nylon velour lined Silastic, Dacron fibrils on Silastic, and Dacron fibril flocked Silastic seeded with bovine fetal fibroblasts (27,28). According to their report (29) the surface coated with either albumin or heparin was not significantly different from Silastic alone. The nylon velour surface was discarded because of its propensity to cause excessive thrombus. In comparative studies between smooth surfaces (Silastic) and rough surfaces (Silastic + Dacron fibril), the rough surface had a tendency to maintain high platelet and fibrinogen levels, and to decrease the incidence of disseminated intravascular coagulopathy, but caused hemolysis (30). Smooth surface Silastic caused less blood damage than the rough surface, but showed greater evidence of embolization (31). The Dacron fibril surface suffered from similar problems that have plagued most rough surface pumps. First, microscopic examination of retrieved blood pumps revealed pitted surfaces, indicative of detachment and loss of fibrils into the circulation. Second, one study reported that detached fibrils were ultimately found in the brain, lung, and kidneys (32). Third, on most Dacron fibril surfaces layer after layer of fibrin was added to the pumping diaphragm. After twenty to thirty days, the fibrin layer began to calcify, resulting in interference with the diaphragm movement and thus compromising cardiac output (33). Fourth, another drawback of the flocked surface which wasn't a factor until survival time increased, was the excellent medium for propagating infectious endocarditis which the fibrin layer presents (34).

Boretos (35) developed a segmented polyether urethane (Biomer) specifically for biomedical applications, and Pierce (36) used this as a smooth surface material to make an implantable roller pump for left ventricular assistance. A roller pump does not provide an ideal model for evaluating biomedical polymers. Not only does it create extremely high shear stresses which of itself is detrimental to the blood, but it also subjects the polymer to elevated stresses and hence a very short flexure life. In spite of these failings, Biomer exhibited a flex life vastly superior to silicone rubber and polyester urethane, and more importantly, thromboemboli and infection were never seen with its use. The relatively good results obtained in these studies restimulated interest in smooth surface materials. More specifically, it destroyed the myth that polyurethanes were part of an homologous family of which Estane was a member and therefore guilty by association.

Another segmented polyether urethane, PU 1025, developed by Lyman was used in a hemispherically shaped TAH in Kolff's group (37). Since the polymer was readily amenable to solvent casting, it was possible to produce a seamless diaphragm-housing junction. Although hemolysis was greatly reduced due to the improved pump design and the seamless fabrication technique, thromboemboli and infarcts were still evident.

In a comparative study between the two kinds of segmented polyurethanes, Biomer and PU 1025, the former material was deemed superior in terms of both durability and compatibility (38). After establishing Biomer as the more suitable segmented polyurethane and making a further improvement in a pump design (ellipsoidal), a comparative study was made between the smooth polyurethane and rough Dacron fibril lined Silastic (39,40). The Biomer surfaces were found clean at autopsy except where the pumping diaphragm joined the housing wall, even without anticoagulation. The localized thrombus at the D-H junction was the result of ventricle design dependent stagnation, which was later remedied (Jarvik 5 pump). Contrarily, in spite of the use of anticoagulants, the rough Silastic surfaces developed thick calcified neointima which severely compromised cardiac output. The average calf survival with the polyurethane pumps was 545 hours in 10 cases (max. 2277 hours) and in the rough Silastic pumps it was 296 hours in 10 cases (max. 862 hours).

Avcothane 51, a complex blend of polyether urethane, dimethylsiloxane, and a block copolymer of the two was developed by Nyilas (41). This material was first used by Engelman (42) who designed and constructed a totally implantable left ventricular bypass pump. Koff's group fabricated an Avcothane TAH and got a 78 day survival (39). This material compared favorably with Biomer fabricated TAHs of the same design.

In Japan, Atsumi's group has been using polyvinyl chloride (PVC) since 1971 (43). The merits of this material are easy handling, easy molding, low cost, and moderately good antithrombogenicity. Recently they have enjoyed improved survival times by coating their PVC pumps with Avcothane and segmented polyurethane (44). They found that the antithrombogenicity of the PVC often became unstable with hypercoagulability of the blood or low blood flow rates. In contrast, Avcothane and polyurethane surfaces did not suffer this same fate. While both these latter materials were found free of thrombus at autopsy, only the polyurethane showed the presence of a fibrin network.

Although many groups were switching over to the smooth surface polyurethanes, one group pursued a viable alternative to a blood compatible surface. Based on the observation that repeated use of formaldehyde stored protheses reduced thrombus formation, the concept of biolized surfaces was born. Biolization refers to the process whereby proteins or tissue are coated and cross-linked on a substrate surface. Initially biolized membranes were made from purified natural rubber latex, Pluronic, and albumin or gelatin (45). Nose (46) took this a step further and fabricated a totally biolized heart using aldehyde-treated pericardium (ATP) backed with hydrophilic aldehyde-treated albumin rubber. ATP demonstrated good blood compatibility but it suffered from several other problems. In vitro testing revealed cracking of the aldehyde-treated tissue within two months, but this was easily solved by cross-linking with glutaraldehyde. Natural rubber, of course, does not have ideal mechanical properties, but most importantly the feasibility of biolized surfaces was demonstrated. The next logical progression was to employ a material of greater flex life as the diaphragm substrate. A polyolefin rubber, Hexsyn, was coated with aldehyde-treated gelatin and supported by Dacron fabric. The housing was constructed of ATP laminated to natural rubber and polyurethane for rigidity. Several design improvements were made (increasing the clearance between the diaphragm and housing, and recessing the outflow valve), which together with an improved fabrication technique (better impregnation of gelatin), led to substantially decreased PNI thickness. A more recent modification has produced a 145 day TAH calf survival (47). This technique called for coating the partially cured diaphragm with a dilute rubber mixture containing screen-graded salt. After curing the salt is dissolved out and the remaining pores are impregnated with gelatin and solubilized with glutaraldehyde. At autopsy this pump was free of thrombus and no thromboemboli were found. Calcium deposits on the diaphragm were the only disconcerting result.

Figure 1 shows the historical trend of the utilization of different plastics toward the fabrication of blood pumps. In this figure one can see that investigators started with smooth materials, which were utilized from approximately five years, changed rather abruptly to porous materials for the next eleven years, and then during the span of a few years virtually every investigator returned to using the smooth surfaces which had been improved in the interim. So the history of utilization of blood contact materials can be divided into three periods. During the first period, only smooth surfaced materials were used. Because of the abbreviated survival time (hours), the most serious problems encountered were hemolysis and thrombus formation. Of these materials, only silicone rubber has been continuously used to date because of its excellent workability and antithrombogenicity. Having tired of struggling with thrombus formation on the surface of these plastic materials, investigators altered their approach in the second period. Since it was extremely difficult to maintain a thrombus free blood pump, they purposely decided to use porous or rough materials to induce fibrin deposition with the expectation of forming a biologically smooth surface of fibrinous membrane and an eventual neointima. The materials used during the first period were relegated to a mechanical supporting role for the rough surfaces. It can not be denied that the survival time of experimental animals was prolonged during this period (from hours to weeks). What has to be realized is that improved pump design deserve as much of the credit for this as did the materials. All these attempts, however, eventually turned out to be less effective than had been originally anticipated. The Dacron fibril flocked polyurethane surface is presently used only in the LVAD developed by Thermo Electron Corporation.

About the time investigators again became discouraged with the material, advanced technology made it possible to synthesize biocompatible, durable smooth polymers, ie. segmented polyurethanes and Avcothane. Blood pumps constructed of these materials have proved to pump continuously without failure for longer than six months. If the material is handled properly during fabrication, use of anticoagulant may not be necessary. Durability longer than one or two years however, still remains to be answered.

What has to be borne in mind, is that the ultimate utility of a polymeric material toward fabrication of blood pumps is intimately related to both the design of the blood pump and the fabrication of the material. Table 1 lists some of the problems encountered with most blood pumps, the source of this problem, and one possible solution to eliminate this problem.

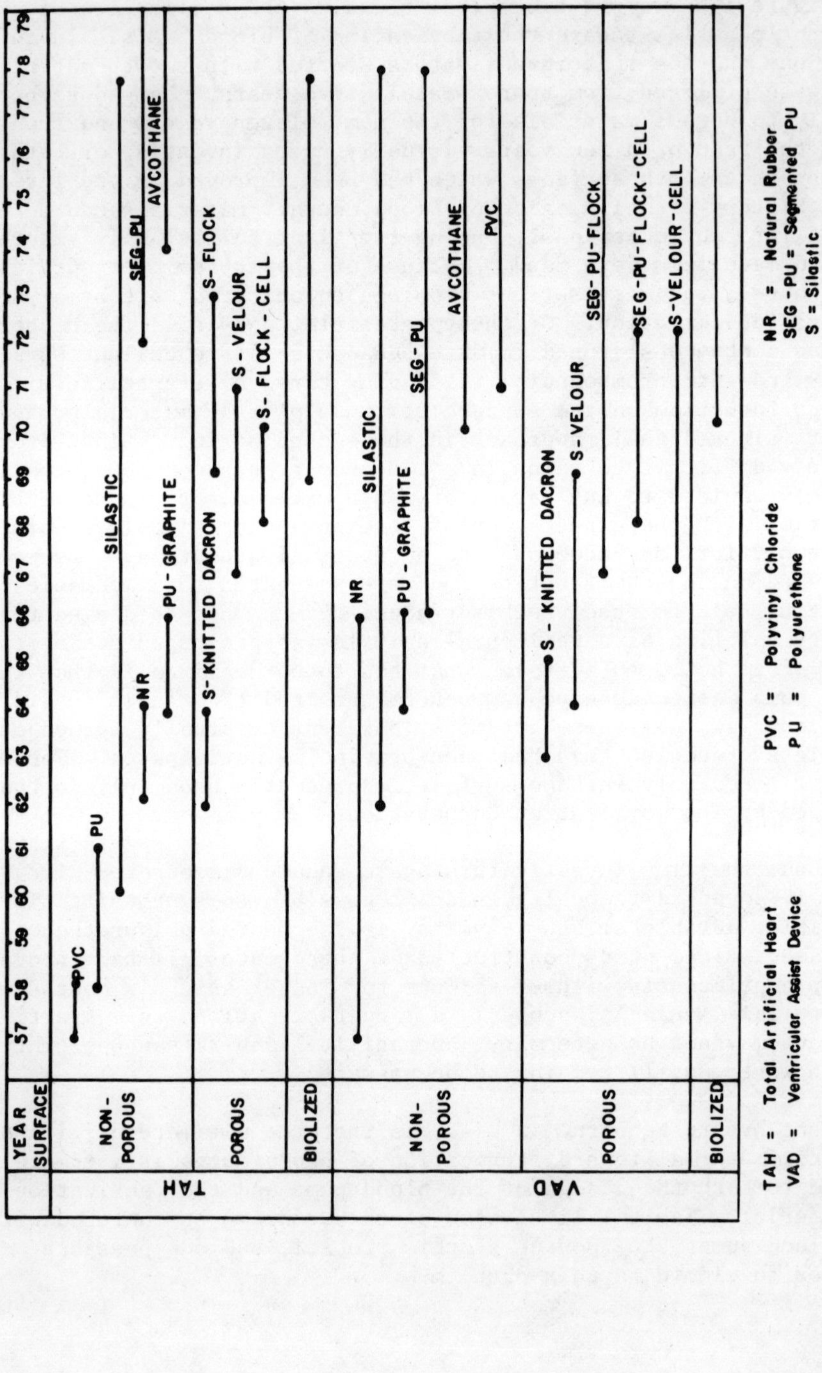
YEAR
SURFACE
57 58 59 60 61 62 63 64 65 66 67 68 69 70 71 72 73 74 75 76 77 78 79
TAH
NON-POROUS
POROUS
BIOLIZED
VAD
NON-POROUS
POROUS
BIOLIZED
PVC
PU
SILASTIC
NR
PU - GRAPHITE
SEG-PU
AVCOTHANE
S- KNITTED DACRON
S - VELOUR
S- FLOCK
S- FLOCK - CELL
NR
SILASTIC
PU - GRAPHITE
SEG-PU
AVCOTHANE
PVC
S - KNITTED DACRON
S - VELOUR
SEG-PU-FLOCK
SEG-PU-FLOCK-CELL
S-VELOUR-CELL
TAH = Total Artificial Heart
VAD = Ventricular Assist Device
PVC = Polyvinyl Chloride
PU = Polyurethone
NR = Natural Rubber
SEG-PU = Segmented PU
S = Silastic

Figure 1. Historical trend of plastic material used for the fabrication of artificial hearts.

Table 1

PROBLEMS ENCOUNTERED WITH ARTIFICIAL HEARTS AND ONE POSSIBLE SOLUTION

PROBLEM	SOURCE	ONE SOLUTION
Seams at Junctions	Material Fabrication	One Piece Construction, Liquid Casts
Transition Tissue to Material	Material	Several Materials
Bladder-Housing Thrombosis	Design	Widen Junction
Pitting of Housing	Fabrication	Control Solvent Evaporation

One of the most perplexing problems to be solved with the early pumps was thrombus at the junctions of different or even similar materials. These seams were virtually impossible to eliminate with materials such as Silastic that require layering and/or compression molding. Figure 2 shows thrombus inside the atrium of a TAH (Silastic) along molding-generated seam. These problems were substantially overcome with the availability of liquid cast polymers. With these easily handled materials, it only depends on a reasonable fabrication regimen to obtain a seamless, continous transition throughout the pump.

The diaphragm-housing (D-H) junction has historically presented two problems: 1) the distance between the flexed diaphragm and the housing and 2) a smooth continous transition between the housing and diaphragm. Initially it was found that because of our pump design, a critical distance was needed to eliminate diaphragm-housing interaction. Even if these (D-H) did not physically contact, the small space between the flexed diaphragm and housing was ideal for thrombus formation (stagnation). In our particular design, a spacing of $\leq$ 2 mm was found to promote thrombus (Figure 3) while an ideal spacing of 3 mm led to a clean junction at autopsy (Figure 4). By this simple design change we were able to significantly reduce thrombus inside the pump. The artificial ventricle shown in Figure 3 suffered from several additional pitfalls. In addition to the diaphragm-housing spacing, rapid solvent evaporation (pitting) and improper prosthetic valve seating were contributing factors to the other thrombus sites.

Smooth surface materials suffer from a distinct disadvantage not encountered with rough surfaces. This is the transitional region between smooth surfaces and natural tissue which presents several problems. First, the junction of these two dissimilar materials will never heal. Second, thrombus usually forms near the junction (Figure 5). One solution to this problem is to interpose a rough surface between the smooth surface and the natural tissue (Figure 6). This strategem is based of course on the success of the rough or porous vascular grafts. Now a problem arises because of the discontinuity

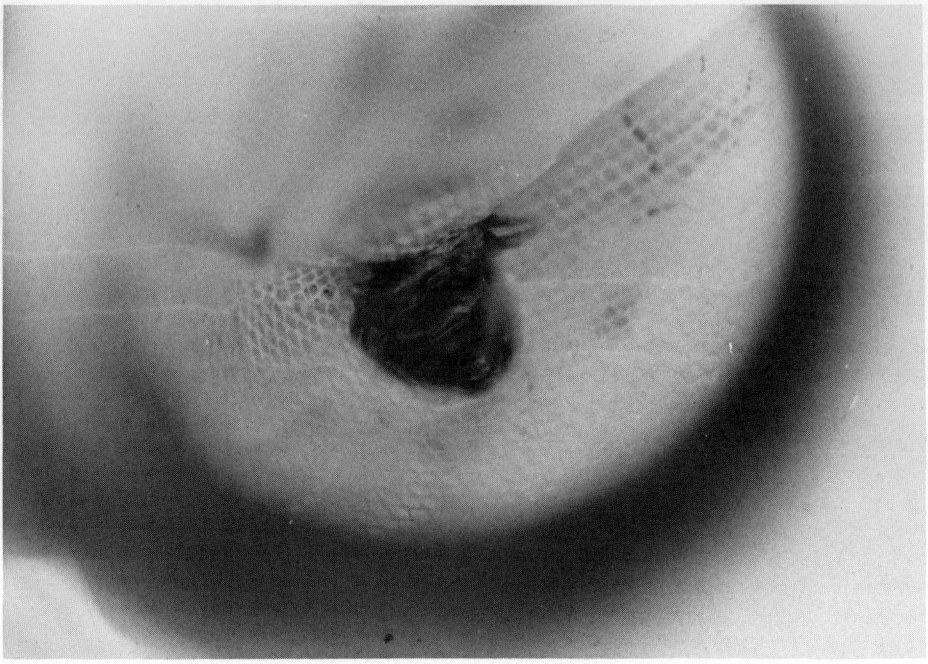

Figure 2. View through the inlet valve region of a Silastic TAH. Thrombus is present at a seam (junction) inside the artificial atrium.

between the rough and smooth surface. To circumvent this, the transitional region can be coated with a solution of Silastic medical adhesive (SMA) plus xylene (Figure 7). This demonstrates how fabrication technique can substantially reduce thrombus formation with the same material and design.

To demonstrate further the effect fabrication technique can have on thrombus formation, Figures 8 and 9 are presented. These LVADs are the same design and material (Silastic). In Figure 8 thrombus is evident inside the pump. These originate from the many tiny surface imperfactions in the Silastic surface generated during the molding process. (It is also possible that either Silastic or mold-borne contamination might have exacerbated the condition). To render the blood contacting side smooth, after the pump is removed from the mold, a solution of SMA and xylene is carefully poured inside. This solution is swirled and carefully removed. The improvement in blood compatibility by this change in fabrication technique is illustrated by Figure 9.

Just as fabrication technique can profoundly effect the experimental results, a design change while retaining similar material and

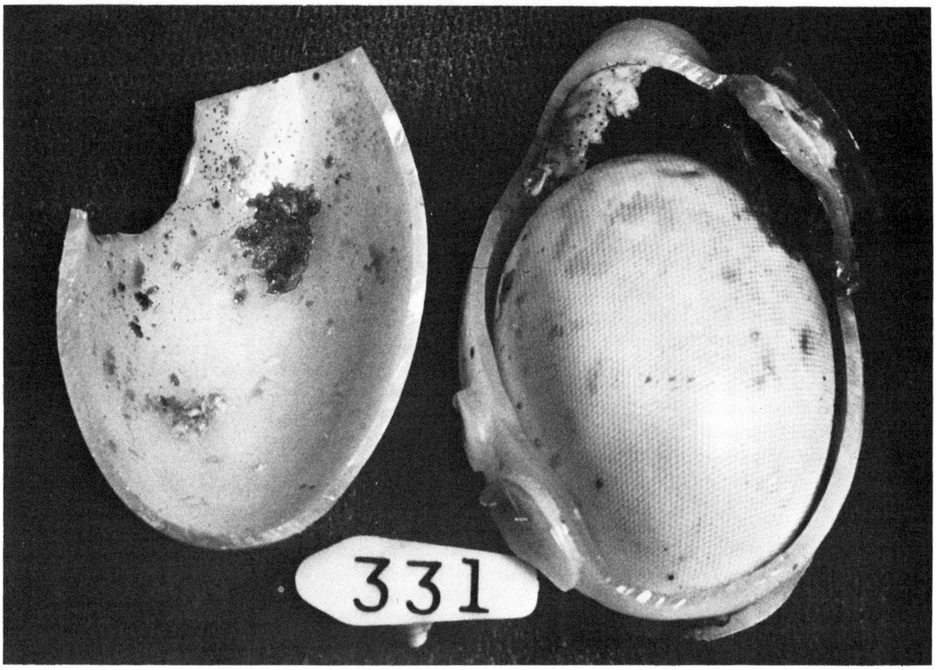
331

Figure 3. TAH, showing thrombus formation in the D-H junction. This is due to the short spacing $\leq$ 2 mm between the diaphragm and the housing. A second thrombus is also visible on the pitted surface of the housing.

fabrication can be just as instrumental. In Figure 10 a series-type LVAD (Dacron mesh reinforced Silastic) is thrombosed. The short inflow length and ovoid shape (transverse section) conspired against the nonthrombogenic Silastic. Again, by retaining the same material and fabrication, a design change (increased inflow length and elliptical shape) improved the results (Figure 11).

With the advent of the new biocompatible liquid cast polymers (Biomer and Avcothane), it is possible to fabricate exceptionally smooth surfaces. Adverse compatibility reactions are minimized because not only are the polymers inherently less thrombogenic but also since they can be liquid case, a continuously smooth, seamless blood pump can be fabricated.

Our laboratory is currently employing Avcothane to fabricate TAH. Table 2 lists the advantages and disadvantages among Avcothane, Biomer, and Silastic. Although Avcothane and Biomer share similar advantages, we feel that Avcothane enjoys a slight advantage because 1) its faster drying time permits the fabrication of more blood pumps and, 2) its enviable track record in intraaortic balloon pumps.

Table 2

ADVANTAGES AND DISADVANTAGES AMONG AVCOTHANE, BIOMER, AND SILASTIC

Material	Advantages	Disadvantages
Avcothane	1. Liquid cast, air cures 2. Fast curing time 3. Thousands hrs use in IABP	1. Photo-oxidation; medical effects unknown 2. Rapid solvent evaporation-bubble formation
Biomer	1. Liquid cast, air cures 2. Slow curing time - less bubble formation 3. Good cement for other polyurethanes	1. Solvent (DMAC) attacks many other polymers 2. Viscosity regulation difficult 3. Unpredictable shrinkage
Silastic	1. Fast fabrication time 2. Compression molding 3. Thickness controlled by sandwiching 4. Curing, mold-independent 5. Medically approved	1. Difficult to form over irregular molds 2. Difficult to eliminate seams at junctions 3. Bonding to certain material is difficult 4. Poor durability

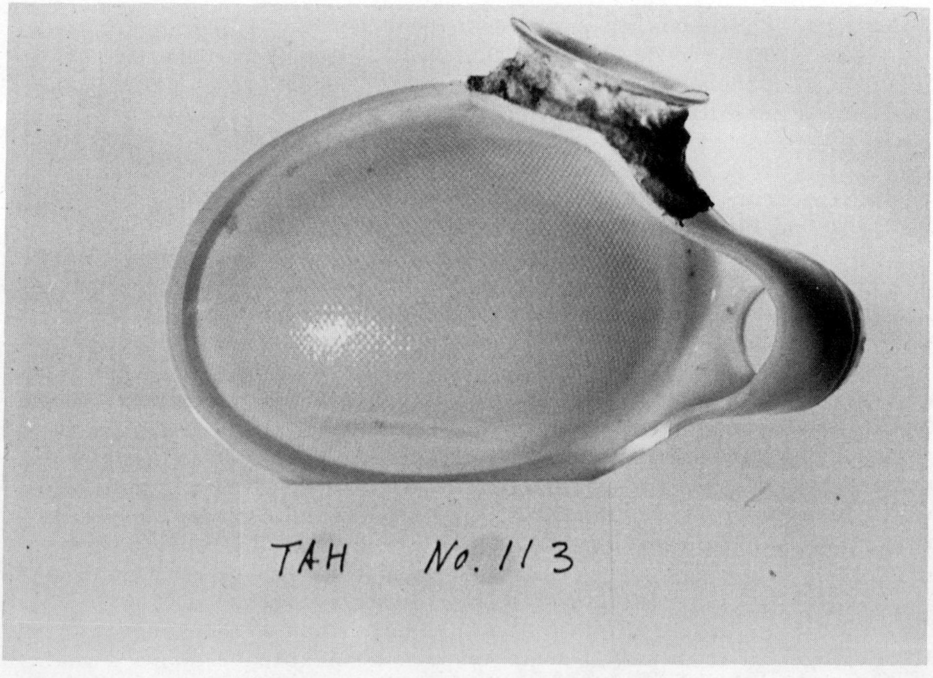
TAH No. 113

Figure 4. A later model of the TAH in figure 3 with a modified D-H junction spacing ($\geq$ 3 mm), showing no thrombus formation.

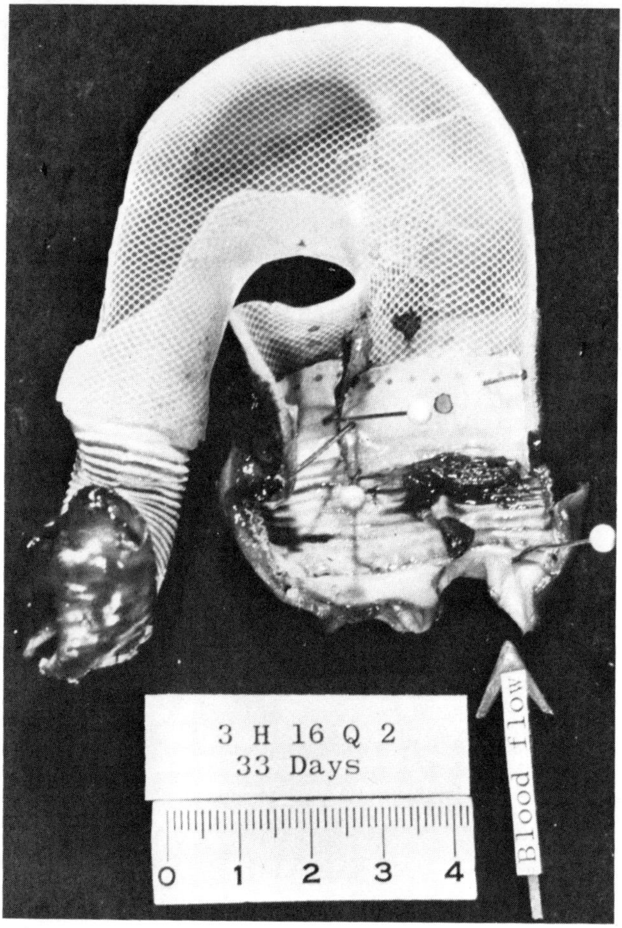
3 H 16 Q 2
33 Days
0 1 2 3 4
Blood flow

Figure 5. A passive (nonpumping) pump bladder retrieved at 33 days. Note the thrombus at the junction of the natural tissue and Silastic.

Our pump housing is first made by repeatedly dipping a dental stone mold into the Avcothane solution. When an adequate thickness is attained a concave mold is inserted into the housing to form the diaphragm (Figure 12). Note the groove around the diaphragm mold which complements the ridge on the inner surface of the housing (Figure 13). This assured reproducible alignment of the diaphragm. Once in place inside the housing, Avcothane is poured through the inflow region, carefully swirled to coat all the blood contacting surfaces and poured out. This technique accomplishes not only the fabrication of a continous, one piece housing and diaphragm, but also the blood contact surface is on the air dried side and not the side in contact with the mold. Figure 14 schematically represents the one piece housing and diaphragm. An assembled pump is shown in

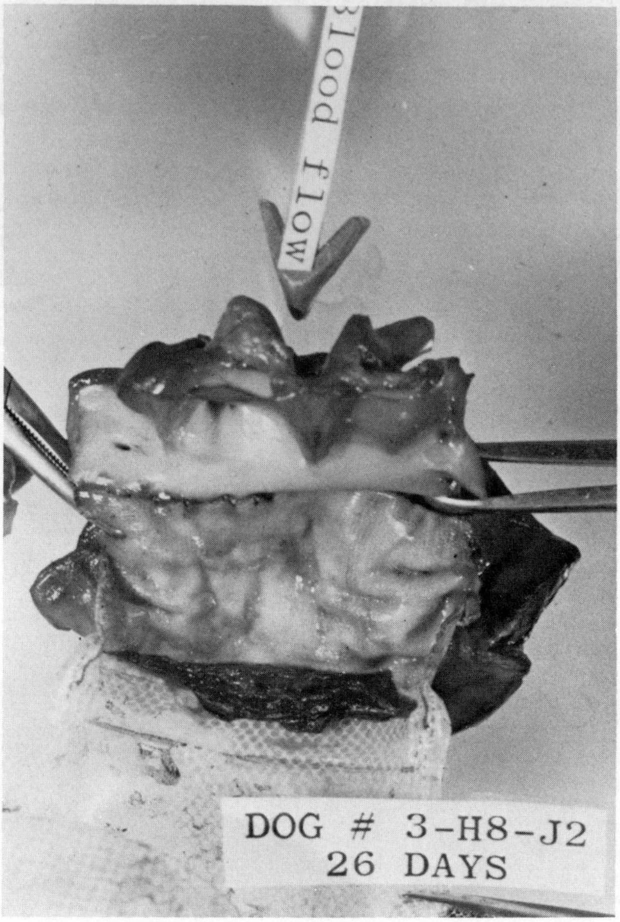
Blood flow
DOG # 3-H8-J2
26 DAYS

Figure 6. Thrombus at the junction of a rough (Dacron graft) and smooth (Silastic) surface. Although a smooth neointima is formed on the graft (below the suture line), thrombus is evident at the graft-Silastic junction.

Figure 15.

Even with liquid cast polymers though, the surface smoothness is a function of the mold surface. A surface imperfaction on the mold can extend its influence through several layers of the dipped polymer. To circumvent this problem our molds are dipped several times in a suitable polymeric releasing agent (Avcomat 40) to mask any survace imperfactions. So when the polymer first contacts the mold surface, it "sees" a smooth regular surface to which it easily conforms.

Surface smoothness is also related to the solvent evaporation

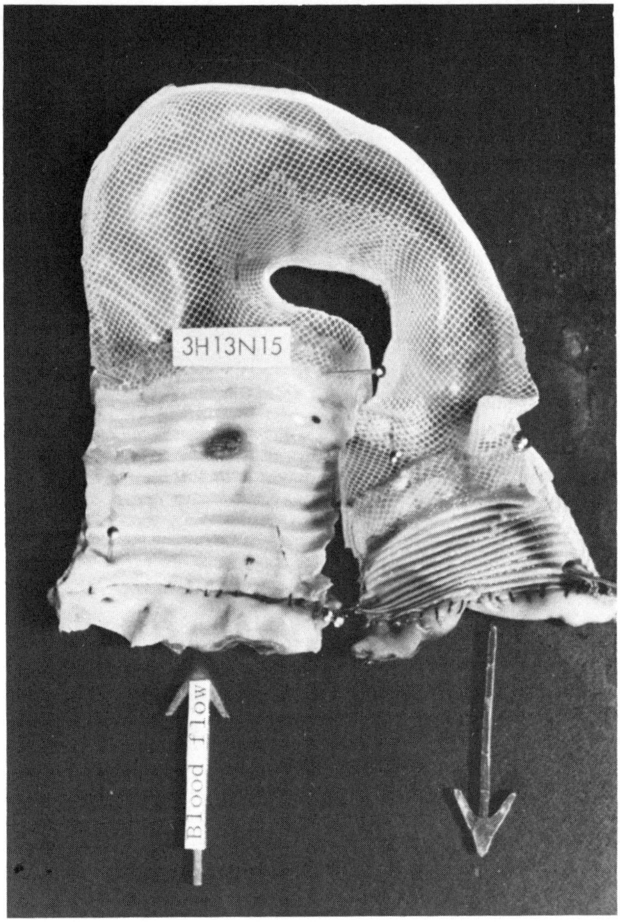
3H13N15
Blood flow

Figure 7. A passive pump bladder recovered virtually free of thrombus. A mistake while coating the graft-Silastic transition region resulted in the small thrombus seen on the graft.

rate in solvent cast polymers. Rapid solvent evaporation from Avcothane produces tiny bubbles in the uncured polymer film. If these bubbles then rise to the surface, micro pits are produced. These present ideal sites for promoting platelet aggregation, hemolysis, and emboli. This pitting of the surface is the reason for the thrombus seen on the diaphragm and housing in Figure 3. Furthermore, these pitted areas also decrease the mechanical strength (durability) of the polymer particularly when they are in the high flex (stress) region of the D-H junctions. There are various ways to regulate the solvent evaporation rate but these all reduce to the following: 1) controlling the temperature in the polymer film and adjoining gas phase and, 2) controlling the mass flow in the

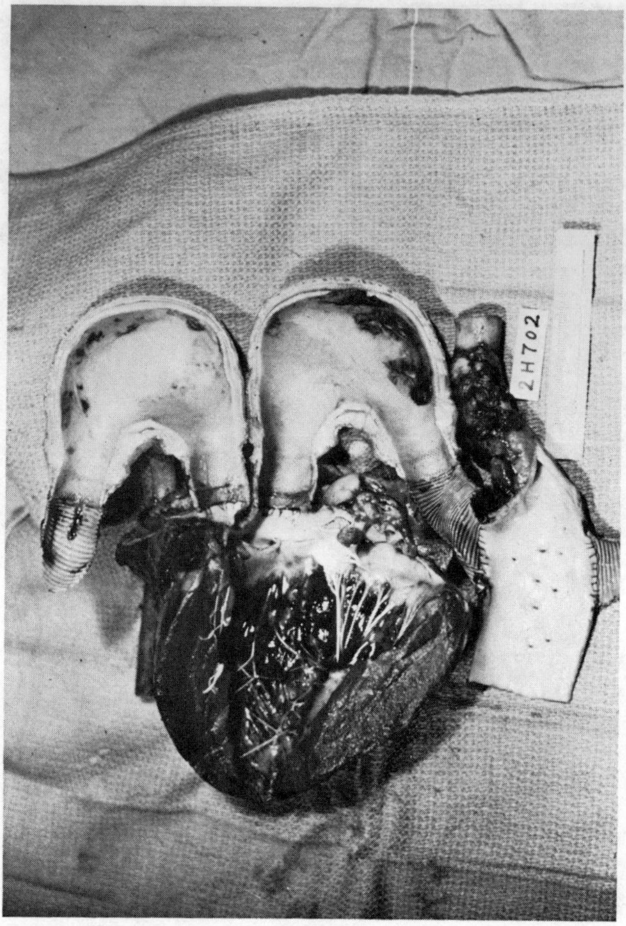
2H702

Figure 8. Recovered Silastic LVAD attached to the natural ventricle.

gas phase. We economically achieve this by maintaining our fabrication room (hence molds and polymer solutions) at 25°C and by covering the molds with a beaker after each dip to establish a small concentration gradient between the gas and liquid phases.

The ultimate physico-chemical characteristics of a prospective biocompatible polymer is intimately related to the surface chemical composition of the casting material. Several recent studies have shown this to be particularly true of segmented polyurethanes, ie. Biomer and Avcothane.

Sung and Hu have demonstrated by a variety of techniques, ESCA (48), Auger electron spectroscopy (49), and (50) Fourier transform IR differences in the surface chemical composition depending on the material at the solvent cast polymer interface. They concluded that

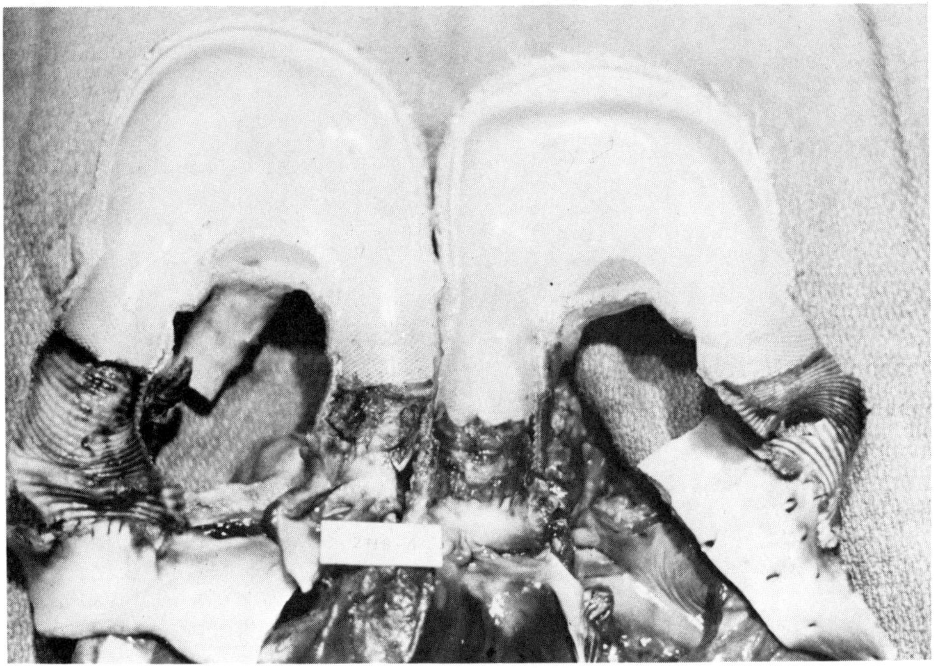

Figure 9. Recovered Silastic LVAD of identical design as that shown in Figure 8. Thrombus-free bladder is the result of a fabrication change.

the air facing surface of Avcothane is covered predominantly with polydimethylsiloxane and very little of the polymer hard segment, whereas the substrate side (cast) contained hard segment, polyether, and polydimethylsiloxane. For Biomer films, the air facing surface contained a greater concentration of polyether soft segments than the substrate surface.

Nyilas (51) has similarly studied these differences in surface chemical composition. More importantly, he has shown significant differences in the Lee-White clotting time and prothrombin time for the air side versus the substrate side. The air-facing side demonstrated superior blood compatibility compared to the substrate side. For this reason fabrication regimens are usually designed so that blood contacting surfaces are from the air side. This is easily accomplished by at least one additional coating of the blood contact surfaces after they are removed from the mold.

Only a couple of different substrate materials (glass and PET) have thus far been studied, but it is possible that some substrates may be found that render the surface chemical composition (blood

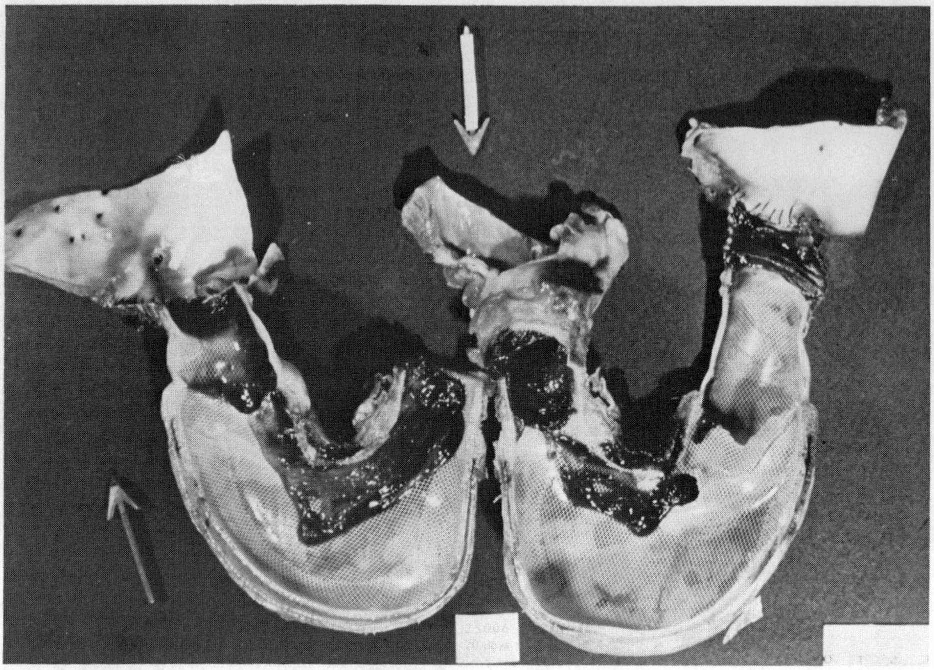

Figure 10. Series-type LVAD (Dacron mesh reinforced Silastic) thrombosed because of poor design.

compatible) superior to that of the air side. The point to be made here is that in preparing a blood contact surface, special attention should be given not only to the material used, but also to the cast material and which surface will be exposed to the blood (air side or cast side).

Contamination of the blood contact surface is a serious problem which can cause thrombus, emboli, and/or hemolysis. Contamination can originate from two primary sources: the solvent cast polymer or its mold and/or coating (releasing agent). These manifest themselves as leachables, solvents, and air-borne dust. The solvents used for Avcothane (THF and dioxane) and Biomer (DMAC) are highly toxic. Their use necessitates a protracted drying time to remove as much of the residual solvent as possible. Dust can convert an otherwise smooth surface to an irregular rough surface with many foci (dust particles) for cellular and protein adhesion. Dust contamination is reduced by performing the various fabrication steps in a clean air room with a special air filtering hood.

Another source of potential contamination is via photo-oxidation of the polymer. This is especially true for most of the segmented

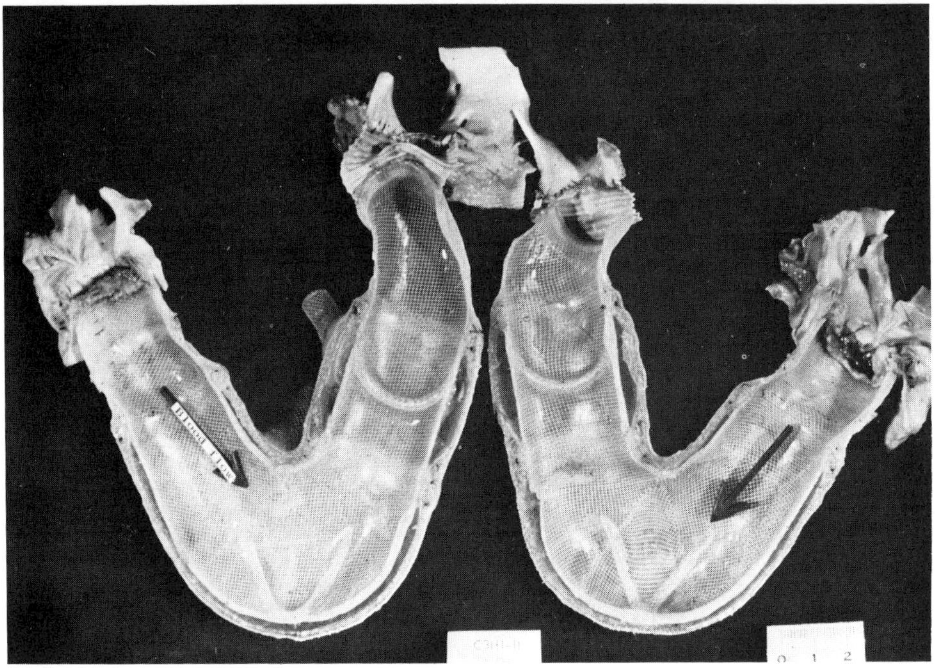

Figure 11. Series-type LVAD of similar material and fabrication as that in Figure 10, but a different design.

poly-urethanes. Avcothane has been no exception. Both the liquid polymer solution and the dried polymer are seen to yellow upon exposure to light. To minimize this effect polymer solutions are kept in tinted bottles covered with aluminum foil. Whenever possible, a nitrogen flush is used to reduce the oxygen concentration in the gas phase above the polymer. Oxidation of the polymer lends to a decrease in flex life and quite possible to a detrimental change in its biocompatibility, although the latter has not been documented.

It should be apparent that even with these new generation biocompatible materials, care must still be exercised in the fabrication and design of artificial blood pumps. Even though survival times have exceeded 10 months we may have reached the mechanical limit of the polymers now being employed. Bucherl (52) recently reported the cause of termination of 10 consecutive TAHs with survival over two months. Seventy (70%) percent of these were related to pump (material) failure. It might be possible to extend the service life of blood pumps somewhat, by design improvements, but long term (greater than 2 years) reliability will probably have to wait for still another generation of biocompatible polymers.

Figure 12. Dental stone coated with releasing agent (left) and a stainless steel (right) mold for fabrication of diaphragms. Note the recessed groove just below the tapered lips of the molds.

ABSTRACT

When artificial heart (AH) research began two decades ago, none of the plastic materials available were intended for medical use. Investigators simply incorporated them without any firm conviction that they might accommodate needs. The history of plastic materials used in AH research can be divided into three periods: 1) use of smooth materials made for industrial purpose, 2) use of porous materials lined on impermeable materials, and 3) use of smooth materials purposely developed for medical implantation.

When evaluating surface properties of any particular plastic material in terms of antithrombogenicity, design and fabrication techniques have to be considered together with the material itself. Ideally the AH including the valves should be fabricated from a single material in one piece. Sometimes however, it may be impossible or at least extremely difficult to develop such a fabrication technique for a particular material which has excellent mechanical durability and nonthrombogenicity.

It has been shown that blood pumps made from a few of the presently available materials can keep the animal alive for seven to eight months with proper use of anticoagulants. Durability longer than one or two years, however, still remains to be answered.

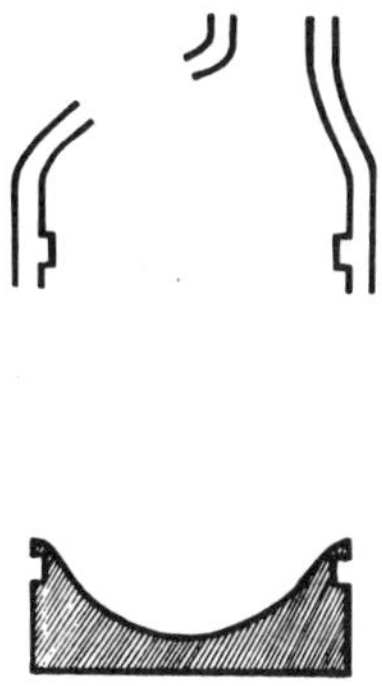

Figure 13. Schematic representation of the pump housing and diaphragm mold.

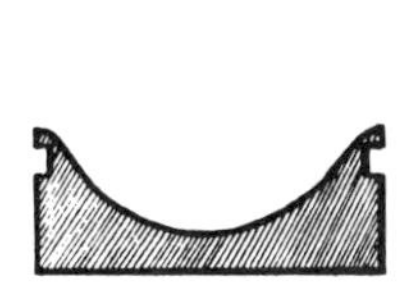

Figure 14. Representation of one piece housing and diaphragm after mold removal.

Figure 15. Assembled artificial artificial ventricles with prosthetic valves.

REFERENCES

1. B. K. Kusserow, Trans. ASAIO, 4, 227 (1958)
2. B. K. Kusserow, Trans. ASAIO, 5, 293 (1959)
3. K. Atsumi, M. Hori, S. Ikeda, Y. Sakurai, Y. Fujimori, and S. Kimoto, Trans. ASAIO, 9, 292 (1963)
4. T. Akutsu, and W. Kolff, Trans ASAIO, 4, 230 (1958)
5. T. Barila, D. Nunn, and K. Woodward, Trans. ASAIO, 8, 30 (1962)
6. Schollenberger, Rubber World (Jan. 1958)
7. W. Kolff, T. Akutsu, B. Dreyer, and H. Norton, Trans. ASAIO, 5, 298 (1959)
8. V. Mirkovitch, T. Akutsu, and W. Kolff, Trans. ASAIO, 8, 79 (1972)
9. W. Seidel, T. Akutsu, V. Mirkovitch, F. Brown, W. Kolff, Trans. ASAIO, 7, 378 (1961)

10. W. Pierce, R. Burney, M. Boyer, R. Driscoll, and C. Kirby, Trans. ASAIO, 8, 118 (1962)
11. T. Akutsu, V. Mirkovitch, S. Topaz, and W. Kolff, J. Thorac. Cardiovasc. Surg., 47, 512 (1964)
12. D. Liotta, T. Taliani, A. Giffoniello, F. Deheza, S. Liotta, R. Lizarraga, L. Tolocka, J. Panano, and E. Bianciotta, Trans. ASAIO, 7, 318 (1961)
13. D. Liotta, E. Crawford, D. Cooley, M. DeBakey, M. deUrquia, and L. Feldman, Trans. ASAIO, 8, 90 (1962)
14. D. Liotta, C. Hall, W. Henly, A. Beall, Jr., D. Cooley, and M. DeBakey, Trans. ASAIO, 9, 182 (1963)
15. C. Hall, D. Liotta, W. Henly, E. Crawford, and M. DeBakey, Amer. J. Surg., 108, 685 (1964)
16. D. Liotta, C. Hall, A. Villanueva, R. O'Neal, and M. DeBakey, Cardiov. Res. Center, Bull., 4, 69 (1966)
17. D. Liotta, C. Hall, W. Akers, A. Villanueva, R. O'Neal, and M. DeBakey, Trans. ASAIO, 12, 129 (1966)
18. C. Hall, D. Liotta, J. Ghidoni, M. DeBakey, and D. Dressler, J. Biomed. Mater. Res., 1, 179 (1967)
19. W. Sharp, A. Finelli, W. Falor, and J. Ferraro, Circulation, 29, 165 (1964)
20. W. Bernhard, C. LaFarge, T. Robinson, J. Yun, K. Shirahige, and S. Kitrilahis, Ann. Surg., 168, 750 (1968)
21. J. Ghidoni, D. Liotta, C. Hall, J. Adams, A. Lechter, M. Barrionueva, R. O'Neal, and M. DeBakey, Amer. J. Path., 53, 375 (1968)
22. J. Ghidoni, D. Liotta, J. Adams, R. O'Neal, and C. Hall, J. Biomed. Mater. Res., 2, 201 (1968)
23. J. Ghidoni, D. Liotta, C. Hall, R. O'Neal, and M. DeBakey, Trans. ASAIO, 14, 69 (1968)
24. M. Adachi, M. Suzuki, and J. Kennedy, J. Surg. Res., 11, 483 (1971)
25. M. Adachi, M. Suzuki, J. Ross, D. Wieting, J. Kennedy, and M. DeBakey, J. Thorac. Cardiovasc. Surg., 65, 778 (1973)
26. W. Bernhard, M. Hussain, and J. Curtis, Surgery, 66, 284 (1969)
27. C. Kwan-Gett, K. Van Kampen, J. Kawai, N. Eastwood, and W. Kolff, J. Thorac. Cardiovasc. Surg., 62, 880 (1971)
28. C. Kwan-Gett, D. Backman, F. Donovan, Jr., N. Eastwood, J. Foote, J. Kawai, T. Kesseler, A. Kralios, J. Peters, K. Van Kampen, H. Wong, H. Zwart, and W. Kolff, Trans. ASAIO, 17, 474 (1971)
29. W. Kolff, Transplantation Proc., 3, 1449 (1971)
30. T. Stanley and W. Kolff, Surg. Forum, 24, 171 (1973)
31. J. Peters, F. Donovan, and J. Kawai, Chest, 63, 589 (1973)
32. D. Olsen, F. Unger, H. Oster, J. Lawson, T. Kessler, J. Kolff, and W. Kolff, J. Thorac. Cardiovasc. Surg., 70, 248 (1975)
33. W. Kolff, and J. Lawson, Trans. ASAIO, 21, 620 (1975)
34. H. Oster, D. Olsen, R. Jarvik, T. Stanley, J. Volder, and W. Kolff, Surgery, 77, 113 (1975)
35. J. Boretos and W. Pierce, Science, 158, 1481 (1967)
36. W. Pierce, M. Turner, Jr., J. Boretos, H. Metz, S. Nolan, and A. Morrow, Trans. ASAIO, 13, 299 (1967)

37. D. Lyman, C. Kwan-Gett, H. Zwart, A. Bland, N. Eastwood, J. Kawai, and W. Kolff, Trans. ASAIO, 13, 299 (1967)
38. T. Kolff, G. Burkett, and J. Feyen, Biomat. Med. Dev. Art. Org., 1, 669 (1973)
39. J. Lawson, D. Olsen, E. Hershgold, J. Kolff, K. Hadfield, and W. Kolff, Trans. ASAIO, 21, 368 (1975)
40. S. Moulopoulos, R. Jarvik, and W. Kolff, J. Thorac. Cardiovasc. Surg., 66, 662 (1973)
41. E. Nyilas, Avco-Everett Res. Lab. No. 327 (1970)
42. R. Engelman, E. Nyilas, H. Lachner, S. Godwin, and F. Sperser, J. Thorac. Cardiovasc. Surg., 62, 851 (1971)
43. J. Kakurai, Artificial Organs, Suppl., 2, 114 (1973)
44. K. Imachi, I. Fujimasa, N. Sato, N. Iwai, H. Miyake, N. Takido, M. Nakajima, A. Kuono, T. Ono, and K. Atsumi, Proc. ESAO, 6, 193 (1979)
45. Y. Imai, K. von Bally, and Y. Nose, Trans. ASAIO, 16, 17 (1970)
46. Y. Nose, Y. Imai, K. Tajima, H. Ogawa, M. Klair, K. von Bally, and D. Effier, J. Thorac, Cardiovasc. Surg., 62, 714 (1971)
47. S. Kasai, I. Koshino, T. Washizu, G. Jacobs, N. Morinaga, R. Kiraly, and Nose, J. Thorac. Cardiovasc. Surg., 73, 637 (1977)
48. C. S. Sung, and C. B. Hu, J. Biomed. Mater. Res., 13, 161 (1979)
49. C. S. Sung, and C. B. Hu, J. Biomed. Mater. Res., 13, 45 (1979)
50. C. S. Sung, and C. B. Hu, J. Biomed. Mater. Res., 12, 791 (1978)
51. E. Nyilas, and R. S. Ward, Jr., J. Biomed. Mater. Res., 11, 69 (1977)
52. W. Krautzberger, M. Behrens, C. Clevert, H. Keilbach, H. Weidemann, C. Grobe-Siestrup, K. Gerlach, K. Affeld, J. Frank, E. Hennig, W. Lemm, A. Mohnhaupt, V. Unger, F. Zartnack, F. Rennekamp, and E. S. Bucherl., ESAO, 6, 19 (1979)

TISSUE CULTURED CELLS: POTENTIAL BLOOD COMPATIBLE LININGS FOR CARDIOVASCULAR PROSTHESES

S. G. Eskin, L. T. Navarro, H. D. Sybers, W. O'Bannon and M. E. DeBakey

Departments of Surgery and Pathology
Baylor College of Medicine, Houston, Texas

INTRODUCTION

Polymers currently used for cardiovascular replacement, both clinically and experimentally, fall into two categories; rough surfaced and smooth surfaced materials. The reaction of the blood upon contact with each type material is different. The rough surfaced materials, e.g., Dacron polyester (polyethylene terephthalate), are porous, and upon blood contact encourage the formation of adherent thrombus on their surfaces. The smooth surfaced materials, e.g., polyether urethanes, are generally impervious to blood and resist the formation of thrombus; although none of the smooth surfaced materials approach the thromboresistance of the natural endothelial lining of the vessel wall (1,2).

In Figure 1, potential problems with blood interactions to these two types of materials are compared to the nonreactive vessel wall. The thrombus that is initially trapped on rough surfaced materials may continue to form, building up into the lumen and possibly occluding it. On smooth surfaced materials, platelets may adhere and aggregate, and then be released as emboli into the blood.

The thrombogenic rough surfaced materials have proven to be successful replacements for large occluded arteries since DeBakey began such replacements in 1954 (3). Clinically, these materials are preclotted with the patient's blood prior to implantation in order to close the pores in the material, otherwise blood loss would be excessive. After implantation, healing of the graft occurs as cells migrate through the thrombus within the pores of the fabric and organize the thrombus into connective tissue (4). This results

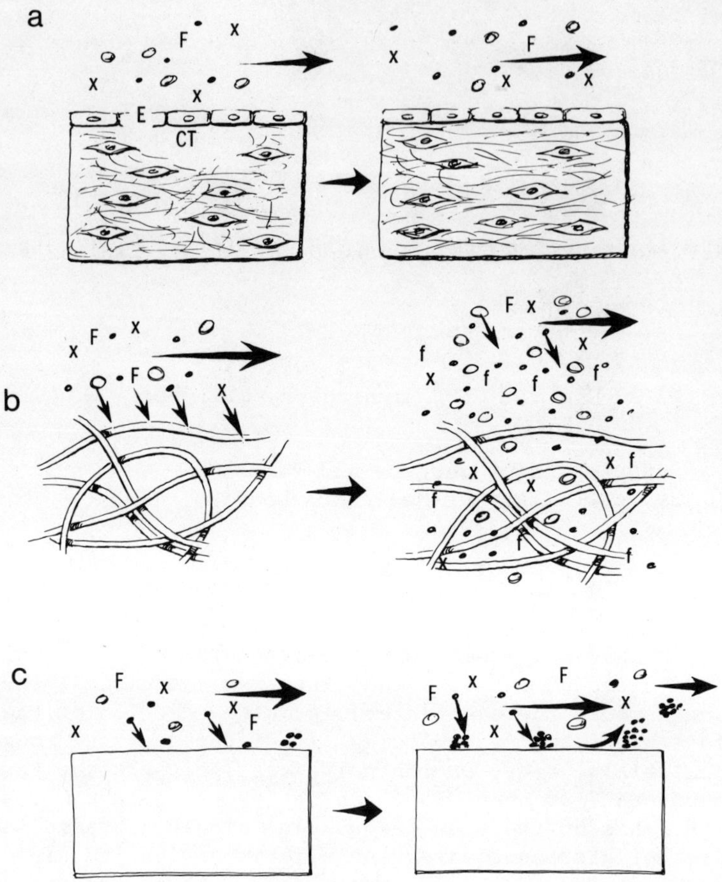
a
F
X
E
CT
b
c

Figure 1. Reaction of blood to different surfaces; a) natural blood vessel wall consisting of endothelium (E) and subendothelial connective tissue (CT) shows no thrombogenic reaction to flowing blood, b) rough surfaced porous polymers entrap and clot blood as it flows past. Fibrinogen in blood precipitates as fibrin entrapping the blood cells. c) smooth surfaced polymers form small platelet aggregates at the interface with the blood, which embolize into the blood stream.
X = leukocytes, o = red blood cells, . platelets,
F = fibrinogen, f = fibrin.

in a vessel wall composed of closely interwoven connective tissue and prosthetic fibers. The success of porous, rough surfaced materials is due to the high blood velocity and relatively simple, steady blood flow patterns found in large arteries, such as the aorta and iliac arteries. In smaller arteries however, for example the coronary arteries, thrombus buildup on rough surfaced materials becomes occlusive.

The patterns of thrombus formation in an artificial heart, or a heart assist pump, in which smooth surfaced as well as rough surfaced materials find application, are considerably more complicated than in a graft. Thrombus formation which is desirable in a graft where it is healed, becomes unacceptable in a pump, where the impermeable surface prevents the ingrowth of cells, and thus prevents healing. In a pump, unhealed thrombus may propagate to a significant extent because 1) the foreign surface area is greater, resulting in more extensive blood-polymer contact and greater potential thrombus formation; 2) the complicated geometry of the device produces complicated patterns of blood circulation which favor the propagation of thrombi; and 3) the blood contact surface must be impervious because pumping the blood requires a barrier between the blood and the pneumatic power source.

The early Baylor heart assist (left ventricular bypass) pumps used Dacron velour as the blood contact material backed with Silastic (Figure 2). These pumps were employed therapeutically in several patients (5), however, transformation of the thrombus into connective tissue on the impervious surface did not occur. In longer term experiments with calves, propagation of thrombus on the surfaces continued, eventually compromising flexion of the diaphragm, and therefore, of pump function.

Smooth surfaced materials may provide sufficiently thromboresistant surfaces for clinical use in blood pumps. However, present day fabrication techniques rarely achieve surfaces which do not form loci of platelet thrombi (Fig. 1c), which embolize and are filtered out by critical organs, causing tissue death and eventually organ failure. A recent vintage Baylor left ventricular bypass pump (Figure 3) utilizes polyether urethane (Biomer) as the blood contact surface (6). Experimental implants on calves showed grossly thrombus free surfaces after 4 weeks (Figure 4). However, examination of these surfaces with scanning electron microscopy revealed thrombus formation (Figure 5).

It was the inability to overcome this problem of thrombogenicity of materials which led to the concept of tissue cultured cells as linings for cardiovascular prostheses. The rationale was that a biological surface might not be recognized as foreign and thus be thromboresistant. Early work showed that the cells must be 1) autologous, to avoid immune rejection (7), 2) probably endothelial, as fibroblasts proved to be thrombogenic (8), and 3) cultured on a substrate which they covered completely and to which they remained firmly attached. The first 2 requirements have proven technically feasible to fulfill in experimental animals. The third problem, that

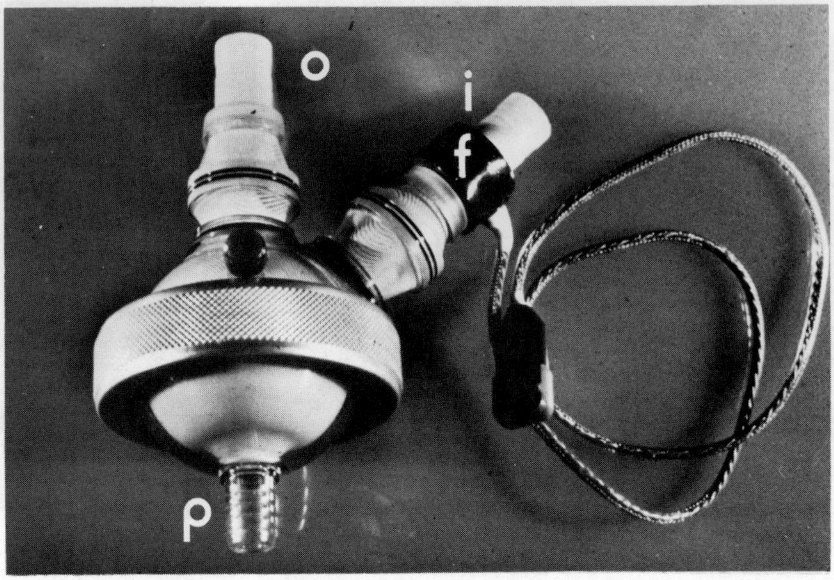
o
i
f
p

Figure 2. Baylor left ventricular bypass pump in which blood contact material is Dacron velour backed with Silastic. The valves are built-in Silastic ball valves. The pump fills with blood from a cannula connecting the right atrium to the pump (i). Blood is ejected from the pump out the top port (o) into a cannula leading to the aorta or other large systemic artery. f = electromagnetic flowmeter, p = pneumatic power supply line connection.

of determining the optimal substrate, is still in question. The substrate must support cell attachment and proliferation and its relationship with the cells must be such that when the cells come into contact with the flowing blood they remain attached. When these requirements are met, cultured cell linings provide a non-thrombogenic surface when implanted (9,10).

EXPERIMENTAL

Materials and Methods

The general procedures for obtaining, culturing and assessing the blood compatibility of the cells are outlined in Figure 6. Autologous endothelial cells were harvested from the thoracic aortae of 100 kg. calves using a modification of an enzymatic method developed for excised aortae (11). A 10 cm. length of thoracic aorta was bypassed with a Dacron graft to provide circulation to the rear of the calf. The blood was withdrawn from the isolated segment, the segment was rinsed with phosphate buffered saline, pH 7.4, (PBS), and then refilled with 0.1% collagenase (Worthington, type CLS) in

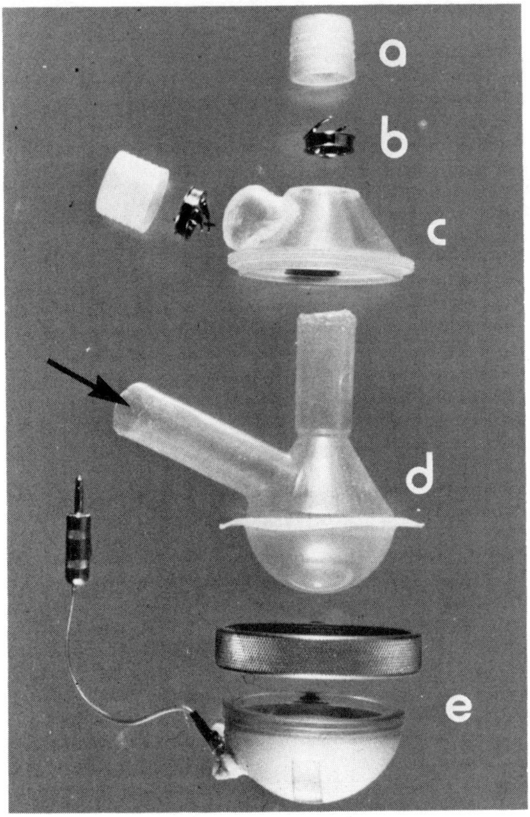
a
b
c
d
e

Figure 3. Exploded view of the components of Baylor left ventricular bypass pump a) polycarbonate connectors, b) Lillehei-Kaster tilting disc valves, c) top of polycarbonate pump housing, d) polyether urethane (Biomer) bladder, e) bottom of polycarbonate pump housing ("dome") with sensor electrode attached which monitors pump filling.

PBS. After 10 min. the enzyme was rinsed out with McCoy's 5 a medium (GIBCO) and the segment was filled with complete medium (McCoy's 5a medium, supplemented with 10% fetal calf serum, 15 mM Hepes buffer, and 0.10mg/ml penicillin, 0.10mg/ml streptomycin, 0.20mg/ml neomycin, and 5.0 mcg/ml Fungizone). The calves were allowed to recover after the cells were removed.

The medium was replenished twice weekly and the cells were subdivided when confluent (Figure 7) with 0.25% trypsin (GIBCO) at a split ratio of 1:2. The cultures were observed daily by phase contrast microscopy to eliminate any contaminating smooth muscle cells.

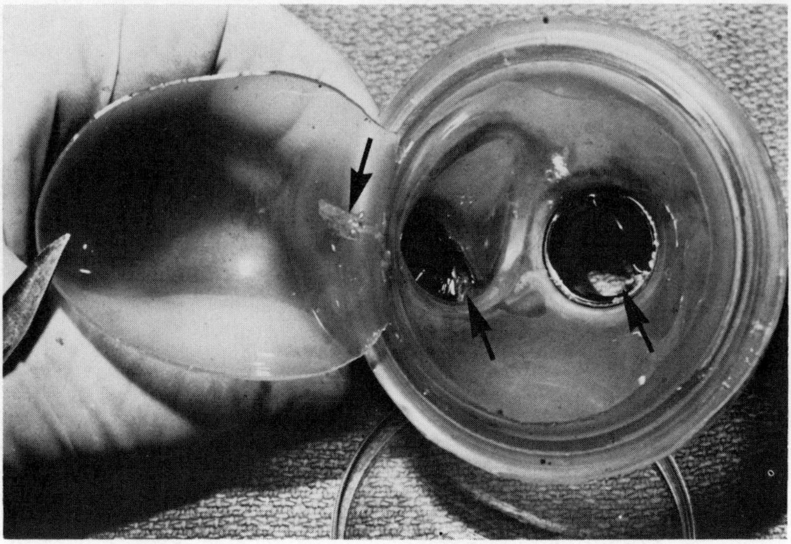

Figure 4. Blood contact surface of left ventribular bypass pump after a 27 day calf implant. Arrows show adherent thrombus.

The substrates considered for use in the in vivo experiments were Dacron knit (U.S. Catheter and Instrument Co., 1624) and polypropylene microfabric (Union Carbide), as these materials proved superior in supporting endothelial cell growth (12). The microfabric consisted of polypropylene fibers extruded onto a 10-12 mil thick backing of polyetherurethane (Estane), then coated with Parylene C, vertically drafted, and microwave descharge treated (Figure 8a). The fibers of the microfabric were 2 um in diameter, the thickness of the material was about 25 um and the spaces between the fibers were 10-20 um. The microfabric was used in the in vivo experiments reported herein because the small fiber size supported a two dimensional (monolayer) growth pattern (Figure 8b), in contrast to the growth pattern on Dacron knit. This provided a better simulation of the morphology of the natural vessel wall and required relatively few cells for complete coverage. The Dacron knit required greater numbers of cells and time for coverage because the large fiber size provided greater surface area for cell growth (Figure 9).

After 5-9 subcultures in polystyrene flasks, the cells were seeded by centrifugation on a 7.6 cm^2 patch of microfabric. The seeded patches were put into 100 mm petri dishes in 10 ml. complete medium and incubated at 37°C. in a humidified atmosphere of 3% CO_2, 97% air. The patch was judged ready for implantation when endothelia coverage was complete using May-Greenwald-Giemsa staining of a 1 cm^2 piece. Thirty minutes before implantation the patch was trimmed to a 2 x 5 cm. oval and rinsed 3 times in PBS. The right

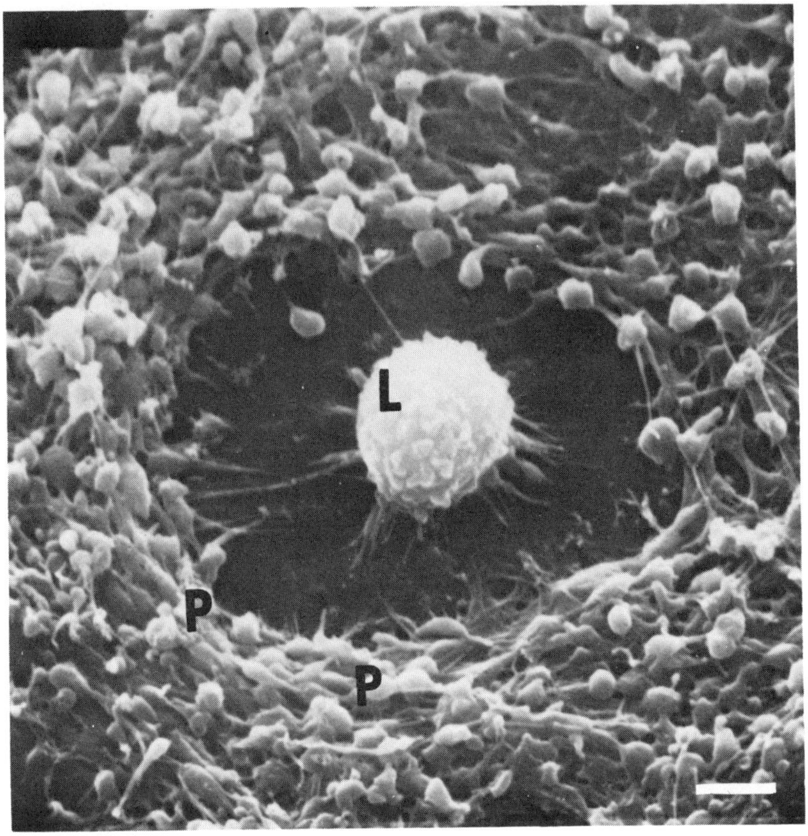
L
P
P

Figure 5. Scanning electron micrograph (SEM) of surface of portions of bladder in Figure 4. L = leukocyte, P = platelets, Bar = 10 µm.

atrium of the autologous calf was clamped in a partial occluding clamp and slit. The patch was then sutured to the cut edges of the atrium so that when the clamp was released the patch unfolded as part of the wall of the atrium. Bare microfabric patches were implanted in calves as controls.

One week after implantation, the calves were heparinized and sacrificed. The patches with the surrounding atria were rapidly removed, and rinsed thoroughly in saline at 37° C. Gross photographs were taken with the patches submerged in saline. Half of each patch was fixed in 2% glutaraldehyde in saline, the other half was fixed in 10% formalin. Samples fixed in glutaraldehyde were processed for scanning electron microscopy (SEM) and transmission electron microscopy (TEM), while those fixed in formalin were processed for light microscopy. SEM observations were made on an ETEC Autoscan. TEM was carried out on a Jeol 100 C. After fixation during trimming of

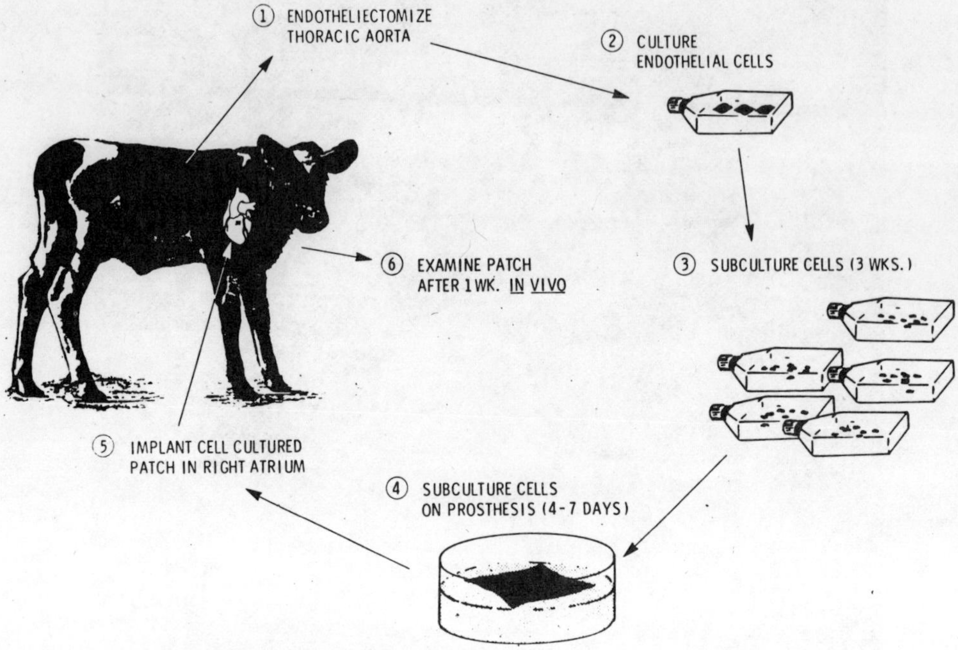
① ENDOTHELIECTOMIZE
THORACIC AORTA
② CULTURE
ENDOTHELIAL CELLS
③ SUBCULTURE CELLS (3 WKS.)
④ SUBCULTURE CELLS
ON PROSTHESIS (4-7 DAYS)
⑤ IMPLANT CELL CULTURED
PATCH IN RIGHT ATRIUM
⑥ EXAMINE PATCH
AFTER 1 WK. IN VIVO

Figure 6. Overall procedure for obtaining the cells and assessing their thromboresistance; 1) endothelial cells are harvested from the aorta 2) cells are cultured, and 3) subcultured to obtain large numbers, 4) cells are subcultured on microfabric surface to be implanted, 5) patch is implanted, 6) calf is sacrificed 1 week later and patch is examined. (Reprinted by permission of Scanning Electron Microscopy, Inc., from Eskin et al. (12).

the samples for processing, the depth of thrombus buildup on the patch surface was measured with calipers using a dissecting microscope. Measurements of depth were made on cross sections which included both the minimum and maximum amounts of thrombus.

Results

Seven implants of autologous cell lined patches and four control implants of bare microfabric were carried out. Table 1 summarizes the range of depth of thrombus on the patches.

Table 1

RANGE OF THROMBUS DEPTH ON PATCHES

Cell Lined Calf No.	Thrombus Depth (mm) Min. Max.	Control Calf No.	Thrombus Depth (mm) Min. Max.
10468	0	10743	0.25 - 0.64
9522	0 - 0.25	11072	1.02 - 1.78
9516	0 - 3.81	9824	2.29 - 3.05
10740	0 - 0.51	10741	7.37 -14.22
10742	0 - 1.02		
10467	2.29- 6.35		
9217	1.27- 4.32		

Five of the 7 cell lined patches had areas on which no thrombus could be measured (Figure 10a); the other 2 were completely covered with thrombus which varied in depth. All the control implants were thrombus-covered, although the thrombus depth varied from 0.25 mm to more than 10 mm (Figure 10b).

Light microscopy (Figure 11) clearly showed the differences between the cell lined and control implants. The controls showed the unorganized multilayered blood cells and fibrin characteristic of thrombus. In contrast, a continuous monolayer of cells on the

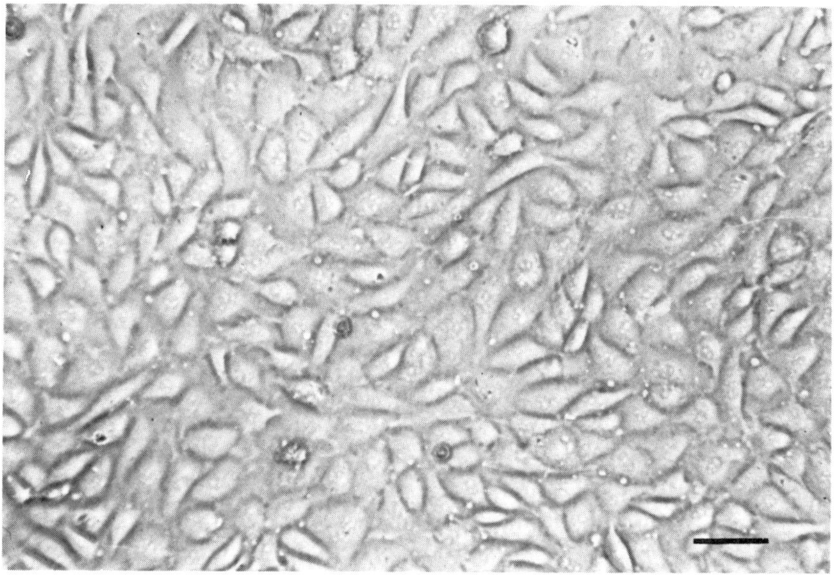

Figure 7. Phase micrograph of a confluent culture of cells on polystyrene, Bar = 100 µm.

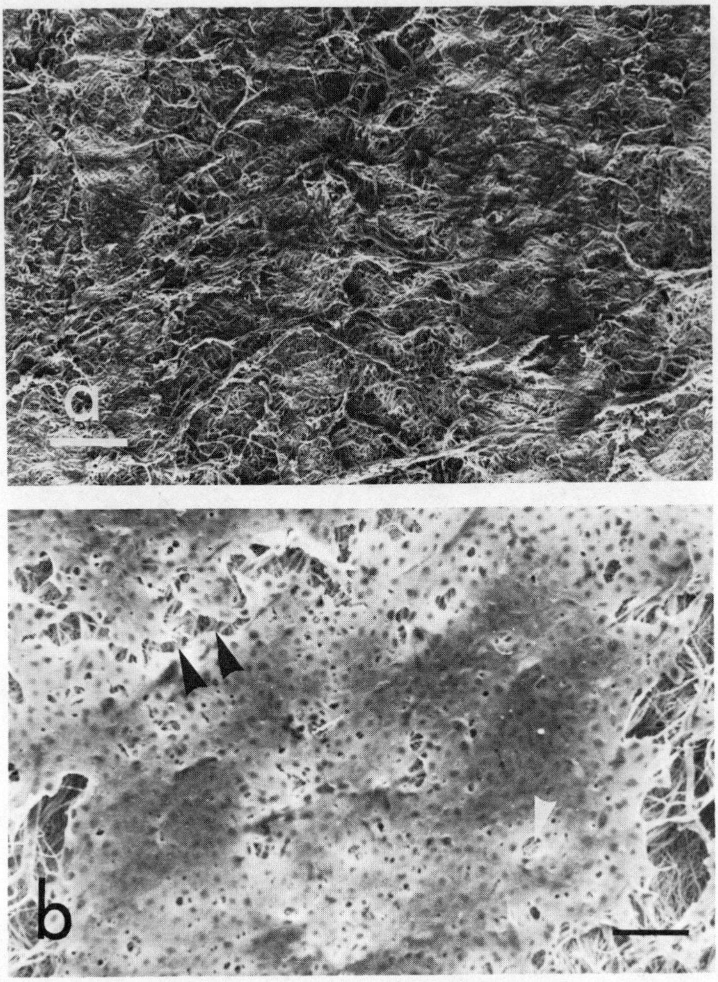
a
b

Figure 8. SEM of a) bare microfabric surface prior to cell culture, b) cultured microfabric prior to implantation. Dark spots are endothelial cell nuclei, arrows show gap in coverage. Bars = 200 μm.

lumenal surface was seen on the precultured patches. Beneath the monolayer scattered leukocytes were observed along with occasional cells which resembled those cells in the surface monolayer. Observations of sections of the cultured patch before implant also showed some cells below the surface enmeshed in the microfabric.

The SEM appearance of the thrombus-free areas of the cell lined patches was that of an intact endothelium, with a smooth surface revealing nuclear profiles of the cells indicating that the patches remained cell lined after implantation (Figure 12a). The surfaces

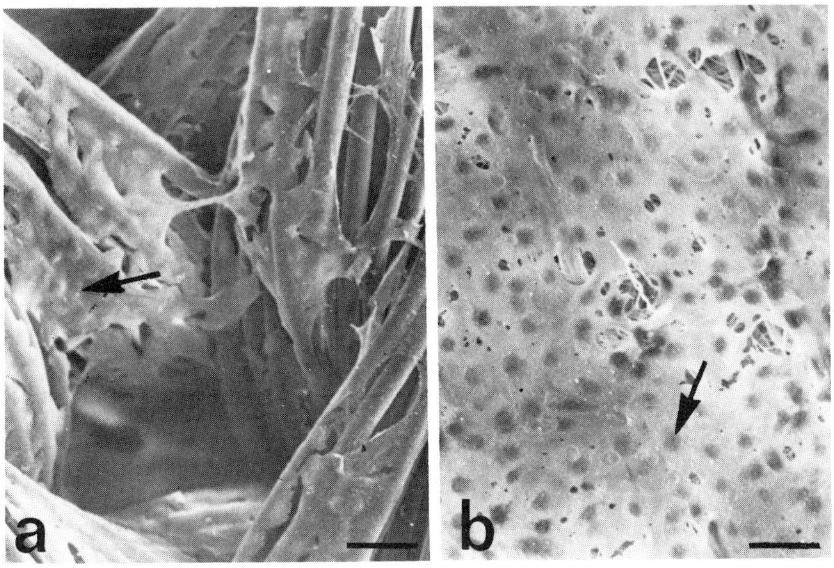
a
b

Figure 9. SEM of endothelial cells cultured on a) Dacron knit, and b) microfabric. Arrows show cell nuclei, Bars = 100 µm.

of bare control implants showed typical thrombus (leukocytes, platelets, red blood cells, and fibrin) adherent to the underlying microfabric (Figure 12b). Areas of the cell lined patches over which endothelial cell coverage was incomplete had foci of thrombus over the uncovered regions (Figures 13 & 14).

Areas were also found where the endothelial cover was regenerating in which cytoplasmic processes stretched between the cells in contrast to the close contact between cells found in a fully endothelialized area.

The TEM observations confirmed the light microscopic and SEM findings. The cell lined patches were covered by a monolayer of endothelial cells which showed overlapping intercellular junctions, with numerous mitochondria and pinocytotic vesicles (Figure 15). Extracellular basement membrane-like material is present on the basal sides of the cells. Often the surface cells had blebs extending into the lumen. Contact between the cells in the lumenal monolayer and the substrate was not observed, nor was any contact observed between the basement membrane material of the surface cells and the substrate. The cells found beneath the surface monolayer had elaborate villous extensions around the cell which often enveloped the substrate fibers (Figure 16). However the cytoplasmic organelles and inclusions of these subsurface cells were similar to those of the cells in the surface monolayer. In addition to the subsurface cells, numerous platelets and many leukocytes (most of

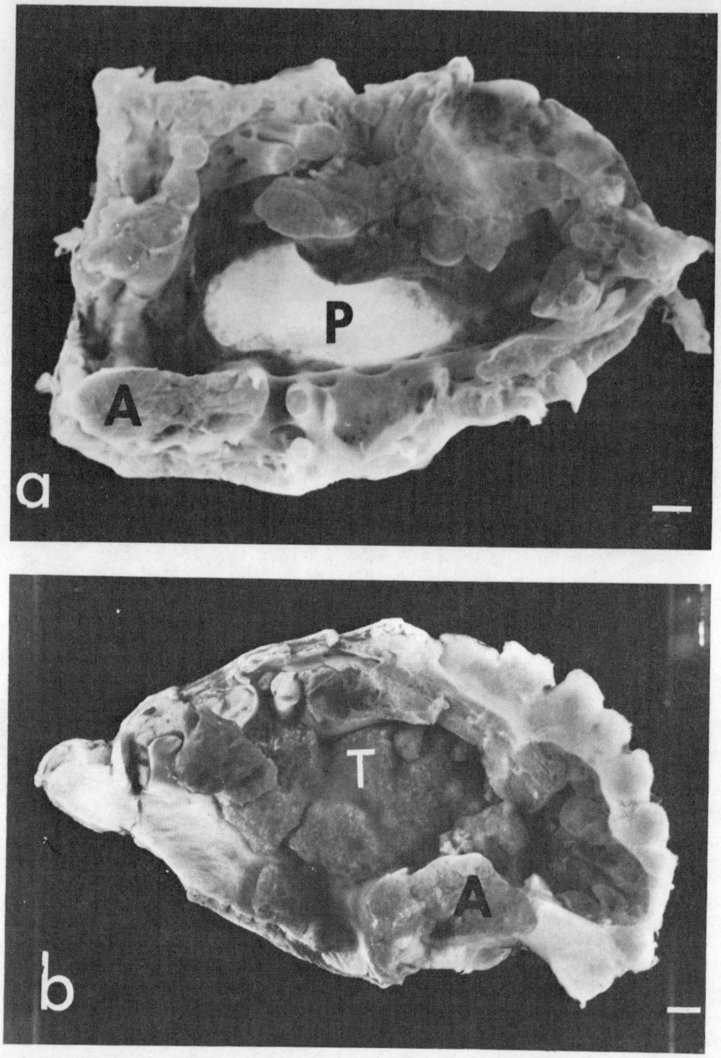

P
A
a
T
A
b

Figure 10. Gross photographs of the blood-contact surfaces of patches with surrounding atrial tissue. a) Cell lined patch; the light area (P) in the center of the picture is the patch. It is translucent and thrombus-free. Around the edges of the patch where anastomosis with the atrial wall is made, slight thrombus buildup is evident. A = cross section through the atrium. b) control patch; thrombus (T) fills the atrial lumen. A = cross section through the atrium Bars = 0.5 cm. (Reprinted by permission of Scanning Electron Microscopy, Inc. from Eskin et al. (12).

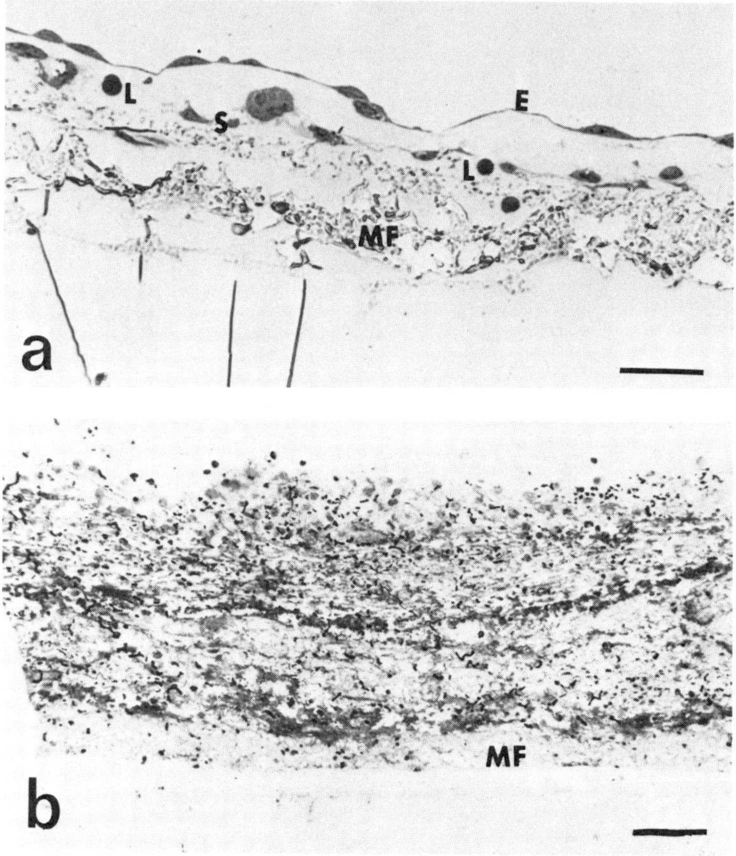
L
S
E
L
MF
a
MF
b

Figure 11. Light microscopic appearance of Paragon-stained plastic sections of implanted patches. a) cell lined thrombus-free patch; lumen is at the top of the picture. E = endothelial cells, L = leukocytes, MF = microfabric, S = subsurface cell. Bar = 50 µm. b) control patch characterized by thrombus buildup; MF = microfabric. Lumen is at the top of the picture. Bar = 100 µm. (Reprinted by permission of Scanning Electron Microscopy, Inc. from Eskin et al. (12).

which were polymorphonuclear leukocytes) were found beneath the endothelial monolayer.

TEM of the control with the least amount of thrombus formation showed leukocytes, erythrocytes, and platelets enmeshed in fibrin (Figure 17). In addition unidentified cells had extruded cytoplasmic processes around the microfabric similar to that which occurred in the subsurface cells on the cell lined patches. In rare cases, cells

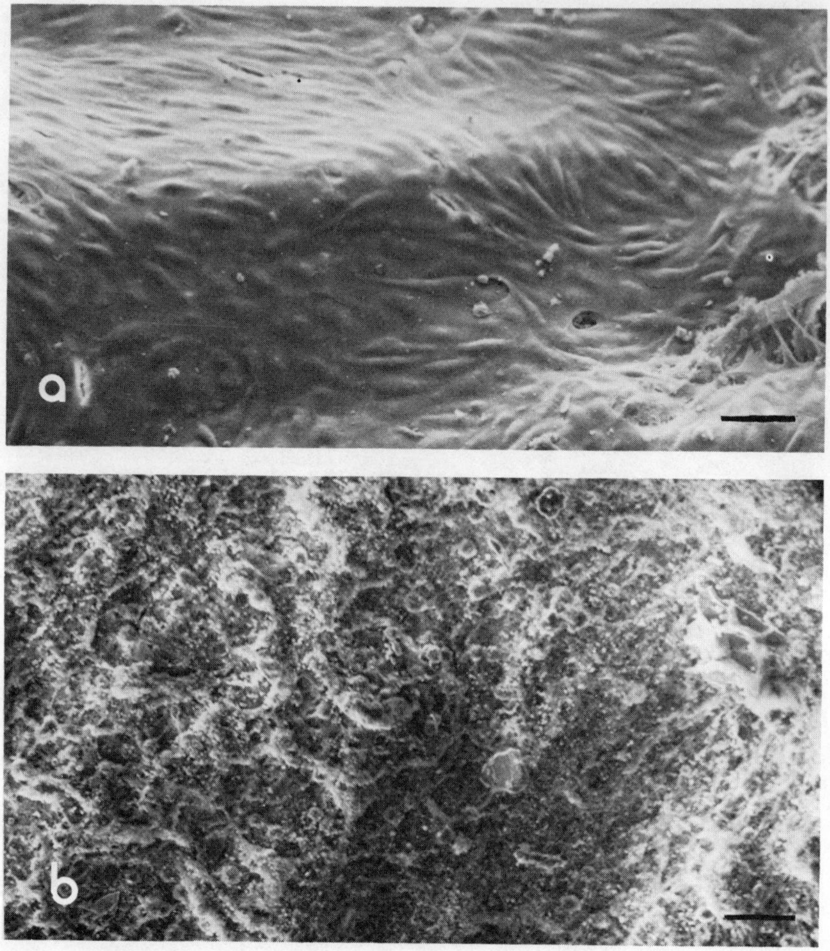

Figure 12. SEM appearance of blood contact surfaces after 1 week implantation a) cell lined patch, regularly spaced bumps are endothelial nuclei. Bar = 100 μm, b) control patch showing rough surfaced thrombus, Bar = 100 μm.

were observed on the lumenal surface of the control which resembled endothelial cells.

DISCUSSION

The ultimate goal of these studies was to explore the feasibility of using cultured endothelial cells to line heart assist pumps. For this reason the cultured surfaces were implanted into the right atrium, which is an area of complicated flow dynamics, which provided a more rigorous test for the cultured surfaces than a straight vessel replacement would provide. The results showed that the cells

E
E

Figure 13. SEM of gap in endothelial cell coverage after 1 week implantation of cell lined patch. E = endothelial cells, arrows = leukocytes adherent to substrate. Bar = 50 μm.

F
BM
L
E

Figure 14. Gap in endothelial cell coverage of a cell lined patch after 1 week in vivo. E = endothelial cell, BM = basement membrane material secreted by E, L = leukocyte, arrow = microfabric substrate, F = fibrin. Bar = 10 μm.

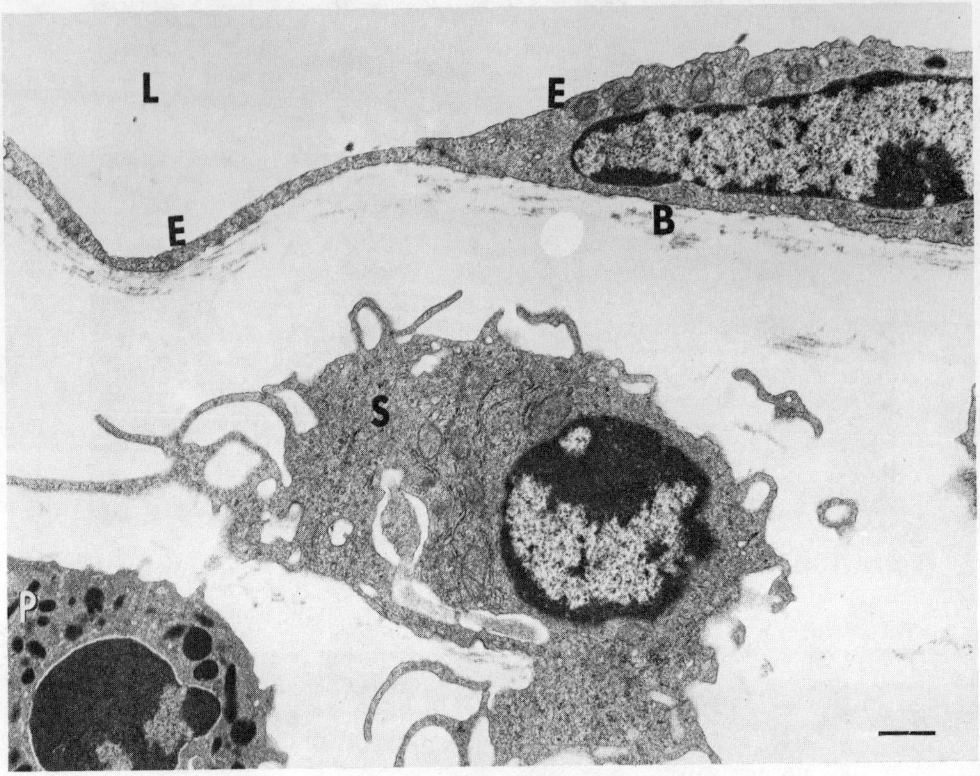
L
E
E
B
S
P

Figure 15. TEM of cell lined patch surface after 1 week in vivo. L = lumen, E = cultured endothelial cells, S = subsurfac cell, P = polymorphonuclear leukocyte, B = basement membrane, Bar = 2 μm.

resisted thrombus formation, where they remained adherent. Adherence problems experienced could have originated from 2 causes; first when the patch was sutured to the beating atrium, the fragile cell monolayer was probably disrupted around the periphery of the patch. Then adherence was probably further compromised when the stationari cultured cells came into contact with the flowing blood. The severity of this adherence problem may be related to the complexity of the blood flow pattern over the experimental surface. Studies reported on aortic grafts showed no difficulties with cell adherenc after implantation (9). However, the same investigators experience removal of a majority of cultured cells upon implantation of cell cultured left ventricular assist devices (13). Our experience with implantation of atrial patches appears to lie intermediate between the graft and the pump results.

The choice of substrate then, as well as the experimental design, is just as critical as the choice of the cells themselves

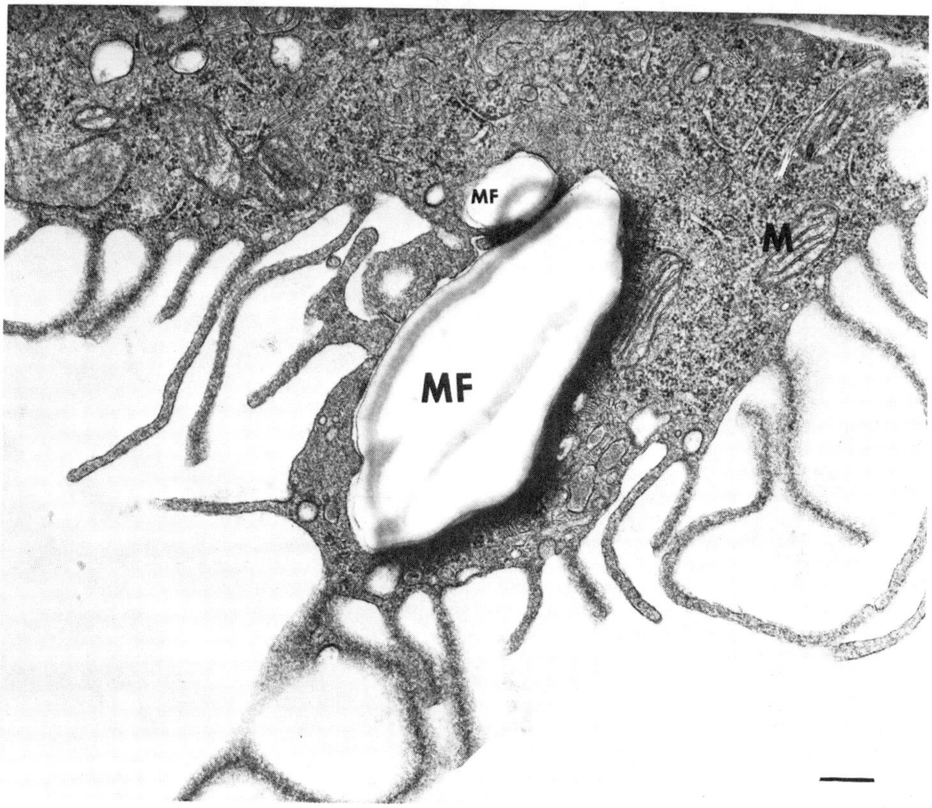
MF
MF
M

Figure 16. TEM of a portion of the cytoplasm of a subsurface cell after 1 week in vivo, showing cytoplasmic processes surrounding the fibers of the microfabric substrate (MF). M = mitochondrion, Bar = 1 μm.

for implantation. The substrate must support cell growth and promote cell adherence, and possess the strength and durability to withstand long term use. The properties which make a material a good cell culture substrate also make it thrombogenic. Thus if the substrate is not covered completely or if the cells do not remain adherent, the thrombogenic substrate is exposed to the blood.

The use of Dacron knit as a substrate may prove more successful than the thin microfabric for 2 reasons; the number of cells covering the Dacron is greater, and many of these cells would not be subject to maximal shear stress because of their positions within the interstices of the knit. Thus although one might expect the cells on the fibers most directly projecting into the flow stream to be removed, they might be replenished by replicating cells deeper in the fabric.

There are other difficulties to surmount prior to clinical

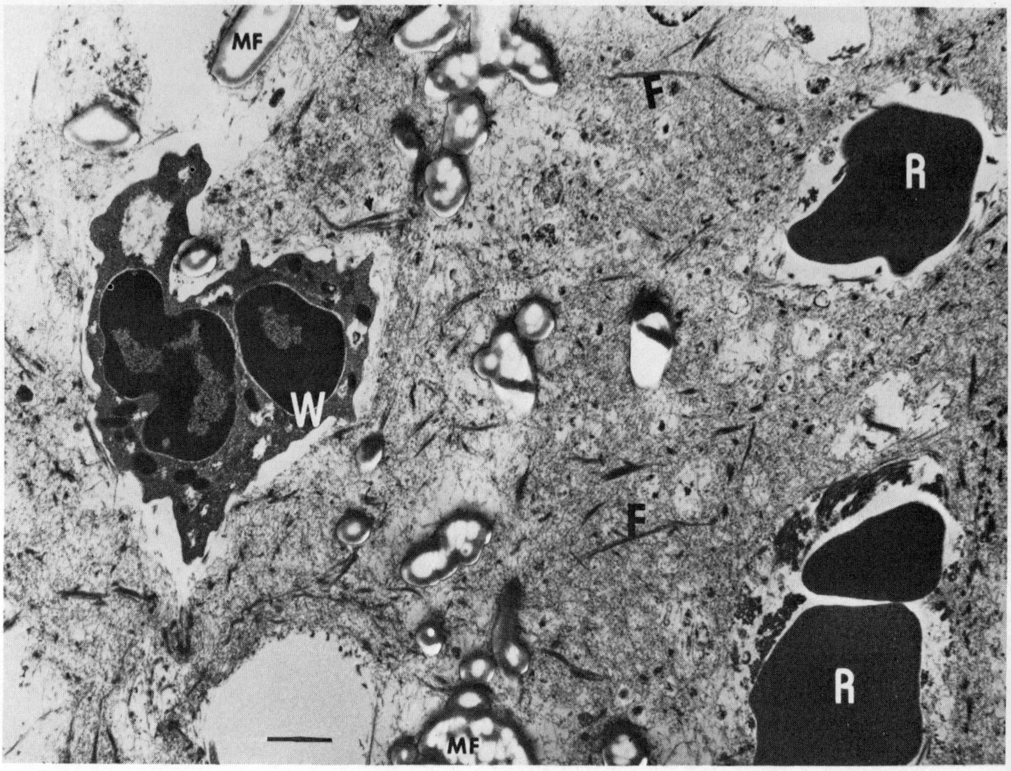
MF
F
R
W
F
R
MF

Figure 17. TEM of control patch after 1 week in vivo. R = red blood cells, W = leukocyte, F = fibrin, MF = microfabric, Bar = 3 μm.

application of this technique. Vascular endothelium must be cultured from a vessel which is amenable to biopsy from high risk patients, unless another source of nonthrombogenic cells is found. The cultured cells could then be stored in liquid nitrogen in the event that the patient required surgery. If vessel grafts were the prosthesis to be cell lined, then culturing and implanting the prosthesis should not be problematic. If the application were lining blood pumps, then the technology for producing cell-substrate complexes that can withstand the flexing of the surface and the complex blood flow patterns generated by the flexion must be developed.

Whether or not clinical use of tissue cultured blood contact linings will become feasible is impossible to judge at present and awaits further study.

ABSTRACT

All materials in current use both clinically and experimentally are thrombogenic. We are attempting to develop techniques for the use of tissue cultured endothelial cells for the blood interface of cardiovascular prostheses, in an effort to prevent thrombus formation. Calf aortic endothelial cells were cultured on a substrate of Parylene C coated polypropylene microfabric backed with polyurethane (Union Carbide). The cell lined materials, in the form of oval patches, were implanted into the right atria of the calves from which the cells were taken. After one week, the calves were sacrificed and the blood contact surfaces examined. Where the cultured cells remained adherent, the surfaces were smooth and thrombus-free. Where the microfabric was exposed to the blood thrombus buildup occurred. The findings shows that cultured endothelial cells can provide a non-thrombogenic surface, but only if problems of cell adherence to the substrate under flow conditions can be overcome.

REFERENCES

1. S. Berger & E. W. Salzman, Prog. Hemost. Thromb., 2, 273 (1974)
2. E. W. Salzman, S. Berger, E. W. Merrill, & P. S. L. Wong, Thromb. diath. haemorrh. Suppl., 59, 107 (1974)
3. M. E. DeBakey, Am. J. Surg., 137, 697 (1979)
4. S. A. Wesolowski, Bull. N. Y. Acad. Med., 48 (2), 331 (1972)
5. M. E. DeBakey, Am. J. Cardiol., 27, 3 (1971)
6. J. E. Chimoskey, W. O'Bannon, J. R. Cant, W. F. Walker, S. G. Eskin, G. P. Noon, and M. E. DeBakey, Biomat., Med. Dev., Art. Org., 5 (4), 361 (1977)
7. P. B. Mansfield & A. Wechezak, in "Organ Perfusion and Preservation", J. C. Norman, ed., Appleton-Century-Crofts, New York, 189 (1968)
8. P. B. Mansfield, Report to Devices and Technology Branch, NIH-NHLI-71-2060-3 & 71-2060-4, NHLI, U.S. Dept. of HEW, PHS, 1975
9. P. B. Mansfield, A. R. Wechezak & L. R. Sauvage, Trans. Am. Soc. Art. Int. Org., 21, 264 (1975)
10. S. G. Eskin, H. D. Sybers, & L. Trevino, Scanning Electron Microscopy/ 1978, 2, 747 (1978)
11. S. G. Eskin, H. D. Sybers, L. Trevino, J. T. Lie, & J. E. Chimoskey, In Vitro, 14 (11), 903 (1978)
12. S. G. Eskin, L. Trevino, & J. E. Chimoskey, J. Biomed. Mat. Res., 12, 517 (1978)
13. A. R. Wechezak & P. B. Mansfield, Scanning Electron Microscopy/ 1979, 3, 857 (1979)

ELASTOMERIC VASCULAR PROSTHESES

Donald J. Lyman, Kenneth B. Seifert, Helene Knowlton, and Dominic Albo, Jr.

Department of Materials Science and Engineering and
Department of Surgery
University of Utah
Salt Lake City, Utah 84112

INTRODUCTION

We have been investigating the effect of chemical structure and surface properties of synthetic polymers on the coagulation of blood. The ultimate goal of these studies is the development of satisfactory prostheses to replace damaged veins, bypass obstructed arteries of small diameter (less than 6 mm) and for use in aortocoronary surgery. Currently, the surgeon is forced to use autogenous tissue such as the saphenous vein in these situations because of the lack of good synthetic polymer prostheses. Although the results using autogenous saphenous veins are very good, this method has certain disadvantages. Some patients (10 to 25%) do not have satisfactory veins either because of prior removal, disease (varicosities), or because the vein is of inadequate size. Even with patients having satisfactory veins, one has to contend with the problem of limited quantity and a significant increase in operating time. The presently available synthetic polymer prostheses are made from relatively thrombogenic polymers (Teflon and Dacron). These materials show patency rates of less than 50% at diameters smaller than 6 mm. Therefore, the development of a successful, small diameter synthetic polymer prosthesis is of utmost importance to both the patient and the surgeon.

Since the current prosthesis materials are thrombogenic, our approach to solving the small diameter vascular prostheses problem was to develop a blood compatible polymer. There appears to be ample evidence from our studies and those of others to support the contention that protein adsorption is the first event occurring as blood comes in contact with a polymer surface *in vivo* (1-10). Studies on the adhesion of platelets to polymer surfaces precoated with a variety of proteins have indicated that albuminated surfaces appear to prevent platelet adhesion and confer nonthrombogenecity to the base polymer (8,9,11-13). In contrast, surfaces precoated with fibrinogen or γ-globulin show increased platelet adhesion and release reactions leading to thrombosis (1,10,12,14-19). Thus, one is led to the hypothesis that the composition of the protein layer formed in situ as whole blood flows over a polymer surface is not the same for every polymer since the degree of platelet adhesion varies from polymer to polymer (11). The data from electrophoretic analyses of proteins desorbed from several polymers exposed to whole blood in situ (1) do show this preferential adsorption. It is also interesting to note that platelet adhesion measurements on these same surfaces using our ex vivo flow-through cell (9,11)(see Table 1) parallel that described above for the preproteinated surfaces; that is if the polymer preferentially adsorbs albumin in situ, it shows less platelet adhesion and the polymer should be nonthrombogenic. Conversely, a polymer preferentially adsorbing globulins or fibrinogen adheres platelets readily and should be thrombogenic. The selectivity of protein adsorption appears to be dictated by the chemical and physical structure of the polymer surface; it is this selectivity which determines the thrombogenic response of the base polymer (1). On the basis of our results, we proposed the following mechanism for the in vivo initiation of a thrombus on a polymer surface (1,20,21).

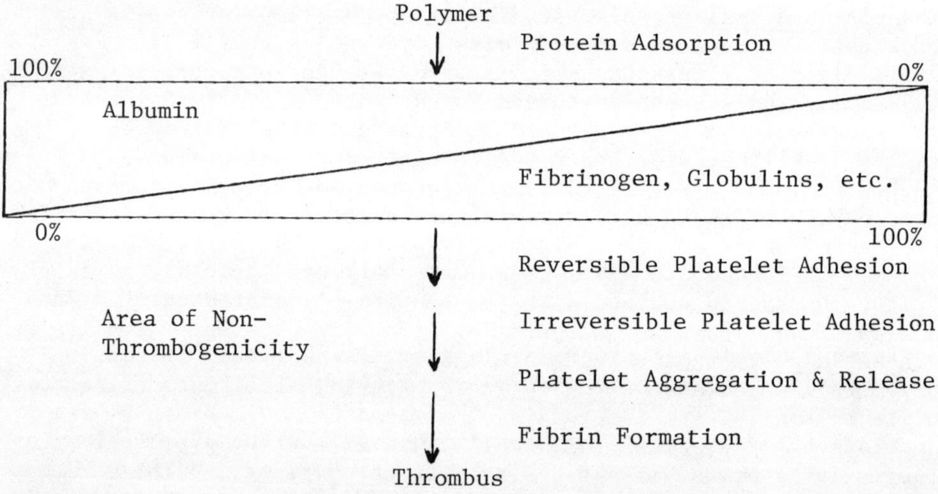
Polymer
Protein Adsorption
100%
0%
Albumin
Fibrinogen, Globulins, etc.
0%
100%
Reversible Platelet Adhesion
Area of Non-Thrombogenicity
Irreversible Platelet Adhesion
Platelet Aggregation & Release
Fibrin Formation
Thrombus

Since Silastic Rubber adsorbs about 70% albumin on its surface and yet is mildly thrombogenic, one would estimate that the albumin content on the surface must be greater than 70% of the total protein adsorbed if the polymer is to be nonthrombogenic.

From these studies, we developed a urethane copolymer which appeared to be nonthrombogenic by in vitro and ex vivo tests (22) and thus was considered as a potential candidate for vascular repair.

TABLE 1

THE RELATIONSHIP BETWEEN *in vivo* PROTEIN ADSORPTION ON POLYMER SURFACES AND PLATELET ADHESION

Polymers	% of Total Protein Present			Platelet Adhesion[2]	
	Albumin	γ-Globulin	Fibrinogen Other Globulins	1 min.	5 min.
Teflon FEP	30	17	53	5.4	clot
710 PEUU	62	30	8	7.5	25.0
Silastic Rubber	70	20	10	4.6	-
Biomer	98	2	0	2.2	5.7
1025 PEUU	98	0	2	0.2	0.9

[1]Exposure time in recirculation was 45 minutes, except for Teflon FEP which was 30 minutes. This timing approximates time necessary for reaching plateau concentration as per *in vitro* experiments. The adsorbed proteins were eluted from the polymer surface, then analyzed by acrylamide gel electrophoresis.

[2]The mean number of platelets adhering to $20,000\mu^2$ surface area exposure to whole blood in the flow through *ex vivo* cell.

However, small diameter arterial prostheses made of this nonthrombogenic material occluded at the anostomosis within 4 to 10 days after implantation. We hypothesized that this failure was due to a difference in wall elasticity between the prosthesis and the artery which caused trauma to the arterial wall and stimulated intimal hypertrophy (23). Initial results using a more compliant form of the prosthesis showed improved patency giving support to this hypothesis. To investigate this hypothesis further and eliminate variables in the initial studies, we developed a technique to vary the elasticities of the prostheses made from this urethane polymer and yet maintain other properties more constant.

EXPERIMENTAL

Urethane Copolymer

The copolyether-urethane-urea used in this study was synthesized from polypropylene glycol (MW 1025), methylene bis(4-phenylisocyanate), and ethylene diamine using a 2-stage solution polymerization technique previously described (24).

Graft Fabrication

A 10% solution of urethane copolymer in N,N-dimethylformamide was coated over a 4 mm O.D. glass mandrel, then immersed in water to extract the solvent and precipitate the copolymer into a coherent substrate. The compliancy of the grafts could be varied by changing the wall thickness by multiple coatings.

The vascular grafts were soaked in deionized water for 24 hours to remove any residual solvent, then removed from the mandrels, trimmed, and steam sterilized for implantation.

Characterization Methods

Stress-strain measurements were made using an Instron Model TM-M Tensile Tester (500 g load cell, cross-head speed was 50 mm/min) with sample holder designed for testing tubular specimens. Test specimens were tubular sections 0.5 cm wide. All measurements were done using wet samples. The copolyurethane samples are shown in Fig. 1 along with the stress-strain curves of samples cut from fresh sections (less than 30 min. old) of the canine thoracic aorta and the inferior vena cava. The samples were maintained in buffered saline from their removal until being measured. Compliancies were calculated as the force (in dynes/test section) required to produce a displacement to 20% increase in cross-sectional area of the lumen.

The porosities of the prostheses to water were measured using a hydrostatic head pressure of 120 mmm Hg.

Histologic studies were made of the removed graft/natural vessel specimens embedded in paraffin. Various stains used to identify tissue and cells were Verhoeff's, Masson's trichrome and hemotoxylin-eosin-phloxine.

Implantation Procedures

Mongrel dogs of either sex, weighing 20-25 kg were anesthetized with intravenous Nembutal (25 mg/kg). Bilateral femoral incisions were made and the femoral arteries exposed from the inguinal ligament distally. A 5 cm segment of artery was excised between 2 vascular clamps, and the precipitated graft sutured in place with 6-0 Prolene, using a continuous suture technique. No heparin was used at any time during or after the implantation procedure. The grafts were not preclotted.

RESULTS

The prostheses prepared in this study had similar porosities, even though they varied considerably in their compliancies (see Table II). Although the prostheses were of sufficient porosity to permit tissue ingrowth, blood loss through the walls was minimal and stopped in less than five minutes in all cases.

Table II

PHYSICAL PROPERTIES AND PATENCY OF 4 mm ID COPOLYURETHANE VASCULAR PROSTHESES

Prosthesis Series	I	II	III	IV	V	Dog Femoral Artery
Elastic Index[1] (x10 dyne/mm)	11.2	10.8	7.5	5.5	5.5	7.4 ± 1.1 (n = 9)
Porosity (ml/cm²/min)	37	33	34	34	27	
Number Implanted	15	9	11	4	5	
Number Patent at One Month[3]	4	3	9	3	3	
Percent Patent	27*	33*	82*	75	60	

[1]Force per unit distance required to increase the cross-sectional area of the lumen 20%.

[2]Measured with deionized water at a hydrostatic pressure at 120 mm Hg.

[3]Determined by re-exposure and direct inspection.

*Indicates values statistically significant ($p<0.005$) by the Cochran-Mentel-Haenzel test.

Patency of the prostheses were determined by re-exposure and direct inspection. The prostheses (series III) in which the elastic index best matched the dog femoral arteries (approx. 7.4 dynes/mm) had the best patency rate at one month. When the prostheses were either more elastic or less elastic, the patency dropped significantly. The decrease in the number patent at one month was statistically significant for those grafts less elastic than the arteries (values significant at <0.005 by the Cochran-Mantel-Haenszel Test); in those prostheses which were more elastic, the low number of samples reduced the significance of the data. However, the trend was toward decreased patency.

Histological examination of specimens removed after one month demonstrated that some intimal hypertrophy had occurred in all arteries near each anostomosis. In the prostheses that failed, the hypertrophy had obliterated the lumen. At the time of this writing, prostheses whose compliances match the natural artery have been patent for 15 months.

DISCUSSION

Current vascular grafts of polyethylene terephthalate and polytetrafluoroethylene have very low compliance. Changes in diameter over the cardiac cycle of pressure are less than 1% as compared to 10% to 15% for the natural artery and saphenous vein (25,26). As a result, both mechanical and hemodynamic stresses occur at the junction of the graft to the host artery. These stresses may be a major contributing factor to suture-line disruption and/or thrombosis, especially for synthetic polymer prostheses of less than 6 mm ID. One solution to this problem of compliance mismatch is to design a graft which simulates the elastic modulus of the host artery.

A variety of elastomeric materials such as latex rubber, Silastic Rubber, fluoroelastomer, hydrocarbon rubbers and copolyurethane materials have been fabricated into vascular prostheses (27-35). The results with latex rubber, Silastic Rubber, fluorocarbon elastomer, and hydrocarbon rubber were not good. Failure in most instances occurred within 24 hours of implantation in dogs. However, these surfaces are relatively thrombogenic. In addition, these grafts were much less compliant than the natural artery. Better results were obtained with the copolyurethane materials, although poor choice of material, impurities, etc. often led to graft fracture or occlusion, pseudoaneurysm formation, embrittlement and degradation. These materials were also less compliant than the natural artery, and this contributed to the failure.

Our implantation studies (22,23) indicated that a major factor contributing to these failures was a mismatch in graft/artery compliance. In vitro hemodynamic studies (36) on compliant-noncompliant tubes also supported the concept of increased stress on the natural tissue side of the junction due to reflected waves. We then developed a method to reduce the wall density of the copolyurethane prostheses thus making it more compliant (Fig. 1) yet maintaining the nonthrombogenic surface. These initial studies showed that in contrast to the noncompliant solid-wall grafts which failed within hours, six of the nine new compliant grafts were patent on sacrifice at times ranging to 77 days.

To test further our hypotheses on compliancy matching, we refined our fabrication method to give variable compliancy with relatively constant porosity of the graft to water. Implantation of these new grafts have shown increased patency (to 82%) for grafts in which the elastic index best matches the dog femoral artery (Table II). As the prostheses became less elastic, patency dropped off rapidly. At 1 month only about 30% of the stiffer grafts were patent. If one makes the prosthesis more elastic, patency also decreases, though apparently not as drastically.

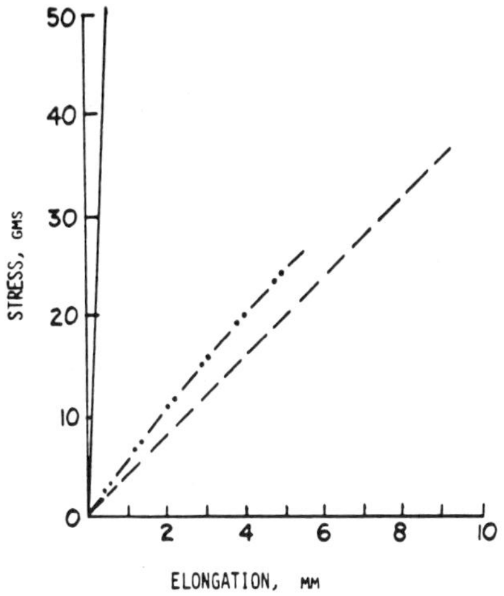
STRESS, GMS
50
40
30
20
10
0
2
4
6
8
10
ELONGATION, MM

Figure 1. Stress-strain curves for thoracic aorta (---); compliant urethane prosthesis (-..-..); and 3.5 mil urethane film (——).

In contrast, a series of polytetrafluoroethylene grafts (Gore-Tex) implanted under identical conditions were not patent at one month.

At the time of this writing, the compliant copolyurethane vascular prostheses are still patent at 15 months implantation. One prosthesis removed at this time showed minimal hypertrophy. Thus, to achieve success in small diameter vascular prostheses, one must balance the surface chemical and physical properties of the material for blood compatibility and the mechanical properties for matching the compliancy of the natural vessel.

ABSTRACT

We have developed a block copolyurethane material with a non-thrombogenic surface. However, a noncompliant small diameter arterial prostheses made of this material soon occluded at the anastomoses. The failure appeared to result from stresses due to a mismatch in wall compliancy between the prosthesis and the natural artery. When this same block copolyurethane material is formed into a compliant small diameter vascular prosthesis, the prosthesis remained patent for long periods of time. Current compliant grafts have been functioning for 15 months, whereas the noncompliant grafts failed within days. Thus, one must incorporate proper chemical, physical and mechanical properties into their material if a workable small diameter vascular graft is to be achieved.

REFERENCE

1. D. J. Lyman, L. C. Metcalf, D. Albo, Jr., D. F. Richards, and J. Lamb, Trans. Amer. Soc. Artif. Int. Organs, 20, 474 (1974)

2. E. W. Salzman, Fed. Proc. 30, 1503 (1971)
3. E. Nyilas, Proc. 23rd Ann. Conf. Eng. Med. Biol., 12, 147 (1970)
4. R. C. Dutton, R. E. Baier, R. L. Dedrick, and R. L. Bowman, Trans. Amer. Soc. Artif. Int. Organs, 14, 57 (1968)
5. R. E. Baier and R. C. Dutton, J. Biomed. Mat. Res., 3, 191 (1969)
6. H. Petscheck, D. Adams, and A. R. Kantrowitz, Trans. Amer. Soc. Artif. Int. Organs, 14, 256 (1968)
7. R. E. Baier, G. I. Loeb, and G. T. Wallace, Fed. Proc., 30, 1523 (1971)
8. A. J. Lande, L. Edwards, J. H. Bloch, R. G. Carlson, V. Subramanian, R. S. Ascheim, S. Scheidt, S. Fillmore, T. Killip, and C. W. Lillehei, Trans. Amer. Soc. Artif. Int. Organs, 16, 352 (1970)
9. D. J. Lyman, K. G. Klein, J. J. Brash, B. K. Fritzinger, J. D. Andrade, and F. S. Bonomo, Thromb. Diath. Haem. Suppl., 42, 109 (1971)
10. D. E. Scarborough, R. G. Mason, F. G. Dalldorf, and K. M. Brinkhouse, Lab. Invest., 20, 164 (1969)
11. D. J. Lyman, K. G. Klein, J. L. Brash, and B. K. Fritzinger, Thromb. Diathes. Haem., 23, 120 (1970)
12. M. A. Packham, G. Evans, M. F. Glynn, and J. F. Mustard, J. Lab. Clin. Med., 73, 686 (1969)
13. V. L. Gott and A. Furuse, Fed. Proc., 30, 1679 (1971)
14. D. S. P. Jenkins, M. A. Packham, M. A. Gucciore, and J. F. Mustard, J. Lab. Clin. Med., 81, 280 (1973)
15. J. F. Mustard, M. F. Glynn, E. E. Nishizawa, and M. A. Packham, Fed. Proc., 26, 106 (1967)
16. S. W. Kim and D. J. Lyman, App. Polymer Symp., 22, 189 (1973)
17. S. W. Kim, R. G. Lee, C. Adamson, and D. J. Lyman, Preprints Div. of Plastics and Organic Coatings, ACS, 165, 36 (1973)
18. M. F. Glynn, M. A. Packham, J. Kirsh, and J. F. Mustard, J. Clin. Invest., 45, 1013 (1966)
19. M. B. Zucker and L. Vroman, Proc. Soc. Biol. Med., 131, 318 (1969)
20. D. J. Lyman and W. J. Seare, Jr., Ann. Revs. Maters. Sci., 4, 415 (1974)
21. D. J. Lyman, K. Knutson, B. McNeil, and K. Shibatani, Trans. Amer. Soc. Artif. Int. Organs, 21, 49 (1975)
22. D. J. Lyman, D. Albo, Jr., R. Jackson, and K. Knutson, Trans. Amer. Soc. Artif. Int. Organs, 23, 253 (1977)
23. D. J. Lyman, F. J. Fazzio, H. Voorhees, G. Robinsons, and D. A.bo, Jr., J. Biomed. Maters. Res., 12, 337 (1978)
24. D. J. Lyman, C. Kwan-Gett, H. J. Zwart, A. Bland, N. Eastwood, J. Kawai, and W. J. Kolff, Trans. Amer. Soc. Artif. Int. Organs, 17, 456 (1971)
25. D. E. Hokanson and E. E. Strandness, Surg. Gynecol. Obstet., 127, 57 (1968)
26. E. R. Gozna, W. F. Mason, A. E. Marble, et al., Can. J. Surg., 15, 176 (1974)
27. W. V. Sharp and W. H. Falor, Am. J. Surg., 105, 802 (1963)

28. W. V. Sharp, A. F. Finelli, W. H. Falor, and J. W. Ferraro, Circulation, 29 (suppl.) 165 (1964)
29. W. V. Sharp, D. L. Gardner, and G. J. Andresen, Trans. Amer. Soc. Artif. Internal Organs, 11, 336 (1965)
30. B. Dreyer, T. Akutsu, and W. J. Kolff, J. Appl. Physiol., 15, 18 (1960)
31. V. Marinescu, E. Pausescu, and S. Carnaru, Thorax, 26, 108 (1971)
32. M. Wagner, G. Reul, J. Teresi, and K. L. Kayser, Am. J. Surg., 3, 838 (1966)
33. H. D. Berkowitz, L. J. Perloff, and B. Roberts, Surgery, 72, 888 (1972)
34. W. V. Sharp, D. L. Gardner, and G. J. Andresen, Trans. Amer. Soc. Artif. Int. Organs, 12, 179 (1966)
35. W. V. Sharp, D. L. Gardner, G. J. Andresen, and J. Wright, Trans. Amer. Soc. Artif. Int. Organs, 14, 73 (1968)
36. D. J. Lyman and R. H. Knowlton, Manuscript in preparation.

ACKNOWLEDGMENTS

This work has been supported by the National Institute for General Medical Science, Grant GM 24487-02.

MORPHOLOGY OF BLOCK COPOLYURETHANES. II.

FTIR AND ESCA TECHNIQUES FOR STUDYING SURFACE MORPHOLOGY

K. Knutson and D. J. Lyman

Department of Materials Science and Engineering
University of Utah
Salt Lake City, Utah 84112

INTRODUCTION

Block copolymers form a domain-matrix morphology due to the chemical and steric incompatibilities of the chemically different blocks or segments. The unusual range of the physical and chemical properties associated with block copolyurethanes result from this unique morphological separation. However, multiple $(AB)_n$ block copolymers such as block copolyurethanes may or may not have as complete phase separation as the simpler AB block copolymers. Bulk morphologies of block copolyurethanes have been studied by many investigators during the last decade via electron microscopy (1,2), temperature-modulus (3,4,5) and infrared spectroscopy (6-11). More recent studies have shown the bulk and surface chemical and morphological structures of block copolyurethanes may be quite different, and are affected by both synthetic alterations of the polymeric repeat unit and by fabrication variables (12-15). Since several block copolyurethanes have shown both compatibility and mechanical properties needed for vascular implants, it is important to determine the chemical and morphological structures of these materials. A block copolyether-urethane-urea and associated model compounds and homopolymers have been studied by Fourier Transform Infrared Spectroscopy (FTIR) and Electron Spectroscopy for Chemical Analysis (ESCA) in order to determine the chemical and morphological structures of the surface of the block copolyurethane as compared to the bulk.

EXPERIMENTAL

Polymer Synthesis

The block copolyether-urethane-urea was synthesized from polypropylene glycol (1000 MW), methylene bis(4-phenylisocyanate) and ethylene diamine using a two-step solution polymerization (16). The repeat unit structure is as follows:

$$[(OCHCH_2(CH_3))_{17}-O-C(=O)-N(H)-C_6H_4-CH_2-C_6H_4-N(H)-C(=O)-N(H)-CH_2CH_2-N(H)-C(=O)-N(H)-C_6H_4-CH_2-C_6H_4-N(H)-C(=O)-]_x$$

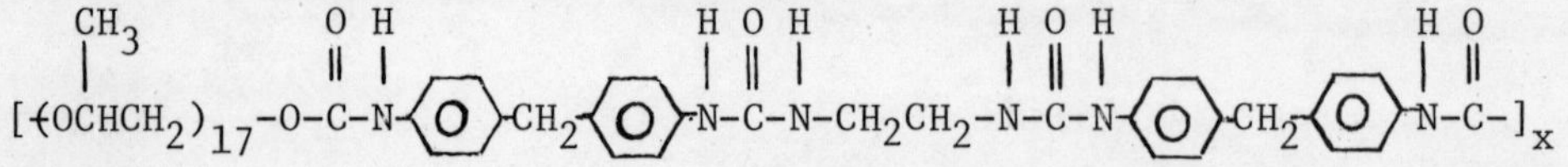

Inherent viscosity was 0.54 in N,N-dimethylformamide (0.5% concentration) at 30°C.

The films used in the studies were prepared by solvent casting a filtered solution (15% solids by weight) of the polymer in distilled N,N-dimethylformamide onto either glass plates or mandrels. The glass plates and mandrels were cleaned by washing with Ivory soap solution, rinsing with deionized water, absolute ethanol and ether before being used. The plates were placed in a forced-air oven at 75°C for 30 minutes to evaporate the solvent and form the film. The mandrels were immersed in deionized water diffusing the solvent out and coagulating or precipitating the polymer. The films were then placed in a vacuum jar (0.1 mm Hg) for 24 hours to insure complete removal of any residual solvent or water.

Model Compound and Polymer Syntheses

The urea domains were modeled with the diurea compound (Urea I) synthesized by reacting a 2:1 molar ratio of p-tolylisocyanate and ethylene diamine in anhydrous toluene to form:

$$CH_3-C_6H_4-N(H)-C(=O)-N(H)-CH_2CH_2-N(H)-C(=O)-N(H)-C_6H_4-CH_3$$

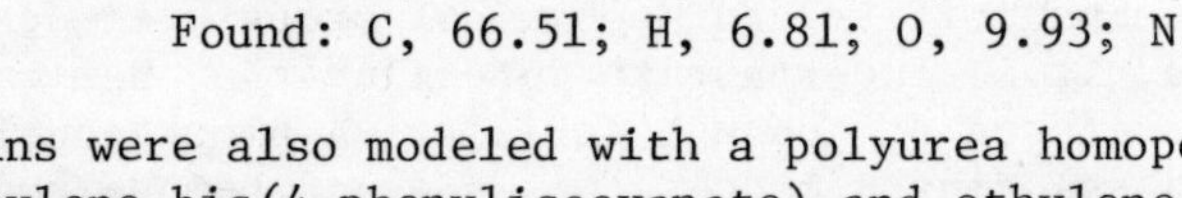

The white powder was recrystallized from hot N,N-dimethylformamide. Melting point was 233-235°C. Elemental analysis was:

Calculated: C, 66.24; H, 6.79; O, 9.80; N, 17.16
Found: C, 66.51; H, 6.81; O, 9.93; N, 17.08

The domains were also modeled with a polyurea homopolymer synthesized from methylene bis(4-phenylisocyanate) and ethylene diamine in dimethylsulfoxide. The repeat unit structure is as follows:

$$[-C(=O)-N(H)-C_6H_4-CH_2-C_6H_4-N(H)-C(=O)-N(H)-CH_2CH_2-N(H)-]_x$$

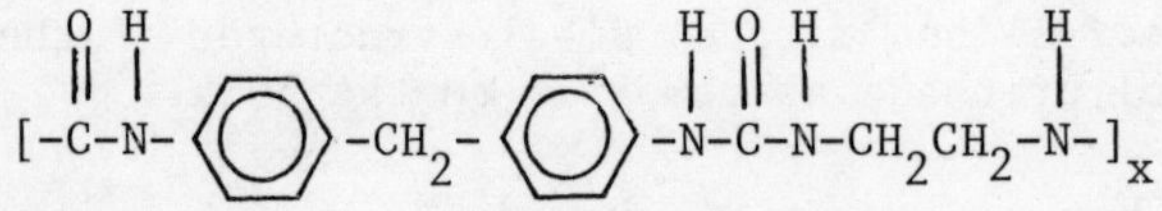

The polymer was slightly soluble in m-cresol. Films were cast from a filtered solution (5% solids by weight) in m-cresol and dried at 75°C. The film was then placed in a vacuum jar for 24 hours (0.1 mm

Hg) to remove residual solvent. The glass casting plates were cleaned as previously described.

The urethane interfacing linkage was modeled with two model compounds. Urethane I was synthesized by reacting p-tolylisocyanate with n-propanol in anhydrous toluene to form:

$$CH_3CH_2CH_2-O-\underset{\underset{O}{\|}}{C}-\underset{\underset{H}{|}}{N}-\langle O \rangle -CH_3$$

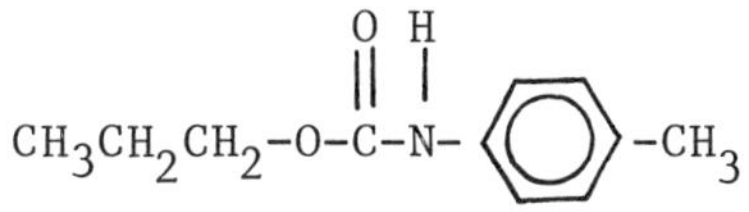

The white powder was recrystallized from hot ethanol. Melting point was 55.5 to 56.0°C. Elemental analysis was:

    Calculated: C, 68.37; H, 7.82; O, 16.56; N, 7.25
    Found: C, 68.59; H, 7.86; O, 16.62; N, 7.22

Urethane II was synthesized by reacting methylene bis(4-phenyliso-cyanate) in anhydrous toluene with n-butanol (1:2 molar ratio) to form:

$$CH_3CH_2CH_2CH_2-O-\underset{\underset{O}{\|}}{C}-\underset{\underset{H}{|}}{N}-\langle O \rangle-CH_2-\langle O \rangle-\underset{\underset{H}{|}}{N}-\underset{\underset{O}{\|}}{C}-O-CH_2CH_2CH_2CH_3$$

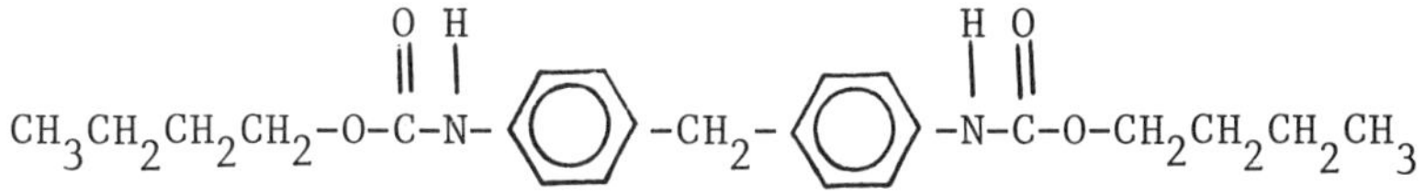

The white powder was recrystallized from hot ethanol. Melting point was 230.5 to 231.5°C. Elemental analysis was:

    Calculated: C, 69.12; H, 7.33; O, 16.36; N, 6.97
    Found: C, 69.35; H, 7.54; O, 16.08; N, 7.04

The urea domains and interfacing urethane linkage were modeled by a copolyurethane-urea synthesized from methylene bis(4-phenyliso-cyanate), propylene glycol and ethylene diamine in a two-step polymerization (16). The repeat unit of the copolyurethane-urea is:

$$[O-CH(CH_3)CH_2-O-\overset{O}{\overset{\|}{C}}-\overset{H}{\overset{|}{N}}-\langle O \rangle-CH_2-\langle O \rangle-\overset{H}{\overset{|}{N}}-\overset{O}{\overset{\|}{C}}-\overset{H}{\overset{|}{N}}-CH_2CH_2-\overset{H}{\overset{|}{N}}-\overset{O}{\overset{\|}{C}}-\overset{H}{\overset{|}{N}}-\langle O \rangle-CH_2-\langle O \rangle-\overset{H}{\overset{|}{N}}-\overset{O}{\overset{\|}{C}}-]_x$$

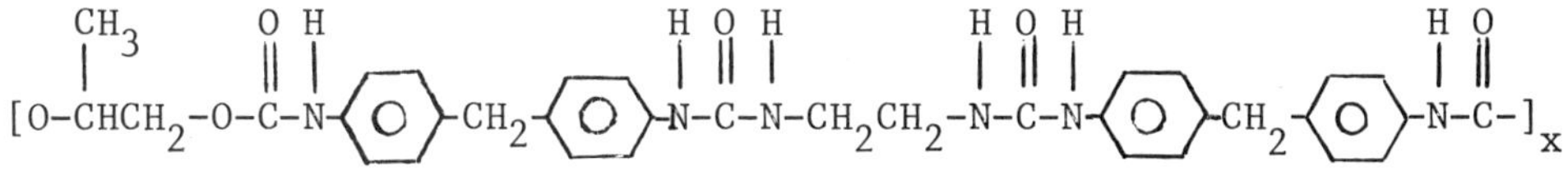

Films were cast from a 6% solution (solids by weight) of polymer in N,N-dimethylformamide onto glass plates cleaned as previously described. The plates were then placed in a forced-air oven at 75°C for 30 minutes, and then placed in a vacuum jar (0.1 mm Hg) overnight to remove residual solvent.

The interface region coupling the urea domains and interacting with the polyether matrix was modeled with a copolyether-urethane synthesized from polypropylene glycol (1000 MW) and methylene bis(4-phenylisocyanate) using a one-step solution polymerization technique (17). The copolymer has the following repeat unit structure:

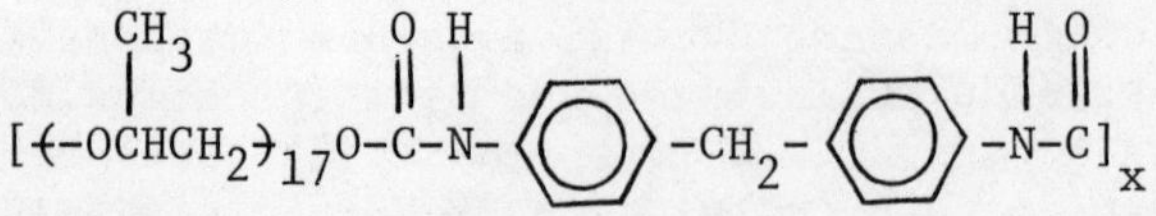

The inherent viscosity was 0.24 in N,N-dimethylformamide (0.5% concentration) at 30°C.

The isolated polyether matrix was modeled through the use of polypropylene glycol (2000 MW) and isotactic polypropylene oxide. The polypropylene glycol was degassed and placed over molecular sieves to remove residual water present in the polyol. The isotactic polypropylene oxide was isolated by repeated crystallization from acetone (18). The inherent viscosity was 1.85 in benzene (0.5% concentration) at 25°C. The isotactic polypropylene oxide films were cast onto glass plates cleaned as previously described from 6% (solids by weight) solution of polymer in N,N-dimethylformamide. They were dried in a forced-air oven for 30 minutes at 75°C and then placed in a vacuum jar (0.1 mm Hg) overnight to remove residual solvent.

Instrumental Methods

A HEWLETT-PACKARD 5950B ESCA spectrometer was utilized in the ESCA studies of solvent cast block copolyether-urethane-urea, copolyurethane-urea, polyurea and isotactic polypropylene oxide films. The samples were allowed to come to equilibrium at 10^{-9} Torr at 300°K prior to data collection. The X-rays from the Al($K_{\alpha 1,2}$) line at 1487 eV were used. Samples were scanned 10 times with a scan width of 20 eV centered around elemental spectra of interest: C(1s) 290-270 eV; N(1s) 405-385 eV; O(1s) 540-520 eV. Radiation damage was evaluated by overlaying the carbon spectrum obtained first for each sample and one obtained after all other elements of interest were scanned. Differences in spectra were within experimental error, therefore radiation damage was considered to be negligible for the qualitative studies. All bands were referenced to the 285 eV band of the C(1s) spectrum for each sample. Peak areas were determined digitally and by planimeter.

A DIGILAB 14B/D Fourier Transform infrared spectrometer was utilized to obtain 1 cm^{-1} resolution spectra over the 4000 to 600 cm^{-1} region for copolyether-urethane-urea, polyether-urethane, polypropylene glycol, Urethane I, Urethane II and Urea I. The sample chamber was allowed to come to equilibrium with a continuous nitrogen purge prior to data collection of 250 scans per sample for transmission studies and 1000 scans per sample for internal reflectance studies.

A HARRICK variable angle internal reflectance attachment and quadruple diamond polarizer were used to obtain depth profiles of copolyether-urethane-urea films. Internal reflectance crystals (25x5x2 mm) made of germanium (Ge) having face cut angles of 45° and 60° were used at incident angles of 30°, 45° and 60° with perpendicular polarization. Absorbance spectra of internal reflectance crystals were obtained prior to placement of the sample films and stored in memory for later subtraction, thus spectra of only the sample films were obtained after the proper arithmetic operations.

Model compounds (Urea I, Urethane I and Urethane II), copolyether-urethane and polypropylene glycol were studied by transmission using a HARRICK transmission cell with 25x2 mm round ZnSe transmission crystals. The viscous polypropylene glycol and copolyether-urethane were spread onto the crystals with spacers separating the two crystals at a uniform thickness. Model compounds were recrystallized onto a transmission crystal from solution. Complete evaporation of the solvent for recrystallization occurred prior to data collection as determined by absence of characteristic solvent bands in the infrared spectra. Spectra were also obtained using the transmission cell of homogeneous mixtures of Urea I and/or Urethane I in either polypropylene glycol or polyether-urethane.

RESULTS AND DISCUSSION

Multiple block copolymers such as copolyether-urethane-ureas may or may not separate into a two phase morphology due to the chemical and steric incompatabilities of the two blocks, block size and fabrication variables. A completely phase separated copolyether-urethane-urea would have the urea domains interfacing through the urethane linkage to the polyether matrix. If no phase separation occurred, the urea domains would be completely dispersed in the polyether matrix giving a homogeneous mixture of the two phases. A partially phase separated structure would show urea domains pure and completely separated, but with some portions of the urea domains dispersed in the polyether matrix. It is also possible that the urethane interface would similarly be mixing with either the urea domains or the polyether matrix. Since the extended length of the urea block in the particular polyether-urethane-urea under study is about 30 A and the polyether segment is about 55 A, the interface region presents an important transition region for determining the completeness and purity of phase separation.

The morphology of the bulk films of the copolyether-urethane-urea was initially studied by transmission spectra of the polymer (the 3600-2600 cm^{-1} region is shown in Figure 1 and the 1800-600 cm^{-1} region is shown in Figure 2). Characteristic bands in Figure 1 include the N-H stretching band centered around 3320 cm^{-1} indicating essentially complete hydrogen bonding of the urethane and urea N-H groups. The CH_3 group of the polyether matrix has an asymmetric C-H stretching band at 2975 cm^{-1} and a symmetric C-H stretching band at 2900 cm^{-1}. The CH_2 group of the polyether matrix and urea domain has an asymmetric

C-H stretching band at 2932 cm^{-1} and a symmetric C-H stretching band at 2872 cm^{-1}. Characteristic bands in Figure 2 include C=O stretching Amide I band of the urethane carbonyl splitting into two shoulders. The 1730 cm^{-1} shoulder is due to the free carbonyls and the 1712 cm^{-1} shoulder is due to the hydrogen bonded carbonyls. The C=O stretching Amide I band of the urea carbonyls is a single symmetrical band at 1635 cm^{-1} indicating complete hydrogen bonding of the carbonyl groups. The C=C stretching band of the aromatic rings and associated N-H bending band of the N-H groups covalently bonded to the ring structures overlap forming the asymmetrical band near 1600 cm^{-1}. The Amide II bands of the urethane and urea groups due to N-H bending, either free or hydrogen bonded, and C-N stretching overlap forming the broad band at 1530 cm^{-1}. The Amide III bands of the urethane and urea groups arising from N-H bending, either free or hydrogen bonded, and C-N stretching overlap with the asymmetric stretching band of the ester portion of the urethane group forming the broad band near 1225 cm^{-1}. The broad ether stretching band is at 1110 cm^{-1}.

The urethane N-H group is capable of forming a dipolar interaction with the ether oxygen similar to a hydrogen bond (9,10), thus resulting in essentially complete N-H hydrogen bonding as reflected in the single N-H stretching band at 3320 cm^{-1} while leaving a significant portion of the urethane carbonyls in the free state as reflected in the intensity of the 1730 cm^{-1} shoulder of the urethane Amide I band. The urethane Amide I band is sensitive to the inter- and intra-molecular interactions occurring between the urethane groups of the interface with either the polyether matrix or the urea domains. The hydrogen bonded shoulder at 1712 cm^{-1} reflects the

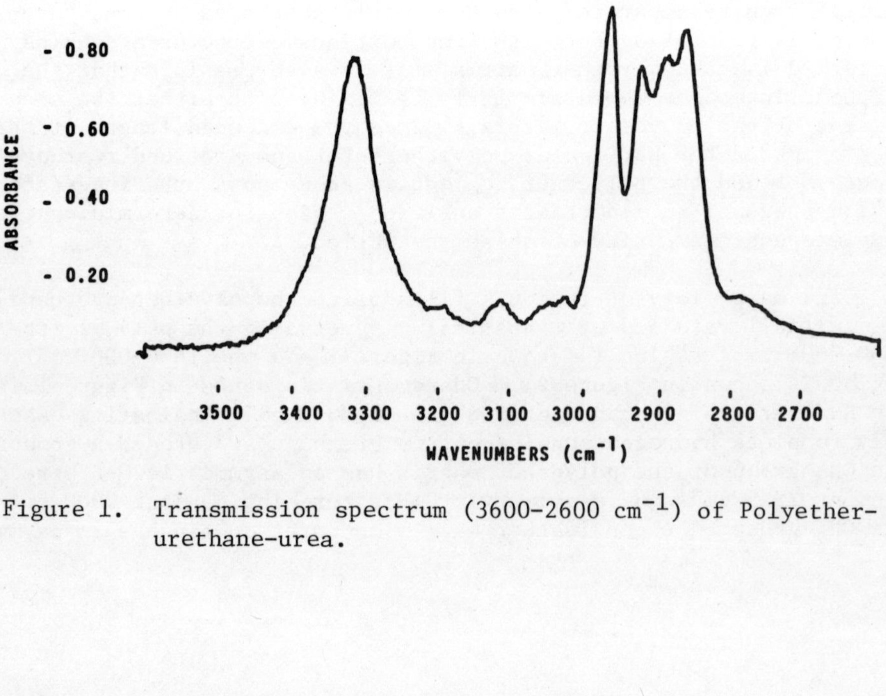

ABSORBANCE
0.80
0.60
0.40
0.20
3500
3400
3300
3200
3100
3000
2900
2800
2700
WAVENUMBERS (cm^{-1})

Figure 1. Transmission spectrum (3600-2600 cm^{-1}) of Polyether-urethane-urea.

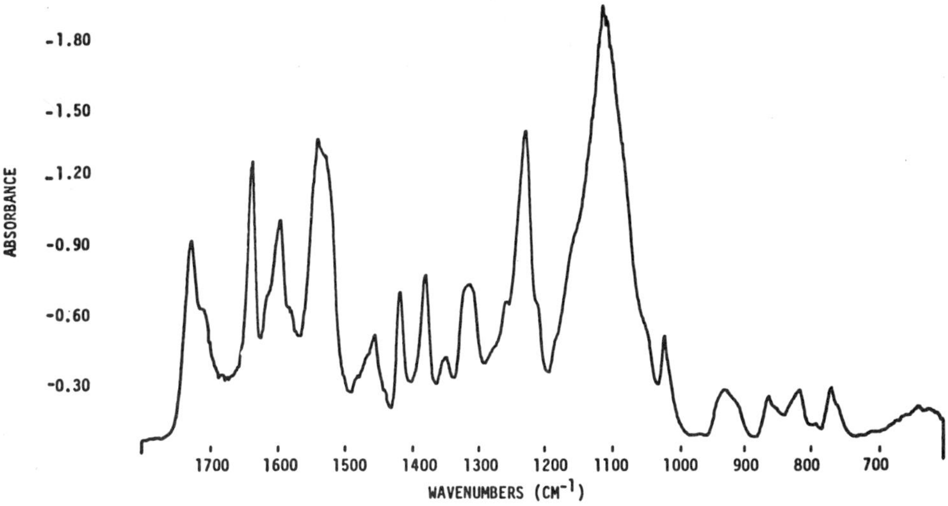
ABSORBANCE
-1.80
-1.50
-1.20
-0.90
-0.60
-0.30
1700
1600
1500
1400
1300
1200
1100
1000
900
800
700
WAVENUMBERS (CM^{-1})

Figure 2. Transmission spectrum (1800-600 cm^{-1}) of Polyether-urethane-urea.

portion of urethane carbonyls hydrogen bonding with either other urethane groups of the interface or urea groups of the domain. The free shoulder at 1730 cm^{-1} reflects the portion of the urethane carbonyls being freed from hydrogen bonding due to the dipolar interaction between the urethane N-H groups and the polyether oxygens. Absorbance ratios of the two shoulders of the urethane Amide I band indicate 60% of the urethane groups were interacting with the polyether matrix.

To determine the degree to which chemical and morphological structures affect secondary bonding involving these sensitive amide and N-H groups, model compounds and polymers were used to study selected secondary bonding environments which mimic various degrees of phase separation and purity possible within the copolyether-urethane-urea.

The individual urethane model compounds were first studied in the crystalline state to identify the characteristic bands of the pure interface region. The 1800-600 cm^{-1} region of the Urethane I spectrum is shown in Figure 3. The characteristic bands include the broad asymmetrical band centered around 1700 cm^{-1} for the urethane Amide I carbonyl stretching, urethane Amide II at 1530 cm^{-1} and the urethane Amide III at 1250 cm^{-1}. Asymmetric stretching of the ester portion of the urethane group overlaps with the Amide III band at 1250 cm^{-1}, while the symmetric stretching gives rise to the 1070 cm^{-1} band. The C=C stretching and N-H bending bands of the aromatic ring and associated N-H group overlap forming the asymmetrical band near

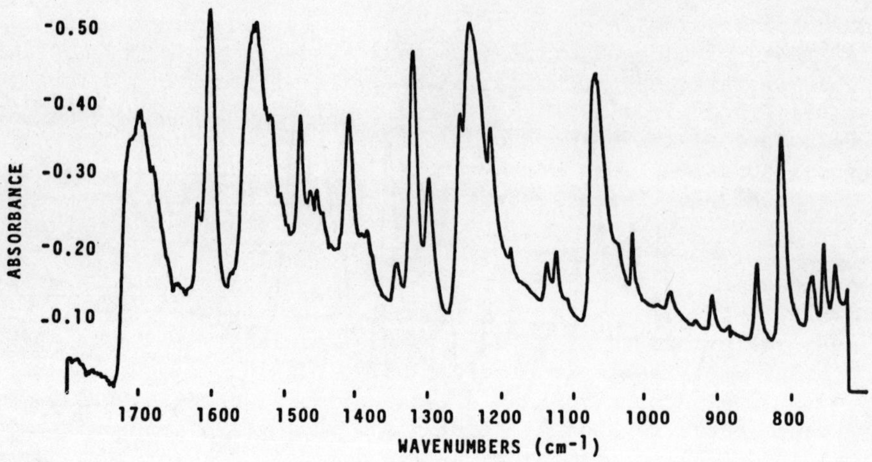

-0.50
-0.40
-0.30
-0.20
-0.10
ABSORBANCE
1700
1600
1500
1400
1300
1200
1100
1000
900
800
WAVENUMBERS (cm-1)

Figure 3. Transmission spectrum (1800-700 cm^{-1}) of Urethane I.

1600 cm^{-1}. The decreased steric hindrances present in the single aromatic ring structure of Urethane I allowed the aromatic ring and urethane group to stack in a coplanar structure. Coplanarity of similar structures have been shown to result in a lower frequency carbonyl stretching Amide I band (19). To determine if the decreased steric hindrances of the single aromatic ring structure allowing coplanar stacking of the Urethane I model compound resulted in the 1690 cm^{-1} shoulder of the Amide I band, the Urethane II model compound was synthesized with a ring-atom-ring structure similar to the poly-ether-urethane-urea. Steric hindrances of the ring-atom-ring structure did not allow coplanar stacking of these types of structures. The Amide I band of the Urethane II model compound was a single symmetrical band centered around 1700 cm^{-1}. Therefore, the low frequency shoulder at 1690 cm^{-1} of the Urethane I was assigned to the portion of the carbonyl groups that are packed in the crystalline state in a coplanar structure.

Complete dispersion of the urethane interface groups in the polyether matrix was modeled by two homogeneous mixtures of Urethane I and polypropylene glycol. The 1800 to 1200 cm^{-1} regions of a 1:2 ratio and a 1:4 molar ratio mixture of polypropylene glycol to Urethane I are shown in Figure 4. The 1:2 molar ratio mixture had 17 ether repeats to one urethane group. Absorbance ratios of the urethane Amide I band indicate 73% of the urethane carbonyls were in the nonhydrogen bonded state due to the dipolar interaction between the urethane N-H groups and the ether oxygen. The 1:4 molar ratio mixture doubled the number of urethane groups to ether groups to what was present in the copolyether-urethane-urea structure. The absorbance ratios of the shoulders of the urethane Amide I indicated only a 5% increase in hydrogen bonding occurs, thus leaving 68% of the carbonyls in a nonhydrogen bonded state. The same percentage of urethane carbonyls are in a nonhydrogen bonded state in the polyether-urethane polymer where the polyether segment was covalently bonded to the urethane linkage, as opposed to the noncovalently bonded free urethane groups found in the mixtures of the urethane model compound with polypropylene glycol. Therefore, the incompatibilities between the polyether matrix and the urethane interface were not sufficient to result in significant phase separation between these two segments, and covalent bonding between the two segments does not increase phase separation. The hydrogen bonded shoulder of the urethane Amide I band was located at 1712 cm^{-1} for the mixtures and for the polyether-urethane polymer. The steric hindrances of the ring-atom-ring structure as found in the polyether-urethane polymer (20) and the associated inter- and intra-molecular interactions as found in either the mixtures or the polymer did not allow coplanar packing of the urethane carbonyl as evidenced by the absence of a lower frequency shoulder.

The urea model compound in the pure state modeled the urea domains as if completely phase separated and pure. The 1800-600 cm^{-1} region of the urea spectrum is shown in Figure 5. Characteristic bands in the spectrum include the single symmetrical band at 1635 cm^{-1} resulting from essentially complete hydrogen bonding of the urea carbonyls. Completeness of hydrogen bonding is confirmed by a single N-H stretching band at 3320 cm^{-1}. The C=C stretching and N-H bending bands of the aromatic ring and associated N-H group of the urea form the intense band near 1600 cm^{-1}.

A homogeneous 1:2 molar mixture of polypropylene glycol to urea has the same ratio of ether repeats to urea groups as found in the polyether-urethane-urea. The urea Amide bands did not broaden nor split into shoulders, thus indicating no freeing of the urea carbonyls due to a similar dipolar interaction between the urea N-H groups and the polyether oxygens as found in the mixtures involving the urethane model compound with polypropylene glycol. Also, the N-H hydrogen bonded band at 3320 cm^{-1} did not broaden towards higher frequencies due to an absence of N-H groups being freed from their hydrogen bonds. Therefore, the urea molecules separate into domain-like structures and are essentially completely hydrogen bonded within themselves.

Tri-mixtures of polypropylene glycol, Urethane I and Urea I were studied to investigate the chemical hindrances present leading to phase separation if the three structures were not covalently bonded into a polymeric chain. A 1:2:2 mixture of polypropylene glycol to Urethane I to Urea I gives 17 repeats of polyether to one urethane group and two urea groups. Absorbance ratios of the urethane Amide I shoulders indicated 73% of the urethane carbonyls were not hydrogen bonding to either other urethane or urea molecules. This was the same percentage of free urethane carbonyls as determined for the

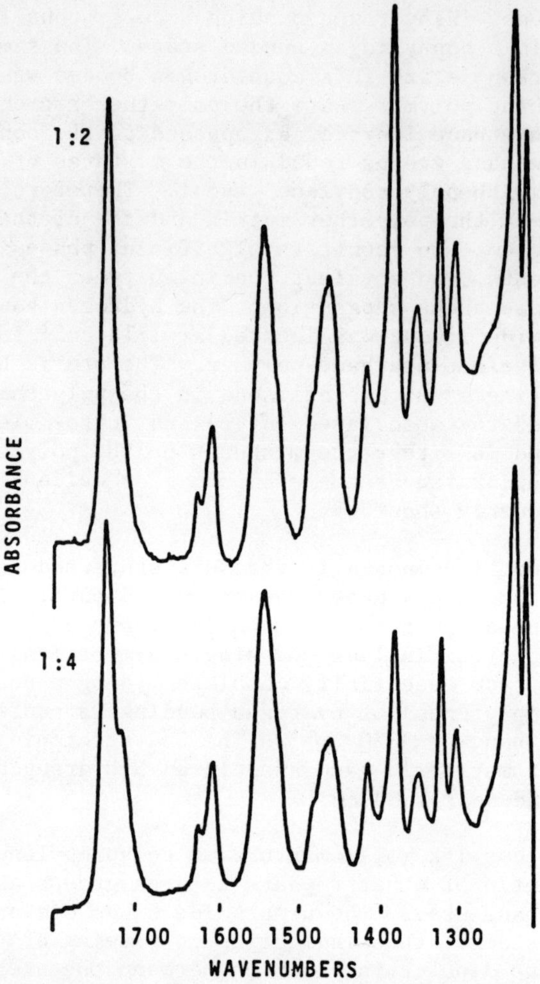
1:2
1:4
ABSORBANCE
1700
1600
1500
1400
1300
WAVENUMBERS

Figure 4. Transmission spectra (1800-1200 cm^{-1}) of 1:2 and 1:4 molar ratio mixtures of Polypropylene Glycol to Urethane I.

1:2 mixture of polypropylene glycol to Urethane I. A 1:4:2 mixture of polypropylene glycol to Urethane I and Urea I gives 17 repeats of polyether to two urethane groups and two urea groups. Absorbance band ratios for this tri-mixture indicated 66% of the urethane carbonyls were in the free state. The urea Amide I band did not broaden nor shift in location. The Amide II and III bands did broaden and change in shape as compared to the mixtures of only polypropylene glycol and Urethane I, but this was due to overlapping of the associated urea and urethane Amide II and III bands. There appears to be little interaction between the urea and urethane molecules with the urea molecules separating into domain-like

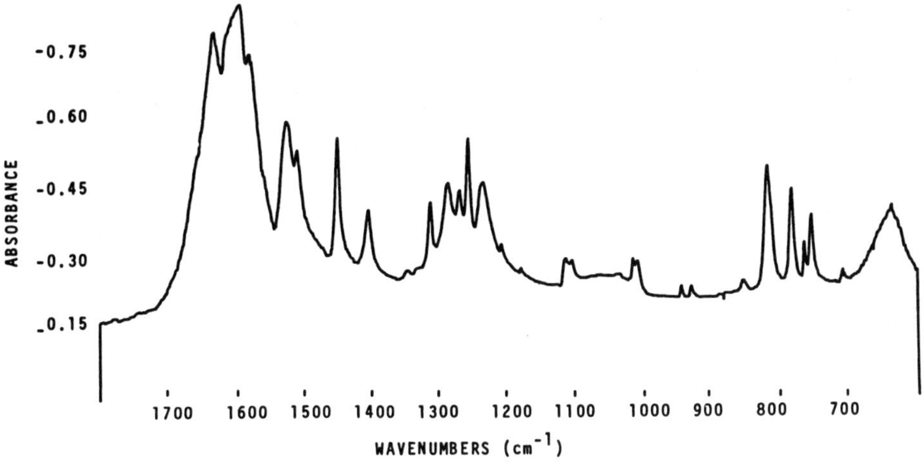
-0.75
-0.60
ABSORBANCE
-0.45
-0.30
-0.15
1700
1600
1500
1400
1300
1200
1100
1000
900
800
700
WAVENUMBERS (cm^{-1})

Figure 5. Transmission spectrum (1800-600 cm^{-1}) of Urea I.

structures, thus leaving the urethane molecules to interact with the polyether matrix as if the urea groups were not present in the mixture. A 1:2 mixture of polyether-urethane to Urea I gave similar results, indicating covalent bonding between the urethane linkage and polyether matrix did not introduce significant steric hindrances to alter the phase mixing of the urethane interface region. However, the covalent bond between the urethane interface and urea domains in the polyether-urethane-urea did increase the steric hindrances of the urethane linkage as reflected in the 8% increase in hydrogen bonding of the urethane carbonyls in the polymer as opposed to the mixtures.

From these studies involving the model compounds individually and in mixtures with either polypropylene glycol or polyether-urethane, the bulk morphology of the polyether-urethane-urea was determined to be a phase separated structure involving essentially complete hydrogen bonded urea domains and a polyether matrix. The interface region is indeed a major transition region of about 10-12 A rather than a simple interfacing linkage of the 3 A urethane linkage alone.

The very surface chemical and morphological structures can be different from that of the bulk structures. Since it is the surface that controls the interaction of a vascular implant with blood, it was important to explore techniques to determine this surface structure as related to the bulk (21). ESCA was originally explored in an attempt to determine the surface structures as compared with the bulk. This technique involves the excitation of inner shell electrons by X-rays and the analysis of the kinetic energies of these excited electrons leaving their orbits from an escape depth of approximately 50 to 100 A. Although electrons throughout the bulk are irradiated, only those excited electrons within the escape depth have sufficient energy to leave the surface for analysis. The low surface depth of penetration and subtle shifts in binding energies which result in peak splittings of the spectra of the elements of interest, except hydrogen, appeared to make this an attractive method to obtain limited knowledge of the chemical and bonding environments of the elements (22-24).

The polyurethane-urea, polyurea and isotactic polypropylene oxide were used to study the peak splittings occurring in the carbon C(1s) spectrum of the polyether-urethane-urea that are particular to either the domain, the matrix or the interface regions of the morphological structure. The isotactic polypropylene oxide modeling the polyether matrix showed the carbon spectrum splits into two distinct shoulders. The carbon atoms covalently bonding to other carbon atoms formed the reference 285 eV shoulder, while those carbons participating in sigma bonding with oxygen atoms formed a shoulder at 286.5 eV. Polyurea modeling the domain structures had a carbon C(1s) spectrum consisting of a shoulder at 285 eV representing carbons bonding to other carbon or nitrogen atoms and a shoulder at 289.5 eV representing carbons pi bonding to oxygen atoms. The polyurethane-urea modeling the domains and interface region had a carbon C(1s) spectrum with three distinctive shoulders. The carbons bonding to either other carbon or to nitrogen atoms formed the reference shoulder at 285 eV, the carbons sigma bonding to oxygen in the ester portion of the urethane linkage forming the 286.5 eV shoulder and the carbon atoms pi bonding to oxygen in either the urethane or urea groups formed the 289.5 eV shoulder. Therefore, the ether portion of the matrix is represented by the 286.5 eV shoulder in the carbon C(1s) spectrum and the carbonyls of either the interface or domains are represented by the 289.5 eV shoulder.

The glass and air surfaces of the block copolyether-urethane-urea film dried in the forced-air oven and the glass surface of the precipitated film were studied using ESCA. The ratios of the peak areas for the three carbon shoulders are given in Table I, along with the theoretical ratios of the carbon atoms participating in the three types of bonds per repeat unit as a representative of a bulk sample. The glass and air surfaces of the film dried in the forced-air oven (air-dried) are quite similar with a higher proportion of ether bonds being present than expected from the theoretical values. The glass surface of the precipitated film has nearly theoretical proportions of the different types of carbon environments present on the surface. However, the information obtained using ESCA does not distinguish if the morphology of the precipitated film surface was completely phase separated or if the domains and matrices were homogeneously mixed. All the information concerning the surface chemical and morphological structures obtained from ESCA concerns only the population of the various chemical structures present, and

Table 1

PEAK AREA RATIOS OF POLYETHER-URETHANE-UREA CARBON C(1s) ESCA SPECTRA

SURFACE	C-O/C=O	C-O/C-C	C=O/C-C
PEUU 1025 PPT	9.00	0.61	0.07
PEUU 1025 GLASS	11.41	1.09	0.10
PEUU 1025 AIR	16.55	1.07	0.06
PEUU 1025 THEORY	9.00	0.78	0.09

not the inter- and intra-molecular interactions leading to various degrees of phase separation and purity.

FTIR coupled with internal reflectance allows one to study the surface structures as well as the inter- and intra-molecular interactions that would occur with different phase separations. Internal reflectance techniques involve the placement of sample films on two sides of an internal reflectance crystal. The infrared beam is reflected through the crystal and interacts with a portion of the surface structures of the films. The depth of these interactions obtained is described by Harrick (25) as the depth at which the amplitude of the electric vector decreases to one half of its original value at the surface. This depth of penetration is determined as a function of the indices of refraction of the sample and the internal reflectance crystal, face cut angles of the internal reflectance crystal, angle at which the infrared beam enters the crystal and the wavelength of the energy. Internal reflectance crystals having higher indices of refraction, higher face cut angles and higher incident angles result in decreasing depths of penetration (25).

Past studies (15) involving the depth profiling of the surface structure of the glass surface of the air-dried film of polyether-urethane-urea by using a germanium internal reflectance crystal having a 45° face cut angle at 30°, 45° and 60° incident angles showed an increase in the hydrogen bonded shoulder of the urethane Amide I band with decreasing depth of penetration. Three surfaces previously studied by ESCA were then studied using FTIR coupled with internal reflectance techniques using a 45° face cut germanium internal reflectance crystal at an incident angle of 45°. The 1800 to 1500 cm^{-1} regions of the three films are shown in Figure 6. These spectra confirm the information obtained from ESCA with the glass and air surfaces of the air-dried films being quite similar, and the glass surface of the precipitated film being different from either air-dried surface. There is an increase in the 1712 cm^{-1} hydrogen bonded shoulder of the urethane Amide I band in the precipitated surface as compared to the air-dried surface. This increase arose from additional interactions between the urethane carbonyl hydrogen bonding with either other urethane groups or urea groups. The information obtained using a 45° germanium crystal at 45° involves only the near-surface with the depth of penetration being approximately 3500 A. When the

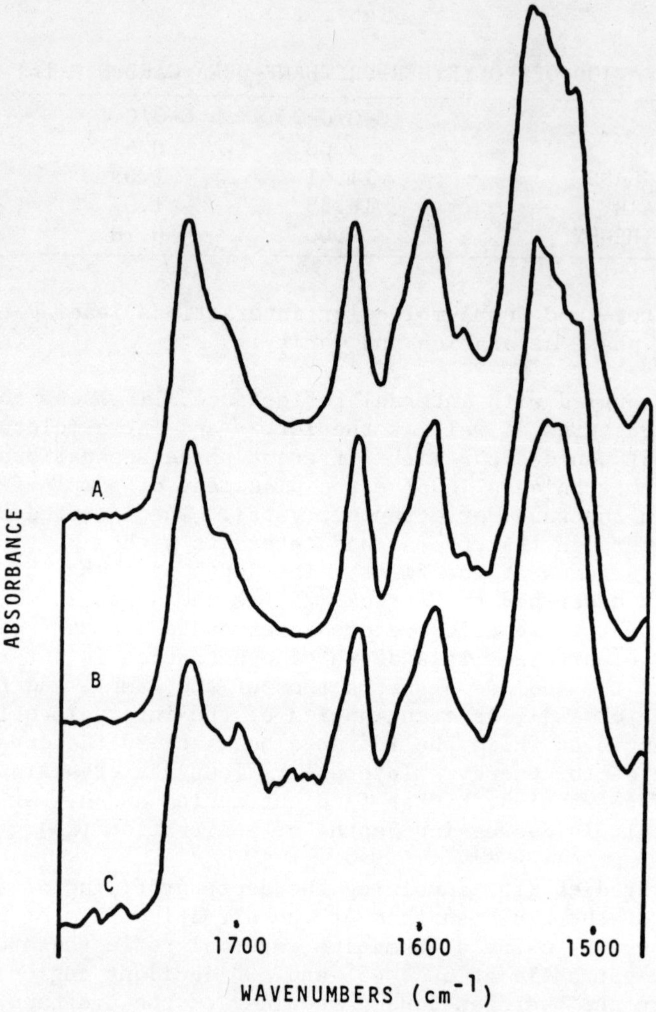
ABSORBANCE
A
B
C
1700
1600
1500
WAVENUMBERS (cm-1)

Figure 6. Internal reflectance spectra (1800-1475 cm^{-1}) of (A) glass surface of air-dried film, (B) air surface of air-dried film and (C) glass surface of precipitated film of Polyether-urethane-urea.

glass surface of the air-dried film was studied using a 60° germanium crystal at 60° incident angle, information for the first 2000 A of the surface was obtained. Data indicated the hydrogen bonded shoulder of the urethane Amide I further increased in absorbance as the surface is approached. Also, the ratio of the hydrogen bond N-H band to the C-H asymmetric stretching band for CH_3 showed an increase in polyether on the surface as compared to the bulk as previously indicated by ESCA. The interface region between the polyether matrix and urea domains decreased in size due to decreased mixing between the urethane linkage and polyether matrix in the near-surface as indicated by the urethane Amide I, although the urea domains remained quite pure on the surface as well as in the bulk.

Thus, both ESCA and FTIR confirm a deviation between surface and bulk chemical and morphological structures that is further altered by fabrication variables. ESCA provides elemental information concerning the first 100 A of the surface with some molecular information concerning population of certain groups. FTIR coupled with internal reflectance techniques provides not only molecular information, but secondary bonding information concerning the near-surface and into the bulk of the polymer structures. Through combining the information obtained from both techniques, the chemical and morphological structures of a polymer film are beginning to be understood on a molecular level from the surface into the inner bulk.

ABSTRACT

Multiple block copolymers form a domain-matrix morphology due to the chemical and steric incompatibilities of the two chemically different blocks. The surface molecular and morphological structures of a series of block copolyether-urethane-ureas have been studied in detail via Electron Spectroscopy for Chemical Analysis (ESCA) and Fourier Transform Infrared Spectroscopy (FTIR) coupled with internal reflectance techniques. ESCA provides elemental information concerning the very surface, while FTIR provides the molecular and secondary bonding information of the surface and into the bulk. Bulk and surface chemical and morphological structures are shown to be quite different, and are affected by synthetic and fabrication variables. A series of model compounds and homopolymers are used to study the possible inter- and intra-molecular interactions for varying degrees of phase separation in controlled concentrations and environments. Together, they are being used to provide the necessary molecular, conformational and secondary bonding information of the surface and into the bulk to allow modeling of these chemical and morphological structures.

REFERENCES

1. D. C. Allport and W. H. Janes, eds., "Block Copolymers", John Wiley & Sons, New York, (1973)
2. J. A. Koutsky, H. U. Hein, and S. L. Cooper, J. Polym. Sci., B-8, 353 (1970)
3. S. L. Cooper and A. V. Tobolsky, J. Appl. Polym. Sci., 10, 1837 (1966)
4. R. L. Bonart, L. Morbitzer, and H. Rinke, Kolloid-Z., 240, 807 (1970)
5. D. S. Huh and S. L. Cooper, Polym. Eng. Sci., 11, 369 (1971)

6. K. Nakayama, T. Ino, and I. Matsubara, J. Macromol. Sci.-Chem., A-3 (5), 1005 (1969)
7. T. Tanaka, T. Yokoyama, and Y. Yamaguchi, J. Polym. Sci., A-1 (6), 2137 (1968)
8. H. Ishihara, I. Kimura, K. Saito, and H. Ono, J. Macromol. Sci.-Phys., B-10 (4), 591 (1974)
9. C. S. P. Sung and N. S. Schneider, Macromolecules, 8, 68 (1975)
10. R. W. Seymour, G. M. Estes, and S. L. Cooper, Macromolecules, 3, 579 (1970)
11. T. Tanaka, T. Yokoyama, and Y. Yamaguchi, J. Polym. Sci., A-1 (6), 2153 (1968)
12. D. J. Lyman, D. Albo, Jr., R. Jackson, and K. Knutson, Trans. Am. Soc. Artif. Intern. Organs, 23, 253 (1977)
13. C. S. P. Sung, C. B. Hu, E. W. Merrill, and E. W. Salzman, J. Biomed. Mater. Res., 12, 791 (1978)
14. C. S. P. Sung, C. B. Hu, and E. W. Merrill, Polym. Preprints, 19, 20 (1978)
15. D. J. Lyman and K. Knutson, "Polymeric Materials and Pharmaceuticals for Biomed. Uses", E. P. Goldberg and A. Nakajima, eds., Academic Press, New York, (1979), in press
16. D. J. Lyman, C. Kwan-Gett, H. J. Zwart, A. Bland, N. Eastwood, J. Kawai, and W. J. Kolff, Trans. Am. Soc. Artif. Intern. Organs, 17, 456 (1971)
17. D. J. Lyman, J. Polym. Sci., 45, 49 (1960)
18. K. Shibatani, D. J. Lyman, D. F. Shieh, and K. Knutson, J. Polym. Sci., Polym. Chem., 15, 1655 (1977)
19. R. T. Conley, "Infrared Spectroscopy", second edition, Allyn and Bacon, Boston, (1972)
20. D. J. Lyman, J. Heller, and M. Barlow, Die Makromolekulare Chem., 84, 64 (1965)
21. D. J. Lyman, K. Knutson, B. McNeill, and K. Shibatani, Trans. Am. Soc. Artif. Int. Organs, 21, 49 (1975)
22. T. A. Carlson, "Photoelectron and Auger Spectroscopy", Plenum Press, New York, (1975)
23. T. H. Schofield, J. Electron Spectrosc. Relat. Phenom., 8, 129 (1976)
24. K. Siegbahn, et al., "ESCA", Almquist and Widsells Publishing, Upsala, Sweden, (1967)
25. N. J. Harrick, "Internal Reflection Spectroscopy", John Wiley Interscience, New York, (1967)

ACKNOWLEDGMENTS

This work has been supported by the National Science Foundation, Grant DMR 76-83681, Polymer Program.

SECTION III

APPLICATIONS OF POLYMERS IN MEDICATION

The use of polymeric materials to control medication is a more recent innovation in biomaterials. Basically the goal of this application is to make drug therapy more specific with a longer period of activity while reducing the toxic side effects. Three different approaches have been developed to achieve this goal: (1) mechanical or diffusion controlled pumps or related devices, (2) controlled release polymeric systems and (3) polymeric drugs. All three types are considered in papers in this section although most papers deal with the newest approach - polymeric drugs. The pumps and related devices are considered fully in Zaffaroni's paper and several examples of this system of approach are either clinically available or undergoing current clinical evaluation. The controlled release delivery system is considered in two papers (Carraher and Mason, et al.) and this system has also found clinical acceptance. The polymeric drug systems are considered in the first six papers of this section (Gebelein, et al; Pitha; Carraher; Brierly/Donaruma et al; Whiteley et al; and Pavlisko/Overberger) and this approach is still in the developmental stages although many promising polymeric drugs have been prepared. While all these approaches have many advantages compared to conventional drug administration methods, the polymeric drug system probably has the greatest potential for high specificity and long duration of activity. These six papers explore many different aspects of what may be one of the most important biomedical advances in the near future.

CHAPTER 11

[illegible]

[illegible]

POLYMERIC DRUGS CONTAINING 5-FLUOROURACIL AND/OR 6-METHYLTHIOPURINE. CHEMOTHERAPEUTIC POLYMERS. XI

Charles G. Gebelein, Richard M. Morgan (1a), Robert Glowacky (1b) and Waris Baig
Department of Chemistry

Youngstown State University
Youngstown, OH 44555

INTRODUCTION

The concept of polymeric drugs and their use in medication is much more recent than some of the other biomedical applications of polymers. In its simplest form, a polymeric drug is a polymer that contains a drug or chemotherapeutic unit as part of the polymer backbone, as a pendant group from the polymer chain or as a terminal group on the polymer chain. Hundreds of such compounds have been synthesized in the past few decades. In their more advanced forms, polymeric drugs could be complex copolymers of polymers attached to some natural substrate such as an imminoglobulin, a histone or a nucleic acid. Numerous reviews and books have appeared on this subject in the past decade (1-12). It would be beyond the scope of this present paper to attempt to review this rapidly developing field. We will concentrate primarily on polymers containing 5-fluorouracil or 6-methylthiopurine and some closely related polymers. Prior to this, however, we will examine the basic concepts involved in polymeric drug medication.

POLYMERIC DRUG MEDICATION THEORY

In medication the objective is usually to treat a specific diseased organ or tissue with a drug agent to affect a cure of the ailment. In actual practice, however, the normal situation involves the drug permeating throughout the entire body and thereby having an effect on many other parts of the body. More often than not, these extraneous interactions lead to undesired side effects such as nausea, dizziness, loss of hair, loss of appetite, skin discoloration, etc. and can even exacerbate the diseased condition of the patient. The basic situation in this conventional medication is diagrammed in Figure 1. The drug or therapeutic agent is introduced into the body by injection and/or oral administration and enters into the body fluids which transport this drug to essentially all parts of the body. While some of the drug does reach the disease target and treat it, most of the drug does not and instead interacts with the other parts of the body resulting in adverse side effects. If it were desired to place a 1.0 mg drug dose at a 500 g disease site in a 70 Kg person, the total drug dose would have to be an 140.0 mg size sample. This means that about 99.3% of this drug does not reach the target and interacts with other areas instead.

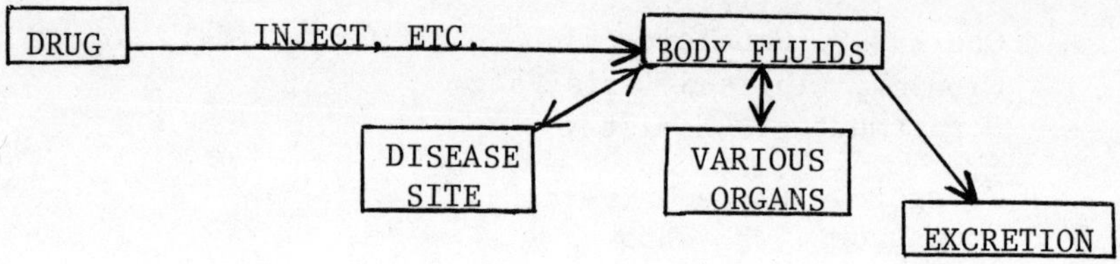
DRUG
INJECT, ETC.
BODY FLUIDS
DISEASE SITE
VARIOUS ORGANS
EXCRETION

Figure 1. Schematic diagram of medication using the conventional drug approach.

Although much research has been directed at making ordinary drugs more specific, there are definite limits to this approach. Most drugs have moderate to low molecular weights and any major change in a compound to change its selective absorption by a tissue or an organ would have a substantial effect on the molecule as a whole. (For example, the simple modification of attaching a carboxylic acid group to the aromatic ring of sulfanilamide would increase the molecular weight by about 25%.) It is not surprising that these changes are not always in the desired direction. While some drugs have been developed that are fairly selective in their migration in the body, it does not appear likely that this approach would be successful in general.

Several alternative medication routes have been proposed in recent years including various controlled or sustained release systems, mechanical mini-pumps and diffusion controlled systems and these have great potential for improving the effectiveness of medication in many cases. The basic mode of operation of these controlled release systems and devices is shown diagrammatically in Figure 2. In these cases, the controlling devise is placed in or near the organ to be treated and a small, relatively constant amount of drug is released reasonably directly to this organ. Only a small amount of drug would thus diffuse away from this organ to other parts of the body and this would help control the deleterious side effects. In some cases, this approach would require surgical implantation which would be a disadvantage. Nevertheless, these approaches can enable the drug level to remain in the therapeutic range for long periods of time and can reduce the toxic side effects of a drug (13-15).

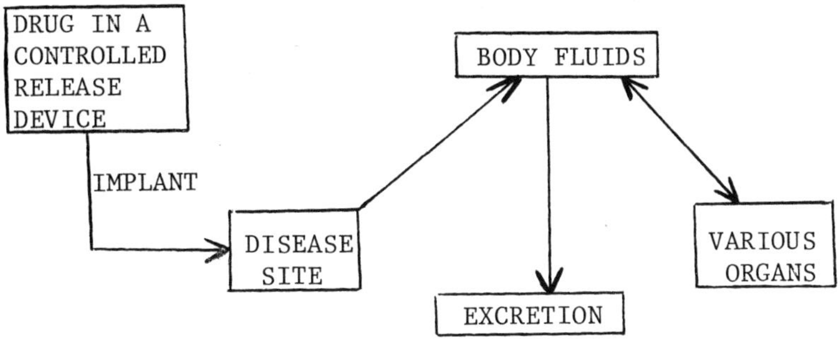
DRUG IN A CONTROLLED RELEASE DEVICE
IMPLANT
DISEASE SITE
BODY FLUIDS
EXCRETION
VARIOUS ORGANS

Figure 2. Schematic diagram of medication using controlled release systems and devices.

Polymeric drugs are another new approach to drug medication control. In this case the drug is attached to or part of a polymer chain. There are essentially three different types of polymeric drugs: (A) insoluble polymeric drugs, (B) soluble polymeric drugs and (C) directed polymeric drugs. A diagram of the mode or operation of these is shown in Figure 3. The primary assumption for all these systems is that the polymers themselves would exhibit biological activity. In some cases, however, polymeric drugs can function by releasing the drug unit to the environment in a manner similar to a controlled release system. The mode of operation of such polymeric drugs is more similar to the system shown in Figure 2 than to those outlined in Figure 3.

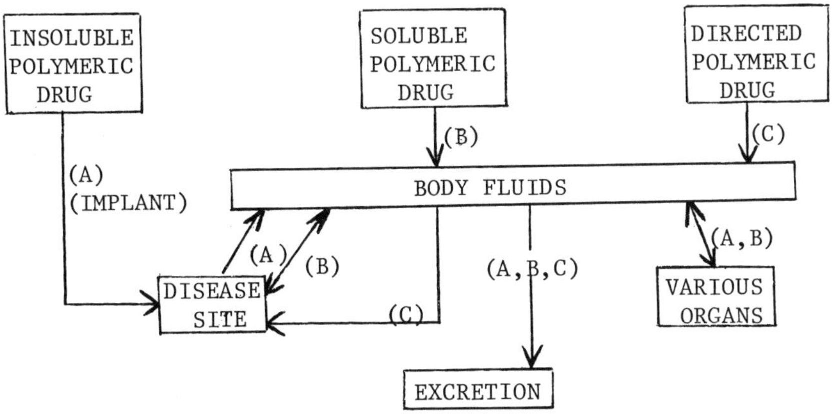
INSOLUBLE POLYMERIC DRUG
SOLUBLE POLYMERIC DRUG
DIRECTED POLYMERIC DRUG
(B)
(C)
(A)
(IMPLANT)
BODY FLUIDS
(A)
(B)
(A,B,C)
(A,B)
DISEASE SITE
(C)
VARIOUS ORGANS
EXCRETION

Figure 3. Schematic outline of polymeric drug medication methods showing the various routes (A, B, C) that each type would take through the body.

As noted in Figure 3, Route A, the insoluble polymeric drug would normally be implanted at, in or near the diseased organ or tissue and would function directly on the disease involved. Some of this material could diffuse to other organs, or be excreted, if this polymer had some low amount of solubility or underwent degradation in the body fluids. These are probably the simplest polymeric drug systems to synthesize and several such systems have exhibited biological activity (4, 9-11). The major drawback to these systems is that they would often require surgical implantation and would lack the ease of application of the other systems. They would, however, be fairly effective in reducing the side effects of the drug and could exhibit long duration of activity. These could be of great value in treating readily accessible parts of the body but would be a less desirable choice for the internal organs or for widespread diseases.

The soluble polymeric drugs differ from the insoluble type in that they are soluble in the body fluids. For this reason, these polymers could be administered orally or injected into the body and would not require surgery. These polymers could be homopolymers, such as poly(acrylic acid) or sodium poly(vinyl sulfonate) which show some antineoplastic activity, or they could be a copolymer which contains a drug unit and a second monomer whose primary function is to confer solubility on the copolymer system in the body fluids. These could be copolymers of a drug unit containing monomer with such comonomers as acrylic acid, N-vinylpyrrolidone, vinylamine oxides, vinyl sulfonate, vinyl imidazole or related monomers. Several examples of such systems have been prepared (5,6, 9-11).

The mode of operation of the simple soluble polymeric drugs is similar to the conventional medication therapy shown in Figure 1, Upon introduction into the body fluids this polymeric drug could go to other sites in the body as well as the diseased organ. The major difference between these and simple drugs would be the molecular weight. This could, however, have a pronounced effect on the biological activity. It has been shown that the antibacterial activity of amino acids increases markedly with the molecular weight of the polymer system (16). In addition, the higher molecular weight could increase the duration of action by (a) causing a high local concentration of a drug and (b) reducing excretion. Side effects might also be reduced since the polymer might not be able to cross some membrane barriers in the body. (Where the polymer could not diffuse into a cell through the cell wall, it could still be placed inside by the process of endocytosis.) While many potentially interesting soluble polymeric drugs have been synthesized and while some may prove to have practical utility, this approach is not the ultimate in polymeric drugs and it does suffer from the general lack of drug specificity which would be typical of the low molecular weight conventional drugs. They could, however, combine drug action with solubility in the body fluids which could be an advantage for some drug systems.

The directed polymeric drugs are a more complicated system which would consist of at least three components: (a) the therapeutic unit, (b) a solubilizing unit and (c) a directing unit. (In some cases these properties could be combined in one or two monomers.) The purpose of the directing unit is to guide the polymeric drug to a specific organ or type of diseased tissue. As shown in Figure 3, Route C, this type polymeric drug would not tend to enter other organs, etc. in the body and would have minimal side effects for this reason. This type polymeric drug would zero in on the target site like a guided missile and effect a cure of the disease without having much effect on the other body functions. The major difficulty is to develop suitable directing units. These directing units could be another monomer, such as a sulfonamide type (17) or a biological unit such as an enzyme, protein, nucleic acid or some related species (18,19). This approach has tremendous potential and promise and several research groups, including ours, are actively pursuing this line of attack. It is this author's opinion that this approach has the greatest potential for highly specific drug action with good duration of activity and lower toxicity and side effects. The method has yet to be fully proven in actual practice, however.

URACIL AND ADENINE DERIVATIVES

While there have been only a few papers published on 5-fluorouracil and/or 6-methylthiopurine based monomers, much work has been reported for uracil and adenine which could be considered the parent compounds. Much of this work has been reviewed recently and it is not necessary to repeat this here in detail (2,8,20,21). The monomers synthesized include 1-vinyluracil (22,23), 9-vinyl-adenine (22,23), 1-(N-2-methacryloyloxyethyl)uracil (24,25), 9-(N-2-methacryloyloxyethyl)adenine (24,25), styrene derivatives of uracil (26,27) and adenine (26,27), 1-(N-vinylcarbamoyl)uracil (28) and various amino acid derivatives of uracil and adenine (29-31).

The vinyl derivatives of uracil and adenine have been studied more extensively than the other monomers. While 1-vinyluracil was found to undergo cyclopolymerization in some initial studies (32), subsequent research was able to prepare linear polymers which were water soluble (33,34). The 9-vinyladenine monomer readily polymerized on free radical initiation to yield water soluble polymers (22,35,36). The solution properties of these systems have been studied and they form complexes with each other in a manner similar to naturally occurring nucleic acids (20,21). These polymers do exhibit some biological activity including the inhibition of murine leukemia virus (37,38) and *E. coli* RNA polymerase (39).

5-FLUOROURACIL MONOMERS AND POLYMERS

The first polymer containing 5-fluorouracil (5-FU) appears to be that of Ballweg, et al (40) which is shown below as (I). This polymer contains the 5-FU as part of the backbone chain and was reported to exhibit biological activity (40). No further work appears to have been reported on this compound since 1969, however.

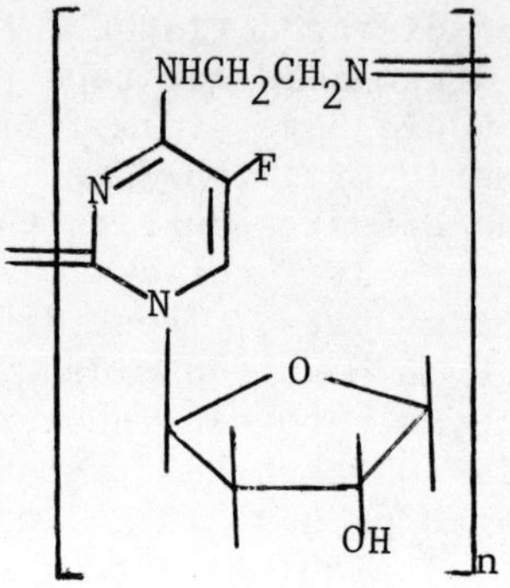

NHCH2CH2N
N
F
N
O
OH
n

(I)

The first reported vinyl-type monomers of 5-fluorouracil appear to be the carbamoyl derivatives which were made as shown in Equation (1). The vinylcarbamoyl derivative (IIa) has been obtained in 42% yield and readily polymerizes under radical initiation. The polymer is active against P388 leukemia but it is not certain whether this is due to activity of the polymer or to a slow release of 5-FU by this polymer in an aqueous system (41-43). The isopropenylcarbamoyl derivative (IIb) also polymerized under radical conditions but not as well as the vinylcarbamoyl derivative. The allylcarbamoyl derivative (IIc) does not appear to polymerize or copolymerize (41).

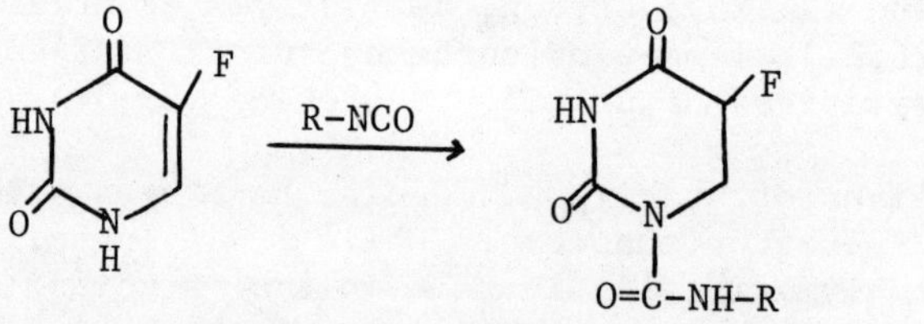

O
F
HN
O
N
H
R-NCO
O
HN
F
O
N
O=C-NH-R

(Equation 1)

(II)

(IIa) R = $-CH{=}CH_2$

(IIb) R = $-C(CH_3){=}CH_2$

(IIc) R = $-CH_2CH{=}CH_2$

Attempts were made by our group to synthesize the acryloyl (III) and vinyl (IV) derivatives of 5-fluorouracil. The acryloyl derivative (III) does appear to form when 5-FU is reacted with acryloyl chloride but this monomer appears to hydrolyze too rapidly to be considered to be of value in a polymeric drug (44). Many attempts were made to prepare the vinyl derivative (IV) via the vinylation reaction (45) under a wide variety of reaction conditions but this was not successful (46).

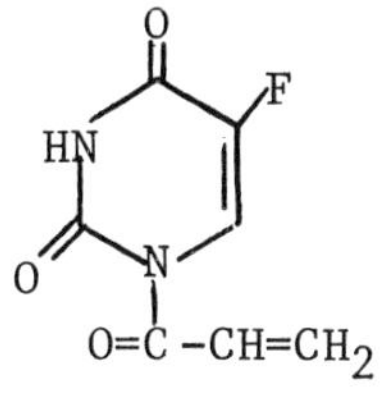
O
F
HN
O
N
O=C -CH=CH2

(III)

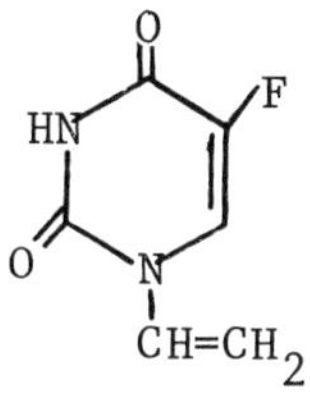
O
F
HN
O
N
CH=CH2

(IV)

Copolymerization of the vinylcarbamoyl derivative of 5-FU (IIa) is currently being studied with various solubilizing and potential directing group monomers. The homopolymer of (IIa) would be an example of an insoluble polymeric drug (Figure 3) or possibly a slow release polymer system (Figure 2). The ultimate goal of this research is to develop directed polymeric drugs for use in treating cancer and other diseases.

Recently Butler, et al have prepared another vinyl-type monomer containing 5-fluorouracil by the reaction of 5-FU with methyl fumaroyl chloride and this is shown below as (V). This monomer hydrolyzes rapidly in water but many of the copolymers hydrolyze more slowly and these polymers show promise as an anti-tumor system (47). Controlled release of 5-FU from various polymeric matrices has also been reported recently (48).

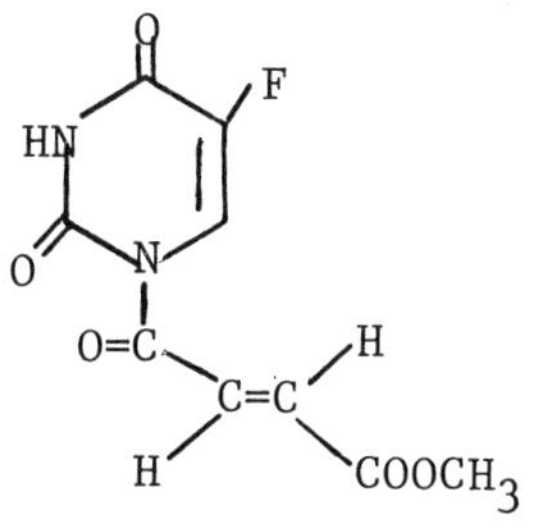
O
F
HN
O
N
O=C
H
C=C
H
COOCH3

(V)

6-METHYLTHIOPURINE MONOMERS AND POLYMERS

The earliest reported example of a polymer containing 6-mercaptopurine or one of its 6-alkyl derivatives is the work of Seita et al in which 6-methylthiopurine was attached as a pendant group to a poly(vinyl alcohol). This structure is shown below as (VI). While no biological data was reported, the water soluble polymer does appear to form complexes similar to the natural nucleic acids (49). Hoffmann et al (50) have prepared the 9-vinyl derivative of various 6-alkylthiopurines (including the methyl derivative) via the vinylation reaction (45) and this structure is shown below as (VII). These monomers do polymerize under free radical conditions but no biological data have been reported.

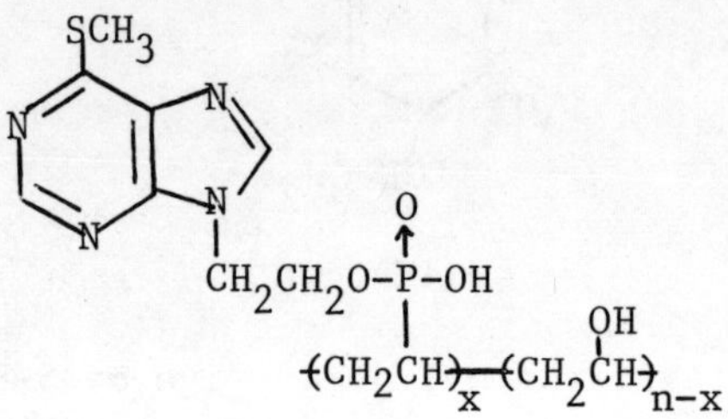

(VI)

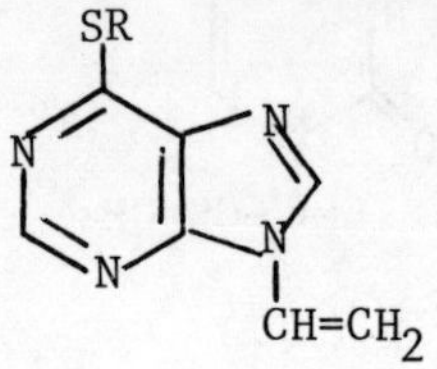

(VII)

The acryloyl derivative of 6-methylthiopurine has been prepared as shown in Equation 2. This derivative does hydrolyze fairly readily but not as rapidly as the corresponding derivative of 5-FU (III) and this monomer can be isolated and polymerized by free radical initiation (51). The carbamoyl derivatives, which can be prepared as shown in Equation 3, are much more stable to hydrolysis and have been studied in some detail (43,51,52). The allylcarbamoyl derivative (IXb) does not appear to homo- or copolymerize but the vinylcarbamoyl derivative (IXa) readily polymerizes under free radical conditions. Monomer (IXa) can be prepared in greater than 60% yield. Copolymerization studies are in progress on (IXa) with various solubilizing monomers and some potential directing groups. The vinylcarbamoyl derivative of 6-methylthiopurine (IXa) does undergo slow hydrolysis in water but less rapidly than the corresponding derivative of 5-fluorouracil (IIa).

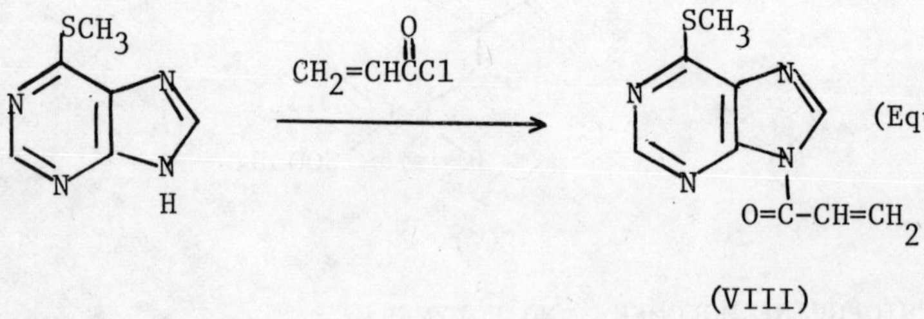

(Equation 2)

(VIII)

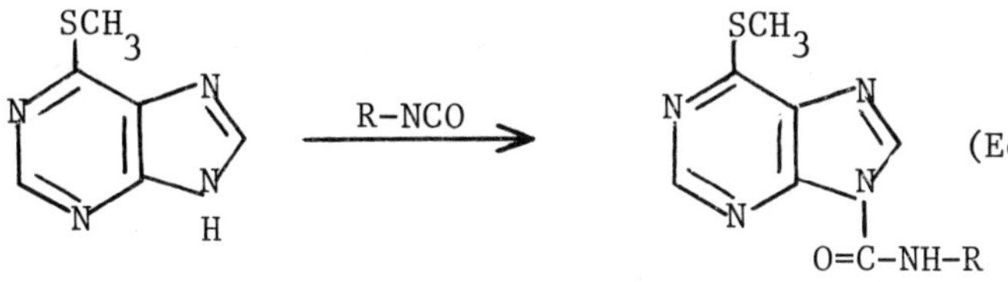
SCH3
R-NCO
SCH3
O=C-NH-R

(Equation 3)

(IX)

(IXa) R = $-CH{=}CH_2$

(IXb) R = $-CH_2CH{=}CH_2$

CONCLUSIONS

Several monomers and polymers have been prepared containing either 5-fluorouracil or 6-methylthiopurine and these materials could be potential antineoplastic polymeric drugs. In several cases anti-tumor activity has been observed with these monomers and/or polymers. The most useful monomers appear to be the vinylcarbamoyl derivatives of 5-fluorouracil (IIa) and 6-methylthiopurine (IXa) but monomers with greater hydrolytic stability may be more desirable for longer duration polymeric drug systems.

ABSTRACT

The basic theory of polymeric drug design and function was discussed with special reference to possible polymeric drugs containing 5-fluorouracil and/or 6-methylthiopurine.

ACKNOWLEDGMENT

This work was abstracted from the Master's Theses of R. Glowacky, R. M. Morgan and M. W. Baig. The research was supported, in part, by a grant from the Youngstown State University Research Council.

REFERENCES

1. (a) Present Address: Dow Chemical Co., Plaquemine, LA
   (b) Present Address: Environmental Protection Agency, Chicago, IL
2. K. Takemoto, J. Macromol. Sci.-Chem., C5, 29 (1970)
3. C. Schuerch, Adv. Polymer Sci., 10, 173, Springer-Verlag, New York (1972)
4. L. G. Donaruma, Prog. Polym. Sci., 4, 1, Pergamon Press, New York (1974)
5. H. Ringsdorf, J. Polymer Sci., Symp. 51, 135 (1975)
6. D. S. Breslow, Pure & Appl. Chem., 46, 103 (1976)
7. V. A. Kropachev, Pure & Appl. Chem., 48, 355 (1976)
8. K. Takemoto, J. Polymer Sci., Symp. 55, 105 (1976)
9. H.-G. Batz, Adv. Polymer Sci., 23, 25, Springer-Verlag, New York

(1977)
10. L. G. Donaruma & O. Vogl, "Polymer Drugs", Academic Press, New York (1977)
11. C. M. Samour, Chemtech, 8, 494 (1978)
12. C. G. Gebelein, Polymer News, 4, 163 (1978)
13. D. R. Cowsar in "Polymers in Medicine and Surgery", R. L. Kronenthal, Z. Oser & E. Martin, editors, Plenum Press, New York (1975), p. 237
14. S. Yolles in "Polymers in Medicine and Surgery", ibid., p. 245
15. A. Zaffaroni, Chemtech, 6, 756 (1976)
16. E. Katchalski, L. Bichowski-Slomnitzki & B. F. Volcani, J. Biochem., 55, 671 (1953)
17. J. Bartulin, M. Przybylski, H. Ringsdorf & H. Ritter, Makromol. Chem., 175, 1007 (1974)
18. E. P. Goldberg in Ref. 10, p. 239
19. J. M. Whiteley, Org. Coatings & Plastics Chem., 42, 529 (1980)
20. K. Takemoto in Ref. 10, p. 103
21. J. Pitha, Polymer, 18, 425 (1977)
22. N. Ueda, K. Kondo, M. Kono, K. Takemoto & M. Imoto, Makromol. Chem., 120, 13 (1968)
23. H. Kaye, Polymer Letters, 7, 1 (1969)
24. K. Kondo, H. Iwasaki, N. Ueda, K. Takemoto & M. Imoto, Makromol. Chem., 120, 21 (1968)
25. M. Akashi, Y. Kita, Y. Inaki & K. Takemoto, Makromol. Chem., 178, 1211 (1977)
26. K. Kondo, T. Sato, Y. Inaki & K. Takemoto, Makromol. Chem., 176, 3505 (1975)
27. K. Kondo, Y. Ohbe & K. Takemoto, Makromol. Chem., 177, 3461 (1976)
28. C. G. Gebelein, R. M. Morgan & R. Glowacky, Polymer Preprints, 18 (2), 513 (1977)
29. M. T. Doel, A. S. Jones & N. Taylor, Tetrahedron Letters, 1969, 2285
30. K. Takemoto, H. Tahara, A. Yamada, Y. Inaki & N. Ueda, Makromol. Chem., 169, 327 (1973)
31. M. Draminski & J. Pitha, Makromol. Chem., 179, 2195 (1978)
32. H. Kaye, Macromolecules, 4, 147 (1971)
33. J. Pitha, P. M. Pitha & P. O. P. Tso, Biochim. Biophys. Acta, 204, 39 (1970)
34. H. Kaye & S. H. Chang, Macromolecules, 5, 397 (1972)
35. J. Pitha, P. M. Pitha & E. Stuart, Biochem., 10, 4595 (1971)
36. H. Kaye, J. Am. Chem. Soc., 92, 5777 (1970)
37. P. M. Pitha, N. M. Teich, D. R. Lowy & J. Pitha, Proc. Nat. Acad. Sci., USA, 70, 1204 (1973)
38. V. E. Vengris, P. M. Pitha, L. L. Sensenbrenner & J. Pitha, Mol. Pharmacol., 14, 271 (1978)
39. H. J. Chou, J. P. Froehlich & J. Pitha, Nucl. Acids Res., 5, 691 (1978)
40. H. Ballweg, D. Schmael & E. von Wedelstaedt, Arzneim. Forsch., 19 (8), 1296 (1969)
41. C. G. Gebelein & R. M. Morgan, Polymer Preprints, 18 (1), 811

(1977)
42. C. G. Gebelein & T. M. Ryan, Polymer Preprints, 19 (1), 538 (1978)
43. C. G. Gebelein, Org. Coatings & Plastics Chem., 42, 422 (1980)
44. C. G. Gebelein & R. Glowacky, unpublished results
45. H. Hopft, Y. Wyss & H. Lussi, Helv. Chim. Acta, 43, 135 (1960)
46. C. G. Gebelein & T. M. Ryan, unpublished results
47. P. P. Umrigar, S. Ohashi & G. B. Butler, J. Polymer Sci., Chem. Ed., 17, 351 (1979)
48. M. Yoshida, M. Kumakura & I. Kaetsu, Polymer J., 11, 775 (1979)
49. T. Seita, M. Kinoshita & M. Imoto, J. Macromol. Sci.-Chem., A7, 1297 (1973)
50. S. Hoffmann, W. Witkowski & H. Schubert, Z. Chem., 14, 14 (1974)
51. C. G. Gebelein & R. Glowacky, Polymer Preprints, 18(1), 806 (1977)
52. C. G. Gebelein & M. W. Baig, Polymer Preprints, 19(1), 543 (1978)

[illegible]
[illegible]
[illegible]
[illegible]
[illegible]
[illegible]
[illegible]
[illegible]
[illegible]
[illegible]
[illegible]
[illegible]
[illegible]

POLYMERIC DRUGS: EFFECTS OF POLYVINYL ANALOGS OF NUCLEIC ACIDS ON CELLS, ANIMALS AND THEIR VIRAL INFECTIONS

Josef Pitha

Section on Macromolecules
Laboratory of Cellular and Molecular Biology
National Institutes of Health
National Institute on Aging, GRC-Baltimore City Hospitals
Baltimore, Maryland 21224

INTRODUCTION

Polymers interact with cells and are distributed in tissues in a manner different from the majority of small molecular weight compounds. These differences may possibly be used in the design and synthesis of new drugs with lower toxicities and diminished side effects. For the successful design of polymeric drugs, in addition to good fortune, it is necessary to acquire a basic understanding of the fate of synthetic polymers in tissues and in the body. This is not an easy undertaking and since there has not been a commercially successful polymeric drug developed, material incentives are minimal. Various classes of polymers have been tested in living systems but there have not been many attempts to investigate their bio-effects systematically. Polyvinyl analogs of nucleic acids are one of the notable exceptions. These polymers were originally synthesized as a tool for the study of interactions on a molecular level, i.e., with nucleic acids and enzymes, but eventually their interactions with cells and tissues were also studied. The interactions of these polymers on the molecular level have been reviewed previously (1). This review contains a list of papers published on these compounds; notably the work of Japanese and German chemists contributing considerably to the understanding of the basic chemistry of these compounds. Presently, after a short summary of the interactions of these compounds on the molecular level, their interactions with cells and tissues will be reviewed.

Summary of Interactions of Polyvinyl Analogs of Nucleic Acids on the Molecular Level

On the molecular level polyvinyl analogs of nucleic acids are capable of nearly all base pair mediated interactions which are known in the field of nucleic acids. In base pairing two heterocyclic moieties form a complex through specific hydrogen binding as illustrated in figure 1. Uracil or thymine residues interact with adenine; cytosine residues interact with guanine or hypoxanthine. Poly-1-vinyluracil and poly-9-vinyladenine (figure 2), which are well water soluble, interact in the same manner with nucleic acids when these contain serially arranged bases capable of base-pairing with the polymers. Nucleic acids carry genetic information and serve as templates for the synthesis of new copies of nucleic acids or as templates for the synthesis of proteins. Polynucleotides, which have the same structure as nucleic acids but contain only one kind of base, can also serve as efficient templates. The templating activity of polynucleotides is completely abolished when these compounds are complexed with polyvinyl analogs carrying bases complementary to those of polynucleotides. This was the result of numerous experiments with various polymerase (i.e., enzymes which synthesize nucleic acids) and

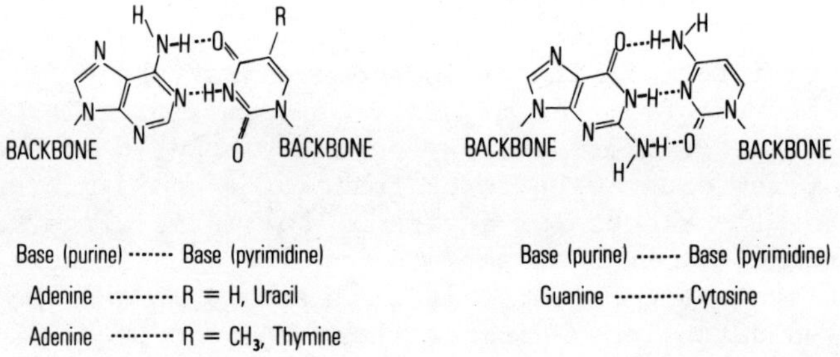
BACKBONE
BACKBONE
BACKBONE
BACKBONE
Base (purine) ······· Base (pyrimidine)
Adenine ·········· R = H, Uracil
Adenine ·········· R = CH3, Thymine
Base (purine) ······· Base (pyrimidine)
Guanine ·········· Cytosine

Figure 1. Formation of base pairs of Watson-Crick type.

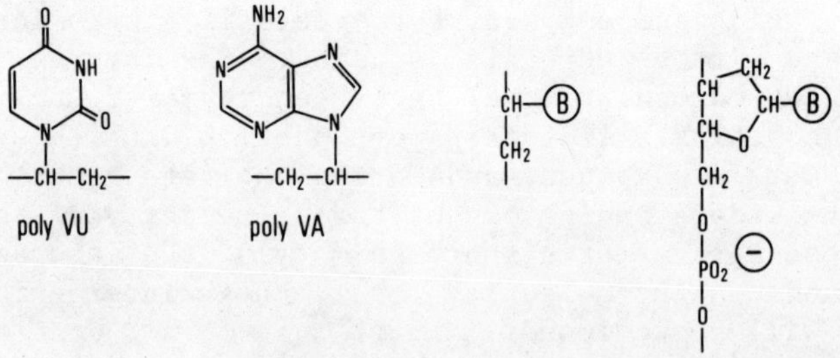
poly VU
poly VA

Figure 2. Formulas of poly-9-vinyladenine (poly vA) and of poly-1-vinyluracil (poly vU) on the left side of the figure. Backbones of polyvinyl polymer and of nucleic acid are compared on the right.

with protein synthesizing systems of bacterial or eukaryotic origin. It is important to realize that when there is no complexing between a polynucleotide and polyvinyl analog, the presence of the analog does not influence enzymatic templated synthesis. A closer investigation of the mechanism of inhibition leads to the conclusion that the enzyme stays immobilized on the template-inhibitor complex rather than being released into the solution. These findings are schematically summarized in figure 3. It may be noted that polyvinyl analogs are capable of some interaction with proteins; the interactions which were detected were nevertheless rather weak and non-specific. Thus, for example, the interaction of poly-9-vinyladenine with an enzyme which processes nucleic acids (reverse transcriptase) could be abolished by the addition of a protein which does not have any known connection with nucleic acid synthesis or metabolism (bovine serum albumin). The studies of the interactions of polyvinyl analogs on a molecular level thus lead to a simple conclusion that these compounds may be potent inhibitors of biosyntheses requiring nucleic acid as templates and such inhibitions occur in a template-specific way.

Polymers and Cells in Culture

Cells grown in culture are the simplest models available for the study of interactions of polymers with living matter. Even then there are two remarkable complications. The cell, from the point of polymer penetration, is a highly compartmentalized system and the polymer distributes in a very uneven way. Thus there are cellular

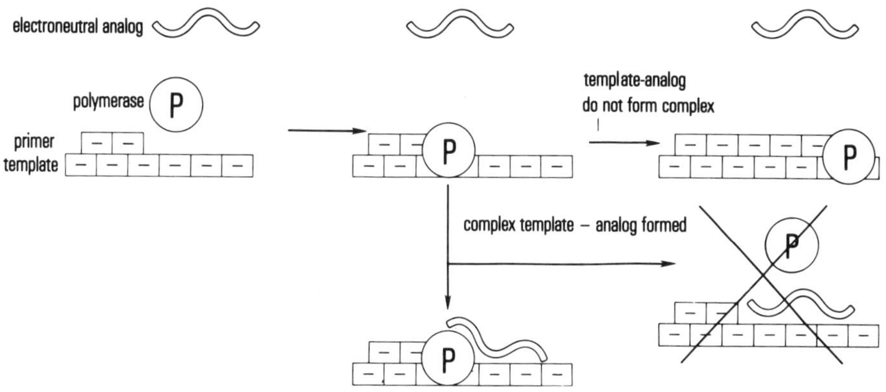
electroneutral analog
polymerase
P
primer
template
template-analog
do not form complex
complex template – analog formed

Figure 3. Scheme summarizing the effects of polyvinyl analogs of nucleic acids on a polymerase (P) using polynucleotide template. When the polyvinyl analog forms a complex with the template the enzyme is immobilized by the complex formed and enzymatic activity is inhibited. When no complexing occurs the enzyme is not inhibited.

spaces which are completely free of polymer and consequently free of any inhibition which the polymer may exert. The second complication stems from an enormous variety of cellular nucleic acids which differ considerably with respect to both the sequence of bases and the structure. Few short, simple sequences are known to occur in natural nucleic acids. For example, polyadenylate sequences are known to occur in many eukaryotic messenger ribonucleic acids, i.e., those nucleic acids which serve as templates for the synthesis of proteins. Furthermore there are several oligothymidylate sequences in eukaryotic deoxyribonucleic acids and a cytidylate sequence in some viral nucleic acids. These segments, in spite of being part of nucleic acids, probably do not carry any immediate genetic information but serve for organizational and regulatory purposes. Thus the presently available polyvinyl analogs cannot be expected to block directly any immediate genetic message; maximally they can interfere in some regulatory processes governing the replication or translation of nucleic acids.

The interior of a cell is separated from its immediate surrounding by a plasma membrane. This complex organ has an overall electronegative charge due to the anionic character of the polysaccharides and proteins located there. The main element barring the polymer from penetrating the cell interior is the lipid bilayer into which

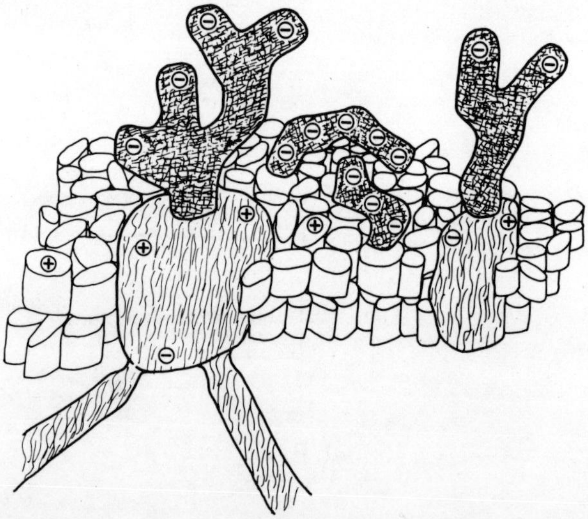

Figure 4. Schematic representation of cytoplasmic membrane, a membrane which surrounds cells and separates the extracellular (on the top) and intracellular (bottom) spaces. Lipids are depicted in light color, proteins are shaded and saccharide residues are crisscrossed.

the proteins and polysaccharide components of the plasma membrane are inserted. This situation is schematically illustrated in figure 4. Synthetic compounds may penetrate into the cytoplasm of cells by phase-phase transfer. From the external aqueous media they are extracted into the non-polar phase of the lipid bilayer and from there they again equilibriate with the aqueous interior of the cell. For this mode of penetration to be efficient a compound must distribute both into the aqueous and non-polar phases. Poly-1-vinyluracil and poly-9-vinyladenine, similarly to other polymers, distribute very unevenly. These polymers dissolve well in water but cannot be extracted into the non-polar media (e.g., n-octanol) and thus this mode of entry into cells is not available to them. In addition to this lack of penetration through the membrane, there is also an absence of strong interaction of such polymers with the membrane as well. Electroneutral polymers do not interact strongly with polysaccharides, proteins or lipids of the membrane and thus there is only relatively weak adsorption of either poly-1-vinyluracil and poly-9-vinyladenine to the surface of various cells in culture. Using highly radioactive polymers in a serum-free media it was estimated that approximate saturation values were around 10 and 1 pg per one fibroblast cell respectively and binding was just about saturated in 30-60 minutes at room temperature (2). In their interactions with polyvinyl analogs both murine and human cells of the fibroblast type show close similarity (2).

The surface of cells is constantly renewed; the plasma membrane is internalized in the endocytotic process (3) which is illustrated in figure 5. During this process the polymer which was adsorbed on

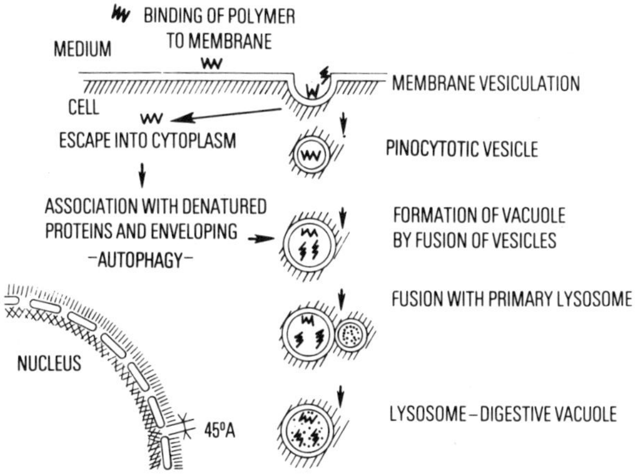
BINDING OF POLYMER
TO MEMBRANE
MEDIUM
MEMBRANE VESICULATION
CELL
ESCAPE INTO CYTOPLASM
PINOCYTOTIC VESICLE
ASSOCIATION WITH DENATURED
PROTEINS AND ENVELOPING
-AUTOPHAGY-
FORMATION OF VACUOLE
BY FUSION OF VESICLES
FUSION WITH PRIMARY LYSOSOME
NUCLEUS
45°A
LYSOSOME-DIGESTIVE VACUOLE

Figure 5. Scheme of endocytosis of foreign macromolecules by cells.

the cell surface and a small portion of the polymer in the surrounding medium enter into cells while fully enclosed in endocytotic vesicles. The internalized polymer stays firmly associated with the cells. When the internalized poly-9-vinyladenine was followed for three cell divisions the amount correlated with the number of viable cells within the experimental error; thus no degradation or excretion from cells could have occurred. The cells containing radioactive polymer were also fractionated and distribution of radioactivity in all these fractions was measured. Results confirm that a substantial amount of the polymer is in an enveloped form, probably lysosomes (2).

The polyvinyl analogs influence the cells grown in culture in an interesting way. Poly-1-vinyluracil at a concentration of 1 mM is toxic to cells that are actively growing and dividing but non-toxic to stationary cultures, i.e., to cells that are not actively dividing. The toxicity of poly-9-vinyladenine is very low and cells can be grown at 10-15 mM concentration of polymer (4).

Polymers and Viral Replication

Polyvinyl analogs of nucleic acids influence replication of some viruses in cells. Poly-9-vinyladenine effectively inhibits replication of murine leukemia virus in mouse cells (4). This virus has an interesting life cycle. The virions of murine leukemia virus contain ribonucleic acid as a genetic material; this material, after infection of cells by virus, is copied into deoxyribonucleic acid (a process called reverse transcription), which is then incorporated into the nucleus of host cells. From here again the genetic information of the virus is transcribed into ribonucleic acid, translated into proteins and new viral particles are formed. Certain mouse cells carry viral deoxyribonucleic acid already in their genome and may be induced by a number of chemicals to produce viral progeny. Poly-9-vinyladenine did not inhibit this induction (5); consequently it may be assumed that the polymer probably inhibits some early event in viral replication, probably the reverse transcription step, which would be in agreement with the results of experiments on this enzyme (6). The inhibition is virus specific, since replication of vesicular stomatitis virus in the same cells was not inhibited (4). A similar compound, poly-9-vinylpurine, also inhibits replication of murine leukemia virus (7). With this polymer it was established that the adsorption and binding of virus particles to cells was not inhibited, but increased in the presence of the polymer. Thus viral replication must be inhibited in some subsequent step in the viral replicative cycle. Still another purine containing polymer, poly-9-vinyl-6-dimethylaminopurine, was investigated; this compound had no effect on the replication of murine leukemia virus. It is interesting to note that this polymer also does not inhibit the reverse transcriptase, thus the ability of the polymer to inhibit the viral enzyme coincides with its potency inhibiting viral replication, a coincidence which suggests that the virus replication is indeed inhibited in this step (7).

The observed selective toxicities (i.e., to virus compared to cells), when considered together with the results obtained on the effects of polyvinyl analogs on individual polymerases, present an intriguing picture. Polyvinyl analogs are template-specific rather than polymerase-specific inhibitors; all tested cellular and viral polymerases were inhibited to about the same extent (8). Thus, polyvinyl analogs in cell free systems do not distinguish between cellular and viral enzymes. Nevertheless, in cellular systems recognition occurs. This is probably due to the uneven distribution of the polymer in the cell. Viral replication occurs in the cytoplasm of cells while cellular replicative processes occur in the cell nucleus. Nuclear materials are isolated from the cytoplasm by an additional membrane which has limited permeability; it has pores the size of which were estimated, using macromolecular probes, to be 45 Å (figure 5). It is important to note that we do not know how some fractions of the polymer get into the cytoplasm. Polymers, as already mentioned, are unable to penetrate the lipid bilayer to the cytoplasm. Endocytosis leads to the uptake of the polymer into the cytoplasm, but the polymer is in an enveloped form (figure 5) and thus it stays isolated from cytoplasmic processes. Perhaps during the closure of the endocytotic vesicles, when continuity of the membrane must be disrupted, a small fraction of the polymer penetrates into the cytoplasm. Another possible mechanism of entry of polymers into the cytoplasm may be through the creation of localized damage in the membrane.

The entry of the macromolecule into the cytoplasm is a process of very low efficiency but is known to occur with various materials. For example, viral infection may be achieved by exposing cells to some viral nucleic acid in the presence of a basic polymer or at high concentrations of calcium ions in the medium (9). In a latter part of this review we will present evidence to the point that foreign macromolecules free in the cytoplasm slowly undergo the autophagy process in which they are converted into an enveloped form (figure 5).

Replication of various viruses in cells may also be inhibited by exposure of cells to interferon. Interferon is a glycoprotein formed by cells themselves when they are exposed to specific inducers. The most effective synthetic inducers of interferon are found between nucleic acids and polynucleotides. Electroneutral polyvinyl analogs of nucleic acids are incapable of inducing interferon in cells in culture (10). On the other hand, the complex of polyvinyl-1-cytosine with a polynucleotide, polyinosinic acid, is a very effective inducer of interferon (11). A principal reason for this effectiveness may be a high uptake of this complex by cells (10).

Polymers and Cells in vivo

In studies on the effects of polymers on cells in culture only few human and mouse cell types were used. Organisms contain a very large variety of cells which differ sharply in their interaction with soluble macromolecules. Some cells, like macrophages and cells of the reticuloendothelial system have very active endocytosis. Organisms also have several supracellular barriers to polymer penetration, e.g., the barrier between the alimentary tract content and blood and the barrier between blood and brain; such barriers are practically impenetrable to polymers. The distribution of synthetic polymers in the organism has been done on the elementary level. In a majority of the polymer distributions studied we do not know to which actual concentration of the polymer the particular part of tissue was exposed. Furthermore various organs are treated as homogenous entities while in reality they are composed from various cells which differ in their interaction with polymers.

The injection route is the only effective one for introduction of vinyl polymers into the organism. Polymers do not penetrate the skin and the uptake of foreign macromolecules from food is negligible.

The response of cells in culture to exposure to a polymer and the response of an organism to an injected polymer are quite different. Cells in culture bind only very weakly the electroneutral polymers. Thus large amounts of highly radioactive polymers have to be used in binding studies and only a very small fraction of those stay associated with the cultured cells; the majority of unwanted foreign material remains in the solution. If the polymer is injected into the organism it is no longer a material which may be left outside, i.e., in the cell media; the injected polymer represents a circulating internal pollution which the organism must take care of. The injected polymer must be effectively and substantially cleared from circulation. Organisms are equipped to deal with such internal pollutions which in natural life come from injuries, infection and cell death, and handle water soluble polymers in about the same way.

The distribution of injected polyvinyl analogs was studied in mice (2). After injection into the peritoneum the polymers enter the bloodstream in a matter of hours and are subsequently distributed and partly excreted from the organism. About one third of the nondegraded polymers were excreted in the urine during the first two days. Simultaneously polymers are cleared from the bloodstream by endocytosing cells; thus the level of polymer there decreases quite rapidly (2). After the first two days the distribution of polyvinyl analogs in tissues changes only moderately but small changes were observed even after several weeks. Polymers concentrate in organs which have endocytosing cells; considerable amounts of polymers were found in liver, spleen, thymus and bone marrow. Some tissues, e.g., kidney and lungs, were found to clear themselves of the polymer, while it accumulated in the aforementioned tissues. The polymer concentration in the brain was always found very low and the amounts detected there were close to observation limits and were probably derived from the blood present in excised brain. These and other observations are summarized in figure 6.

Mammals apparently do not possess any enzymes capable of the degradation of vinyl polymers. Polyvinyl analogs from liver and spleen two weeks after the injection were compared with the original sample and distribution of molecular weights was found identical within the limits of experimental error (2).

Poly-9-vinyladenine was found non-toxic to mice when administered by intraperitoneal injection at all obtainable doses. Effects of this compound on the immunosystem, on viral leukemia, on chemically induced leukemia and on the infection by lytic virus were investigated in detail (12).

Poly-9-vinyladenine was found to have no influence on both humoral and cell mediated immunity. This finding contrasts with the strong influence of polynucleotides or anionic polymers on the immunosystem; these potentiate humoral responses and also influence macrophages (9).

Poly-9-vinyladenine inhibits replication of leukemia viruses in cells in culture; this polymer also accumulates and is stored in the spleen, the location where leukemia viruses replicate. Consequently rather strong and protracted antiviral effects had been expected. Different protocols of administration of the polymer and

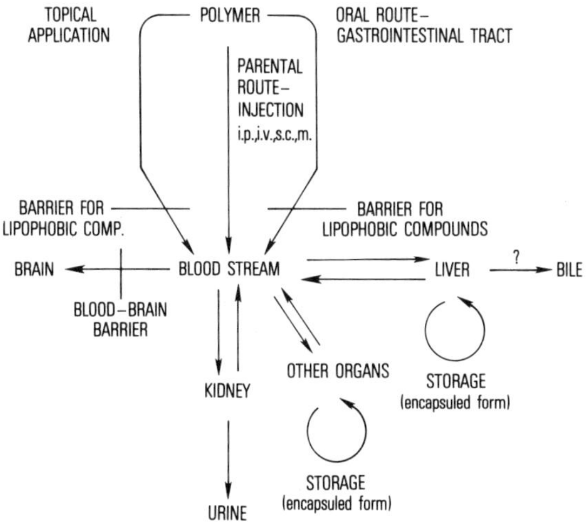
TOPICAL APPLICATION
POLYMER
ORAL ROUTE–GASTROINTESTINAL TRACT
PARENTAL ROUTE–INJECTION i.p.,i.v.,s.c.,m.
BARRIER FOR LIPOPHOBIC COMP.
BARRIER FOR LIPOPHOBIC COMPOUNDS
BRAIN
BLOOD STREAM
LIVER
?
BILE
BLOOD–BRAIN BARRIER
KIDNEY
OTHER ORGANS
STORAGE (encapsuled form)
STORAGE (encapsuled form)
URINE

Figure 6. Scheme of distribution fate of polyvinyl analogs of nucleic acids in mice.

virus were tested and results were more intriguing than impressive. The polymer had to be administered daily to suppress viral replication; large doses administered once only were ineffective (12). This interesting finding was explained in the following way. The polymer is taken by the spleen cells by endocytosis, the majority of it in enveloped form, thus it is isolated by a membrane from the cytoplasm where viral replication occurs (figure 5). But some fraction of the polymer which is taken up by cells enters the cytoplasm in a free, non-enveloped form and can there inhibit replication of the virus. Nevertheless after some time this free polymer is converted into a membrane-enveloped form; thus at the end all of the polymer is isolated from viral replication. There is no direct evidence for intracellular circulation of synthetic material, but the phenomenon in question, called autophage, is well known to occur with natural compounds. During cell life cytoplasmic proteins are constantly renewed, the old being enveloped into membranes, then transferred into lysosomes and there degraded to amino acids.

The removal of the polymer from the site of its antiviral activity to an insulated site is an interesting and unusual process leading to a loss of beneficial drug effects. Small molecular weight drugs are excreted or metabolically transformed into inactive compounds; a polymer which cannot be excreted nor degraded is converted into an inactive form by autophagy.

Poly-9-vinyladenine was found inactive in mice against transplanted leukemia of chemical origin. Also no inhibition of the replication of Semliki forest virus in mice was observed (12). This virus is a lytic one and contains, similarly to leukemia viruses, ribonucleic acid as a genetic material, but in difference to leukemia viruses the life cycle of this virus does include reverse transcription of ribonucleic acid into the corresponding deoxyribonucleic acid. These results suggest that the effects of poly-9-vinyladenine on the replication of leukemia viruses in animal may also be due to direct inhibition of reverse transcription, which is a vital and unique step in the life cycle of this particular virus group. These plausible agreements nevertheless do not form much of a basis for extrapolation as the following example indicates. Encephalomyocarditis virus contains ribonucleic acid as genetic material which is, during viral infection, replicated by viral polymerase into multiple copies. Poly-9-vinyladenine is a very effective inhibitor of this replicative process when it occurs in a cell-free system; nevertheless the viral infection/replication in mice is not affected by this polymer (13).

CONCLUSION

Relatively little work has been done on the biological effects of water soluble polymers (14-16). The polyvinyl analogs of nucleic acids described here and in previous reviews (1,17) together with polymers containing carboxylate groups (18-20) seems to be the most systematically investigated. There are some similarities between these groups of polymers in their distributions in various tissues of an organism and in their overall biological fates; they differ distinctly in their toxicities, effects on membranes and effects on the immunosystem. Hopefully systematical work on the biology of water soluble polymers will result in a mapping of all unknown territories and help in the development of better drugs.

REFERENCES

1. J. Pitha, M. Akashi & M. Draminski, in "Biomedical Polymers", E. P. Goldberg, ed., Academic Press, New York, in press (1980)
2. L. Noronha-Blob, V. E. Vengris, P. M. Pitha & J. Pitha, J. Med. Chem., 20, 356 (1977)
3. S. C. Silverstein, R. M. Steinman & Z. A. Cohn, Ann. Rev. Biochem., 46, 669 (1977)
4. P. M. Pitha, N. M. Teich, D. R. Lowy & J. Pitha, Proc. Natl. Acad. Sci. USA, 70 (4), 1204 (1973)
5. P. M. Pitha, J. Pitha & W. P. Rowe, Virology, 63, 568 (1975)
6. J. Pitha, K. Kociolek & C. A. Apffel, Cancer Res., 39, 170 (1979)
7. J. Pitha, S. H. Wilson & P. M. Pitha, Biochem. Biophys. Res. Commun., 81 (1), 217 (1978)
8. J. Pitha & S. H. Wilson, Nucleic Acids Res., 3 (3), 825 (1976)
9. J. Pitha, in "Polymers in Biology and Medicine", L. G. Donaruma, R. M. Ottenbrite & O. Vogl, eds., John Wiley & Sons, Inc. Publishers, New York, in press (1980)
10. L. Noronha-Blob & J. Pitha, Biochim. Biophys. Acta, 519, 285 (1978)
11. J. Pitha & P. M. Pitha, Science, 172, 1146 (1971)
12. V. E. Vengris, P. M. Pitha, L. L. Sensenbrenner & J. Pitha, Mol. Pharmacol., 14, 271 (1978)
13. A. G. Stewart, J. Pitha & N. Stebbing, unpublished results
14. L. G. Donaruma, Progr. Polym. Sci., 41, 1, (1974)
15. H. Ringsdorf, J. Polym. Sci. Polym. Symposium, 51, 135 (1975)
16. J. Kalal, Makromol. Chem. Suppl., 2, 215 (1979)
17. A. S. Jones, Int. J. Biolog. Macromolecules, 1, 194 (1979)
18. R. M. Ottenbrite, in "Polymers in Biology and Medicine", L. G. Donaruma, R. M. Ottenbrite & O. Vogl, eds., John Wiley & Sons, Inc. Publishers, New York, in press (1980)
19. W. Regelson, J. Polym. Sci. Polym. Symposium, 66, 483 (1979)
20. D. S. Breslow, Pure Appl. Chem., 46, 103 (1976)

ORGANOMETALLIC POLYMERS AS DRUGS AND DRUG DELIVERY SYSTEMS

Charles E. Carraher, Jr.

Department of Chemistry
Wright State University
Dayton, Ohio 45435

INTRODUCTION

The use of metal and organometallic containing polymers in medical applications is widespread focusing on siloxane polymers and to a lesser degree on polyphosphazenes (for instance 1). This work is focused on the use of these polymers as medical materials in applications such as biomedical implants in catheters, blood pumps and breasts.

Here we will concentrate on a new, emerging area - the use of organometallic polymers as drug delivery (controlled release or direct) agents. Two topics will be dealt with here. First, a general, brief description of some of the philosophical considerations involved with the use of organometallic polymers as drug delivery agents. Second, specific examples emphasizing the use of platinum containing polymers as antitumor agents, tin containing polymers in topical applications and arsenic containing polyamines, will be described.

The toxicity as well as therapeutic value of metals is well known. The interaction of metal ions with biological macromolecules such as proteins and nucleic acids is a continuing area of research. The appearance of metal containing macromolecules in the human body is extensive including the metals of iron (transferrin), molybdenum (xanthine oxidase), vanadium (hemovanadin), zinc (carbonic anhydrase) and copper (hepatocuprein). The use of organometallic medicinals is widespread and includes merbromine (mercurochrome; mercury), meralein (mercury; antiseptic), silver sulfadizine (prophylactic treatment for severe burns), arsphenamine (antimalarial; arsenic), 4-ureidophenylarsonic acid (therapy of ameblasis, tryparsamide (Gambian sleeping sickness; arsenic), and antimony dimercaptosuccinate (schistosome).

EXPERIMENTAL

The synthesis of a wide variety of metal containing polymers was accomplished utilizing either or both interfacial and solution techniques. Table 1 contains a list of polymers referred to in this paper along with other pertinent data. Where possible, syntheses were effected utilizing systems which resulted in the production of polymers within less than a minute reaction time. The combination of short reaction time and mild reaction conditions allows the use of a wide variety of potentially unstable reactants where thermal or solution induced rearrangements are possible.

For the antimony polypyrimidines, bacteria were chosen for the preliminary biological studies. Representative bacteria were chosen for study. Dimethylsulfoxide, DMSO, was chosen as the solvent since the polymers are soluble in it and DMSO is not appreciably toxic to man nor to most of the bacteria tested. The stock polymer solution was prepared by dissolving the polymer in DMSO and the solution filtered through a sterile Zeitz filter. Microorganisms were grown in a liquid medium of nutrient broth containing 10 grams of dehydrated nutrient broth per liter of distilled water. Nutrient agar was prepared as above except 1.5 grams of agar was added. Inhibition assays were conducted using sterile absorbent paper disks, 0.25 cm square, soaked in polymer solutions of designated concentrations. The disks were placed on agar plates which had been seeded with a lawn of the particular bacteria. Slant culture assays were conducted using standard saline solutions containing 0.9 ml of saline plus 0.1 ml cells and 0.8 ml saline, 0.1 ml cells and 0.1 ml DMSO. Test solutions were prepared using 0.8 ml saline, 0.1 ml cells and 0. 1 ml DMSO containing 2.75 to 0.171 ug polymer. Nutrient broth solutions were prepared as above except that nutrient broth replaced the saline.

Two types of tests were utilized in the evaluation of the modified cellulose products. The paper disk assays were conducted with each test organism suspended in sterile water (10^6 spores per ml). Appropriate dilutions were made in Sabourand's dextrose agar and poured into Petri plates. The best confluent growth resulted when employing 10^3 spores per ml which is the concentration employed for subsequent studies. Protein analysis of the samples suspended by continuous vibration were conducted. Cold perchloric acid was added to each culture tube after seven days of growth. Pellets were collected by centrifugation and subsequently suspended in NaOH. After autoclaving, the solubilized protein was determined using bovine serum albumin, fraction V, as the standard.

The effect of select platinum polyamines on viral replication in HeLa cells, human cancer cells, was studied as follows. The HeLa cells were treated with 5 ml of solution (polymer or nonpolymer containing). The cells were removed and infected with Polio Type 1 virus with subsequent incubation for 24 hrs. The virus was harvested from the resulting fluids.

Mice were injected with DMSO solutions containing varying amounts of platinum polyamine and observed for 24 hours. Injection occurred in the intraperitoneal cavity region of the mouse.

PHILOSOPHY

Desirable objectives in the design of chemotherapy drugs include the synthesis of materials which have a. good specificity, b. good activity, c. a long duration of activity and d. a wide concentration range of biological activity (i.e. if it is to act as a toxin, then its toxicity to the desired object as a tumor is great compared with its toxicity to healthy cells and organs in which it will come in contact with). Thus variability and specificity are key ingredients. Organometallic polymers, because of the variety of metal size, chemical environment and electronic structure, offer a wide variability in biological activity and the opportunity for good specificity. Thus far we have synthesized polymers containing the following metals: U, Pt, Mo, Ti, Zr, Hf, Fe, Co, Ru, Bi, Sb, As, Pb, Sn, Rh, Ge, Si and Mn. In theory these metals can be included within polymers through reaction with a wide variety of alkyl and/or aryl containing Lewis bases. For reaction to occur, the Lewis bases must be in what can be referred to as "active" or in an "active form". There should be a match between the number of "active" Lewis base sites such that the final structure of the product will be linear or crosslinked, depending on the particular intent. For instance, xanthene dyes typically exhibit three Lewis bases (two phenols and one carboxylic acid) but through a wide pH range (often from about 2 through 13) only two of these Lewis bases is active forming "polydyes" of form 1 from condensation with Group IV A and B organodihalides (1).

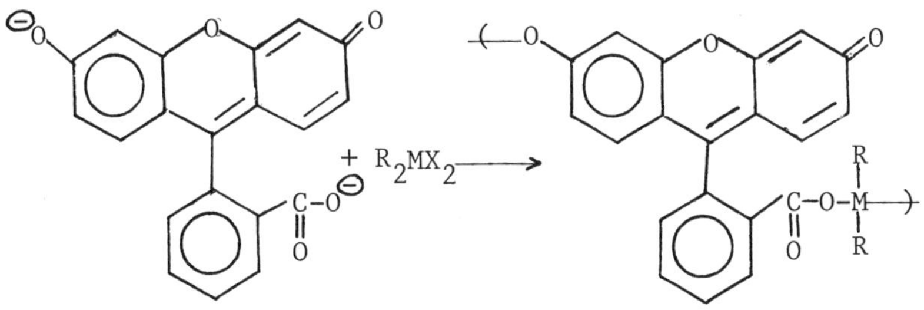
+ R_2MX_2

I

Organometallic dihalides are typically more reactive than general organic diacid halides and will undergo certain condensations for which analogous products are not formed through attempted condensation with acid chlorides utilizing mild conditions. For instance a number of organometallic dihalides have been successfully condensed with salts of dicarboxylic acids forming organometallic polyesters yet the analogous reaction with diacid chlorides producing the analogous polyanhydrides does not occur under similar reaction conditions. Thus reactions employing organometallic halides can be utilized in condensations employing potentially useful drugs which contain one or two acid groups whereas analogous reactions with acid chlorides do not occur.

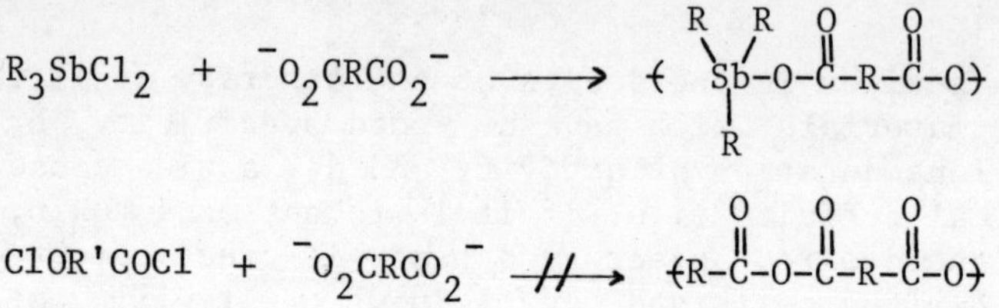

The placement and chemical environment of the drug varies as to intended use. Permanent attachment can occur through use of highly polar attachments which contain highly hydrophobic substituents, through attachment by slightly or nonpolar bonding, through attachment at a highly sterically hindered site, or any combination of the afore. Temporary attachment of the drug results from attachment at a polar site containing little or no hydrophobic nature, etc. Rate of drug release can be controlled by varying the above noted factors. Organometallic polymers offer a wide variability of bonding kind and the potential to couple desired activity with necessary "performance".

Desirability of the location of the drug also will vary as to its intended use. For instance the drug may be an integral part of the polymeric backbone as in II or within a side chain near the backbone III or within a side chain removed from the backbone IV (2,3).

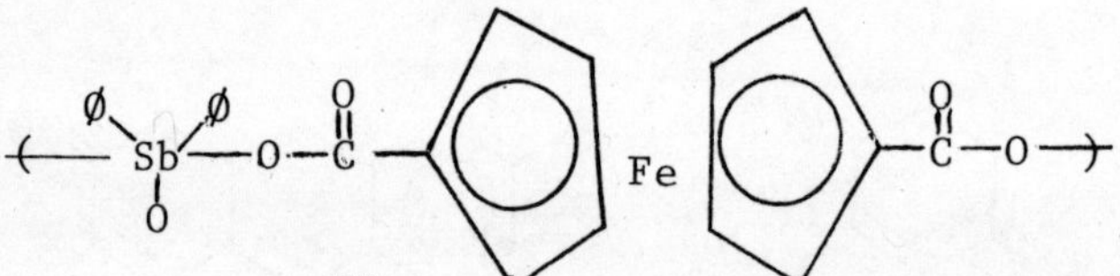

II

III

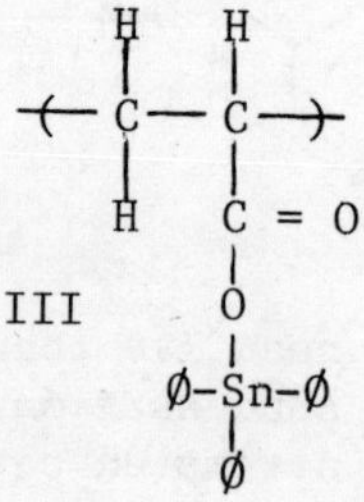

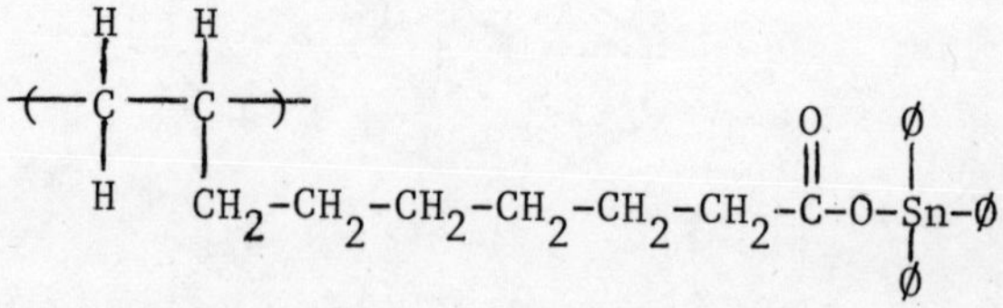

IV

Again, as noted for forms II, III and IV organometallic polymers offer a wide variety of delivery sites.

We have been actively including metal-containing moieties into polymers for a number of reasons including use of such compounds as drug delivery agents. The desired portion to be delivered (i.e. the drug) may be either or both the metal containing moiety or the organic comonomer. For instance, our purpose in the synthesis of a number of Group IVB polymers containing steroids and derivatives of vitamin K, is to use such polymers as delivery agents for the nonmetal comonomer portion utilizing the relatively biologically inactive Group IV B Cp_2M moiety (4).

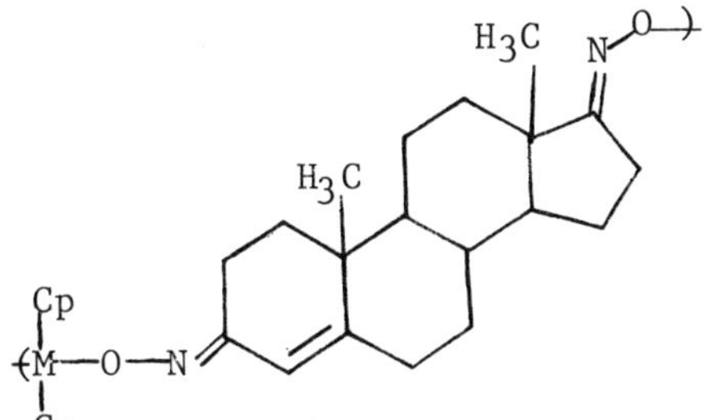
H3C
N
O
H3C
Cp
M
O
N
Cp

V

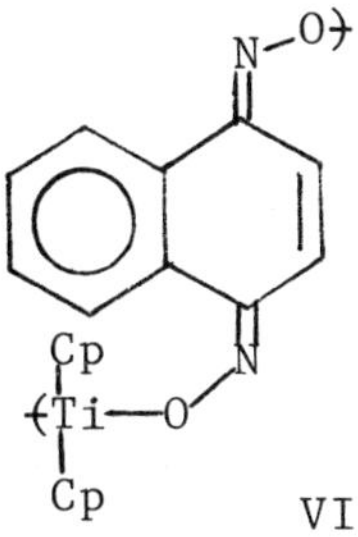
N
O
Cp
Ti
O
N
Cp

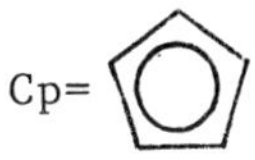
Cp=

VI

Polymer VII was synthesized to deliver both the manganese moiety, which is an essential meal, and the pyrimethamine portion which is an antimalarial and antimicrobial agent (5). Here the delivery of both portions may be advantageous not necessarily through "toxic" routes but through "assisting" routes. Counter to this is the synthesis of VIII where the cobalticinium moiety is suggested as a potential hampton, yet the arsentic moiety is a known toxin to many biological species (6).

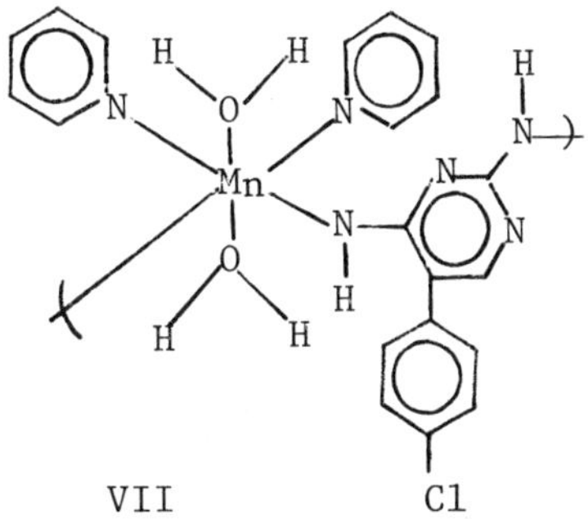
H
H
O
N
N
Mn
O
H
H
N
H
H
N
N
N
Cl

VII

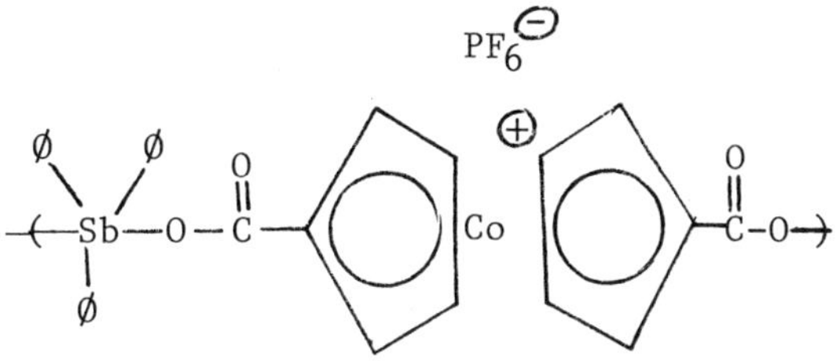
PF6
Ø
Ø
Sb
O
C
O
Ø
Co
O
C-O

VIII

Thus there exists the capability to "tailormake" a number of potential organometallic drug delivery polymers.

TIN-CONTAINING POLYMERS

A wide variety of organotin-containing polymers have been made along with the modification of a number of commercially available polymers including polyvinyl alcohol, polyethyleneimine, polyacrylonitrile and cellulose derived from cotton and dextran. In all cases the resulting tin containing polymer exhibited biological toxicity towards some bacteria and/or fungi at concentrations permitting their use in topical applications such as certian burns and athletics foot.

A number of organotin containing polymers have been synthesized and their biological properties partially studied. Here we will briefly describe only two systems which illustrate biological behavior typical of these polymers.

The biological toxicity is quite varied depending on the substituents on the tin. For instance, polyethyleneimine modified through condensation with diphenyltin dichloride (61% inclusion) showed no activity against Aspergillus fumagatus and only a small inhibition against a combination Penicillin species yet the analogous polymer derived from triphenyltin chloride (also in solid form) showed good activity toward both to a concentration of 3×10^{-2} μg/ml (9-11).

A majority of the organotin-modified cellulose derived from cotton showed good inhibition towards all tested fungi. For instance, cellulose derived from dipropyltin dichloride, dibutyltin dichloride and triphenyltin chloride all exhibited good inhibition towards Aspergillus flavus, Aspergillus niger and Aspergillus fumagatus for paper disks sprinkled with several granules of the modified cellulose (7,8). Only products from dioctyltin dichloride failed to show a good response towards the fungi.

Protein assays were determined on the modified cellulose samples in Trichoderma reesei and Chaetomium globosum using saline and dextrose solutions. The modified cellulose samples were active against the fungi in both solutions in agreement with inhibition being the result of the toxicity of the modified cellulose rather than an inability of the fungi to degrade the test compound.

The fungi tested in these studies are typical and widespread representing a good cross sectional test for the applicability of such modified products as retarders of fungi related maladies.

ARSENIC POLYPYRIMIDINES

The synthesis and biological evaluation of arsenic-containing polymers is not new. Wang and Sheetz reported on the synthesis of 10-(alkenyloxy)phenoxarsine based monomers and polymers. The polymers and monomers were both active against a variety of fungal and bacterial agents (12).

The rational behind the synthesis of many of the potential drugs by us is to supply a host with a metabolizable portion within the polymer, which when metabolized, will release a moiety toxic to the host.

Only the compound derived from the condensation of triphenylarsenic dichloride with 4,6-diamino-2-mercaptopyrimidine (poly-(AsDASP)) has been extensively studied (13). It was picked for study because of the presence of the thiol group which is both a potential modifying moiety towards the toxicity of arsenic and a typically desirable metabolite for a number of biological organisms - thus encouraging both select toxicity and metabolism.

The effect of poly(AsDASP) on a number of bacteria was studied. It was found that activity was limited to Pseudomonas fluorescens at concentrations of 27.5 μgram per disk and above. At the highest concentration tested (220 μgram) no inhibition was apparent suggesting resistant colonies may be appearing. To check for this, colonies growing around the highest concentration disks were transferred to a fresh plate and the 55 μgram inhibiting dilution applied. The cells were not inhibited suggesting higher concentrations may be selected by resistant organisms. In fact toxicity is so selective that only certain strains of Ps. fluorescens are affected by poly(AsDASP).

Inhibition tests in nutrient broth and saline solution were undertaken. Inhibition was greatest for the nutrient broth mixtures being consistent with activite metabolism being involved in inhibition of the Ps. fluorescens (concentration range of 27.5 μgram to 1.71 μgram of polymer studied). The inhibition of triphenylarsenic dichloride is 50% or less than that of the polymer and is similar in both nutrient broth and saline solutions. Thus combining of the arsenic with the thiopyrimidine as a polymer appears to be advantageous in inhibiting the growth of at least certain strains of Ps. fluorescens.

PLATINUM CONTAINING ANTITUMORAL AGENTS

Malignant neoplasms are the second leading cause of death in the U.S. In 1964 Rosenberg and coworkers discovered by chance that bacteria failed to divide, but continued to grow giving filamentous cells in the presence of platinum electrodes (12). A major cause of this inhibition to cell division is the cis-dichlorodiammineplatinum II, IX, (c-DDP). Much work centered about the clinical use of c-DDP leading recently to the licensing of it as an antineoplast drug. It is currently successfully used in conjunction with other drugs in the treatment of a wide variety of tumors. The importance of the discovery that c-DDP is a valid antineoplast drug includes the reawakening of the potential and actual use of metal containing compounds as drugs, here as neoplast drugs.

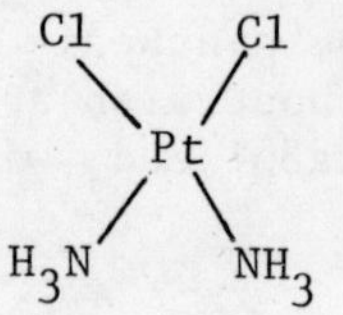

IX

A number of derivatives have been synthesized and partially tested in regard to antineoplastic activity. Briefly the derivatives of XII must be a. neutral, b. contain two inert and two labile ligands and c. must have the ligands cis to each other.

The use of c-DDP has been complicated because of negative side effects including gastrointestinal, hematopoietic, immunosuppressive, auditory and renal disfunction. One method of overcoming many of these negative side effects is to limit the movement of the c-DDP through inclusion of it into a polymer prohibiting the filtration of the polymer by the kidneys.

Two general approaches can be taken. First, attachment of the c-DDP derivative onto a preformed polymer. This was done by Allcock and co-workers. Poly[bis(methylamino)phosphazene], XIII, is a water soluble polymer that bears coordination sites on both the side group and chain nitrogen atoms (15, 16). Compound X reacts with K_2PtCl_4 and 18-crown-6-ether in organic media to yield a coordination complex of structure XI (15,17). Compound XI shows tumor inhibitory activity against mouse P388 lymphocytic leukemia and in the Ehrlich Ascites tumor regression test (17). Work is continuing related to this.

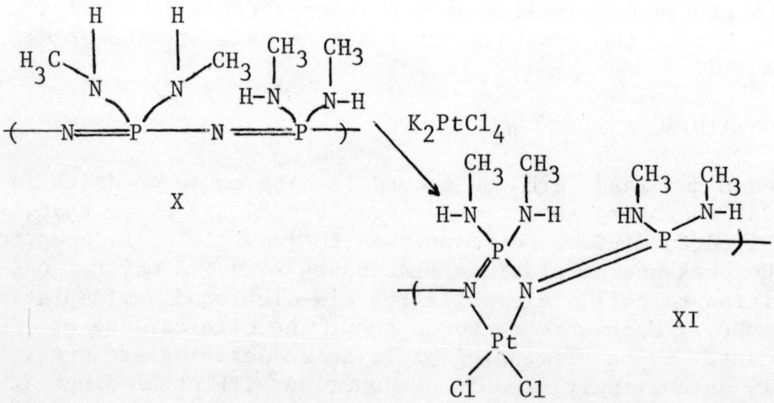

X XI

A second approach has the c-DDP derivative actually included as part of the polymer backbone chain. In 1977 we initially synthesized the first poly(cis-dichlorodiammine platinum II) compound

(18). Since then we have synthesized a wide variety of similar compounds including aromatic, aliphatic and pyrimidine diamines and other biologically active diamines.

$$K_2PtCl_4 + H_2N\text{-}R\text{-}NH_2 \rightarrow \left(\overset{Cl \quad Cl}{\underset{}{Pt}}\text{-}NH_2\text{-}R\text{-}NH_2 \right)$$

XII

These compounds show a wide variety of biological activities (18). At concentrations of 10 to 20 μg/ml they can suppress, enhance or have no effect on the replication of Polio Virus type I, a L RNA virus, but none of the polymers tested have any activity toward L929 (mouse) of HeLa (human) tumor cells at these concentrations. At concentrations of 30 μg/ml and greater all of the polymers tested inhibit both L929 and HeLa tumor cells. Further mice tolerate a dose of 400 μg of a methylthio pyrimidine containing polymer with no apparent ill effects (highest dosage tested). Thus the polymers show specific, differential activity in the μg/ml level.

ABSTRACT

The use of organometallic polymers as controlled release or direct agents in medical applications is reviewed with two major sections. One, a brief philosophical consideration as to potential advantages in utilization of metal containing polymers as drugs. Second, specific examples are presented in three areas: a. tin containing polymers for topical applications; b. arsenic polypyrimidines; and c. platinum polyamines as antitumoral agents.

TABLE 1

DATA FOR COMPOUNDS NOTED IN PAPER

Reactants	Synthesis[a.]	% Yield (range)	$\bar{M}_w$ [b.]	Reference
Cp_2TiCl_2, Bromophenol Blue	IF	20-70	5×10^5	1
Cp_2TiCl_2, Δ'-Pregnane-3,20-dione dioxime	IF	30-60	3×10^5	4
Cp_2TiCl_2, 2-Methyl-1,4-Naphthoquinone dioxime	IF	30-60	9×10^4	4
Triphenylantimony dichloride, 1,1'-ferrocene dicarboxylic acid disodium salt	IF	30-60	7×10^4	2
Poly(acrylic acid), triphenyltin chloride	IF	30	--	3
Poly(ethyleneimine), triphenyltin chloride	IF	50-100	--	7,8
Poly(ethyleneimine), diphenyltin dichloride	IF	50	--	7,8
Cellulose, dibutyltin dichloride	IF	10-90		5,6
Cellulose, dioctyltin dichloride	IF	70-90	--	5,6

TABLE 1 (CONTINUED)

Reactants	Synthesis[a.]	Yield (range)	$\bar{M}_w$ [b.]	Reference
Triphenylarsenic dichloride, 4,6-Diamino-2-mercaptopyrimidine	IF	10-80	6×10^5	13
K_2PtCl_4, 1,6-Hexanediamine	Soln.	80-100	1×10^6	18

a. Synthesis: IF = interfacial; Soln = solution

b. Weight Average Molecular Weight via Light Scattering Photometry

REFERENCES

1. C. Carraher, R. Schwarz, M. Schwarz and J. Schroeder, Organic Coatings and Plastics Chemistry, 42, 23 (1980)
2. C. Carraher, H. Blaxall, J. Schroeder and W. Venable, Organic Coatings and Plastics Chemistry, 39, 549 (1977)
3. C. Carraher and J. Piersma, J. Applied Polymer Sci., 16, 1851 (1972)
4. C. Carraher and L. Torre, Organic Coatings and Plastics Chemistry, 42, 18 (1980)
5. C. Carraher, V. Foster, H.M. Molloy and J. Schroeder, Organic Coatings and Plastics Chemistry, 41, 203 (1979) and unpublished results.
6. J. Sheats, C. Carraher and H. Blaxall, Polymer Preprints, 16, 655 (1975)
7. C. Carraher, W. Burt, D. Giron, J. Schroeder, M. Taylor, H.M. Molloy and T. Tiernan, J. Applied Polymer Sci., submitted.
8. C. Carraher, J. Schroeder, C. McNeely, D. Giron and J. Workman, Organic Coatings and Plastics Chemistry, 40, 560 (1979)
9. C. Carraher, D. Giron, W. Woelk, J. Schroeder, M. Feddersen, J. Applied Polymer Science, 23, 1501 (1979)
10. C. Carraher, J. Schroeder, W. Venable, C. McNeely, D. Giron, W. Woelk and M. Feddersen, "Additives for Plastics", Vol 2, Edited by R. Seymour, Academic Press, N.Y., 1978, page 81.
11. C. Carraher and M. Feddersen, Angew. Makromolekulare Chemie, 54, 119 (1976)
12. C. Wang and D. Sheetz, Dow Chemicals, U.S. Pat. 3,660,353, May 2, 1972
13. C. Carraher, W. Moon and T. Langworthy, Polymer Preprints, 17, 1 (1976)
14. B. Rosenberg, L. VanCamp and T. Krigas, Nature, 205, 698 (1965)
15. H. Allcock, R. Allen and J. O'Brien, J. Chem. Soc., Chem. Comm., 717 (1976)
16. H. Allcock, W. Cook and D. Mack, Inorg. Chem., 4, 2584 (1972)
17. H. Allcock, "Organometallic Polymers", (Edited by C. Carraher, J. Sheats and C. Pittman), Academic Press, N.Y., 1978, page 283-288
18. C. Carraher, W. Scott and D. Giron, J. Macromolecular Science - Chemistry, in press; and unpublished results

POLYTHIOSEMICARBAZIDES AS ANTIMICROBIAL POLYMERS

James A. Brierley, L. Guy Donaruma, Steven Lockwood, and Robert Mercogliano - Departments of Biology and Chemistry, New Mexico Institute of Mining Technology, Socorro, NM 87801
Shinya Kitoh - The Lion Co., Ltd., Odawara-Shi, Kanagawa-Ken, JAPAN
Robert J. Warner - The Johnson Wax Co., Racine, WI 53403
J. V. Depinto and J. K. Edzwald - Department of Civil and Environmental Engineering, Clarkson College of Technology, Potsdam, NY 13676

Over a number of years, the antiparasitic and antibacterial properties of a large number of synthetic polymers have been examined by our research group. Among the materials examined were: sulfonamide-formaldehyde copolymers (1,2), sulfonamide-dimethylolurea copolymers (1,2), tropolone-formaldehyde copolymers (1,2), N-methacrylyl-1-aminoadamantane-methacrylic acid copolymers (1), and piperazine-dibasic acid copolymers (1). These materials have been tested against reasonably large numbers of parasite and bacterial species (1) in order to see if any structure-activity relationships could be observed. From this work it seemed that a number of potential structure-activity relationships could be noted (1). That is to say, observed activities apparently could be correlated with molecular weight, copolymer composition, and changes in substituents on the monomers from which the polymers were derived (1). Further, upon searching the literature, other polymer properties such as polyelectrolyte character, stereochemical configuration, crosslinking, etc. could be seen perhaps to exhibit trends toward being related to biological activities of various sorts (2). In addition, the molecular weight, copolymer composition, and substituent change presumed correlations which we observed in our own work also could be observed from independent results in the literature (2).

In the studies cited above for our group, in order to maximize chances for observing biological activity, the polymers employed were largely prepared using known drugs as monomers or comonomers (1). We wanted to look at some polymers which were prepared from materials not really considered to be drugs. Thus, an antibacterial screen was set up which enabled us to test all polymers which we prepared for whatever purpose. It is one of these polymer systems which we would like to discuss further.

In other work designed to prepare and test ion exchange materials for metal ion recovery and oxidation-reduction properties (3), we had occasion to prepare a series of polythiosemicarbazides. The initial compounds worked with were prepared as follows:

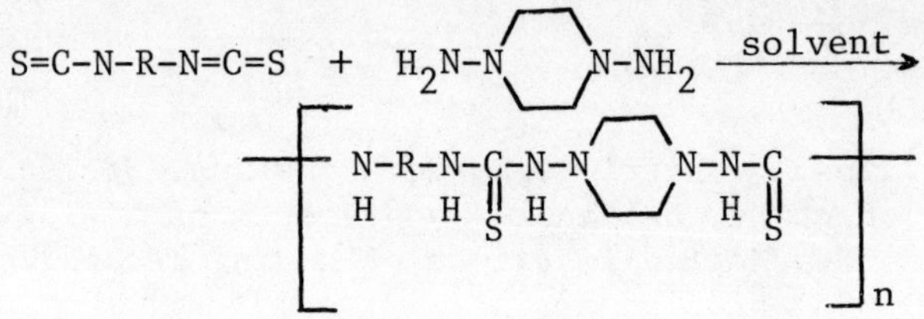

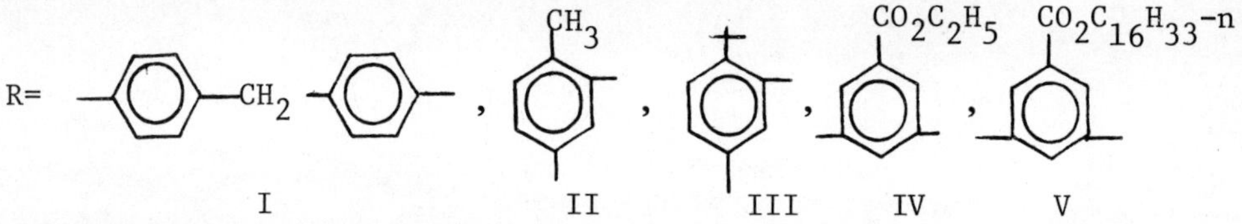

The reaction products were quite water insoluble as were their respective copper(+2) complexes. Having made these materials which, incidentally, were highly selective for copper(+2), we tested some of the polymers(I and II) and the respective polymeric copper(+2) complexes as antibacterials against both gram positive and gram negative bacteria. The test results are shown in Table 1. Evidently, there is reason to suspect that copper(II) may impart or enhance antibacterial activity in these two systems.

Table 1

ANTIBACTERIAL ACTIVITY OF SOME POLYTHIOSEMICARBAZIDES AND THEIR RESPECTIVE COPPER(+2) COMPLEXES

Polymer System	Antibacterial Activity: Free Polymer		Antibacterial Activity: Copper(II) Complex	
	B. subtilis*	E. coli**	B. subtilis	E. coli
I	Inactive	Inactive	Active	Inactive
II	Active	Inactive	Active	Active

*gram positive
**gram negative

Also, the copper(+2) complexes of polymers I-V were tested as copper(+2) release agents for schistosomiasis prophylaxis. Schistosomiasis is a serious, often fatal, viral disease which attacks humans where aquatic snails are the disease carrier (4). Such snails live in irrigation ditches, streams, swamps, and other natural waters in the warmer climates. The disease carrying snails can be killed by adding copper(+2) to their aquatic habitat, and the classic source of the needed copper(+2) is cupric sulfate. However, the salt rapidly migrates away in moving waters or is exchanged into the surrounding soil and is very rapidly lost. Thus, frequent additions are required to provide effectiveness. One way tried to retard loss of copper(+2) has been to mill copper sulfate into rubber pellets from which water slowly leaches out copper(+2) to provide the needed ions to kill snails. However, in natural waters these pellets float away or sink into the channel soil. Of course, any ion exchange resin could provide a substrate to hold copper(+2) so that natural water soluble salts would exchange away the copper(+2) while the regenerable ion exchanger was in a fixed accessible site. However, the copper must be slowly released and not exchanged away from the ion exchanger too quickly. Thus, an ion exchanger which binds copper(+2) tightly is needed. Yet, the binding of the copper(+2) must not be so tight that it will not exchange away from the polymer. The desired rate of copper(+2) release should be 1-2% per day (5). Knowing that copper(+2) was tightly bound to I-V, the rate of release from the polymer in water was observed over a period of time. Table 2 displays these results.

As can be seen from Table 2, the preliminary data indicate that a slow release of copper(+2) may be occurring over a fourteen day period. This release apparently is characterized by an initial surge followed thereafter by release of smaller amounts as time goes on. Over a fourteen day period after the initial surge, the release for polymers I, III, and IV averages about in the neighborhood of 1% per day.

We also prepared a number of very similar polymers which had the capability of being crosslinked (6). These polymers were water soluble. Also, they could be crosslinked through the hydroxymethyl group via various etherification reactions using concentrated sulfuric acid or concentrated sulfuric acid in the presence of ethylene glycol, 1,2-propylene glycol, and glycerol. Crosslinking also could be obtained by esterification reactions with acid chlorides and acid anlydrides. Naturally, the crosslinked products were water insoluble.

Table 3 shows the antibacterial test results obtained when these materials, both linear and crosslinked were tested against both gram positive and gram negative bacteria.

Table 2

RELEASE OF COPPER(+2) FROM POLYMERS I-V AS A FUNCTION OF TIME

	% Release Over Time*							
Polymer	1 da.	2 da.	3 da.	4 da.	7 da.	9 da.	11 da.	14 da.
I	14.6(11.9)	15.6(20.4)	17.4(21.8)	18.8(23.2)	17.3(27.4)	21.6(28.8)	23.3(31.1)	23.2(33.5)
II	4.3(5.8)	5.8(8.6)	7.2(11.6)	8.7(14.4)	10.1(17.3)	10.1(20.2)	9.8(25.6)	12.9(25.8)
III	11.6(18.9)	14.5(20.3)	14.5(21.7)	14.4(24.6)	14.4(21.4)	19.4(28.8)	19.4(32.3)	24.5(34.8)
IV	11.6(13.1)	13.1(14.5)	11.6(15.9)	13.0(17.4)	11.5(18.9)	11.5(20.2)	19.4(19.4)	18.1(23.2)
V	<0.1(0.6)	<0.1(0.6)	0.3(0.6)	0.3(0.6)	<0.1(.84)	<0.1(1.1)	<0.1(4.9)	0.5(5.0)

*Figure in parentheses is with agitation of the samples.

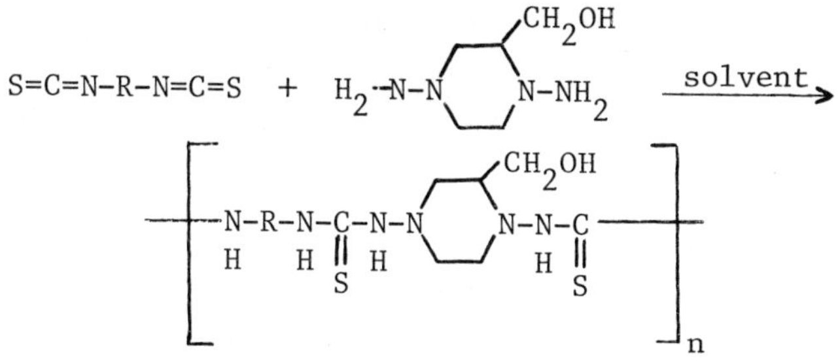

R= 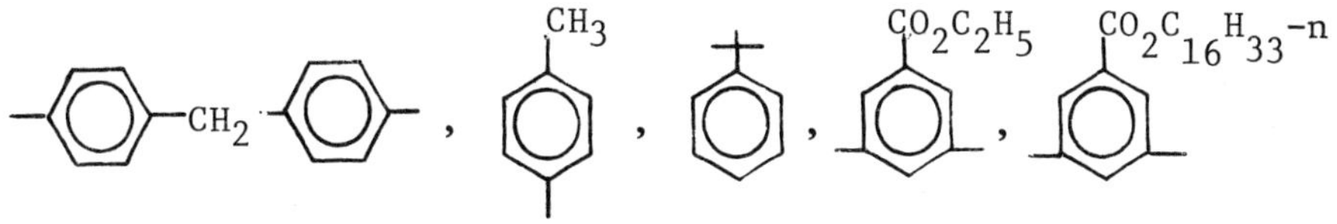

The interesting observations to be made from looking at Table 3 are that (1) the polythiosemicarbazides seem to be more active against gram positive bacteria and (2) crosslinking seems to enhance activity. The reasons why these phenomena apparently occur are not clear at all. However, possible enhancement of biological activity by crosslinking is not unreported (2) in the literature. Most certainly, the results in Table 3 having to do with crosslinking are somewhat surprising in view of the fact that the crosslinked materials are water insoluble.

Table 3

ANTIBACTERIAL TESTS FOR

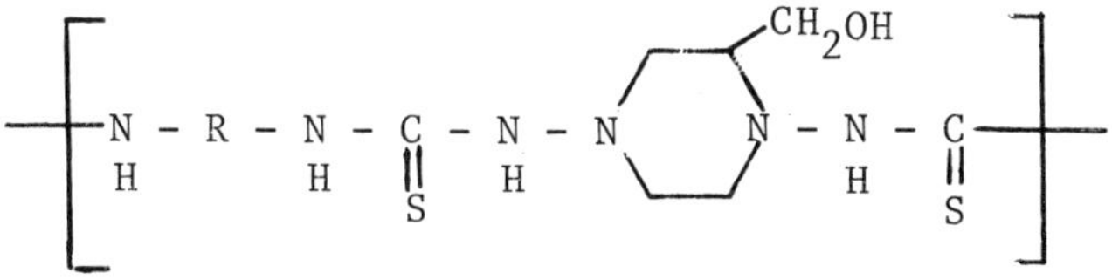

R	S. aureus (g.pos.)	E. coli (g.neg.)
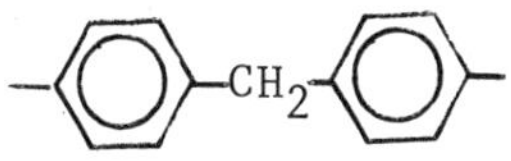	inactive	inactive
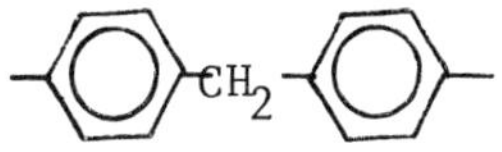 succinic anhydride*	active	active

Table 3 (Continued)

R	S. aureus (g.pos.)	E. coli (g.neg.)
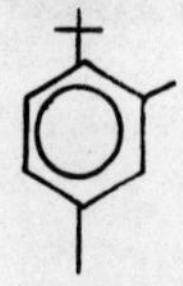	inactive	inactive
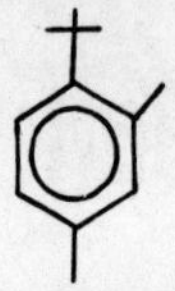 succinic anhydride*	active	inactive
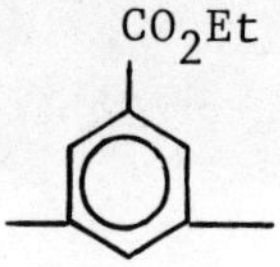	inactive	inactive
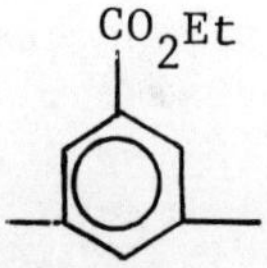 H_2SO_4	inactive	inactive
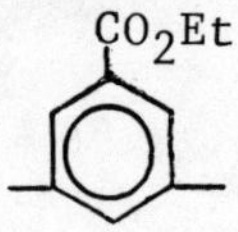 succinic anhydride*	active	inactive
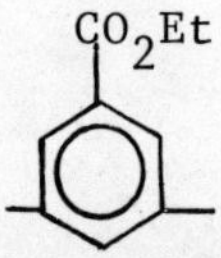 ethylene glycol*	active	inactive
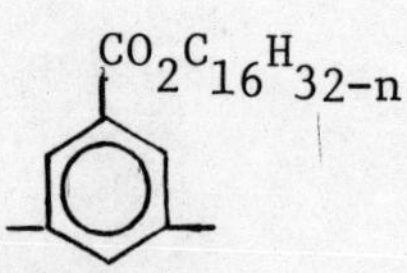	inactive	inactive
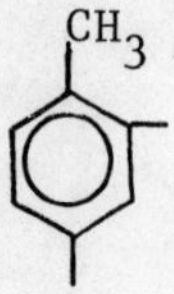	inactive	inactive

Table 3 (Continued)

R	S. aureus (g.pos.)	E. coli (g.neg.)
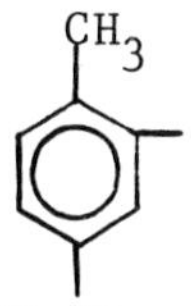 H_2SO_4	active	inactive
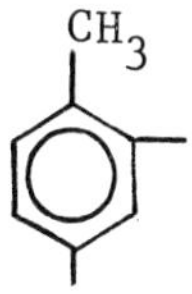 succinyl chloride*	active	inactive
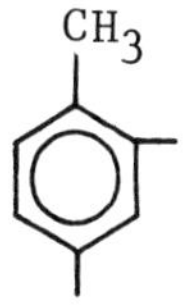 ethylene glycol*	active	inactive
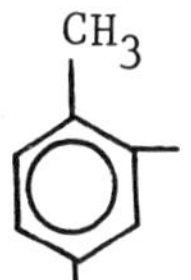 1,2-propylene* glycol	active	inactive
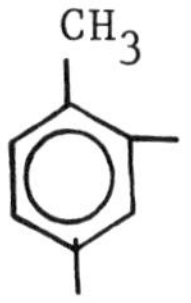 glycerol*	inactive	inactive

*Denotes crosslinking agent

Continuing the studies on polythiosemicarbazides, preparation of a copolymer (VII) from vinyl alcohol containing thiosemicarbazide units was attempted via Scheme A using VI (6). What we actually got was VIII due to dehydration of some vinyl alcohol mers. VIII had markedly decreased water solubility over poly[vinyl alcohol] due no doubt to the introduction of the water insoluble mer derived from VI and perhaps some crosslinking via intermolecular etherification between vinyl alcohol mers. Poly[vinyl alcohol] alone then was treated with sulfuric acid to provide IX which appeared to be crosslinked by etherification.

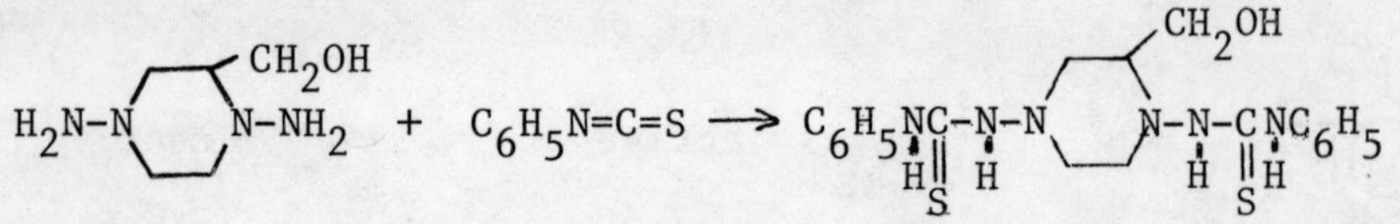

VI

Scheme A

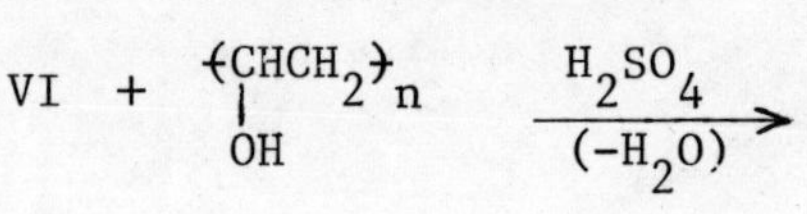

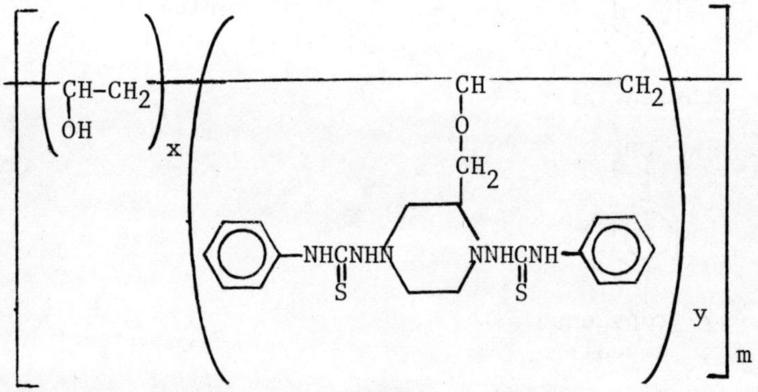

VII

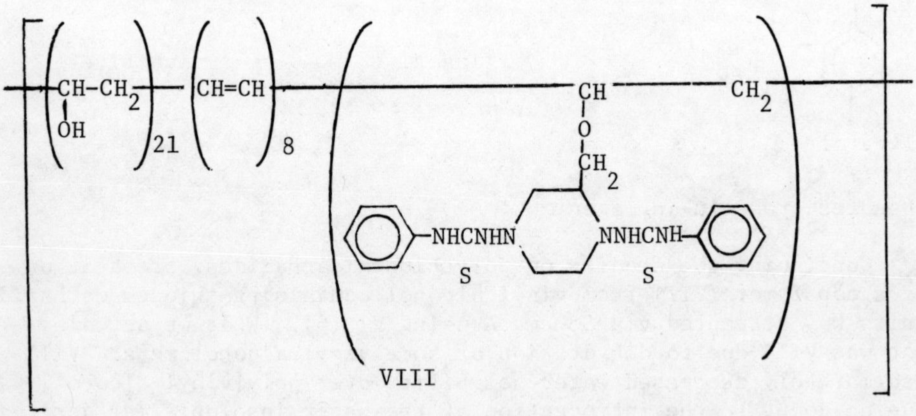

VIII

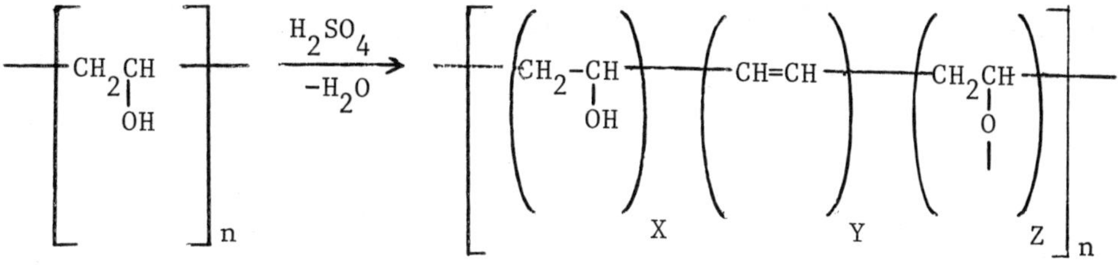
H_2SO_4
$-H_2O$
X
Y
Z
n

IX

Antibacterial test results for VIII and IX are shown in Table 4. Note that IX is active against both gram positive and gram negative bacteria, but VIII is active only against gram positive bacteria. As in Table 3, it can be observed that the polythiosemicarbazides are seemingly more active against gram positive bacteria.

Table 4

ANTIBACTERIAL TESTING FOR VIII AND IX

Polymer	S. aureus (g.pos.)	E. coli (g.neg.)
IX	active	active
VIII	active	inactive

Preparation of a copolymer (X) derived from poly[vinyl chloride] and VI was attempted (6) by displacement of halogen (Scheme B). Actually, the lithium alkoxide of VI was employed as the nucleophile. Again, some olefinic mers were present and what was obtained as a product was XI. Z as shown in XI is an oxygen containing mer type also present in the poly[vinyl chloride] employed (6). Both XI and poly[vinyl chloride] are, of course, water insoluble. Since olefinic mers were present in XI, poly[vinyl chloride] was subjected to the identical reaction conditions employed to prepare XI, sans the presence of VI, to provide XII. Antibacterial test results for XI and XII are shown in Table 5. Again, the thiosemicarbazide polymer seems to be active against gram positive bacteria only.

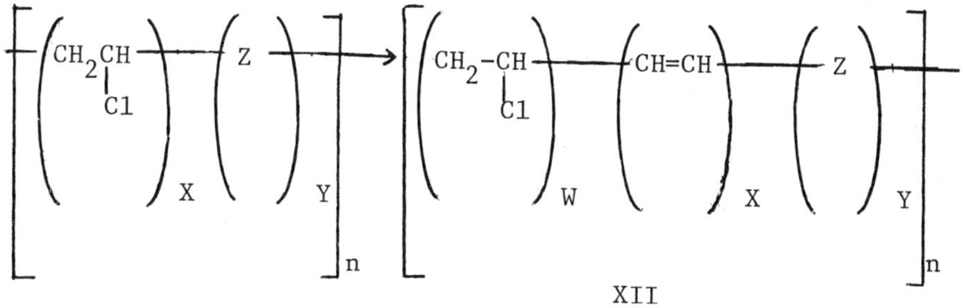
Z
X
Y
n
W
X
Y
n

XII

Scheme B

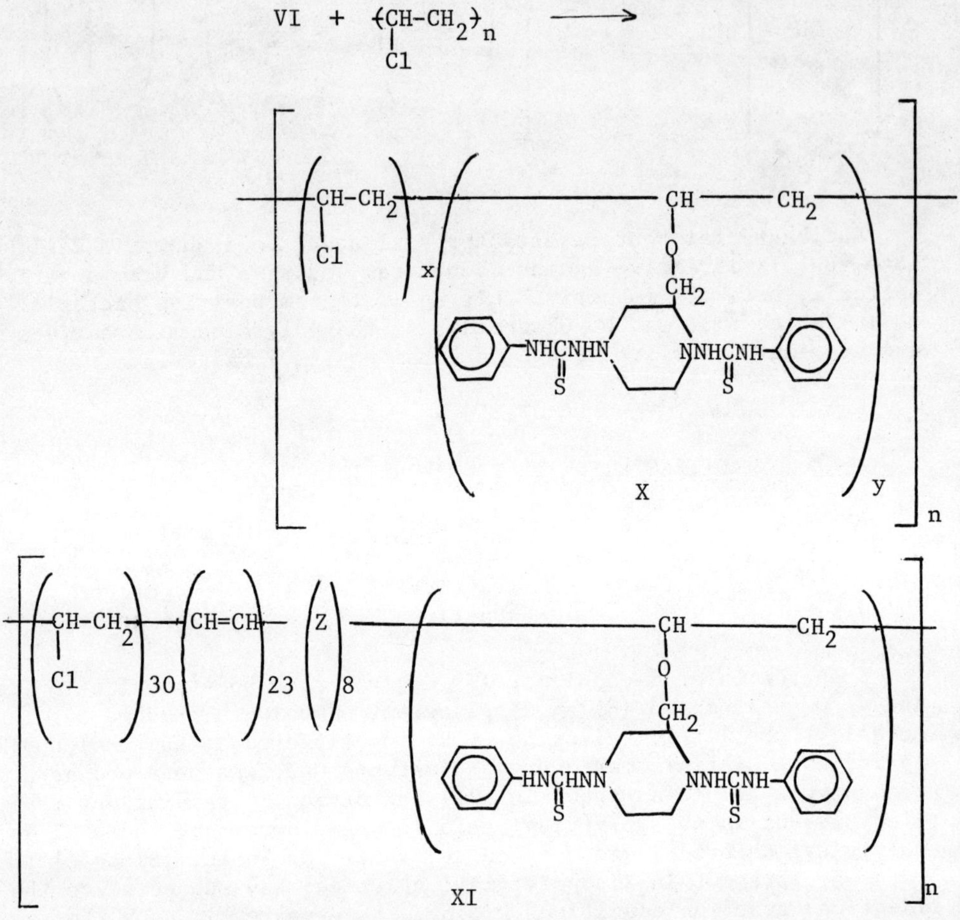
VI
CH-CH2
Cl
n
CH-CH2
Cl
x
CH
CH2
O
CH2
NHCNHN
S
NNHCNH
S
X
y
n
CH-CH2
Cl
30
CH=CH
23
Z
8
CH
CH2
O
CH2
HNCHNN
S
NNHCNH
S
XI
n

Finally, thiosemicarbazide containing polymers (XIV and XVI) were prepared from poly[glycidyl methacrylate] (XIII) and VI as well as poly[glycidyl methacrylate-co-methyl methacrylate] (XV) (Scheme C) (6).

Table 5

ANTIBACTERIAL TESTING FOR XI AND XII

Polymer	A. aureus (g.pos.)	E. coli (g.neg.)
XII	inactive	inactive
XI	active	inactive

XIV and XVI were tested against both gram positive and gram negative bacteria as were products XIII_a and XV_a which were prepared from XIII and XV which had been subjected to reaction conditions identical to those employed to prepare XIV and XVI with no VI being present in the reaction mixtures. The antibacterial test results shown in Table 6, again may indicate a preferential antibacterial activity of polythiosemicarbazides for gram positive bacteria.

Scheme C

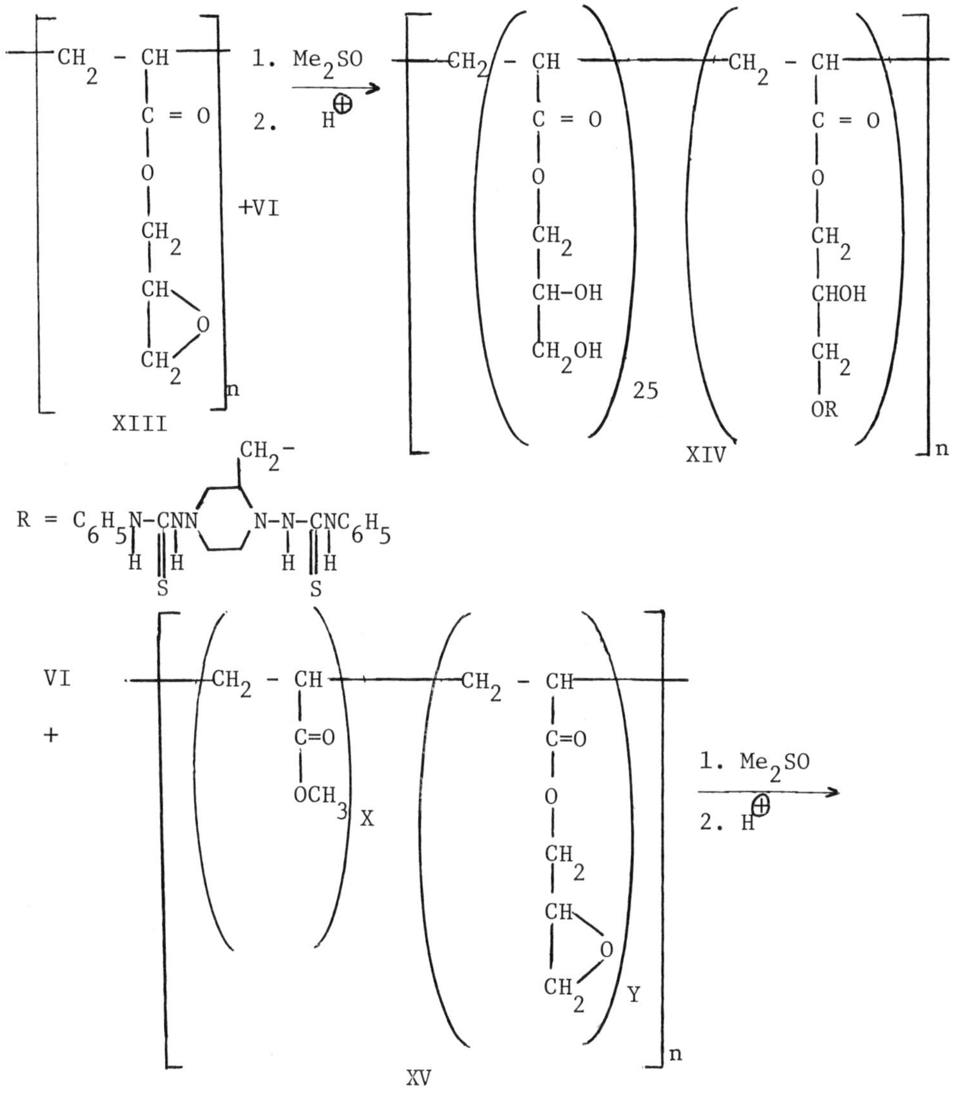

1. Me_2SO
2. $H^{\oplus}$
+VI
XIII
XIV
25
n
R =
VI
+
X
Y
XV

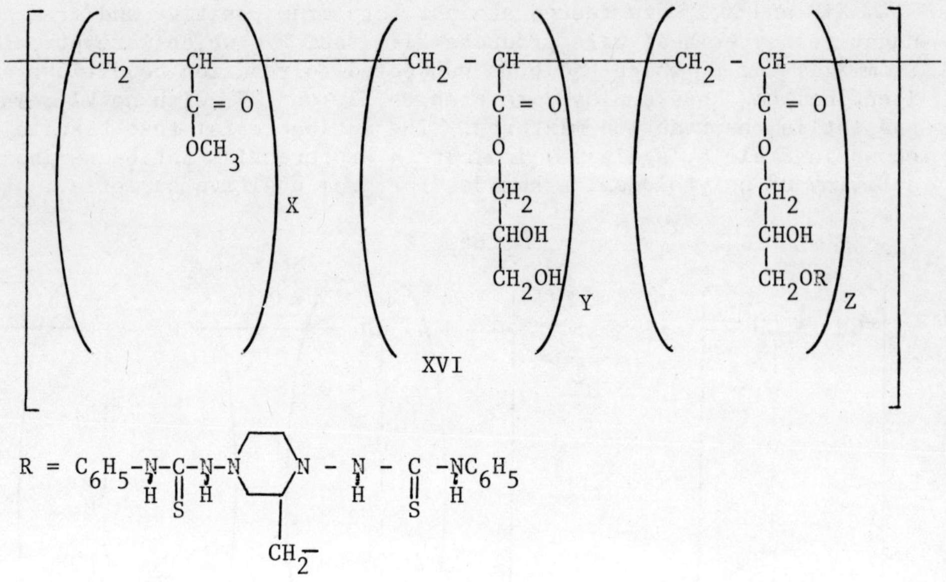

XVI

Table 6

ANTIBACTERIAL TESTING FOR XIII_a, XIV, XV_a, AND XVI

Polymer	S. aureus (g.pos.)	E. coli (g.neg.)
XIII_a	active	inactive
XIV	inactive	inactive
XVI	active	inactive
XV_a	inactive	inactive

We are indebted to the International Copper Research Association for their generous financial support of this research.

REFERENCES

1. L. G. Donaruma and O. Vogl eds., "Polymeric Drugs", Academic Press, New York, NY, 1978, p. 349
2. L. G. Donaruma, "Progress in Polymer Science", Vol. 4, Pergamon, Oxford, 1974, p. 1
3. L. G. Donaruma, S. Kitoh, G. Walsworth, J. V. Depinto, and J. K. Edzwald, Macromolecules, 12, 435 (1979)
4. T. C. Cheng, "Molluscicides in Schistosomiasis Control", Academic Press, New York, NY 1974

5. International Copper Research Association, Private communication
6. R. L. Warner, "Synthesis of Some Novel Polymeric Thiosemicarbazides", Doctoral Thesis, Clarkson College of Technology, December 1978

THE BIOCHEMICAL PROPERTIES OF CARRIER-BOUND METHOTREXATE

John M. Whiteley, Scripps Clinic and Research Foundation, La Jolla, CA 92037
Barbara C. F. Chu, Department of Medicine T-006, University of California, San Diego, CA 92093
John Galivan, Division of Laboratories and Research, New York State Department of Health, Albany, New York 12201

From antiquity mankind has attempted to soothe ailments and control bodity functions with palliatives which usually on closer examination have contained some form of chemical compound. Examples include the use of laudanum and opiates for upset stomachs, the somewhat gross medieval use of leeches, which secrete fluids now known to contain heparin, to alleviate bloodflow problems, the discovery during the past century of the anesthetic properties of halogenated hydrocarbons, and the bacteriocidal properties of phenols, the more current usage of sulfonamides, tetracyclines, penicillins and other antibiotics, and presently the use of specific antimetabolites to control or quell aberrant enzymatic activity. Throughout this long period of development one can perceive the achievement of greater purity of the particular chemical compound or drug, greater understanding of its specific action, and a clearer delineation of the function it is being asked to perform.

Specificity until recently has usually been achieved by the discovery or design of molecules which carry out a limited and selective destructive function. The antibiotics again provide one of the best examples in this area. Currently, however, considerable effort is being directed towards both antiviral and antineoplastic therapy, and in these instances, selective cell destruction has become a much more difficult goal to attain. In fact, in the former case, prevention via vaccination is still the preferred procedure.

Several obstacles to success in the latter case occur because rapidly proliferating cell populations, often found in malignancies, can develop resistance to an antimetabolite in a variety of ways: by the elevation in concentration of key enzymes with which the agent interacts, by utilizing alternate biosynthetic routes for key components, and by developing transport deficiencies which block the passage of the drug into the cell. Additionally, in whole animals, the drug may be rapidly excreted, or metabolized and excreted, and moreover, because of the specific drug's intrinsic physical properties, there can be impermeable blocks such as the "blood-brain barrier" which can prevent the passage of the therapeutic agent to its goal.

To attempt to circumvent some of these problems, myriad structural variations on a basic efficacious molecule have been used to secure continuous successful therapy for the patient. Examples are the many tetracyclines, the antihistaminics, the prostaglandins, etc., and in the field of chemotherapy, the alkylating agents, the nucleoside analogs and the many synthetic variants on aminopterin of which methotrexate (MTX) or amethopterin is the best known example. (Abbreviations: MTX, methotrexate; pCMS, p-chloromercuribenzenesulfonate; 5-formylTHF, 5-formyl-5,6,7.8-tetrahydrofolate; BSA and MSA, bovine and murine serum albumin, respectively.)

Although simple variations in molecular structure, such as altered alkyl, aryl, or halogen substituents, isomeric displacement of hydroxyl groups, etc., led in mnay of these instances to greater drug specificity, recent interest has been stimulated by the concept of carrying the drug to, and securing its release at, the required site of action. Examples of such carrier-bound drug systems which have already been investigated include: DNA (adriamycin and daunomycin) (1); lipid vesicles (actinomycin D and cytosine arabinoside) (2); antibodies (chlorambucil and N,N-bis(2-chloroethyl)-p-phenylenediamine) (3); bovine serum albumin and high molecular weight dextrans (MTX) (4). Proteins have been of special interest because of the possibility that a tumor-specific antibody might be isolated which could act as a carrier and target its drug directly to the tumor site. However, the formidable obstacle of producing specific antibodies against the different antigens of each human tumor has not yet made this a feasible prospect.

Interest in this laboratory in constructing carrier-bound drugs did not arise from a comprehensive synthetic program designed to overcome deficiencies in current drug delivery protocols, but from investigations into the development of materials suitable for the specific isolation of dihydrofolate reductases from bacterial and mammalian cell preparations (i.e., affinity chromatographic materials (5)). During the course of this work, it became apparent that MTX bound to an inert support did, in fact, retain specificity for the reductase, suggesting that such derivatives might demonstrate altered and perhaps useful biological properties in whole animals. In addition, there was a wealth of data on drug therapy employing MTX in the literature, which could form a basis for comparative studies. Therefore, MTX bound to an inert carrier appeared to be a good model for investigating this concept.

EXPERIMENTAL

Materials

The following were obtained from commercial sources: bovine chymotrypsinogen, poly-L-lysine (MW ∿ 30,000), pCMS, 5-formylTHF, folic acid and crystalline BSA (Sigma Chemical Co.); MTX (B. L. I. Biochemicals); bovine IgG (Schwarz/Mann Chemical Co.); Dextran T70 and Ficoll Hypaque (Pharmacia Chemical Co.); [3',5'-^{3}H]MTX (7.5 Ci/mmole) and [^{3}H]Dextran T70 (21 mCi/g, average MW 82,000) (Amersham Searle); 6-[^{3}H]UdR (242 Ci/mmole) (New England Nuclear). MTX-BSA, [^{3}H]MTX-BSA, [^{3}H]MTX-Dextran T70, MTX-[^{3}H]Dextran T70, Folate-BSA, MTX-chymotrypsinogen, MTX-IgG and MTX-poly-L-lysine were synthesized according to a previously published procedure (4), in which MTX is coupled, via one of its terminal carboxyl groups, in a carbodiimide-promoted reaction to either lysine ε-amino groups contained in the protein or to aminoethyl groups chemically inserted into the dextran. MTX-[^{125}I]BSA was obtained from MTX-BSA using the procedure of McConahey and Dixon (6). BDF_1 and DBA_2 female mice (19-22 g) were obtained from Simonsen Laboratories, Gilroy, CA.

Methods

MTX and its derivatives were tested for anti-tumor activity in mice using the protocol recommended by the Drug Evaluation Branch, National Cancer Institute (7). Animals were inoculated with 10^6 L1210 cells on day zero, and the drugs were injected intraperitoneally on day 1.

Fragments (2-4 mm) of the Lewis lung carcinoma from female C57BL/6 mice were transplanted once every 14 days into female BDF_1 mice by sc injection into the axillary region, with puncture in the inguinal area. Drugs were injected ip beginning on either day 1 or day 7 after tumor transplant and continued every 4 days for a total of 4 doses. Tumor sizes were measured with calipers on days 8, 12, 15, and 19 after transplantation, and their weights were calculated with the use of the conversion $w = (a \times b^2)/2$, where a = length (mm), b = width (mm), and w = weight (mg). We assumed that the width of the tumor was equivalent to its depth and that its specific gravity was unity (8). Survivors were recorded daily.

For measurement of lung metastases, untreated control animals and drug-treated animals were killed on day 19 after tumor transplant. According to the method of Wexler (9), the trachea was isolated, and via an incision made above the carina, approximately 2 ml of 15% India ink in water was injected through an 18-gauge needle into the lungs. The lungs were dissected, washed free of excess ink, and placed in a solution of glacial acetic acid: formaldehyde:70% ethanol (1:2:17). The metastatic foci, visible as white nodules on the black surface of the lungs, were counted and measured.

To determine plasma drug levels, BDF_1 mice were injected intraperitoneally with 7.5 mg/kg of $[^3H]$MTX-BSA, $[^3H]$MTX-MSA, and $[^3H]$MTX (specific activities, 650 μCi of MTX per millimole) or with 2.5 mg/kg of MTX-$[^{125}I]$BSA (1.5 μCi/mg). Blood samples (30 μl) were removed from the eye with heparinized blood-collecting microtubes 0, 0.12, 0.25, 0.5, 1, 2, 4, 7.5 and 24 hr after injection. Experiments in which uncomplexed $[^3H]$MTX was injected were terminated after 1 hr as the blood levels were no longer measurable. Erythrocytes were separated by centrifugation for 5 min in a Beckman 152 Microfuge, and the radioactivity in a 20-μl aliquot of the plasma was measured. Larger samples were mixed with nonradioactive MTX as a carrier and were analyzed by paper chromatography. MTX-albumin remained at the origin, whereas MTX migrated with R_f of 0.66.

Analytical chromatography was carried out with $MeOH:NH_4OH:H_2O$ (7:1:2) on Whatman No. 1 paper, radioactivities of tritiated and iodinated materials were measured with Beckman LS-233 liquid scintillation and Nuclear-Chicago 1185 gamma counters, respectively, and SDS polyacrylamide gel electrophoreses were performed employing the procedures of Weber and Osborn (10).

L1210 and L1210 R6 cells were grown in RPMI medium 1640 (Flow Labs, Inglewood, CA) supplemented with 5% fetal calf serum and 1% penicillin-streptomycin, starting with a concentration of 0.4-0.5 x 10^5 cells/ml. Aliquots of the culture were removed 24, 48 and 72 hr after start of growth and counted in a hemocytometer. Dihydrofolate reductase inhibition experiments were kindly performed by Dr. Chinan Fan of this department using the enzyme isolated from the MTX-resistant L1210 R6 strain of ascites tumor cells (11). The assays were performed in a system that contained, in 1.0 ml, 56 μM dihydrofolate, 150 μM NADPH, 0.1 M K-phosphate, 0.1 M K-Hepes and 0.5 M KCl, pH 7. Initial velocities (at 37°) were determined via absorbance changes at 340 nm after a 10 min, 25° pre-incubation period of the enzyme with MTX or the MTX derivative (12).

H-11-E-C3 cells derived from the Reuber H35 hepatoma (13,14) (abbreviated to H35 cells) were grown as monolayers on 60-mm Falcon dishes in a 5% CO_2 atmosphere. They were subcultured weekly and plated at a density of 2 x 10^5 cells/dish. The medium consisted of Swims Medium S-77 supplemented with 20% horse serum, 5% fetal calf serum (Grand Island Biological Co., Grand Island, NY), and 4 mM glutamine. Media changes were done routinely at 72 and 120 hr after plating. Cells were released from the dishes with 0.05% trypsin, and cell counting was carried out with a Z_{BI} Coulter counter. Cell stocks of normal and resistant lines were stored in liquid nitrogen. Assays for Mycoplasma contamination (15) were conducted routinely on cultures and frozen samples and were negative.

Resistant lines were developed by initially adding MTX at 10^{-9} M through a culture period of one week. The concentration was then raised by 3-fold, and cells were grown for 2 to 3 weeks in the elevated level of MTX until their growth curve equaled that of normal cells. This was repeated until the MTX concentration reached 3×10^{-7} M, and the cells that grew normally at this level are designated H35R. A more resistant cell line was also developed by a similar procedure which could be maintained in 2×10^{-6} M MTX.

DISCUSSION

The carrier-bound MTX derivatives retained their ability to interact with dihydrofolate reductase, indicating a freely accessible pteridine component. From previous knowledge of folate chemistry, it was assumed that substitution in a carbodiimide promoted reaction would occur primarily between the α- or γ-carboxyl groups of MTX and the lysine ε-aminoalkyl groups of the protein molecules or the aminoalkyl substituents inserted into the dextrans. No distinction was made between α- or γ-substitution in the therapeutic protocols. The derivatives showed absorbance spectra similar to a composite of free MTX and carrier chromophores, were essentially homogeneous on polyacrylamide gel electrophoresis, and were freely soluble in aqueous media. Using the known extinction coefficient of MTX at 370 nm (16), uptakes of 5 to 15 moles of MTX per mole of carrier were observed with the different polymers.

In experiments with BDF_1 mice, measurements of the free and bound MTX levels which developed after treatment with a single ip dose of either free MTX or MTX covalently bound to high molecular weight carriers indicated that the animals treated with the latter complexes demonstrated a higher, more prolonged extracellular concentration of the drug (MTX bound to carrier) than animals treated with the free agent (Fig. 1). In addition, administration of MTX covalently bound to proteins was comparable to free MTX when used in treatment of the ascitic L1210 leukemia carried by the mice (Table 1). In contrast, the various molecular weight dextran derivatives were essentially ineffective, and MTX-MSA proved too toxic a compound for therapy.

The effectiveness of a single dose of MTX-BSA in prolonging survival times, elevating drug levels and increasing body retention times, however, suggested that the compound might also be useful for the treatment of solid tumors. In particular, the prolonged high levels of circulating drug might offer increased cytotoxic effects against metastatic tumor cells. To examine this proposal BDF_1 mice carrying implants of the Lewis lung carcinoma were treated with MTX-BSA and MTX alone under a variety of conditions. Groups of mice were treated every 4 days for a total of 4 doses beginning either 1 or 7 days after receiving the tumor transplant. Drug-treated animals were killed after 19 days, and the lungs were examined for malignant nodules. The average numbers of metastatic centers found on the lungs of the treated animals are compared in Table 2 with

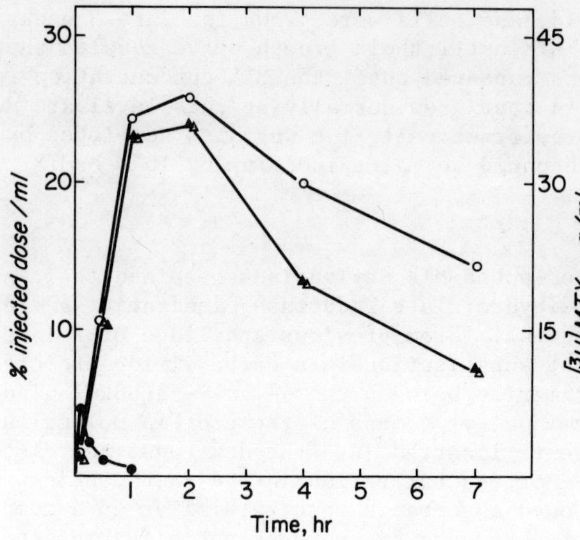
% injected dose / ml
[3H] MTX, µg/ml
Time, hr
30
20
10
45
30
15
1
2
3
4
5
6
7

Figure 1. Time-dependent variation of plasma drug concentrations found in BDF_1 mice treated with high molecular weight MTX derivatives. The left ordinate shows the time-dependent variation of the percentage of injected radioactivity and the right ordinate shows the concentration of MTX found per milliliter of plasma in BDF_1 mice after intraperitoneal injection of 7.5 mg/kg of [^{3}H]MTX (●), [^{3}H]MTX-BSA (▲), or [^{3}H]MTX-MSA (Δ) (specific activity, 650 μCi/mmole) or 2.5 mg/kg of MTX-[^{125}I]BSA (O) (1.5 μCi/mg).

those observed in untreated controls. Both MTX and its BSA derivative effectively reduced the number of metastases when therapy was begun on day 1 after tumor implantation. The average number of nodules was reduced from 28.1 to 20.0 after treatment with 10 mg MTX/kg, and this figure declined to 15.6 when the MTX dose was increased to 20 mg/kg. MTX-BSA was much more effective than either concentration of non-derivatized MTX, and when used at 10 mg/kg, it allowed the formation of only 7.9 nodules. Treatment with BSA alone (2.5 mg/animal) appeared to have no effect: however, when MTX (10 mg/kg) and BSA (2.5 mg/animal) were injected simultaneously into the mice, the MTX effect was enhanced, and the number of metastases was reduced to 14.7. When treatment was begun later (on day 7), MTX-BSA was still the more effective agent for reducing malignant proliferation, although early treatment was preferable.

Table 1

Survival times of L1210 tumor-carrying BDF_1 mice treated with high molecular weight derivatives of MTX. BDF_1 mice were inoculated on day 0 with 10^6 L1210 cells and treated on day 1 with either MTX or a series of high molecular derivates of MTX.

Antagonists		No. of Mice	Mean Survival Time
	(mg/kg)		(days)
Control	0	44	8.0 ± 1.2
MTX	15	11	15.2 ± 1.9
MTX-BSA	15	22	15.1 ± 2.3
MTX-MSA	15	11	14.3 ± 2.6[a]
MTX-chymo-trypsinogen	15	12	17.3 ± 3.2
MTX-IgG	15	8	20.8 ± 3.7
MTX-Dextran 10	15	14	10.1 ± 2.6
MTX-Dextran 40	15	11	9.5 ± 2.3
MTX-Dextran 70	15	6	8.7 ± 1.3
MTX-Dextran 150	15	6	10.3 ± 4.0

[a]In this experiment seven mice died of drug toxicity, and are excluded from this figure.

The percentage of metastatic nodules larger than 1 mm in diameter is also shown in the table. MTX and MTX-BSA both significantly reduced the number of large nodules when treatment was begun on day 1. If treatment was started on day 7, the number of metastases occurring with each regimen was also reduced, but the number of larger metastases was not reduced, because these malignant centers would have been implanted before drug treatment began.

The effect of the drug on the growth rate of the primary tumor was less impressive. In the group of mice receiving early treatment with MTX or MTX-BSA, a smaller tumor was evident in the MTX-BSA-treated animals (50% of the control weight) compared to animals treated with MTX alone (72% of the control weight) by the eighth day after transplant. These differences continued as tumor growth progressed. On day 12 the average weight of the control tumors was 980 mg, and the average weights of the tumors of the MTX-treated and MTX-BSA-treated mice were limited to 83% and 54% of this size, respectively. This trend continued through days 15 and 19. For

Table 2

Number and size of metastatic nodules found on the lungs of mice carrying the Lewis lung carcinoma after treatment with MTX, MTX-BSA, BSA, and MTX plus BSA[a].

Drug	No. of mice	Dose, mg/kg	Day of treatment	Metastatic nodules found on lungs, per mouse[b]	Percent of nodules larger than 1 mm in diameter
Control	28	--	--	28.1 ± 10.6	14.5
MTX	17	10	1	20.0 ± 9.9[c]	9.2
MTX	13	20	1	15.6 ± 8.9[d]	8.8
MTX-BSA	25	10	1	7.9 ± 4.1[d]	4.0
BSA	16	2.5[e]	1	27.9 ± 16.9	18.0
MTX + BSA	14	10 2.5[e]	1	14.7 ± 11.1[d]	6.7
MTX	11	20	7	13.4 ± 6.2[d]	14.1
MTX-BSA	21	10	7	11.7 ± 8.5[d]	12.1

[a]Groups of mice were treated every 4 days for a total of 4 doses, with each agent at the levels indicated, beginning either on day 1 or on day 7 after tumor transplant.

[b]Results are expressed as mean values $\pm$SP.

[c]$P<0.05$, statistically significant, Student's *t*-test.

[d]$P<0.001$, statistically significant, Student's *t*-test.

[e]Dose expressed in mg/mouse.

mice treated with BSA alone, the observed slight inhibition of growth was not significant. If treatment was delayed until day 7 after tumor transplant, differences in tumor growth between the controls and mice treated with 10 mg MTX/kg were not significant; however, therapy with MTX-BSA at this dose level was still effective, although differences in tumor size were less than those observed with early treatment and were not apparent until day 12 after tumor transplant. To obtain significant reduction of tumor size with late therapy using MTX alone, it was necessary to increase the dose to 20 mg/kg.

In order to discover how the carrier-bound drugs interacted with cells, experiments were conducted _in vitro_ with cultured L1210 and Reuber H35 hepatoma cells. Figure 2 shows the growth curves of L1210 cells grown in RPMI media in the presence of MTX, MTX-BSA, or MTX-Dextran T70, starting with a cell concentration of 0.3 x 10^5 cells/ml. In the presence of 10^{-7} M MTX cell growth is almost completely inhibited, whereas growth in the presence of 10^{-7} M MTX-BSA or 10^{-6} M MTX-Dextran T70 is comparable to that of the untreated control cultures. At 10^{-6} M, MTX-BSA begins to exert some growth-inhibiting effects and after 72 hr the cell concentration is 10-fold lower than that observed in the control cultures; cells grown in the presence of MTX-Dextran T70 are not severely inhibited until a concentration of 10^{-5} M is used. The amounts of drug found in the L1210 cells when grown _in vitro_ in the presence of 4 x 10^{-6} M [^{3}H]MTX, [^{3}H]MTX-BSA or [^{3}H]MTX-Dextran T70, at 37° are indicated in Table 3. The growth-inhibiting properties of the various drugs illustrated in Fig. 2 appear to follow the intracellular concentrations achieved with each agent. For example, after a 1-hr incubation at 37°, the internal drug concentrations of cells incubated with 4 μM MTX, MTX-BSA and MTX-Dextran T70 are 2.2, 1.12 and 0.42 nmoles/10^9 cells, respectively. The differing intracellular levels of MTX and MTX-BSA observed when cells are incubated in the presence of each drug at 0° instead of 37°, and in the presence of two known inhibitors of MTX transport, pCMS and 5-formylTHF, are also shown in the table. At 0° MTX transport is almost completely inhibited, whereas that of MTX-BSA is only reduced by 46%. In the presence of 100 μM pCMS at 37°, MTX transport is inhibited by 65% in contrast to MTX-BSA where drug transport is only reduced by 32%, moreover, at 0° 100 μM pCMS has little effect on MTX-BSA transport. When the medium contains 10:1 excess 5-formylTHF over MTX, again the transport of MTX alone is inhibited to a greater extent (55%) than that of MTX-BSA (22%). When cells are incubated with [^{3}H]MTX and a ten-fold excess of non-radioactive MTX-BSA, or with [^{3}H]MTX-BSA and a ten-fold excess of non-radioactive MTX, 12% inhibition is observed in each case.

Table 3 also shows that when the lysate and membrane fractions are separated, almost all the radioactivity of cells incubated with MTX is found in the lysate, and that when this fraction was dialyzed, 68% of the radioactivity passed out of the sac, leaving a residual 0.7 nmole; possibly that bound to dihydrofolate reductase. With

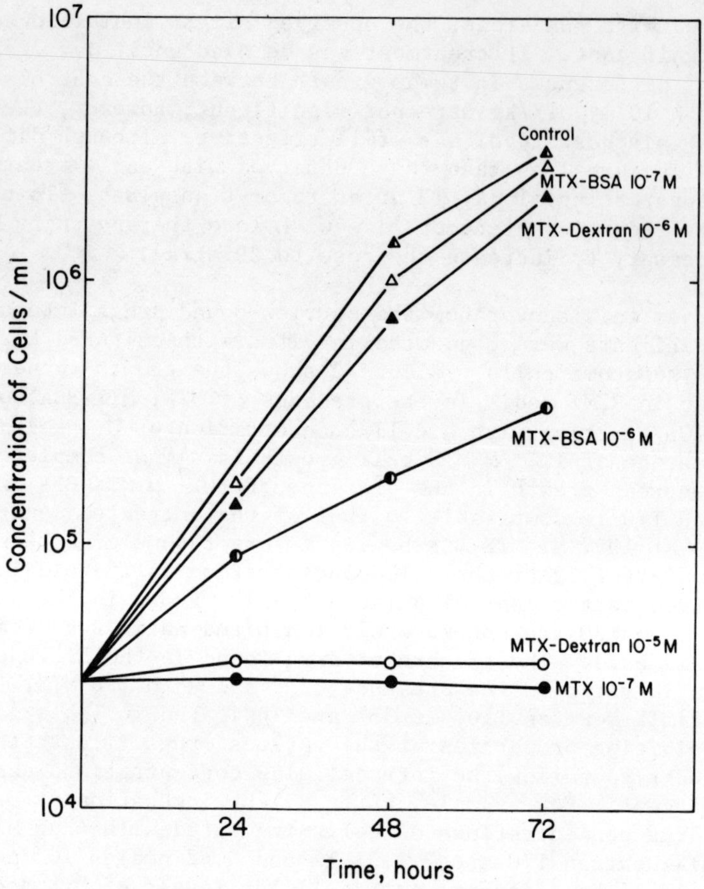
Concentration of Cells / ml
10^7
10^6
10^5
10^4
Control
MTX-BSA 10^-7 M
MTX-Dextran 10^-6 M
MTX-BSA 10^-6 M
MTX-Dextran 10^-5 M
MTX 10^-7 M
24
48
72
Time, hours

Figure 2. The comparative growth curves of L1210 cells in RPMI 1640 medium which occurred in the presence of MTX, MTX-BSA and MTX-Dextran T70 at the concentrations indicated.

MTX-BSA, 34% of the lysate radioactivity was lost by dialysis, leaving a similar quantity, 0.6 nmole, of drug within the dialysis membrane.

In order to determine if hydrolysis of MTX-BSA had occurred on the external surface of the cell membrane, the supernatant fractions of cells incubated with [^{3}H]MTX-BSA, a 1:10 ratio of [^{3}H]MTX-BSA: 5-formylTHF, and an unincubated sample of [^{3}H]MTX-BSA were analyzed by paper chromatography. Excess 5-formylTHF was included in the one case to restrict the cellular uptake of any MTX that appeared in the medium should MTX-BSA be hydrolyzed on the outer membrane surface. In each case 2-3% of the label appeared as MTX. The lack of a

Table 3

Intracellular drug concentrations of L1210 cells incubated with [^{3}H]MTX, [^{3}H]MTX-BSA and [^{3}H]MTX-Dextran T70 at 0° and 37°. 2 x 10^7 cells were incubated for 1 hr with either 4 x 10^{-6} M drug alone or with the addition of the transport inhibitor or competitor indicated. After lysis, lysate and membrane fractions were separated by centrifugation and the lysate dialyzed against water.

Drug	Conc^n.	Inhibitor	Conc^n.	Intracellular Drug Conc^n. 37°	0°	Inhibition	Percentage of total radioactivity in lysate	Percentage of lysate radioactivity remaining after dialysis
	(μM)		(μM)	(nmoles MTX/10^9 cells)		(%)	(%)	(%)
MTX	4			2.24±0.51			94-100	32
					<0.1	∿100		
		pCMS	100	0.78±0.13		65		
		5-formylTHF	40	1.00±0.20		55		
		MTX-BSA	40	1.98±0.23		12		
MTX-BSA	4			1.12±0.33			75-85	66
					0.60±0.13	46		
		pCMS	100	0.76±0.19		32		
					0.52±0.07	54	64-75	
		5-formylTHF	40	0.84±0.12		22		
		MTX	40	0.98±0.29		12		
MTX-Dextran T70	4			0.42±0.02	<0.05			

significant increase in MTX in the medium when cells were incubated with [^{3}H]MTX-BSA suggests that hydrolysis of MTX from its carrier does not occur on the outer membrane, but that the originally synthesized sample of MTX-BSA might contain a small percentage of non-covalently bound MTX which resisted previous separation by column chromatography and dialysis. Alternatively it is possible that certain of the MTX-BSA amide linkages are more labile than others.

When Reuber H35 hepatoma cells were treated with MTX-BSA little effect could be observed on the growth of an MTX-resistant cell line, however, both 10 and 40 μM of the carrier-bound agent showed toxicity to the normal cells (Table 4, Fig. 3). Since only 0.03 μM free MTX is necessary to kill these cells ∿ 0.3% of the bound sample need be cleaved to kill the cells. In contrast, the resistant cells can grow in 5 μM MTX, so > 10% degradation would be necessary to give an effect. It, therefore, appears that the hepatic cells are killed either by free MTX occluded in the carrier-bound sample or by that which is liberated by slow chemical or enzymic hydrolysis in the cell growth medium. Although similar effects no doubt could occur in the L1210 cell-uptake experiments, normal cell growth is less sensitive to MTX concentration. Moreover, the use of [^{3}H]MTX-BSA and MTX-[^{125}I]BSA with L1210 cells indicated that both components of the drug entered these cells, suggesting a pinocytotic effect was predominant in this instance.

With MTX bound to poly-L-lysine, however, the situation is more complex (Fig. 4). In this instance both resistant and non-resistant cell lines submit to attack by the carrier-bound agent suggesting that the nature of the carrier is a critical factor in cellular response to carrier-bound drugs. At the levels used the carrier itself is essentially non-toxic to the cells.

Table 4

The number of cells/plate observed after 72 hr when resistant and non-resistant H35 cells were grown in the presence of the indicated levels of MTX-BSA.

Cell type	Concentration of MTX-BSA	Cells/plate at 72 hr
	(μM)	
H35	0	2.7×10^{6}
H35	10	9.4×10^{4}
H35	40	7.3×10^{4}
H35R	0	2.2×10^{6}
H35R	10	2.2×10^{6}
H35R	40	1.8×10^{6}

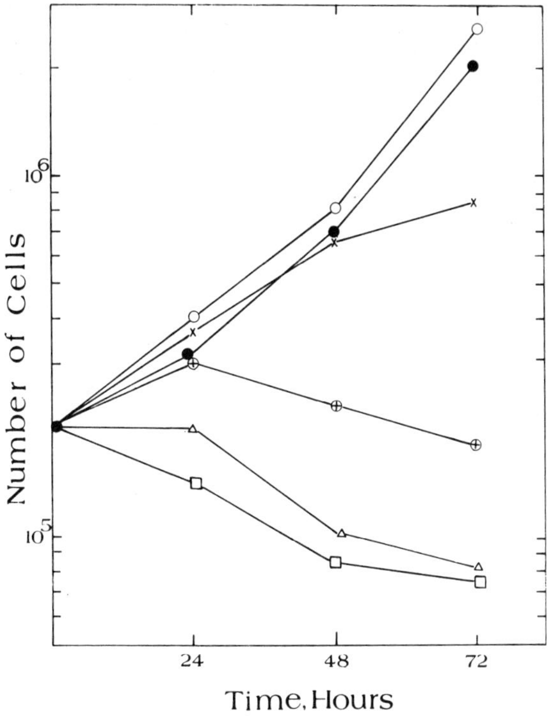

10^6
10^5
Number of Cells
24
48
72
Time, Hours

Figure 3. The comparative growth curves of H35 cells plated at a density of 2 x 10^5 cells/dish in Swims Medium S-77 supplemented with 20% horse serum, 5% fetal calf serum, and 4 mM glutamine, when treated with 0 (O——O), 2 (X——X), 5 (⊗——⊗), 20 (Δ——Δ), and 40 μM (□——□) MTX-BSA. (●——●) The response of the MTX-resistant H35R cells to treatment with 40 μM MTX-BSA.

The above experiments indicate that carrier-bound drugs do have altered properties from those exhibited by the free entity, however, they do not necessarily lead to enhanced drug action in a particular host. The greater retention time of the MTX-BSA in vivo affords comparable control of the L1210 ascitic tumor to that observed with free MTX, and superior control of metastases with the Lewis lung carcinoma, however, in vitro experiments with the lymphocytic tumor indicate lower uptake of carrier-bound drug, with differences attributable to carrier structure (e.g., polysaccharide of polypeptide). With the hepatoma cells there is no uptake of albumin-carried drug, yet poly-L-lysine is an effective carrier. These observations suggest further experimentation with other carriers, other hosts, and many more cell types should be accomplished before the full potential of carrier-bound agents can be established.

ABSTRACT

When methotrexate (MTX) covalently bound to bovine or murine serum albumin (BSA or MSA), bovine IgG, chymotrypsinogen and various molecular weight dextrans is injected ip into BDF_1 mice, higher, more prolonged serum concentrations and decreased rates of excretion

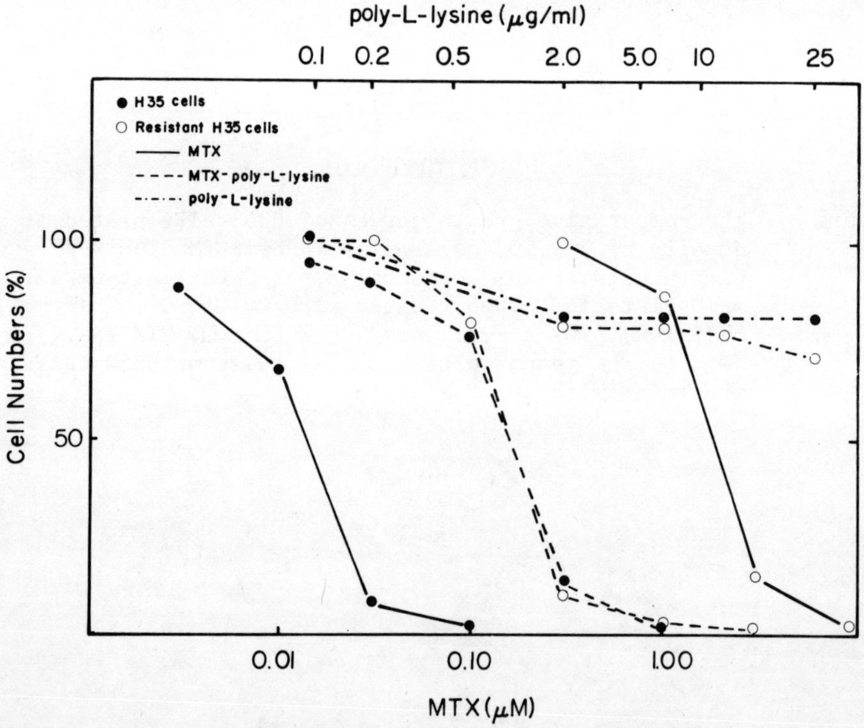
poly-L-lysine (µg/ml)
0.1 0.2 0.5 2.0 5.0 10 25
H35 cells
Resistant H35 cells
MTX
MTX-poly-L-lysine
poly-L-lysine
Cell Numbers (%)
100
50
0.01 0.10 1.00
MTX (µM)

Figure 4. The comparative growth curves of resistant and non-resistant Reuber H35 cells when treated with MTX, MTX-poly-L-lysine and poly-L-lysine alone at the levels indicated. The growth rates are expressed as a percentage of those observed with untreated cell samples.

of MTX are observed compared with a similar group of mice treated with free MTX. Treatment of mice carrying the L1210 ascitic tumor with MTX, or MTX bound to a carrier protein, leads to a similar increment in life-span with each drug. In contrast, the dextran derivatives are ineffective. In mice carrying the Lewis lung carcinoma, MTX-BSA is markedly superior to free MTX in the control of metastases. Experiments with L1210 and Reuber H35 hepatoma cells *in vitro* indicate, in the former case, that the carrier-bound drug is taken up to a lesser extent than the free agent and measurements with radioactively labeled carrier indicate uptake possibly by a pinocytotic mechanism. The hepatoma cells differ by being unresponsive to albumin-bound drug but take up MTX bound to poly-L-lysine. The observations suggest that the response to carrier-bound MTX therapy in whole animals may be caused primarily by the altered pharmacokinetic properties of the derivatized drug, but suggest also that in some instances the carrier may play more than a passive role.

ACKNOWLEDGMENTS

This work was supported by USPHS grants CA-11778, AM-07097 and CA-25933 and by grant CH-31 from the American Cancer Society. J. M. W. is the recipient of a USPHS Research Career Development Award, CA-00106, from the National Cancer Institute. The authors wish to thank Ms. Zenia Nimec for her excellent technical assistance.

REFERENCES

1. A. Trouet, D. Deprex-de Campeneere and C. De Duve, Nature New Biology, 239, 110 (1972)
2. G. Gregoriadis and E. D. Neerunjun, Res. Commun. in Chem. Pathol. and Pharmacol., 10, 351 (1975)
3. G. F. Rowland, G. J. O'Neill and D. A. L. Davies, Nature, 255, 487 (1975)
4. B. C. F. Chu and J. M. Whiteley, Mol. Pharmacol., 13, 80 (1977)
5. G. P. Mell, J. M. Whiteley and F. M. Huennekens, J. Biol. Chem., 243, 6074 (1968)
6. P. J. McConahey and F. J. Dixon, Int. Arch. Allergy, 29, 185 (1966)
7. R. I. Geran, N. H. Greenberg, M. M. MacDonald, A. M. Schumacher and B. J. Abbott, Cancer Chemother. Rep., Part 3, 3, 7 (1972)
8. J. G. Mayo, Cancer Chemother. Rep., 3, 325 (1972)
9. H. Wexler, J. Nat. Cancer Inst., 36, 641 (1966)
10. K. Weber and M. Osborn, J. Biol. Chem., 244, 4406 (1969)
11. R. C. Jackson, D. Niethammer and F. M. Huennekens, Cancer Biochem. Biophys., 1, 151 (1975)
12. C. Fan, G. Henderson, K. Vitols and F. M. Huennekens, "Antimetabolites in Biochemistry, Biology and Medicine (Eds. J. Skoda and P. Langen) Pergamon Press, Oxford and New York (1979) p. 315
13. H. Pitot, C. Periano, P. Morse and V. R. Potter, Natl. Cancer Inst. Monogr., 13, 229 (1964
14. M. D. Reuber, J. Natl. Cancer Inst., 26, 891 (1961)

15. E. M. Levene, Exp. Cell Res., 74, 99 (1972)
16. D. R. Seeger, D. B. Cosulich, J. M. Smith, Jr. and M. E. Hultquist, J. Amer. Chem. Soc., 71, 1753 (1949)

ESTEROLYTIC ACTION OF WATER-SOLUBLE IMIDAZOLE CONTAINING POLYMERS

J. A. Pavlisko (Chemistry and Life Sciences Group, Research Triangle Institute, Research Triangle Park, NC 27709)
C. G. Overberger (This review is taken from the work of the following students and postdoctoral fellows: M. Morimoto, T. W. Smith, K. Dixon, S. Mitra, A. Guterl, and Y. Kawakami)
Department of Chemistry, The University of Michigan, Ann Arbor, Michigan 48109

This review article summarizes our more recent investigations of the esterolytic action of water soluble imidazole containing polymers. In our previous work we have explained increased reactivity of polymeric reactants in terms of electrostatic and cooperative effects. Recently, we have directed our efforts in emphasizing the apolar interaction as it has reflected dramatic rate enhancements and proven a predominant factor in determining the maximum catalytic efficiency. Polymeric catalysts which are included in this review are poly[1-alkyl-4(5)-vinyl-imidazoles] and their copolymers, hydrophobic terpolymers of 4(5)-vinylimidazole, copoly[vinylamine/4(5)-vinylimidazole], and poly(ethylenimine-g-L-histidine). All of these catalysts share the common property of water solubility and the capability of attaining high catalytic efficiencies which is attributed to apolar interactions.

INTRODUCTION

Attempts have long been made to use synthetic polymers as model compounds for biologically active macromolecules. An approach to biopolymers from the field of polymer science can be depicted as follows (1):

$$\text{biopolymer} \underset{\text{elucidation}}{\overset{\text{simplification}}{\rightleftarrows}} \text{model compound} \xrightarrow{\text{development}}$$

$$\text{biologically important polymer}$$

The main purpose of this approach is the investigation of the mechanism of the biological activity of biopolymers using model compounds to explain the function of the biopolymer. The step "development" is evidenced in application of biopolymers and their model compounds in the areas of biomaterials, collagen-glycosaminoglycan membranes as artificial skin (2), and chemotherapy, synthetic polynucleotides (3) and polyanions (4) as potent antiviral agents.

Hydrolytic enzymes, characterized by high specificity and high catalytic reactivity have been the most frequently modeled biopolymers. At the active site of the enzyme there are usually several functional groups responsible for the overall catalytic reaction which are covalently bound to remote areas of the enzyme. Collectively these interactions, which are termed as intramolecular multiple catalyst reactions are closely related to an enzyme's specificity and efficiency.

The histidine residue, of which the imidazole group is a component, in particular is involved in the intramolecular multiple catalyst reaction of hydrolases. The catalytic activity of vinyl polymers containing imidazole groups was initiated by Overberger (5) with the synthesis of poly[4(5)-vinylimidazole].

Enhanced catalytic activity of polymeric bound imidazole has been investigated by numerous researchers who have used various types of macromolecular backbones and imidazole derivatives as biopolymer models. A summary of our work in the area of polymer containing imidazole has been presented in several review articles (6-8). In polymer catalysts containing imidazole there has been defined three important factors which contribute to the overall enhancement of the catalyst's efficiency: cooperative interactions, electrostatic interactions and hydrophobic or apolar interactions.

Cooperative interactions between catalytically active imidazole groups and other functional groups has been well documented (6). The total rate equation for the polymeric catalyst can be expressed by the following equation:

$$k_{cat} = k_1\alpha_1 + k_2\alpha_1^2 + k_3\alpha_2 + k_4\alpha_1\alpha_2$$

where k_1 is the single nucleophilic rate constant, k_2 is an imidazole catalyzed imidazole nucleophilic constant, k_3 is the rate constant for imidazole anion, and k_4 the rate constant for anionic imidazole-catalyzed imidazole catalysis, α_1 and α_2 as fractions of neutral and anionic imidazole, respectively.

Since the fraction of anionic imidazole present is generally extremely small in the pH regions investigated, the cooperative interactions have been mainly attributed to neutral imidazole-imidazole interactions ($k_2\alpha_1^2$). Overberger (9) presented three possible mechanisms for neutral imidazole cooperative interactions which are

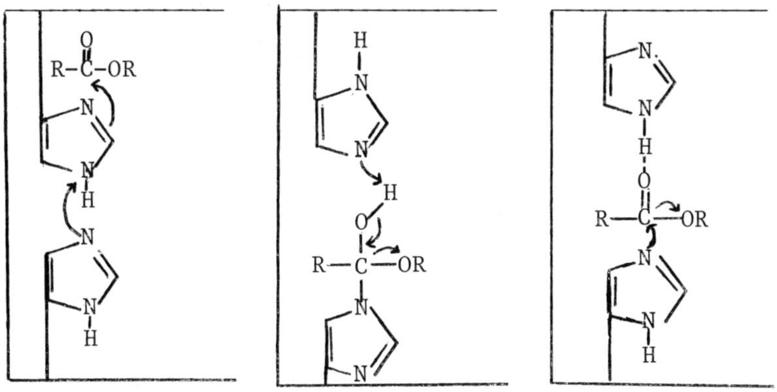
I
II
III

Figure 1. Cooperative interactions of two neutral pendant imidazoles (Overberger et al (9)).

shown in Figure 1. The first two mechanisms (A and B) are general base catalysis reactions and the third mechanism (C) entails a general acid catalyzed nucleophilic imidazole reaction.

The role of electrostatic interactions involving a partially charged polymer catalyst and an oppositely charged substrate has also been extensively investigated (10).

Hydrophobic interactions are more a property of the solvent than a particular nonpolar solute. Water molecules surrounding each individual nonpolar solute molecule orient in such a way as to maximize hydrogen bonding. This results in the ordering of the solvent in the vicinity of the solute corresponding to a dramatic increase in the entropy of the total system. The phenomena of apolar bonding is favorable because of the entropy factors derived from the strong tendency of the water to hydrogen bond.

Recently, we have directed our efforts to emphasizing the apolar interaction as it has reflected dramatic rate enhancements and proven a predominant factor in determining maximum catalytic efficiency. In this article we summarize our most recent activities in the preparation of imidazole catalysts which are soluble in highly aqueous solvent systems with an emphasis on their related apolar binding properties.

CATALYSIS BY VINYL POLYMERS CONTAINING IMIDAZOLE

As described in the introduction, the imidazole groups are involved in the catalytic action of most hydrolases. Unfortunately, nucleophilic catalysis by chymotrypsin and related enzymes occurs with the intermediate formation of an ester, formed from the acyl group of the substrate and hydroxyl group of a serine residue. There is also no evidence that imidazole acts as a nucleophilic catalyst in any enzymatic acyl transfer reaction.

Nucleophilic catalysis of acyl transfer in nonenzymatic reaction is illustrated by imidazole catalysis of the hydrolysis of activated acyl compounds. The reaction proceeds as shown:

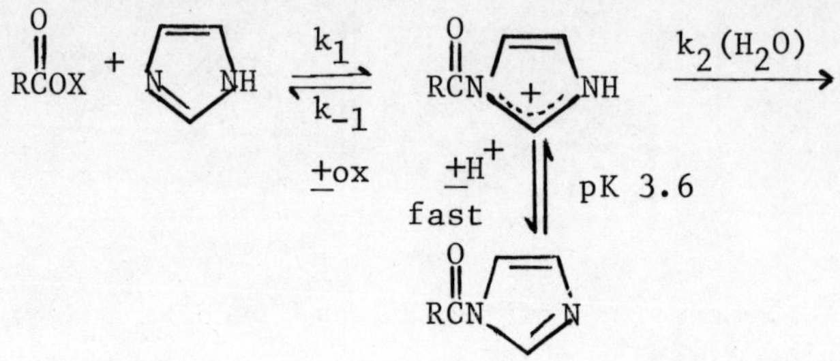
RCOX
NH
k1
k-1
±ox
RCN
+
NH
k2(H2O)
RCO2H + N
NH
±H+
fast
pK 3.6
RCN
N

If N-methylimidazole is substituted for imidazole, the reactive acylimidazolium ion intermediate cannot lose a proton and undergoes rapid hydrolysis (11).

Numerous investigators have recognized the importance of polymer catalysis and there has appeared several comprehensive review articles (1,12) which are inclusive of imidazole catalysis.

Efficiencies of vinyl polymers containing imidazole catalysts can be increased by enhanced apolar binding between substrate and either the catalyst's apolar polymer backbone, a specific apolar group incorporated for the purposes of enhanced binding, or a long-chain N-acylimidazole intermediate of the hydrolysis reaction.

In the proceeding sections we describe several polymer catalysts which utilize these individual or multiple properties for effecting apolar binding and increased efficiencies.

Poly[1-alkyl-4(5)-vinylimidazoles]

The most important physical properties of poly[1-alkyl-4(5)-vinylimidazoles] is their water solubility. The alkylation of the imidazole ring reduces intramolecular H-bonding which causes the insolubility of the polymer in aqueous media. The apolar character of the poly[1-alkyl-4(5)-vinylimidazoles] is easily varied by the nature of the alkyl substituent. The preparation of 1-alkyl-5-vinylimidazole monomer and polymer is shown in Scheme I.

Poly(1-methyl-5-vinylimidazole)

Poly(1-methyl-5-vinylimidazole) catalyzed hydrolysis of 3-nitro-4-acyloxybenzoic acid substrates (Sn^-) exhibit substantial rate

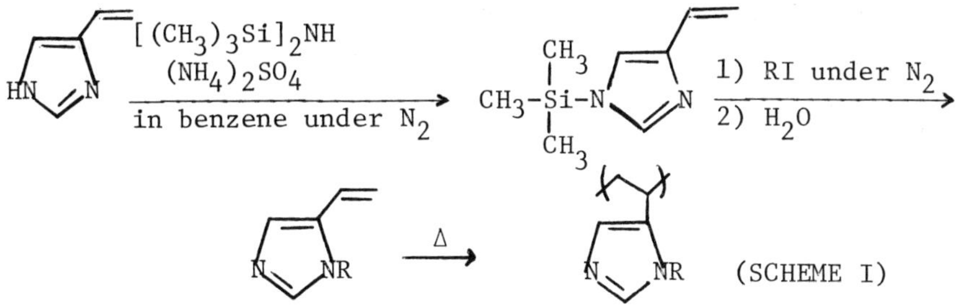

[(CH3)3Si]2NH
(NH4)2SO4
in benzene under N2
CH3-Si-N
1) RI under N2
2) H2O
Δ

(SCHEME I)

enhancements due to hydrophobic interactions. Figure 2 shows the relative rates of hydrolysis of Sn^- as for the poly(1-methyl-5-vinylimidazole) catalyst relative to its model compound 1,5-dimethylimidazole in an aqueous medium and in an ethanol-water mixture. The large influence of the solvent composition on the catalytic behavior is related to the confromational change of the catalyst polymer. Overberger and Morimoto (14) presented a schematic illustration of the conformational change (Figure 3) for poly[4(5)-vinylimidazole]. The shrinkage of the polymer chains at low and high ethanol compositions is attributed to increased hydrophobic interactions and hydrogen bonding, respectively. Poly(1-methyl-5-vinylimidazole) in a solvent of low ethanol content is similarly in a tightly coiled conformation due to an increased hydrophobic interaction. This shrinkage in polymer conformation has been attributed by us to the apolar nature of the polymer's hydrocarbon backbone.

The hydrophobic domain formed by the vinyl backbone represents an apolar binding site for the more apolar substrates. Mirejovsky (15) has recently proposed that the conformational changes are attributed to an inherent tendency of imidazole to aggregate resulting in an inefficient utilization of imidazole and a decrease in catalytic efficiency. Therefore, it is difficult to conceive, as suggested by the author, how the apolar interactions of poly(1-methyl-5-vinylimidazole) and substrate increases the observed catalytic efficiency.

Copoly[1-alkyl-4(5)-vinylimidazole/4(5)-vinylimidazole]

Copoly[1-methyl-4-vinylimidazole/4(5)-vinylimidazole], copoly-[1-methyl-5-vinylimidazole/4(5)-vinylimidazole], copoly[1-ethyl-5-vinylimidazole/4(5)-vinylimidazole], were synthesized according to Scheme I and their catalytic activity towards 3-nitro-4-acyloxybenzoic acid substrates (Sn^-) measured in 28.5% ethanol-water and in water (17,18).

Water soluble poly[1-methyl-5-vinylimidazole] efficiently catalyzed the hydrolysis of Sn^- substrates because of the hydrophobic nature of the polymer backbone. Unsubstituted imidazole catalysts such as poly[4(5)-vinylimidazoles] are capable of accelerative kinetics by accumulation of a more hydrophobic acylated polyvinyl-

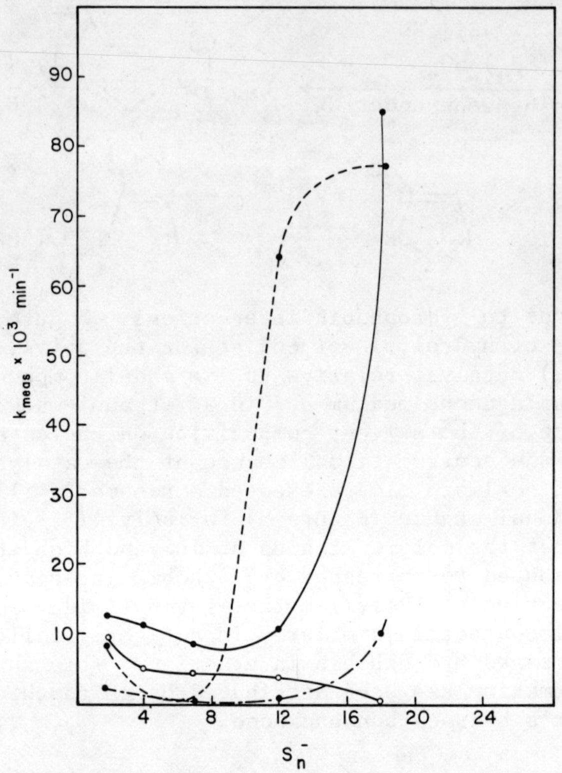
90
80
70
60
50
40
30
20
10
4
8
12
16
20
24
$k_{meas} \times 10^3$ min^{-1}
S_n^-

Figure 2a. Solvolysis of the Sn^- series by poly(N-alkylimidazoles). Hydrolysis of Sn^-, 26.7% ethanol, 3.3% CH_3CN, 70% H_2O (v), pH 6.85 buffer, $\mu = 0.02$, 26°C; [catalyst] 5×10^{-4} M; $[Sn^-] = 5 \times 10^{-5}$ M; (O) 1,5-DMIm; (●) poly(1-N-VIm); (◐) poly(1-Me-5-VIm); (◒) poly(1-Bu-5-VIm) (Overberger and Smith (13)).

imidazole esterolytic intermediate caused by a slow rate determining deacylation step shown in Scheme II.

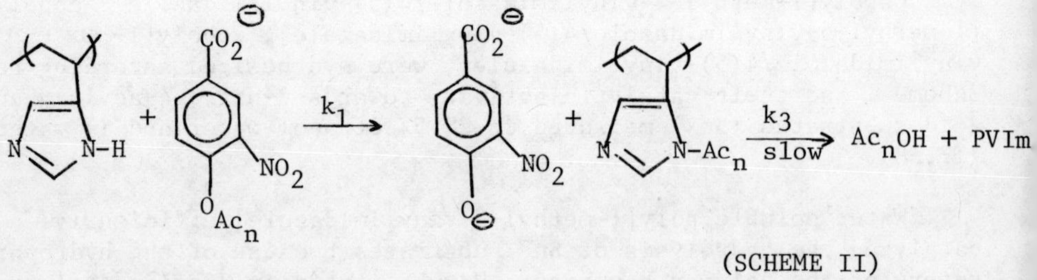
N
N-H
+
CO_2
NO_2
OAc_n
k_1
CO_2
NO_2
O
+
N
N-Ac$_n$
k_3
slow
Ac$_n$OH + PVIm

(SCHEME II)

The esterolytic intermediate for poly(1-methyl-5-vinylimidazole) is a positively charged acylonium ion, which is very reactive and

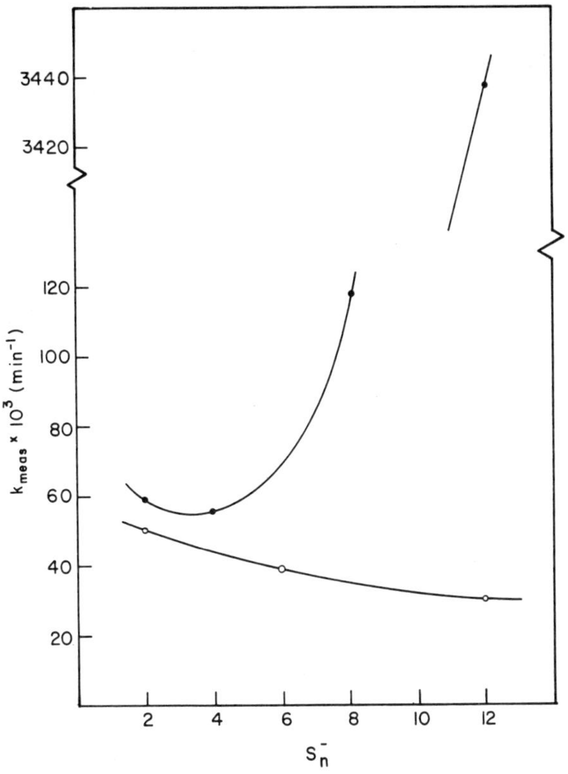

3440
3420
120
100
80
60
40
20
k meas x 10³ (min⁻¹)
2
4
6
8
10
12
Sn⁻

Figure 2b. The poly(1-Me-5-VIm)-catalyzed solvolysis of the Sn^- series. Hydrolysis in pH 7.90 aqueous buffer, $\mu = 0.02$, 26°C, [catalyst] = 5×10^{-4} M, [Sn^-] = 5×10^{-5} M; ●, poly(1-Me-5-VIm); O, 1,5-DMIm (Overberger and Smith (16)).

decomposes upon formation to regenerate catalytically active imidazole residues.

These copolymers, by exhibiting differences in their ester hydrolysis mechanisms, permit an increase in the apolar nature of the catalyst by incorporation of the poly[4(5)-vinylimidazole] groups while simultaneously incorporating the 1-alkly substituted imidazole capable of rapid substrate turnover.

The data in Figure 4 shows little difference in the catalytic activity of copoly[1-methyl-4-vinylimidazole/4(5)-vinylimidazole] and a mixture of the respective homopolymers towards S_2^- and S_4^-. Hydrolysis of S_{12}^- and S_{18}^- was more efficiently catalyzed by the copolymers than the mixtures due to the increased hydrophobic interaction between copolymer and substrate. The hydrolysis of S_{12}^- proceeded in an accelerative manner resulting from accumulation of acylated vinylimidazole intermediate.

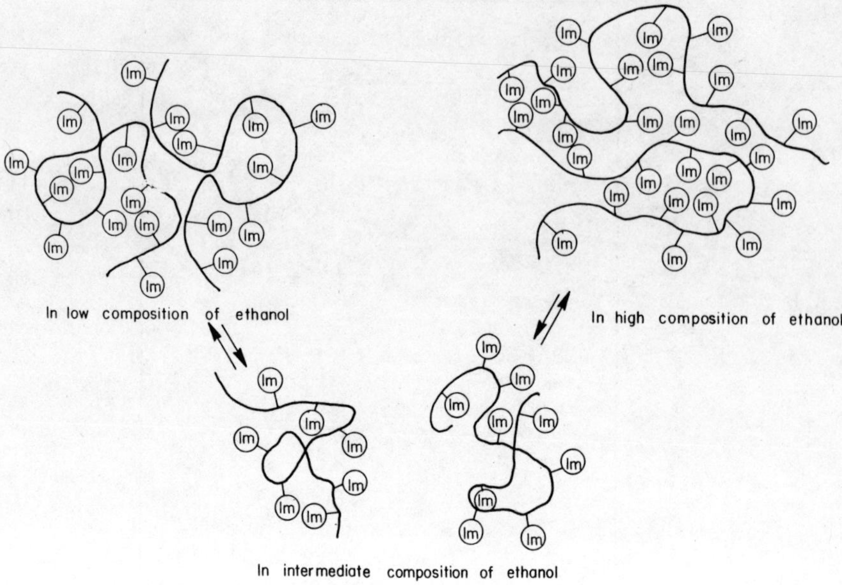
In low composition of ethanol
In high composition of ethanol
In intermediate composition of ethanol

Figure 3. A schematic illustration of the conformation of poly(4-vinylimidazole) in ethanol-water (Overberger and Morimoto (14)).

The effect of the length of the polymer n-alkyl side chain on the rate of hydrolysis of S_2^- and S_{12}^- are shown in Figures 5 and 6. The effect of the side chain length of one to three carbon atoms on the catalytic activity is minimal. However, higher reactivity towards S_n^- was obtained as the length of the alkyl side chain was increased.

Although there were few exceptions, there were two important features. (1) The hydrolysis rates for S_2^- and S_4^- at each pH increased with 4(5)-vinylimidazole content; (2) the hydrolysis of S_7^- and S_{12}^- at intermediate pH 4.83 and pH 8.44 was dependent on the 4(5)-vinylimidazole content. The hydrolysis of S_7^- and S_{12}^- at intermediate pH and the hydrolysis of S_{18}^- at pH 7.11 displayed maxima for the dependence of k_{cat} on the content of 4(5)-vinylimidazole in the polymers. The k_{cat} values for S_{18}^- were larger in the presence of poly[4(5)-vinylimidazole] than any of the copolymers.

These data indicate that poly[4(5)-vinylimidazole] was a better catalyst than the copolymers for S_n^- in 28.5% ethanol-water when deacylation was not the rate determining step. The larger pKa value for poly[4(5)-vinylimidazole] leads to a higher nucleophilicity of the imidazole groups towards ester hydrolysis and increased electrostatic attraction of the negatively charged substrates by the greater number of protonated pendant imidazole groups. When deacylation was

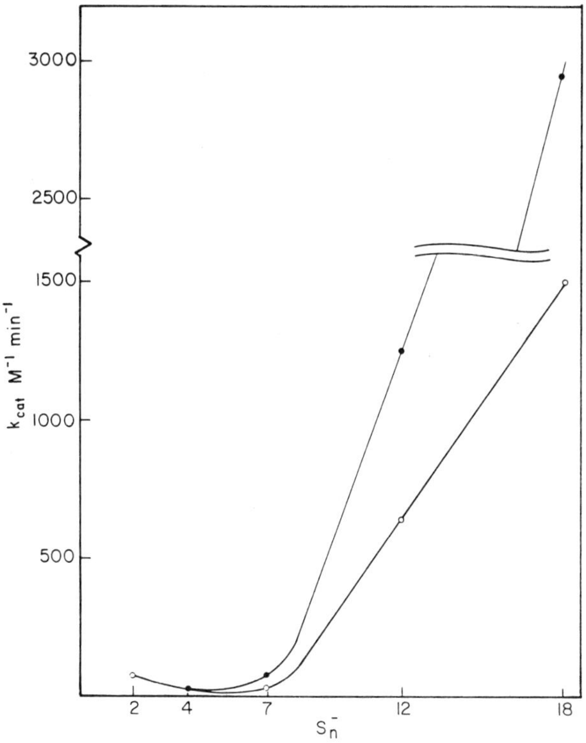
3000
2500
1500
1000
500
k_{cat} M^{-1} min^{-1}
2
4
7
12
18
S_n^-

Figure 4. Second-order rate constants for solvolysis of Sn^- series of 4(5)-VIm containing polymers: (●) cop[VI/1-Me-4-VI]$_{50}$; (O) [PVI-p(1-Me-4-VI)]$_{50}$. In 28.5% ethanol-water, pH 7.11, 26°C, μ = 0.02 (Overberger and Kawakami (18)).

the rate determining step, as for S_{12}^- hydrolysis in low ethanol-water compositions, then the copolymers were more catalytically efficient.

The most efficient hydrolysis rates for the hydrolysis of S_n^- by the copolymers occurred for long chain substrates. Because of the copolymer's water solubility, N-alkylimidazole homopolymers and copolymers with 4(5)-vinylimidazole are capable of exhibiting enhanced catalytic efficiencies due to stronger hydrophobic effects. Table I contains the second order rate constants for the hydrolysis of S_n^- at pH 5.48 and 7.14 in 96.7% water-acetonitrile. The second order rate constants increase with increasing chain length and with increasing length of the polymer side chain. The increase was more dramatic than that resulting in various ethanol-water compositions where the hydrophobic interactions are weaker. The hydrolysis of S_{12}^- by the copolymers was about five times faster in water than 28.5% ethanol-water.

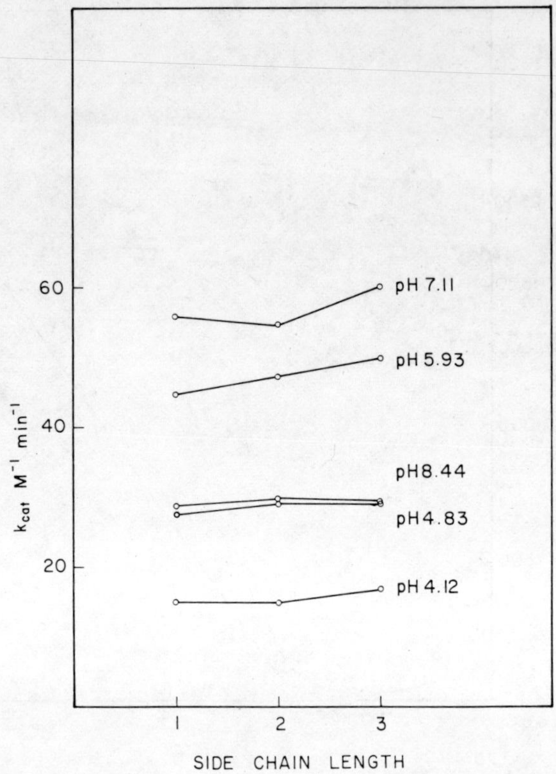
60
40
20
k_{cat} M^{-1} min^{-1}
pH 7.11
pH 5.93
pH8.44
pH4.83
pH 4.12
1
2
3
SIDE CHAIN LENGTH

Figure 5. Second-order rate constants for solvolysis of S_2^- as a function of polymer side-chain length. Hydrolysis in 28.5% ethanol-water catalyzed by copoly[4(5)-VI/1-alkyl-5-VI]; 26°C, μ = 0.02 (Overberger and Kawakami (18)).

Table 1

k_{cat} IN 96.7% WATER AT 26°C, μ = 0.02

	k_{cat}					
	S_2^-		S_7^-		S_{12}^-	
	pH 5.48	pH 7.14	pH 5.48	pH 7.14	pH 5.48	pH 7.14
cop[VI/1-Me-4-VI]$_{75}$	220	276	69	1220	216	4321
cop[VI/1-Me-4-VI]$_{50}$	212	379	155	1700	214	5695
cop[VI/1-Me-4-VI]$_{25}$	160	184	55	600	151	3987
cop[VI/1-Me-5-VI]$_{75}$	121	394	167	1812	183	5233
cop[VI/1-Me-5-VI]$_{50}$	130	380	138	1612	194	5821
cop[VI/1-Me-5-VI]$_{25}$	115	240	120	544	180	5089
cop[VI/1-Me-5-VI]$_{25}$	120	316	133	1271	187	5233
cop[VI/1-Prop-5-VI]$_{25}$	123	(a)	141	(a)	194	(a)

[a]The copolymer is insoluble at this pH (18).

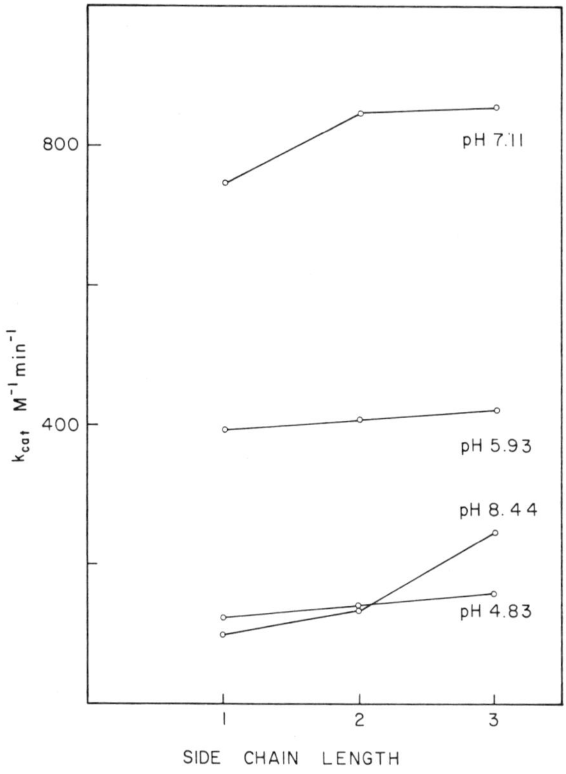
800
400
k_{cat} M^{-1} min^{-1}
pH 7.11
pH 5.93
pH 8.44
pH 4.83
1
2
3
SIDE CHAIN LENGTH

Figure 6. Second-order rate constants for solvolysis of S_{12}^- as a function of polymer side-chain length. Hydrolysis in 28.5% ethanol-water catalyzed by copoly[4(5)-VI/1-alkyl-5-VI]; 26°C; μ = 0.02 (18).

Hydrophobic Terpolymers of 4(5)-Vinylimidazole

Water soluble, hydrophobic terpolymers of 4(5)-vinylimidazole were prepared in order to investigate hydrophobic effects attributed to random, covalently bound pendant apolar side chains in the hydrolysis of the charged apolar ester substrates (S_n^-). The rationale for the preparation of these terpolymers is provided in the analogy to the poly[4(5)-vinylimidazole] catalysts. The apolar nature of the catalyst poly[4(5)-vinylimidazole] is derived from its hydrocarbon backbone and formation of the transient and hydrophobic N-acylimidazole intermediate. Accumulation of long-chain acylated imidazole residues, which increases hydrolysis rates dramatically, inspired the concept of utilizing hydrophobic polymer side chains.

Terpolymers of 4(5)-vinylimidazole, acrylamide (AA), and either 3-buten-2-one, 1-penten-3-one, 1-undecen-3-one, or 1-hexadecen-3-one were evaluated for esterolytic activity with several substrates of differing hydrophobic chain lengths and the results are shown in Table II.

Table II

SECOND-ORDER RATE CONSTANT (k_{cat}) VALUES (M^{-1} min^{-1})

In Water	S_{18}^-	S_{12}^-	S_7^-	S_2^-
Poly[4(5)VIm-C_{13}-AA*]		1,030	266	42.3
Poly[4(5)VIm-C_3-AA]		882	152	37.3
Poly[4(5)VIm-C_3(C=O)-AA]		872	173	37.2
Poly[4(5)VIm-C_2-AA]		517	44.5	37.3
Poly[4(5)VIm-C_1-AA]		291	23.9	26.1
In 30% Ethanol				
Poly[4(5)VIm-C_{13}-AA]	513	294	20.9	13.4
Poly[4(5)VIm-C_3-AA]	385	143	8.8	10.4
Poly[4(5)VIm-C_3(C=O)-AA]	283	138	8.3	10.3
Poly[4(5)VIm-C_2-AA]	(a)	(a)	5.7	9.7
Poly[4(5)VIm-C_1-AA]	(a)	(a)	3.5	6.5

[a] Substrate not soluble in these solutions (19).
*Acrylamide.

An analysis of the kinetic data suggests the following: (1) Increasing the size of the side chain increases the rates faster than the corresponding increase in total apolar weight of terpolymer; (2) comparable apolar weights are more effective as a few long side chains than numerous short side chains; (3) an increase of the acyl chain length of the substrate increases kinetic rates; (4) the substrate S_7^- shows the largest relative increases in rates when the transition from a 30% ethanol-water solvent to a 100% water solvent is made.

Figure 7, which is a graph of the second-order rate constant (k_{cat}) versus aliphatic chain length of substrate in water, shows the increase in rates as longer chain substrates are employed. The substrate chain length has previously been proven quite effective in increasing hydrolysis rates. The substrate is one of the two reacting species in the bimolecular hydrolysis reaction with its own inherent apolar qualities. The effectiveness of the terpolymer side chain is shown in Figure 8. Hydrolysis of the least affected substrate S_2^- shows a 60% increase in rates as the terpolymer side chain is increased from 1 to 13 carbons. The most affected substrate, S_7^-, shows an increase of 1110% with the same terpolymer side chain increase. The rate contribution from the side chain obviously exceeds that from the backbone in this instance. These observed increases are greater than the total apolar weight increase, per monomer unit, of the terpolymers in going from 1 to 13 carbon side chains. This apolar weight increase is a lesser 47%

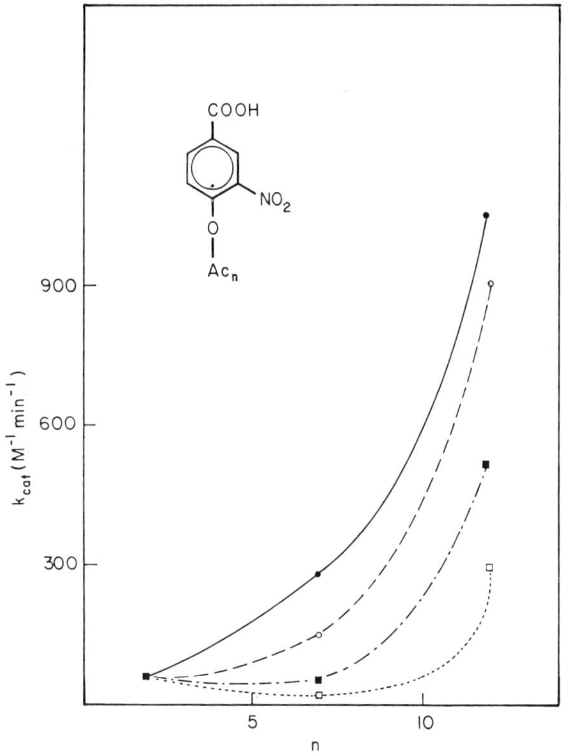
COOH
NO2
O
Acn
900
600
300
kcat (M-1 min-1)
5
10
n

Figure 7. Substrate chain length vs. k_{cat} (M^{-1} min^{-1}) in 3.2 vol% acetonitrile and 96.8 vol% water at pH 7.9, μ = 0.02, TRIS = 0.02 M, 26°C; (-●-) poly[4(5)VIm-C_{13}-AA], (--O--) poly[4(5)VIM-C_8-AA], (-·-□-·-) poly[4(5)VIm-C_2-AA], (---□---) poly[4(5)VIm-C_1AA] (19).

The observation that the same weight of apolarity is more effective when associated with a few long pendant chains than with a larger number of shorter apolar chains is evidenced by the 1-undecen-3-ol terpolymer. This terpolymer, which has 41% of its total apolar weight in side chains, is 1.7 times more effective for the hydrolysis of S_{12}^- in water and 3.4 times more effective for the hydrolysis of S_7^- in water than the terpolymers with 1-penten-3-ol which has a comparable 36.7% of its apolar weight in side chains.

Copoly[vinylamine/4(5)-vinylimidazole]

Copoly[vinylamine/4(5)-vinylimidazole] extends the studies of poly[4(5)-vinylimidazole] catalyzed ester hydrolysis to an aqueous environment. Copoly[vinylamine/4(5)-vinylimidazole] were prepared as shown in Scheme III.

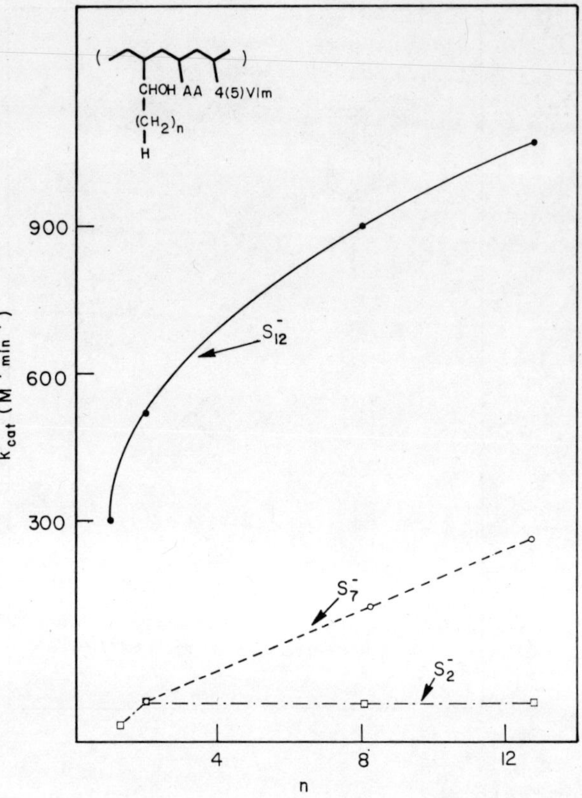
CHOH AA 4(5)VIm
(CH2)n
H
900
600
300
kcat (M-1 min-1)
S12-
S7-
S2-
4
8
12
n

Figure 8. Terpolymer side chain length vs. k_{cat} (M^{-1} min^{-1}) in 3.2 vol% acetonitrile and 96.8 vol% water at pH 7.9, μ = 0.02, TRIS = 0.02 M, 26°C (19).

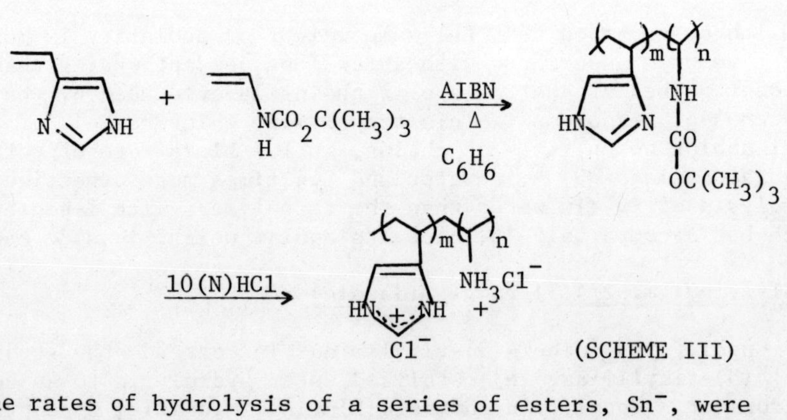
AIBN
Δ
C6H6
10(N)HCl

(SCHEME III)

The rates of hydrolysis of a series of esters, Sn^-, were determined and are shown in Figure 9 for two vinylamine/4(5)-vinylimidazole copolymers: IA-1 with 25% vinylamine units and IA-7 with

33% vinylamine units. As in the previous cases, very large rate enhancements are achieved in the hydrolysis of the long chain esters. Thus, with S_{12}^- the copolymers were more than 300 times as efficient as imidazole in catalyzing the hydrolysis. As opposed to the previous studies, the catalytic rate constants increased in going from S_2^- to S_4^-, indicating that the hydrophobic interactions begin to play a dominant role with these copolymers even with S_4^-. This dominance is even more significant when considering the large steric factors associated with the copolymer catalyzed hydrolysis in aqueous systems where the polymer chains are more tightly coiled.

Decelerative kinetics were observed for the copolymer catalyzed hydrolysis of S_4^-, S_7^- and S_{12}^- which further evidenced the significant apolar contributions of substrates starting from S_4^-. Such decelerative kinetics have previously been observed with other imidazole catalysts but do not generally appear until S_{12}^- hydrolysis.

Of the two copolymers studied, IA-7 was about 50% more reactive than IA-1 for the substrate S_7^-. This trend has been attributed to the fact that IA-1, with less amine content, is more tightly coiled in solution, and steric effects decrease the accessibility of the bulky S_7^- to the active imidazole sites. With S_{12}^-, however, hydrophobic effects mask other factors and IA-7 is only 5% more reactive than IA-1.

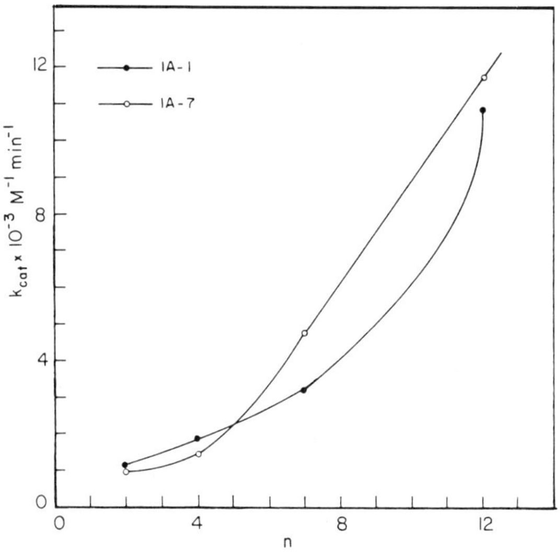
IA-1
IA-7
$k_{cat} \times 10^{-3}$ M^{-1} min^{-1}
0
4
8
12
n

Figure 9. Second-order rate constants for the hydrolysis of the Sn^- series by [4(5)-VIm-VAmine] copolymers: (●) IA-1, (O) IA-7 in 96.7% water-3.3% acetonitrile, pH = 8, $[S_n^-] = 5 \times 10^{-5}$ M, [catalyst] = 5×10^{-4} M, $\mu = 0.02$, 26°C (20).

Hydrophobic interactions were also studied with the same copolymers and substrates in 20% ethanol-water systems in order to establish if the observed large rate enhancements could be attributed to the high water content of the solvent and an apolar effect. The results are shown in Table III and a plot of rate constants against the substrate chain length is shown in Figure 10. Comparison of these data with those obtained in water reveal very large effects of the water content of the solvent on the k_{cat} values for the copolymers. Hydrolysis of S_2^- by the copolymers is lower in 20% ethanol-water largely due to the reduction in bifunctional catalysis because of the extended nature of the catalyst. With longer substrate chains the value of "r" [r = k_{cat}(polymer)/k_{cat}(imidazole)] changes dramatically until S_{12}^- where they are about 55-60% of the value in water. This large rate decrease in 20% ethanol-water solutions is accountable only by major reductions in hydrophobic interactions. Upon comparison of the copolymers' k_{cat} or "r" values with poly[4(5)-vinylimidazole] values it becomes apparent that apolar effects are reduced by "dilution" with the vinylamine units. However, where apolar interactions are not predominant, as in S_2^- hydrolysis, the more highly charged copolymer proves to be a more effective catalyst than the homopolymer at the pH of the kinetic studies. This can be attributed to strong electrostatic effects.

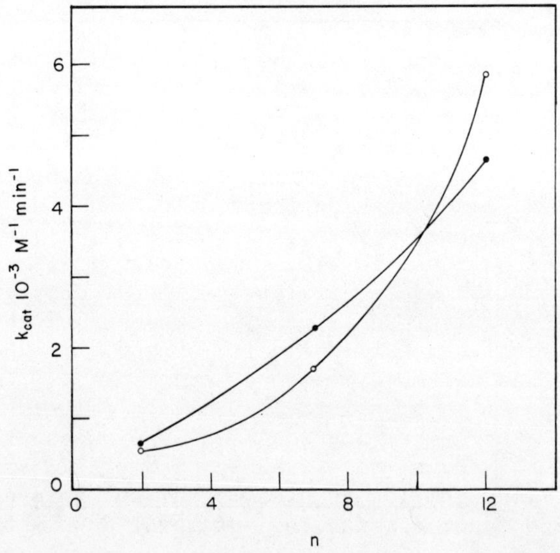
6
4
2
0
k_{cat} 10^{-3} M^{-1} min^{-1}
0
4
8
12
n

Figure 10. Second-order rate constants for the hydrolyses of the S_n^- series by the [4(5)-VIm-VAmine] copolymers: (●) IA-1, and (O) IA-7 in 20% ethanol-80% water, pH 8, $[S_n^-] = 5 \times 10^{-5}$ M [catalyst] = 5×10^{-4} M, $\mu = 0.02$, 26°C (20).

Table III

SOLVOLYSES OF THE S_n^- SERIES IN WATER[a] (20)

Catalyst		n = 2	4	7	12
1A-1[e]	$k_{cat}(M^{-1}min^{-1})$	1,101	1,788	3,228	10,706
IA-7[e]	r^b	16.2	29.8	64.6	315
	$k_{cat}(M^{-1}min^{-1})$	984	1,451	4,670	11,409
Poly[4(5)-VIm][c]	r^b	14.5	24.2	93.4	336
	$k_{cat}(M^{-1}min^{-1})$	338	---	2,038	10,326
	r^d	5.8	---	67.0	392
Imidazole	$k_{cat}(M^{-1}min^{-1})$	68	60	50	34

[a]Solvent 96.7% H_2O-3.3% CH_3CN; pH = 806; μ = 0.02; [cat] = 5×10^{-4} M; $[S_n^-] = 5 \times 10^{-5}$ M.

[b]$r = k_{cat}(pol)/k_{cat}(imid)$ under conditions in (a).

[c]Solvent 20% EtOH-80% H_2O; pH = 8.0; μ = 0.02; [cat] = 5×10^{-4} M; $[S_n^-] = 5 \times 10^{-5}$ M.

[d]$r = k_{cat}(pol)/k_{cat}(imid)$ under conditions in (c).

[e]Polymer solutions became turbid on standing.

CATALYSIS BY POLYETHYLENIMINES CONTAINING IMIDAZOLE

The use of modified polyethylenimines was initiated by Klotz et al (21) in binding studies of acylated polyethylenimine with bovine serum albumin and methyl orange. This same group in 1971 applied a similarly modified polyethylenimine containing imidazole groups to the solvolysis of phenyl esters. Because of polyethylenimine's different chemical nature and related solubility properties, polyethylenimine was expected to provide a microenvironment dissimilar to that associated with a vinyl polymer catalyst. The second-order rate constant for the hydrolysis of p-nitrophenylacetate with the modified polyethylenimine, D(10%)-PEI-Im(15%) [10% dodecyl groups, 15% methylimidazole], was 45 M^{-1} sec^{-1} which was 100 times more effective than imidazole itself (Table IV).

More recently Klotz's group (23,24) have reported esterolytic catalysis by polyethylenimine derivatives containing pendant diethylamino groups and functionalized triazine derivatives which exhibit increased catalytic effectiveness due to cooperative effects and increased apolar binding.

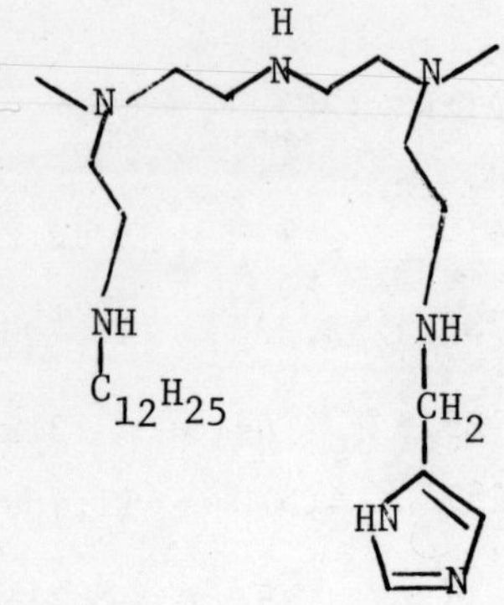

D(10%)-PEI-Im(15%)

Table IV

HYDROLYSIS OF p-NITROPHENYL ACETATE[a]

Catalyst	k_{cat} ($M^{-1}min^{-1}$)
Imidazole	10
PEI(600)D(10%)-Im(15%)	2,700
α-Chymotrypsin	10,000

[a]From Klotz et al (22).

Overberger and Dixon (25) reported the preparation and the catalyzed hydrolysis of activated esters by a L-histidine graft copolymer of polyethylenimine. A major difference in the polyethylenimine catalysts employed by Klotz and Overberger was in the structural characteristics of the polyethylenimine. Klotz had employed commercial polyethylenimine prepared by the ring opening polymerization of ethylenimine. The resulting polymer is a highly branched, relatively compact gossamer type of network. The water-soluble polymer contains approximately 25% primary, 50% secondary and 25% tertiary nitrogen atoms (26). Modified polyethylenimine catalysts of Overberger's group are a linear crystalline polyethylenimine prepared by the ring opening polymerization of 2-oxazoline as first described by Saegusa (27). Linear polyethylenimines are of a characteristically lower molecular weight than their ethylenimine derived counterparts and exhibit a limited solubility in cold water.

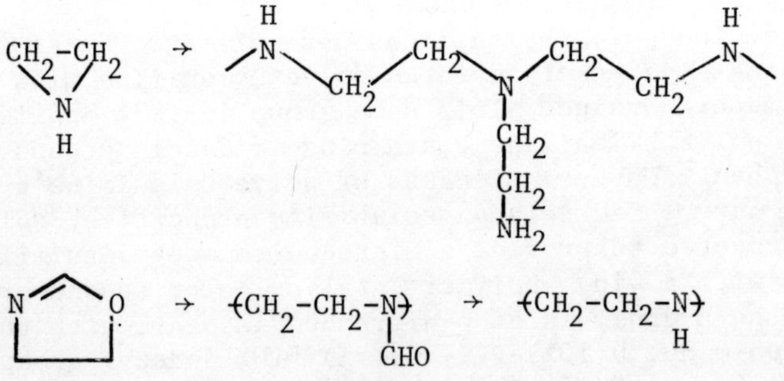

Branched PEI

Linear PEI

Poly(ethylenimine-g-L-histidine)

The catalyst used in this study was prepared by grafting a protected L-histidine onto linear polyethylenimine (DP ca. 40) by using dicyclohexylcarbodimide as shown in Scheme IV.

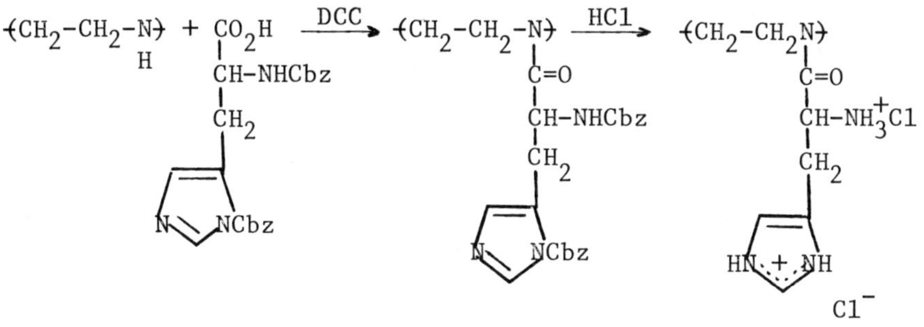
DCC
HCl
H
CH-NHCbz
C=O
NCbz
Cl⁻

(SCHEME IV)

Analogous to the vinyl polymers which contain imidazole, poly-(ethylenimine-g-L-histidine) exhibits large rate enhancements due to hydrophobic interactions. This effect may be seen if one examines the rate of hydrolysis of a series of 4-acyloxy-3-nitrobenzoic acids (S_n^-). As the length of the acyloxy group increases, the rate initially drops from S_2^- to S_4^- and then increases dramatically (Figure 11). The kinetics of the hydrolysis of S_{12}^- were complicated by decelerative behavior. This behavior, decreased rate from S_2^- to S_4^- and decelerative behavior for S_{12}^- hydrolysis, indicates that the apolar interactions between substrate and catalyst are not as strong as those observed in copoly[vinylamine/4(5)-vinylimidazole] catalyzed ester hydrolysis. The copolymer exhibited a rate increase from S_2^- to S_4^- and decelerative kinetics from S_4^- to S_{12}^- hydrolysis. The greater catalytic activity of the copolymer could be attributed to an increased apolar effect associated with its vinyl backbone.

The graft copolymer possesses an optically active center in the vicinity of the nucleophilic catalytic site. It was therefore of interest to determine if the poly(ethylenimine-g-L-histidine) catalyst would hydrolyze one enantiomeric ester substrate in preference to the other. A hydrophobic substrate was chosen as it was felt that the presence of a strong apolar interaction would provide for a strong substrate-catalyst complex.

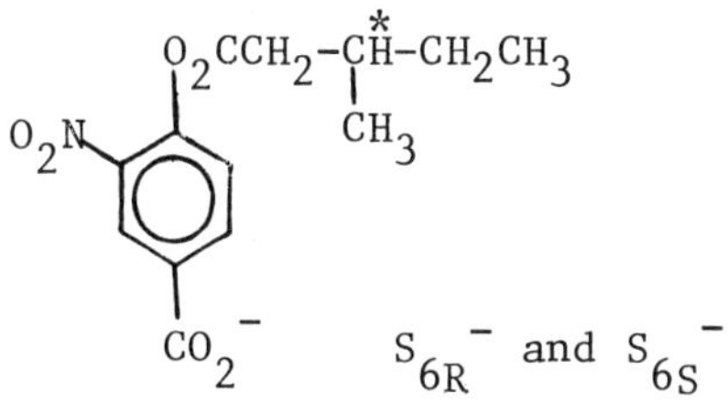

S_{6R}^- and S_{6S}^-

Unfortunately, preferential hydrolysis of either ester was not observed as is evident from Table V. The rate constants for the branched substrates (S_{6R}^- and S_{6S}^-) compared to the straight chain

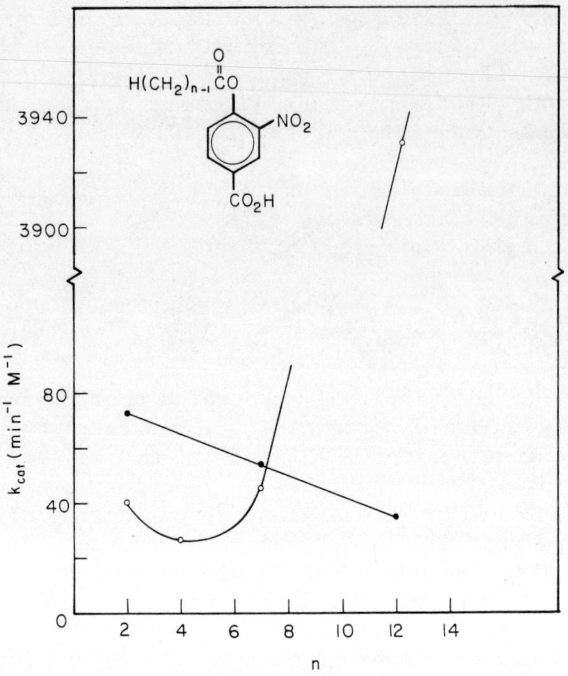

O
H(CH2)n-1CO
NO2
CO2H
3940
3900
80
40
0
kcat (min-1 M-1)
2
4
6
8
10
12
14
n

Figure 11. Second-order rate constants for the hydrolysis of the S_n^- series by catalyzed (●) imidazole and (O) by the histidine graft copolymer in 96.7% water, 3.3% acetonitrile, pH 8.06, $[S_n^-] = 5 \times 10^{-5}$ M, $\mu = 0.02$, 26°C (Overberger and Dixon (25)).

substrates (S_4^- and S_7^-) are considerably smaller. Therefore, steric and not hydrophobic interactions may be the dominant factor and a more apolar substrate may be required before any differences could be expected to appear. Alternatively, the apolar interactions may mask any specificity of the hydrolytic reaction, similar to the steric interactions, and a different chiral substrate may be required.

Table V

VALUES OF RATE CONSTANT FOR SUBSTRATES STUDIED

Substrate	$k_{meas} \times 10^3$, min^{-1}[a]
S_{6R}^-	3.94
S_{6S}^-	3.92
S_4^-	14.4
S_2^-	23.4

[a]Conditions: pH 8.06, $\mu = 0.02$, 26°C, 96.7% water, 3.3% acetonitrile, [catalyst] = 5×10^{-4} M, [substrate] = 5×10^{-5} M (Overberger and Dixon (25)).

ACKNOWLEDGEMENTS

This work was supported, in part, by the Macromolecular Research Center, and Grants DMR76-22246 and DMR-7813400 from the National Science Foundation.

REFERENCES

1. Y. Imanishi, J. Polym. Sci., Macromol. Rev., 14, 1 (1979)
2. I. V. Yannas, J. F. Burke, P. L. Gorden, C. Huang, and R. H. Rubenstein, J. Biomed. Mater. Res., 14, 107 (1980)
3. J. Pitha, Polymer, 18, 425 (1977)
4. W. Regelson, J. Polym. Sci., Symp., 66, 483 (1979)
5. C. G. Overberger, T. St. Pierre, N. Vorchheimer, N. Yaroslavsky, and S. Yaroslavsky, J. Am. Chem. Soc., 85, 3513 (1963)
6. C. G. Overberger and J. C. Salamone, Accts. Chem. Res., 2, 217 (1969)
7. C. G. Overberger, T. W. Smith, and K. W. Dixon, J. Polym. Sci., Symp., 50, 1 (1975)
8. C. G. Overberger, A. C. Guterl, Y. Kawakami, L. J. Mathias, A. Meenakshi, and T. Tomono, Pure and Appl. Chem., 50, 309 (1978)
9. C. G. Overberger, T. St. Pierre, C. Yaroslavsky, and S. Yaroslavsky, J. Am. Chem. Soc., 88, 1184 (1966)
10. H. Morawetz and H. Landenheim, J. Am. Chem. Soc., 81, 4860 (1959)
11. W. P. Jencks, "Catalysis in Chemistry and Enzymology", McGraw-Hill, Inc., New York, 1969, pp. 67-68
12. T. Kunitake and Y. Okahata, Adv. Polym. Sci., 20, Springer-Verlag, Berlin, Heidelberg, New York, 1976, pp. 159-221
13. C. G. Overberger and T. W. Smith, Macromolecules, 8, 407 (1975b)
14. C. G. Overberger and M. Morimoto, J. Am. Chem. Soc., 93, 3222 (1971)
15. D. Mirejovsky, J. Org. Chem., 44, 4881 (1979)
16. C. G. Overberger and T. W. Smith, Macromolecules, 8, 401 (1975c)
17. C. G. Overberger and Y. Kawakami, J. Polym. Sci., Polym. Chem. Ed., 16, 1237 (1978)
18. C. G. Overberger and Y. Kawakami, J. Polym. Sci., Chem. Ed., 16, 1249 (1978)
19. C. G. Overberger and A. C. Guterl, J. Polym. Sci., Symp., 62, 13 (1978); Erratum, J. Polym. Sci., Chem. Ed., 17, 1887 (1979)
20. C. G. Overberger and S. Mitra, Pure and Appl. Chem., 51, 1391-1404 (1979)
21. I. M. Klotz, G. R. Royer, and A. R. Sloniewsky, Biochemistry, 8, 4752 (1969)
22. I. M. Klotz, G. P. Royer, and I. S. Scarpa, Proc. Natl. Acad. Sci., USA, 68, 263 (1971)
23. I. M. Klotz and M. Nango, J. Polym. Sci., Polym. Chem. Ed., 16, 1265 (1978)
24. I. M. Klotz, M. Nango, and E. P. Gamson, J. Polym. Sci., Polym. Chem. Ed., 17, 1557 (1979)
25. C. G. Overberger and K. W. Dixon, J. Polym. Sci., Polym. Chem. Ed., 15, 1863 (1977)

26. I. M. Klotz and T. W. Johnson, Macromolecules, 7, 149 (1974)
27. T. Saegusa, H. Ikeda, and H. Fujii, Macromolecules, 5, 108 (1972)

Figures are reprinted with permission from:

Figure 1 J. Am. Chem. Soc., 88, 1188 (1966).
Figure 2a Macromolecules, 8, 410 (1975).
Figure 2b Macromolecules, 8, 418 (1975).
Figure 3 J. Am. Chem. Soc., 93, 3226 (1971).
Figure 4 J. Polym. Sci., Polym. Chem. Ed., 16, 1255 (1978).
Figure 5 J. Polym. Sci., Polym. Chem. Ed., 16, 1259 (1978).
Figure 6 J. Polym. Sci., Polym. Chem. Ed., 16, 1260 (1978).
Figure 7 J. Polym. Sci., Polym. Chem. Ed., 17, 1887 (1979).
Figure 8 J. Polym. Sci., Polym. Chem. Ed., 17, 1888 (1979).
Figure 9 Pure Appl. Chem., 51, 1398 (1979).
Figure 10 Pure Appl. Chem., 51, 1399 (1979).
Figure 11 J. Polym. Sci., Polym. Chem. Ed., 15, 1867 (1977).

HYDROLYTIC DEGRADATION OF POLY DL-(LACTIDE)

N. S. Mason, C. S. Miles, and R. E. Sparks

Department of Chemical Engineering
Washington University
St. Louis, Missouri 63130

INTRODUCTION

Poly (DL-lactide) or poly (DL-lactic acid) (PLA), has been proposed as a component of several controlled-release drug forms because it degrades in living tissue to harmless products (1,2). Thus, devices or particles made of this material could be slowly metabolized and thus leave the body harmlessly after the drug is released. The solubility of PLA in many organic solvents (ketones, esters, aromatics, and chlorinated aliphatics) gives considerable flexibility in device fabrication. PLA has been proposed for use as fibers (3), microcapsules (4), and larger implants (2). Experimenters have made a variety of drug forms and devices to deliver drugs ranging from birth-control agents and narcotic antagonists, to anticoagulants and depressants.

A drug to be delivered may be dispersed throughout the polymer to give monolithic devices, or the polymer may envelop a drug-rich phase as in reservoir devices. Syringe-injectable microcapsules have been fabricated by several groups (4,5,6). Figure 1 shows an SEM of the surface of PLA microcapsules, originally containing 30% progesterone, after 140 days in human plasma. For these applications it is necessary that the polymer remain intact for a predictable period of time.

PLA has had an interesting history. Bischoff (7) and Carothers (8) noted the ease with which the polymer and cyclic dimer (dilactide) were interconverted. However until techniques were developed for purifying the monomer, the cyclic dimer in this case, only low molecular-weight, brittle polymers could be obtained (9). The purification of dilactide is facilitated by the formation of a

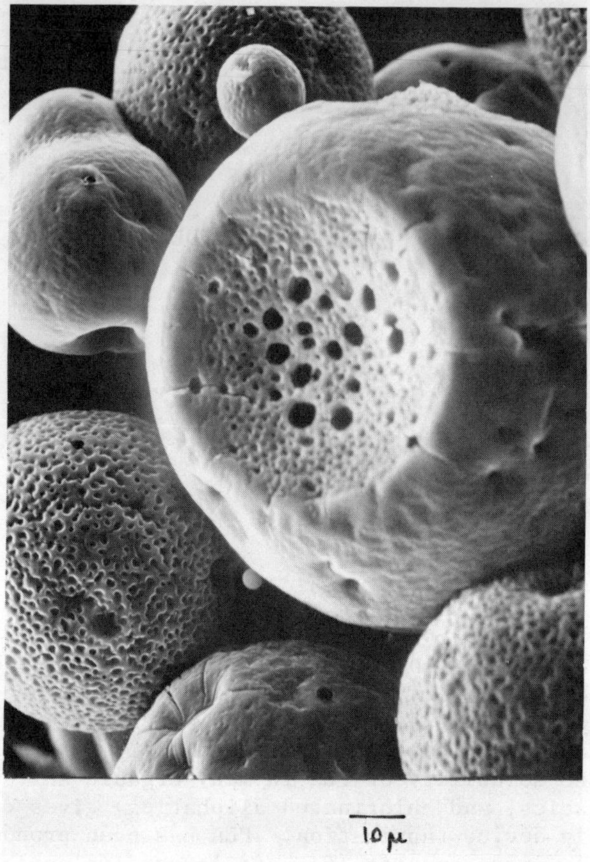
10μ

Figure 1. PLA capsules containing 30% by weight Progesterone after contacting with plasma for 140 days.

racemic compound melting at 127°C consisting of an equal number of D-dilactide and L-dilactide units (10). The purification is normally carried out by multiple recrystallization from solvents such as acetone and benzene. The meso-dilactide, which consists of one D- and one L-lactic acid, melts at 41°C, is much more difficult to purify (11), and is normally discarded.

The bulk polymerization of the racemic dilactide may be carried out under vacuum in the presence of a tin or zinc salt. Under these conditions the configuration of the asymmetric carbons is maintained (12). In each growth step two dilactide units are added. The resulting polymer is considered atactic although there are restrictions in the possible stereosequences (13). Poly (DL-lactide) is amorphous by X-ray diffraction measurement and has a glass transition temperature of 57°C. While PLA is said to be biodegradable in vivo, the chain scission is primarily hydrolytic and does not involve enzymes. This is believed to be the case since PLA is a completely synthetic substance which does not exist in nature. Schindler et al (13) have reported on the kinetics of hydrolysis of poly-lactides, caprolactones and their copolymers. In their view the in vivo and in vitro degradations follow the same mechanism. However, in vivo, ester interchange may be involved. This contribution should be small in view of the low concentration of free fatty acids in the body relative to water. This study was stimulated by the observation that relatively small changes in experimental conditions, have a large effect upon the rate of degradation. It is necessary to take such effects into account in the design of drug-release devices.

For random degradation of PLA chains (disregarding changes in molecular weight distributions), the decrease in number average molecular weight, $\bar{M}$, may be written

$$-\frac{d\bar{M}}{dt} = -k\bar{M} \quad \text{or} \quad \ln\frac{\bar{M}_o}{\bar{M}_t} = kt. \qquad \text{(Equation 1)}$$

Thus a graph of log $\bar{M}$ versus time would yield a straight line. The Mark-Houwink relationship between molecular weight and intrinsic viscosity $\{\eta\}$ (Equation 2)

$$\{\eta\} = K\bar{M}^a \qquad \text{(Equation 2)}$$

permits equation (1) to be written as shown in Equation 3.

$$\ln\frac{\{\eta_o\}}{\{\eta_t\}} = akt. \qquad \text{(Equation 3)}$$

Hence, if the ln $\{\eta\}$ is plotted vs. time the slope is ka, where a is the exponent of the Mark-Houwink equation (13).

If it is assumed that the Huggins and Kreamer equations are valid for this system, i.e.

$$\frac{\ln \eta_{sp}}{c} = \{\eta\} + k'\{\eta\}^2 c, \qquad \text{(Equation 4)}$$

$$\frac{\eta_r}{c} = \{\eta\} - k''\{\eta\}^2 c \qquad \text{(Equation 5)}$$

and further, k' + k" = 0.5, then the intrinsic viscosity can be obtained from one viscosity determination by means of the following expression:

$$\{\eta\} = \frac{\sqrt{2(\eta_{sp} - \ln \eta_r)}}{c} \qquad \text{(Equation 6)}$$

EXPERIMENTAL

PLA was prepared according to a procedure described by Kulkarni et al (14). After dilactide was recrystallized four times, it was polymerized under vacuum with a programmed temperature rise from 145°C to 175°C, with 0.01% tetraphenyl tin as the catalyst. The polymerization was terminated when the charge solidified. The polymer was stored in a dessicator. Films approximately 5 mils thick were cast on glass from 20% benzene solution and dried under a high-velocity hood for 48 hours. Separate 200 mg samples of polymer were exposed to 18 ml of medium in 20 ml glass scintillation bottles shaken at constant temperature. At time intervals up to a month a bottle was removed from the bath and the film removed, washed off and dried under vacuum to constant weight. The intrinsic viscosity of a portion of the film was determined in benzene at 30°C.

The degradation behavior in deionized water is shown in Figure 2. In Figure 3 the rates of degradation in pH 7 phosphate buffer at 25 and 37°C are compared. The data in Figure 4 show the effect of pH. Phthalate buffer (0.1 molar) was used to obtain pH 3 and HCl (0.1 molar) to obtain pH 1.

To provide a comparison with a biological fluid, a study was made of the effect of degradation in pig plasma (containing 0.01% sodium azide and 0.1 thymol to inhibit bacterial growth) at 25 and 37°C. The results are given in Figure 5. Figure 6 illustrates the rate of hydrolysis in benzene solution. Benzene dried by passing through a molecular sieve (3A) and silica gel column was compared to benzene saturated with water and to as-received benzene. To determine whether autoxidation, in addition to hydrolysis is involved in PLA degradation, 2.5 parts of antioxidant (Shell Ionox 330) was added to another set of bottles. Finally Figure 7 shows the degradation of the polymer while simply being stored in laboratory air.

DISCUSSION

It may be seen that in a liquid medium, whether aqueous or non-aqueous, the change in the intrinsic viscosity is gradual and fits the predictive equation. The weight of the sample remains nearly constant until a critical intrinsic viscosity is reached (approximately 0.2). Then the weight loss becomes rapid. Apparently this is the point where some of the chain fragments are short enough to become water soluble.

The first-order degradation constants are summarized in Table 1 data.

The rate constant is relatively insensitive to pH and to salt content since it does not vary significantly whether the sample is in deionized water of in buffers. The constant is highly sensitive

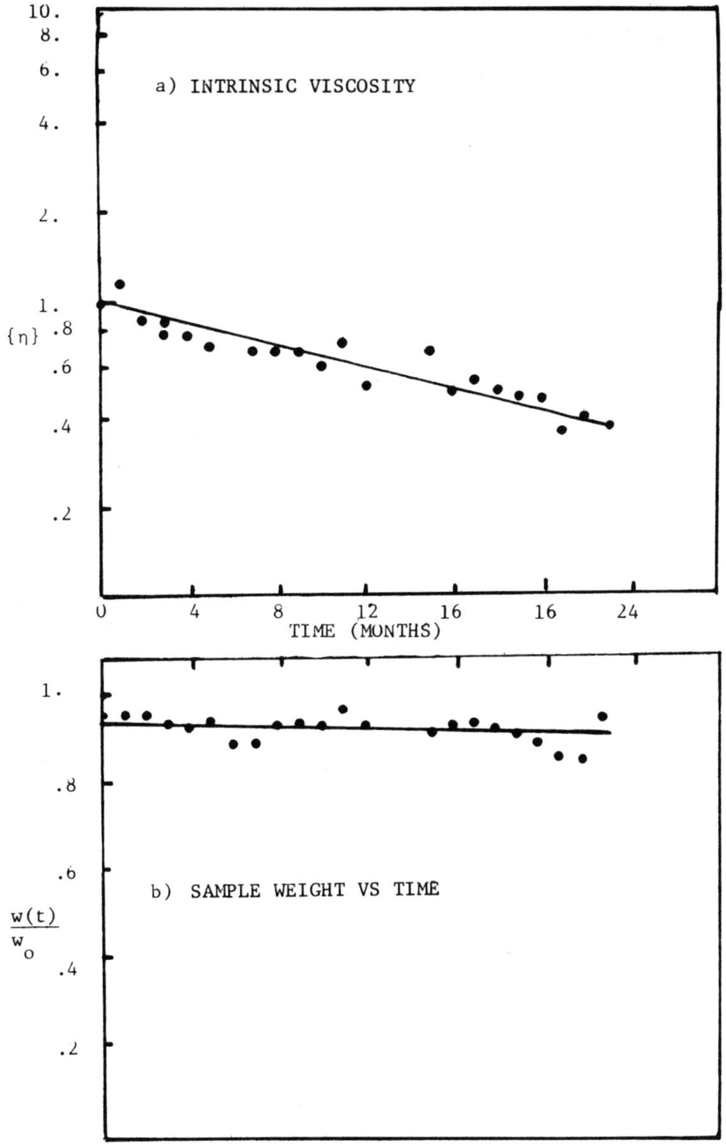
a) INTRINSIC VISCOSITY
10.
8.
6.
4.
2.
1.
{η}
.8
.6
.4
.2
0
4
8
12
16
16
24
TIME (MONTHS)
b) SAMPLE WEIGHT VS TIME
1.
.8
.6
w(t)
w o
.4
.2

Figure 2. Aging in Deionized Water

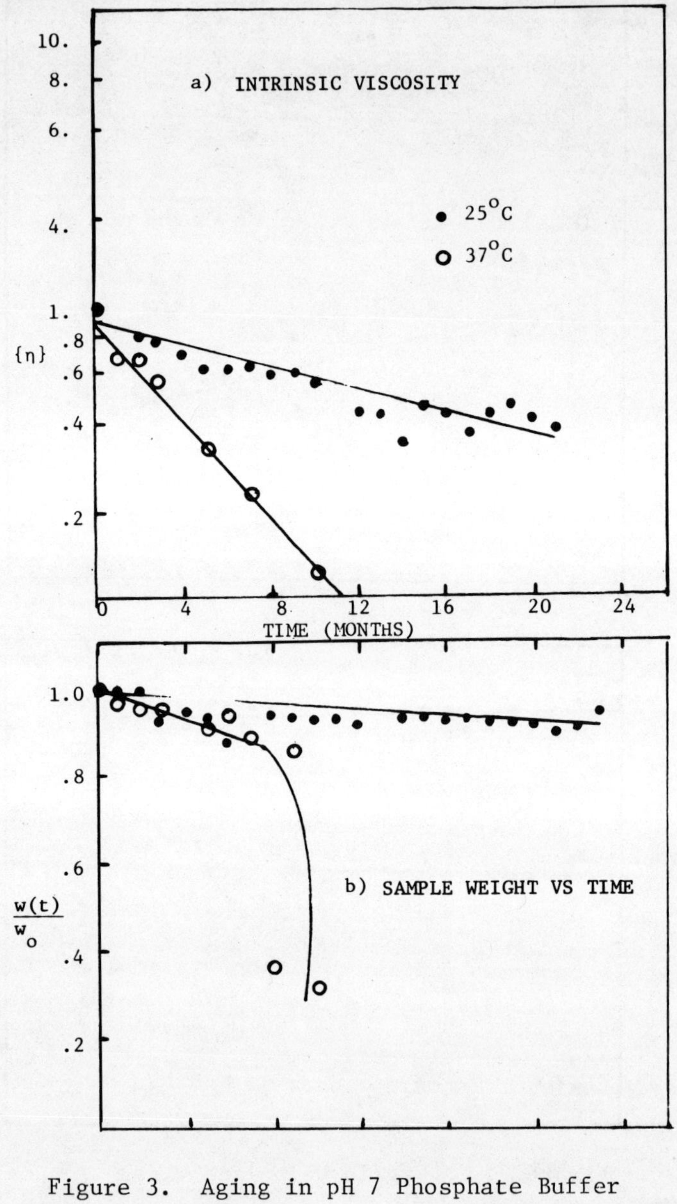
10.
8.
6.
4.
1.
.8
.6
.4
.2
a) INTRINSIC VISCOSITY
● 25°C
○ 37°C
{η}
0
4
8
12
16
20
24
TIME (MONTHS)
1.0
.8
.6
.4
.2
w(t)/w₀
b) SAMPLE WEIGHT VS TIME

Figure 3. Aging in pH 7 Phosphate Buffer

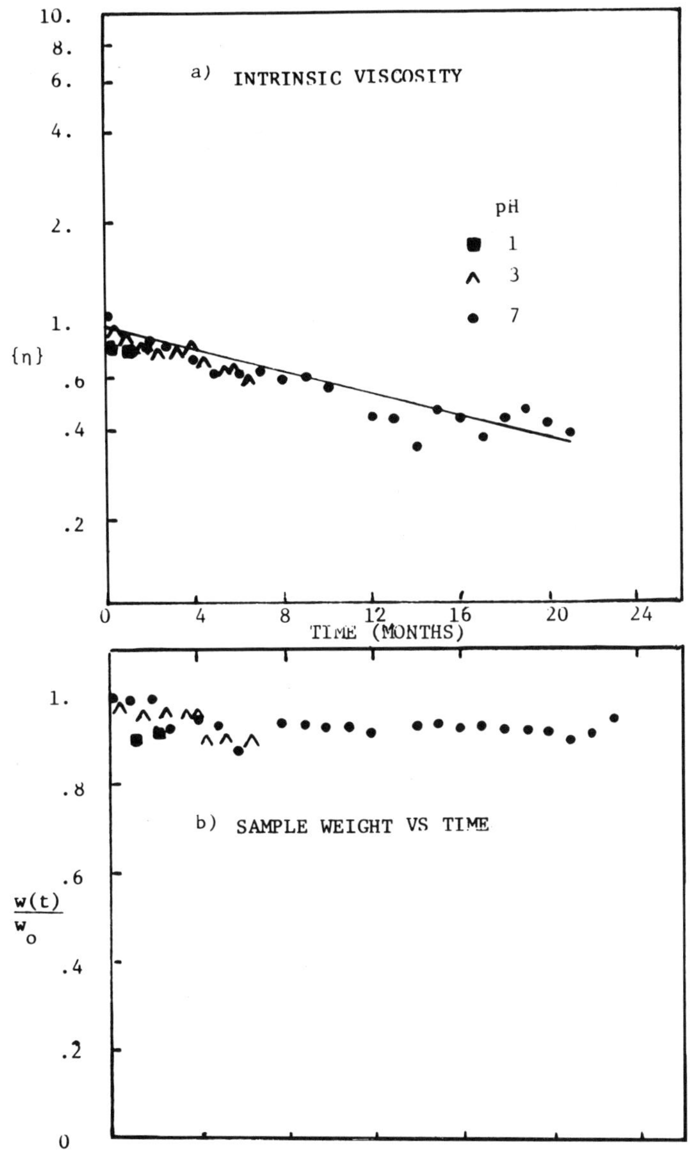
10.
8.
6.
4.
2.
1.
{η}
.6
.4
.2
a) INTRINSIC VISCOSITY
pH
1
3
7
0
4
8
12
16
20
24
TIME (MONTHS)
1.
.8
.6
w(t)
w
o
.4
.2
0
b) SAMPLE WEIGHT VS TIME

Figure 4. Effect of pH

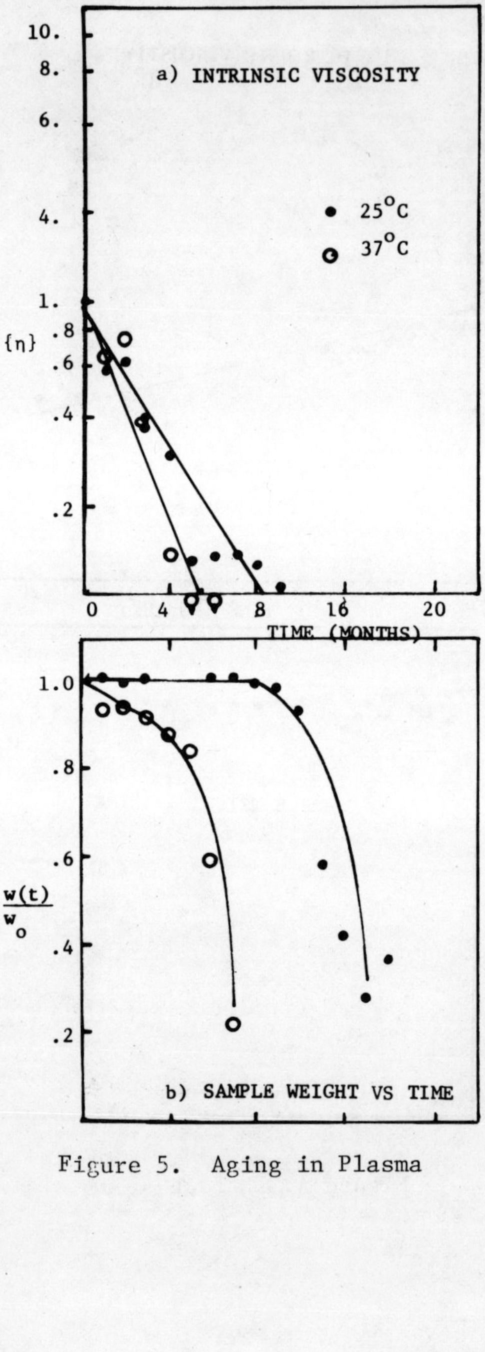
10.
8.
6.
4.
1.
.8
.6
.4
.2
{η}
a) INTRINSIC VISCOSITY
25°C
37°C
0
4
8
16
20
TIME (MONTHS)
1.0
.8
.6
.4
.2
w(t)
w
o
b) SAMPLE WEIGHT VS TIME

Figure 5. Aging in Plasma

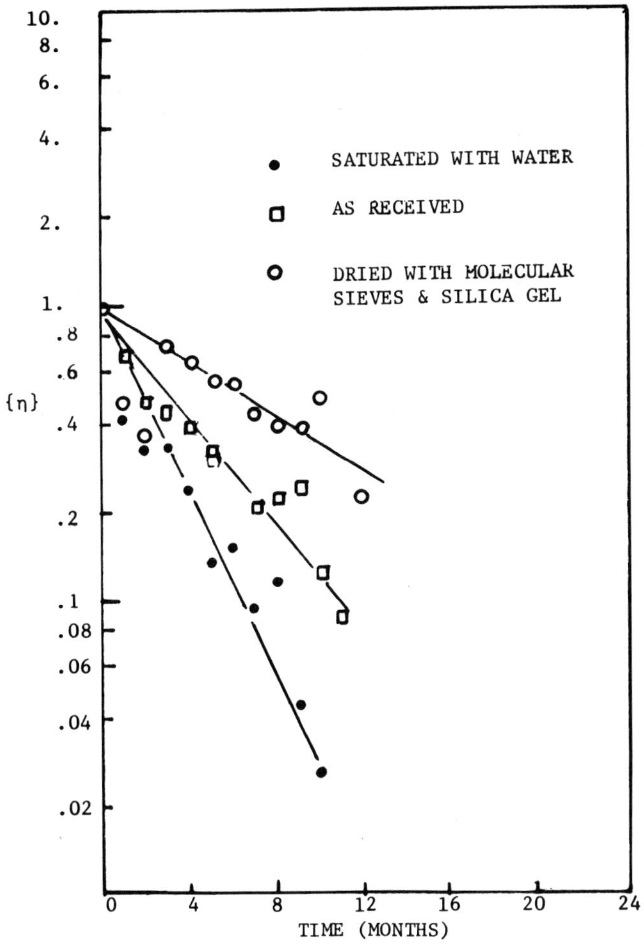
10.
8.
6.
4.
2.
1.
.8
.6
.4
.2
.1
.08
.06
.04
.02
{η}
SATURATED WITH WATER
AS RECEIVED
DRIED WITH MOLECULAR
SIEVES & SILICA GEL
0
4
8
12
16
20
24
TIME (MONTHS)

Figure 6. Aging in Benzene Intrinsic Viscosity Vs Time

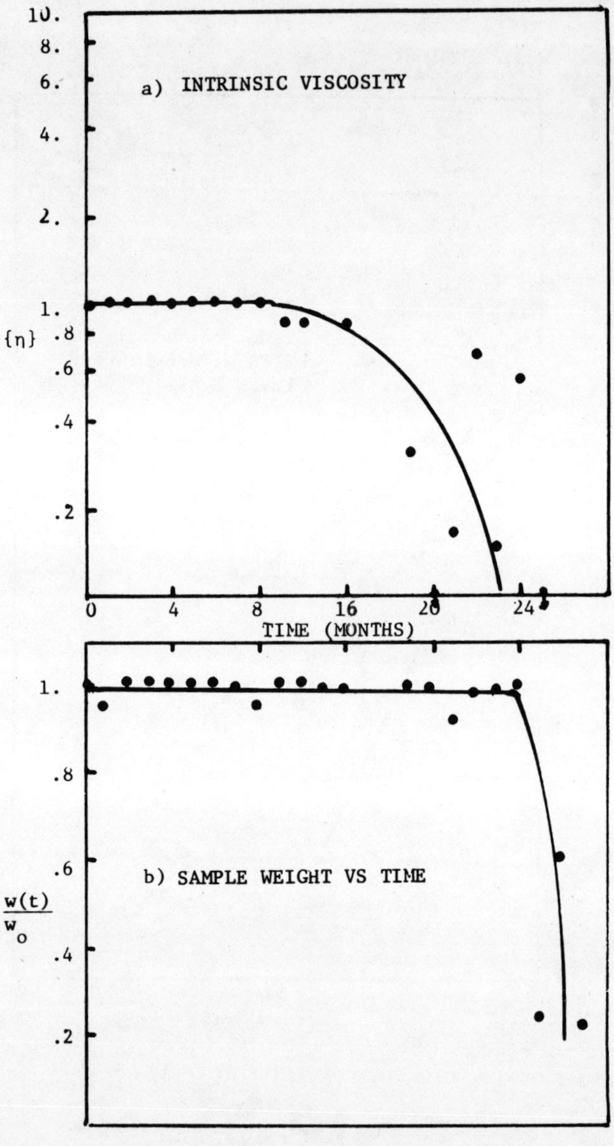

10.
8.
6.
a) INTRINSIC VISCOSITY
4.
2.
1.
{η} .8
.6
.4
.2
0
4
8
16
20
24
TIME (MONTHS)
1.
.8
.6
b) SAMPLE WEIGHT VS TIME
w(t)/w_o
.4
.2

Figure 7. Aging in Laboratory Air

Table 1

FIRST-ORDER DEGRADATION RATE CONSTANTS

Medium	ka ($day^{-1} \times 10^3$) 25°C	37°C	50°C
Deionized Water	1.13 ± .10	-	123. ± 13
pH 7 Buffer	0.97 ± .12	9.0 ± 1.4	
pH 3 Buffer	1.18 ± .5	-	
Inhibited Plasma	6.2 ± .73	18.7 ± 2.4	
Water-Saturated Benzene			
without antioxidant	8.53 ± 1.1	-	
with antioxidant	8.33 ± 2.0	-	
Dried Benzene			
without antioxidant	.833 ± 1.1	-	
with antioxidant	.600 ± .77		
Benzene (as received)	5.73 ± 0.63	-	

Activation Energy Buffer 29.6 kcal
Plasma 16.9 kcal

to temperature, exhibiting an activation energy of 29.6 kcal for buffer and 19.6 kcal for plasma.

The rate of degradation in plasma is from 2 to 6 times higher than in buffers. This could be caused by an increase in chain mobility due to lipid components present in plasma which are absorbed by PLA.

When the mobility of the chain is invreased as much as possible, e.g. by dissolving it in benzene, the rate constant increases further. It should be noted that the solubility of water in benzene at 25°C is approximately 53 mg/100 ml, hence the water concentration is only about 1/2000 of that in dilute aqueous media, yet the hydrolytic degradation is more rapid. Even in molecular-sieve-dried benzene the degradation rate is significant. The effect of antioxidant does not appear to be significant.

In laboratory air the degradation rate was not linear with time (Figure 7). Rather, there was an induction period and then a relatively rapid deteriorization. The decrease in the intrinsic viscosity, which began to be detectible after 9 months, reached the critical intrinsic viscosity corresponding to weight loss in about 24 months. This might be caused by the decrease in polymer molecular weight. This would give more hydrophilic end groups and increase hygroscopic moisture uptake. Schindler (13) also alluded to the possibility of autocatalytic effects which might result from the presence of short open-chain oligomers of lactic acid. Such water soluble species could be in lower concentration in the polymer when it is immersed in water instead of air. When the rate of degradation was compared in volumes of deionized water which differed by a factor of 10, the samples in the smaller volume of water degraded more rapidly by 15%. This was statistically significant since it represented 5 times the sum of the standard deviations of the two rates. However, even in the smaller volume of water (2 ml) a linear degradation rate was obtained, indicating that the curvature that was observed in moist air likely results from absorption of moisture by the polymer.

The rate constant in buffer at 30°C, with 0.75 chosen for the value of a, was 4×10^{-3} day $^{-1}$ vs 1.2×10^{-2} day $^{-1}$ reported by Schindler (13). This difference may be due to small temperature errors in view of the high sensitivity of the rate to temperature.

Analyses were also carried out on a polymer sample which had been stored for approximately 2 years in a dessicator. These data are presented in Table II. During this time the intrinsic viscosity decreased from 1.35 to 1.05.

Table II

CHARACTERISTICS OF PLA USED IN DEGRADATION STUDIES

Molecular Weight	
by Osmometry, $\bar{M}_n$	52,600
by Light Scattering, $\bar{M}_w$	231,200
Calculated from $\{\eta\}$, $\bar{M}_n$	41,100
Molecular Length by GPC (polystyrene calibration)	
number average	1120Å
weight average	5507Å
$\frac{M_w}{M_n}$ by GPC	4.92

In conclusion, it appears that the hydrolytic degradation of PLA in liquid media fits a model based on random chain-scission. The rate increases when the polymer is softened or dissolved. The rate is relatively insensitive to pH of the medium.

ABSTRACT

The rate of hydrolytic degradation of poly DL-lactic acid was studied in a variety of media at 25°C and 37°C. The rate was linear in liquid media, but showed autoacceleration in moist air. The rate was found to be sensitive to factors which increased chain mobility but insensitive to pH or ionic strength of the medium.

REFERENCES

1. A. G. Boswell and R. E. Scribner, U.S. Patent 3,773,919 (1973)
2. J. H. R. Woodland, S. Yolles, J. Med. Chem., 16, 897 (1973)
3. R. J. Ruderman et al, Trans. A.S.A.I.O., XVIII, 30 (1972)
4. L. R. Beck et al, Fertil. & Steril., 31, 545 (1979)
5. C. Thies in "Microencapsulation", T. Kondo ed., Techno, Inc., Tokyo (1979), p. 426
6. N. Mason, C. Thies, T. J. Cicero, J. Pharm. Sci., 65, 847 (1976)
7. C. A. Bischoff, Chem. Ber., 26, 263 (1903)
8. W. H. Carothers et al, J. Am. Chem. Soc., 54, 761 (1932)
9. A. K. Schneider, U.S. Patent 2,703,316 (1955)
10. C. H. Holten et al, "Lactic Acid", Verlag Chemie Weinheim, (1971), p. 232
11. D. D. Deane and E. G. Hammond, J. Dairy Sci., 43, 1421 (1960)
12. J. Kleine and H. H. Kleine, Makromol. Chem., 30, 23 (1959)
13. A. Schindler et al, "Biodegradable Polymers for Sustained Drug Delivery, Continuing Topics in Polymer Sci.", Vol. II, Plenum 1977
14. R. K. Kulkarni et al, Arch. Surg., 93, 839 (1966)
15. A. Seidel, "Solubilities of Organic Compounds" 3rd ed., D. Van Nostrand Co., Inc., New York (1941)

APPLICATIONS OF POLYMERS IN RATE-CONTROLLED DRUG DELIVERY

Alejandro Zaffaroni

ALZA Corporation

Palo Alto, California 94304

In recent years, polymers have found functional uses as components of new dosage forms that precisely control the therapeutic administration of drugs. These developments, which came about because of the drawbacks of conventional dosage forms, promise advances in many areas of medical treatment.

LIMITATIONS OF CONVENTIONAL DOSAGE FORMS

The life-saving hormone insulin illustrates the drawbacks of conventional dosage forms. Everyone recognizes what a problem it is that the only effective dosage form for insulin is the injectable, and how much a better dosage form for this hormone could improve both therapy and the quality of life for diabetics.

Insulin injections, like all conventional dosage forms--including tablets, eyedrops, ointments, and even sustained-release capsules--deliver their active substance(s) in what is called first-order fashion: delivery occurs at rates that are highest initially, and decline steadily thereafter. This pattern confers no known therapeutic benefits but is simply the result of the inability of conventional dosage forms to control drug release and thus drug concentration in blood and tissues. In fact, the sole role of the conventional dosage form is to convey a unit dosage from a container to the body. Then, just when therapy begins, the dosage form disintegrates! This inability of dosage forms to maintain a therapeutic drug level means that potent agents have to travel through the bloodstream in much larger concentrations than are required at the target organ or tissue.

More explicitly, the results of repetitive dosing with such

first-order dosage forms is a saw-tooth pattern of peaks and troughs in the concentrations of agent in blood and tissues (Fig. 1). This pattern produces a time-dependent mixture of side effects and desired effects, with side effects tending to predominate early in the interval following the dose. Later, the concentration of drug falls too low to be effective; the need then arises to remedicate. Elapsed time from dosing until loss of effect can be prolonged by putting more drug into the dosage form, but the penalty is even higher peaks in drug level in blood and tissues, and thus more and/or more severe side-effects.

For some agents, effect and side-effect occur at widely different concentrations in blood or tissue, but for many others the difference in these levels is small. With the latter category of drugs, rate-controlled delivery is essential to maintain concentrations within the ideal therapeutic "corridor", where the drug's selectivity of action is greatest and its side-effects are least.

FORCES FOR CHANGE IN PHARMACEUTICAL DEVELOPMENT

The limitations of conventional dosage forms exist because of a strange anomaly in the development of medicine. Over the years, pharmaceutical companies have developed increasingly powerful agents to fight disease. But the methods available for the delivery of these drugs did not keep pace. Conferring selectivity of drug action

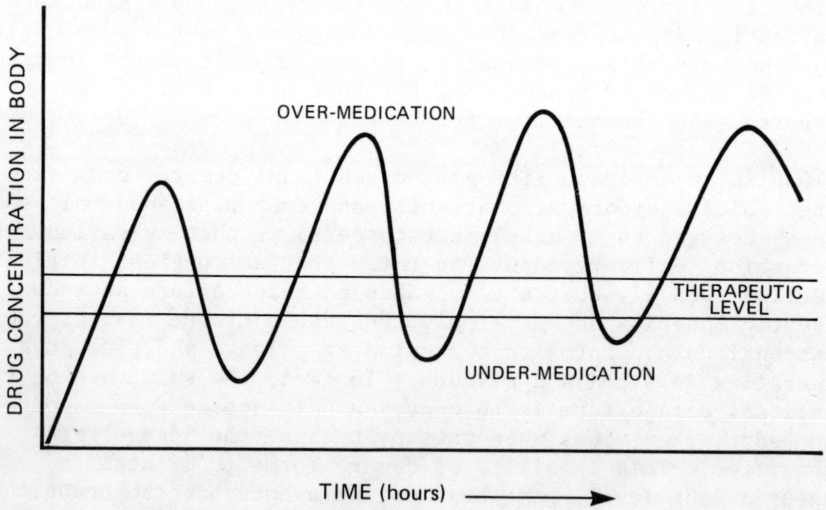
DRUG CONCENTRATION IN BODY
OVER-MEDICATION
THERAPEUTIC LEVEL
UNDER-MEDICATION
TIME (hours)

Figure 1. Schematic representation of plasma drug levels when drug is administered repeatedly from a first-order dosage form. (Copyright 1979 ALZA Corp.)

generally was seen not as an important function of the dosage form but as a goal to be achieved by chemical means: either the synthesis of new drug molecules or alterations of old ones to divest them of undesired actions.

In recent years this approach has--for a variety of reasons--proved less and less fruitful and increasingly expensive. Moreover, advances in pharmacology, pharmacokinetics, and the technology for controlling the rate of drug delivery have shown that a new parallel approach is feasible to attaining the ideal of selective drug action.

Central to this new approach is the growing recognition that the actions of many drugs are related to their concentrations in blood (1). Controlling those concentrations can therefore be a method of improving the selectivity of a drug's beneficial actions, provided that the beneficial actions are elicited by lower concentrations than those needed to elicit unpleasant or adverse effects. A growing body of data is becoming available on the concentration-effect relations of important drugs.

Moreover, rate-control is now recognized as a physiologic principle. That is, the body's own control systems for regulating hormonal secretions are known to operate on the basis of continuous rate-control. In fact, the ultimate physiological selectivity of hormone action depends upon the concentration of hormones in blood staying within a certain range. Administering many of the same hormones by conventional dosing, e.g., insulin injections, is thus a poor imitation of nature. That is why improving the management of day-to-day insulin requirements has evoked great interest, giving rise to various efforts to provide continuous insulin administration, which necessarily implies rate-control. In acute diabetic keto-acidosis, for example, rate-controlled I.V. infusion of insulin has replaced the practice of giving insulin by injection, because rate-controlled infusion is safer and more effective treatment (2).

In fact, rate-controlled continuous drug delivery is now the rule rather than the exception in almost every phase of acute, intensive care--e.g., after major surgery, coronary occlusion, and serious trauma. Rate control makes both safe and effective the use of potent agents whose concentrations in blood must be maintained within a narrow range. Nitroprusside, dopamine, heparin, dobutamine, lidocaine, norepinephrine, and nitroglycerine are examples of such agents. The growing ability to use them safely and effectively was to some extent due to technological advances in the early 1970s that converted the large and cumbersome infusion pumps of the physiology laboratory into the compact, convenient, and reliable infusion pumps now widely used for drug administration in acute, intensive care.

These developments, however, left unfilled the needs for the same precision of drug delivery to make outpatient treatment safer, more reliable, and more comfortable. It would be belaboring the obvious to point out the problems of compliance associated with prescribing medication to be taken several times daily--for a lifetime--to treat chronic illness such as hypertension or glaucoma. Noncompliance may be a particular hazard if the patient finds the prescribed drug's side-effects more onerous than the symptoms of the disease, which is not infrequently the case.

THE THERAPEUTIC SYSTEM CONCEPT

Thus, at the beginning of the 1970s a unique opportunity existed to strive for a quantum jump in the benefits of drug therapy through the creation of a technology that would permit controlled delivery of drugs to the body in easily self-administered dosage forms. The aim was to maintain drug levels within a range that provided maximum therapeutic effect and minimal side effects. ALZA was organized to accomplish this in 1968; its approach has been to develop methods of membrane-controlled, molecular movement to create entirely new rate-controlled pharmaceutical dosage forms, which we call therapeutic systems. These dosage forms have attained--or are in the process of attaining--registration with both static specifications of drug content and kinetic specifications of *in vivo* delivery rate. The kinetic specification of rate is not an alternative to the static specification of content, but complementary to it.

The very existence of this technology is a force for transition: it has begun to unleash the imagination of researchers in pharmacology and medicine, stimulating them to explore the actions of drugs when drug concentrations in blood and tissues are controlled. Such control is aimed at either maintaining constant concentrations or causing concentrations to follow particular time-patterns.

ZERO ORDER PHARMACOLOGY IN PRACTICE

ALZA's new types of dosage forms are based on the principles of zero-order pharmacology (3) rather than first-order pharmacology (4-8). A drug delivered at zero-order is released at a constant rate from the dosage form, rather than in a surge. Depending on release rate, its spectrum of actions can vary greatly. In fact, it is now generally recognized that a drug's full potential remains unknown until it has been studied with extended-duration, rate-controlled delivery.

Understanding a drug's zero order pharmacology is to know the rank order of its actions elicited during delivery at a series of constant rates ranging from the lowest rate sufficient to elicit any detectable effect, to the highest rate allowed by common sense and the ethics of experimentation. As one progresses stepwise upwards through that range of delivery rates, successively more actions of the drug appear. Thus, if we can deliver drug at zero-order, we can accomplish selectivity in drug action by peeling the top action--or top several actions--from the array of delivery-rate-dependent actions, and excluding the rest.

The principles of zero-order pharmacology have been reduced to practice in therapeutic systems. Such a system--unlike a conventional dosage form--does not disintegrate after application; its functional life-time closely parallels the duration of drug therapy. Therapeutic systems are described by their duration and rate of drug release--for example, so many micrograms of pilocarpine per hour for seven days; conventional dosage forms simply specify the bulk amount of drug, such as 25 mg, placed in the body.

Therapeutic systems have another attribute that enhances the use-effectiveness of drugs in everyday life. No matter how inherently beneficial a drug's action may be, it fails as a therapeutic agent if patients shun its use because the prescribed regimen is too complex. Therapeutic systems can be expected to encourage compliance because they provide therapy continuously from one application for much longer periods--ranging from once daily to three years at present--than conventional dosage forms.

EXAMPLES OF THERAPEUTIC SYSTEMS

Some achievements or work in progress on therapeutic systems for localized therapy include:

One-week-duration OCUSERT® ocular therapeutic systems (Figure 2) in the form of a thin, elliptical film, worn beneath the eyelid, for delivery of the antiglaucoma drug pilocarpine (9,10) (an alternative to taking the drug in eyedrops).

One- and three-year-duration, T-shaped PROGESTASERT® systems (Figure 3), providing intrauterine delivery of progesterone for contraception (11), with concomitant reduction of menstrual blood loss (12,13).

These two systems are now generally available. Both demonstrate three values of controlled drug delivery: a decrease in the amount of agent required for a full therapeutic effect; reduction of side-effects (12-14); and a lessened frequency of dosing (once weekly vs four times daily for pilocarpine eyedrops and once yearly vs once daily for the progestational hormone).

In addition, the contraceptive exemplifies two other values of continuous, controlled delivery. One is the achievement of targeted drug delivery to a single organ (the uterus) of a type of agent (a birth control hormone) that previously had to be delivered systemically to the whole body. The other is the ability to use--because of continuity of delivery--a biological substance (progesterone)

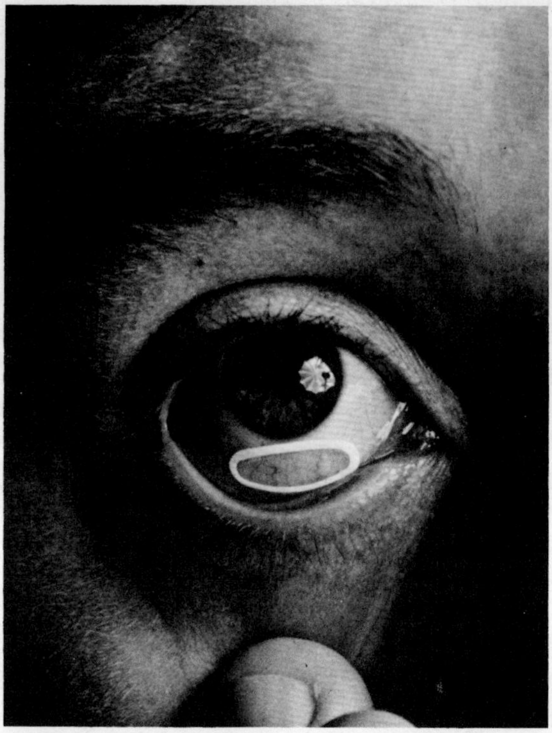

Figure 2. OCUSERT® ocular therapeutic system in the lower cul-de-sac of the eye. (Copyright 1975 ALZA corp.)

that has too brief a half-life in the body to permit its administration in conventional dosage forms. A huge array of such substances exists that could prove valuable in medicine.

Because of the overriding importance of systemic drug delivery, several therapeutic systems for systemic treatment are under development. They also can overcome many disadvantages of the corresponding conventional dosage forms. Three of these systems are:

Half-year-duration erodible systems, which are injectable, for systemic delivery of a progestational steroid for contraception (15).

Day-long-duration, osmotically actuated tablets, for oral administration of many rapidly metabolized/excreted agents (16).

Three-day-duration system in the form of an adhesive disc, worn on a few square centimeters of skin surface, for transdermal systemic delivery of a drug for motion-induced nausea (17,18).

The transdermal system is at the marketing stage. The osmotic

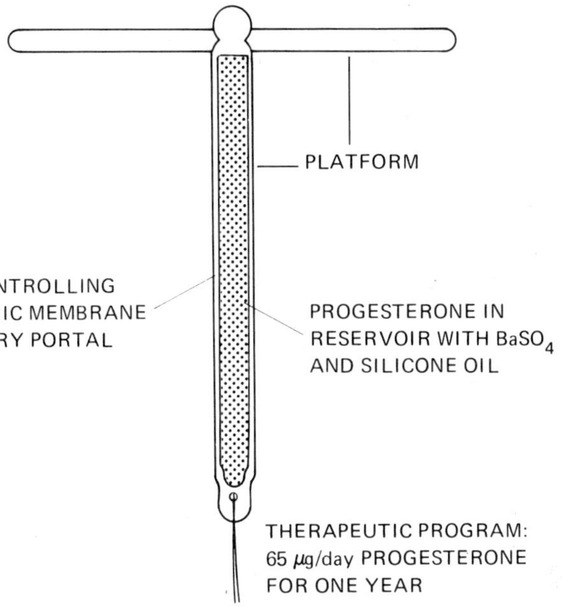
PLATFORM
RATE-CONTROLLING
POLYMERIC MEMBRANE
AND ENTRY PORTAL
PROGESTERONE IN
RESERVOIR WITH BaSO4
AND SILICONE OIL
THERAPEUTIC PROGRAM:
65 μg/day PROGESTERONE
FOR ONE YEAR

Figure 3. Schematic diagram of the PROGESTASERT® system. (Copyright 1980 ALZA Corp.)

technology is in clinical trials as an oral dosage form for therapy, but is already available in two forms usable for studies in animals and human subjects.

STRUCTURE AND COMPOSITION OF THERAPEUTIC SYSTEMS

Each therapeutic system consists of a structural platform, a drug or other bioactive agent, a reservoir, and a rate-controlling element. The various systems are small enough to be taken orally or by injection, or to be placed in a body orifice without discomfort. Since therapeutic systems, by design, remain in the body for long periods of time, their constituent materials must have a high level of biocompatibility, to prevent them from causing the user any harm or discomfort.

The chief materials used to make system components (other than the drug) have been a variety of polymers, and thus polymer technology has been a cornerstone of our work (19). Some therapeutic systems provide controlled administration of medication through a nonporous, nonhomogeneous, polymeric membrane by a process of dissolution of the permeating drug in the polymer at one interface and diffusion through the membrane under the driving force of a concentration gradient (5). Many polymers, in the form of thin films, can operate as diffusion membranes for permeants of low molecular weight (mol. wt. <499.) Rate of drug release is inversely proportional to thickness of the membrane.

OCUSERT® ocular therapeutic systems use diffusional control over drug delivery (Figure 4). These systems are elliptical, 3-layer units made by laminating a drug reservoir core between two rate-controlling membranes of ethylene/vinyl acetate (EVA) copolymer. The drug reservoir is a film of pilocarpine base with a minor amount of polymeric gelling agent, alginic acid. A ring of EVA with TiO_2 surrounds the core to provide a white border. The 40 μg/hr units contain twice as much drug as the 20 μg/hr units, and also di(2-ethylhexyl) phthalate, which increases the release rate of pilocarpine. The two systems are 0.5 mm and 0.3 mm thick, respectively, and both measure approximately 6 mm x 13 mm axially.

PROGESTASERT® systems also utilize diffusion through polymeric membranes to provide controlled administration of the therapeutic agent--in this case progesterone delivered to the uterus at an essentially constant rate of 65 μg/day for one year (Figure 5). The

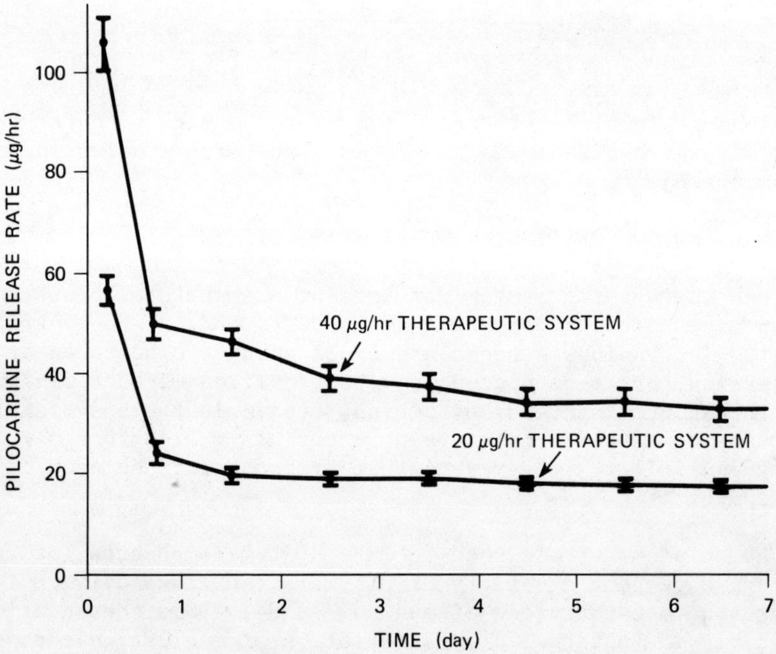
PILOCARPINE RELEASE RATE (μg/hr)
100
80
60
40
20
0
40 μg/hr THERAPEUTIC SYSTEM
20 μg/hr THERAPEUTIC SYSTEM
0
1
2
3
4
5
6
7
TIME (day)

Figure 4. *In vitro* release rate/time profiles (at 37°C) for the 20 μg/hr and 40 μg/hr ocular therapeutic systems. (Copyright 1974 ALZA Corp.)

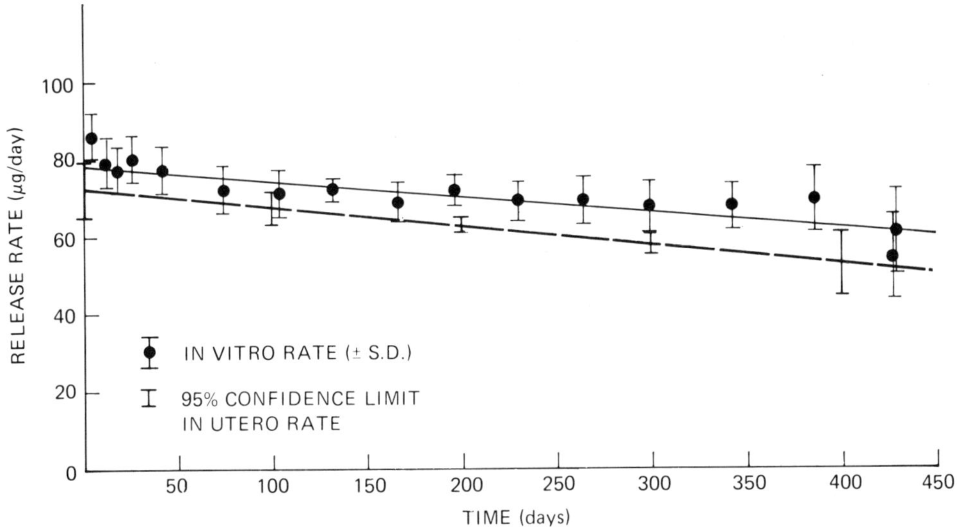
RELEASE RATE (μg/day)
100
80
60
40
20
0
50
100
150
200
250
300
350
400
450
TIME (days)
IN VITRO RATE (± S.D.)
95% CONFIDENCE LIMIT
IN UTERO RATE

Figure 5. *In vivo*/*in vitro* release rates of the intrauterine progesterone contraceptive system. (Copyright 1977 ALZA Corp.)

vertical stem of this flexible T-shaped system is a tube, with wall thickness approximately 0.5 mm. The tube lumen is filled with a dispersion of fine-particle crystalline progesterone, in medical-grade silicone oil.

The EVA copolymer used in these two therapeutic systems is particularly versatile. It is available in broad ranges of comonomer ratios and offers a correspondingly wide choice of drug permeability. EVA is chemically stable, offers a broad range of softness and flexibility, and can be sterilized by gas or ionizing radiation with little or no change in properties. A system for rate-controlled transdermal drug delivery uses a microporous membrane, while the oral therapeutic system--actuated by an osmotic energy source--utilizes a membrane with a high tensile modulus and permeability to water. The structure and functioning of these systems are described in subsequent sections.

Many considerations enter into the selection of polymers as structural elements in these therapeutic systems. The processability considerations shown in Table 1 apply across the broad range of therapeutic systems, although each system does not require all the capabilities listed.

Concern for patient safety transcends all other considerations.

Table 1

CONSIDERATIONS IN SELECTING POLYMERS FOR THERAPEUTIC SYSTEMS

Safety	Primary functional characteristics	Polymer supply	Processability
Nontoxic	Permeability to specific drugs	Minimum intralot variability	Amenability to purification--extraction or reprecipitation
Nonirritating to tissue	Sorption and permeability to water	Minimum interlot variability	Moldable--injection, transfer, compression
Noncarcinogenic	Chemical stability *in vivo* and *in vitro*	Assured long-range availability	Extrudable--profile, sheet blowfilm
Nonallergenic	Morphological stability		Solvent castable
Nonthrombogenic and nonhemolytic (in polymers contacting whole blood or where residues can induce adverse cardiovascular reaction)	Minimum chemical interaction with drug		Thermoformable
	Minimum interaction with body lipids		Sealable--hot bar, ultrasound, RF, solvent cement
	Hardness and flexural modulus		Sterilizable--minimum chemical or morphologic disturbance
	Mechanical integrity *in vivo*		Minimum reliance on antioxidants
			Minimum chemical change during processing
			Minimum sensitivity of final morphology to processing conditions

It is necessary to establish the safety of polymers by extensive animal and subsequent human testing. All polymers must pass Class VI U.S.P. tests for polymers in contact with drugs. Chemical interaction with the drug is ruled out and interaction with body lipids should be minimal to avoid shifts of permeability and loss of mechanical integrity _in vivo_.

Materials used must be exceedingly well characterized; specifications should include: melt index, hardness, tensile and flexural properties, and other mechanical and physical properties; data on residual monomer, catalysts, modifiers, antioxidant content; and kinetics of chemical changes during processing. Chemical stability is important, since changes in composition, either in processing or _in vivo_, could alter the permeation rate or even mechanical integrity of the system. Ordinarily an initial purification process is required for polymers obtained commercially to remove impurities or additives that would be prone to leach out _in vivo_.

RESEARCH TOOLS FOR EXPERIMENTAL STUDIES OF RATE-CONTROLLED DELIVERY

The OCUSERT® and PROGESTASERT® systems described previously were developed when no routine means existed for testing the action of drugs delivered at various constant rates for either local or systemic administration. Thus, prototype systems had to be developed to find the optimal release pattern for therapy--first, systems for testing such patterns in animals and then for testing them in small groups of human subjects. This empirical approach to designing the therapeutic program was both time-consuming and costly.

Now rate-controlled systems--which are membrane-controlled and osmotically actuated--are available to carry out every phase of rate-controlled pharmaceutical research and development from animal toxicology to incorporation of the drug of interest into a dosage form for routine use in human therapy. The experimental tools developed make it practical to study--in both animals and man--the distribution and actions of drugs in the body when their concentrations are maintained at a virtually steady level. Data from such studies are basic to the logical design of rate-controlled pharmaceutical products for medical treatment. For such research, two classes of osmotically driven pumps have been developed--one for animals and one for clinical pharmacology. Within each class, two delivery rates are available (20-22).

ALZET® MINIPUMPS FOR ANIMAL STUDIES

Although it has always been possible to administer drugs and other bioactive agents to small animals in a rate-controlled manner, the advent of ALZET® osmotic minipumps has made it reliable and simple to do so. The minipumps (Figure 6) have been designed with delivery rates of 1.0 or 0.5 μl/hr, and delivery durations, respec-

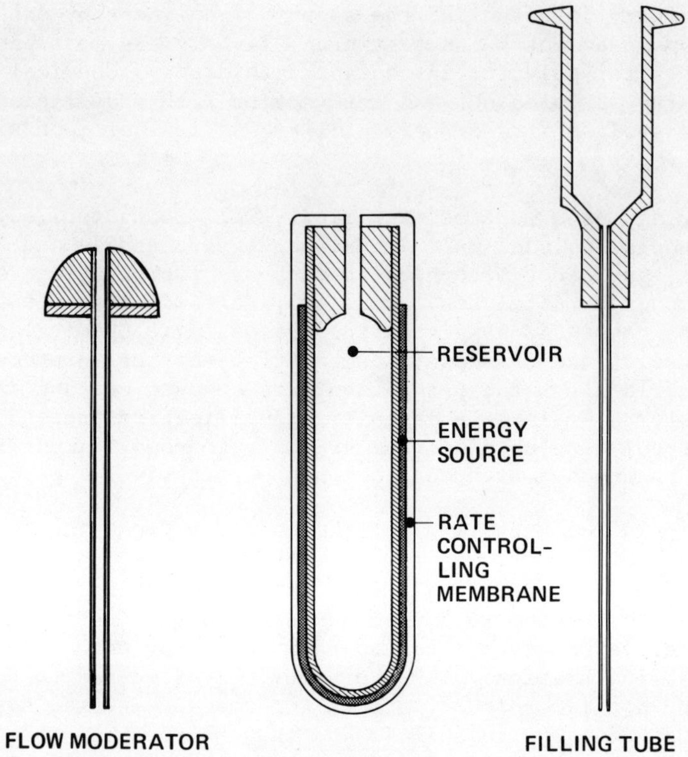
RESERVOIR
ENERGY SOURCE
RATE CONTROL-LING MEMBRANE
FLOW MODERATOR
FILLING TUBE

Figure 6. ALZET® osmotic minipump with flow moderator. (Copyright 1980 ALZA Corp.)

tively, of 1 or 2 weeks. The pumps are capsular in shape, and have the same cross-section, internal fill volume (200 μl), and external dimensions (2.5 cm long and 0.6 cm in diameter), regardless of rate/duration specifications. Internally each pump consists of: 1) an inert, impermeable, flexible drug reservoir open to the exterior via a single portal; 2) a thin sleeve of osmotic agent surrounding the reservoir; and 3) a semipermeable membrane surrounding the sleeve of osmotic agent. The researcher fills the reservoir with a solution of the drug to be administered. When the filled minipump is implanted, generally subcutaneously or intraperitoneally (21), water from surrounding tissues begins to move osmotically through the membrane into the osmotic sleeve. Water moves into the osmotic sleeve at a rate controlled by the permeability of the membrane; the rigidity of the membrane causes the swelling of the osmotic sleeve to displace the liquid drug formulation within the reservoir out through the portal in a continuous and controlled manner.

The user fills the minipump with drug solution through the portal, using a special filling tube attached to a small syringe. The flow moderator is then inserted through the portal, to restrict the diffusional mode of solute exit, assuring that the osmotic process will control the delivery rate. If the transparent overcap is removed from the flow moderator, a catheter (PE 50 or 60 size) can be attached for localized administration of drugs to areas remote from the site of implantation or to target organs.

OSMET[T.M.] MODULES FOR HUMAN PHARMACOLOGICAL RESEARCH

OSMET[T.M.] drug delivery modules are designed to be orally administered to human subjects and to deliver drug formulation at 8 μl/hr over 24 hours or at 15 μl/hr over 12 hours. Each module has the same internal volume, cross-section, and operational principles and procedures as ALZET® osmotic minipumps.

The OSMET[T.M.] drug delivery modules are available by special arrangement on a drug-by-drug basis to pharmaceutical companies interested in assessing the feasibility of developing rate-controlled, orally administered products. The modules have been used in clinical investigational studies to deliver anti-inflammatories, antihypertensives, vitamins, and receptor-blocking agents.

By providing a rate-specified source of drug moving through the gastrointestinal tract, the modules can be used to define the absorption window of a drug, which follows from measured discrepancies between: A) the extent of drug absorbed vs extent delivered; and B) the time course of absorption rate vs that of delivery rate. The validity of these important comparisons depends on the reliable kinetic performance of the OSMET[T.M.] modules *in vivo*.

IN VIVO KINETIC PERFORMANCE OF EXPERIMENTAL OSMOTIC PUMPS

For osmotic minipumps and the 24-hour OSMET[T.M.] module -- all delivering a triphenylmethane dye solution in isotonic saline at 37°C (22)--Figure 7 shows the pumping rate normalized to the pumping rate specified on the label, vs time normalized to the duration specified on the label. Beyond 0.1 label duration, regardless of system type or duration, performance shows essential constancy; steady-state precision of delivery rate is ± 5%. All systems have a startup lag of 1-4 hours (depending on label duration) but this transient is large enough to show only for the OSMET[T.M.] pump (22).

Figure 8, which plots ratio of *in vitro*/*in vivo* pumping rates of the same pumps vs label duration, shows that their precision is practically unchanged with respect to the *in vitro* results in isotonic saline at 37°C (22). The results qualify the *in vitro* test conditions as bioanalogous, in that they correspond to the *in vivo* situation.

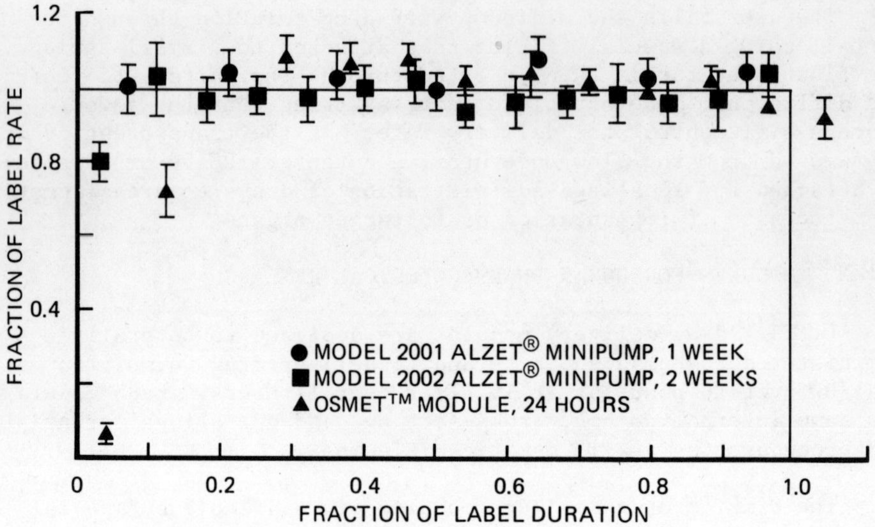
1.2
0.8
0.4
FRACTION OF LABEL RATE
● MODEL 2001 ALZET® MINIPUMP, 1 WEEK
■ MODEL 2002 ALZET® MINIPUMP, 2 WEEKS
▲ OSMET™ MODULE, 24 HOURS
0
0.2
0.4
0.6
0.8
1.0
FRACTION OF LABEL DURATION

Figure 7. In vitro performance of ALZET® osmotic minipump and 24-hour OSMET T.M. drug delivery module.

Model 2001 ALZET® Osmotic Minipump
Theoretical Pumping Rate: 1.0 μl/hr
Actual Pumping Rate: 1.03 ± 0.04 μl/hr

Model 2002 ALZET® Osmotic Minipump
Theoretical Pumping Rate: 0.5 μl/hr
Actual Pumping Rate: 0.49 ± 0.02 μl/hr

24 Hour OSMET T.M. Module
Theoretical Pumping Rate: 8 μl/hr
Actual Pumping Rate: 8.17 ± 0.49 μl/hr
Data points represent the average, and error bars the standard deviation

(Copyright 1980 ALZA Corp.)

RATE-CONTROLLED ORAL DOSAGE FORMS FOR THERAPY

Oral dosage forms have been the focus of intensive research aimed at simplifying regimens and more closely controlling drug concentrations in blood. Partly because of the gastrointestinal tract's wide variations in acidity, motility, and lumenal contents, much controversy surrounds the question of whether in vivo drug release from oral dosage forms can be predicted from in vitro data. The oral dosage form recently developed (16), however--the elementary osmotic pump (EOP or OROS® system)--has the following attributes: 1) predictability of drug release rate in vivo from simple in vitro tests; 2) capability of delivering, for 12-24 hr, drugs having a wide

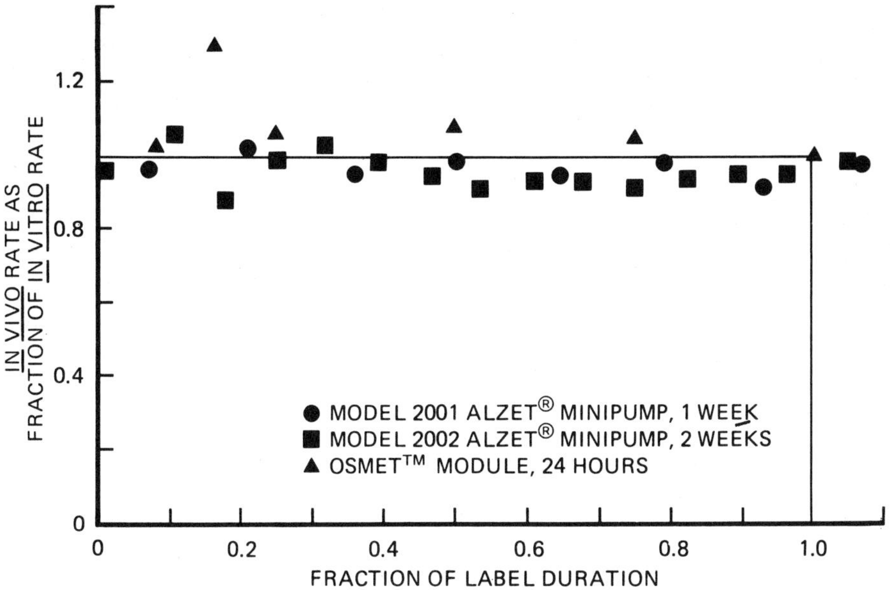
IN VIVO RATE AS FRACTION OF IN VITRO RATE
1.2
0.8
0.4
0
● MODEL 2001 ALZET® MINIPUMP, 1 WEEK
■ MODEL 2002 ALZET® MINIPUMP, 2 WEEKS
▲ OSMET™ MODULE, 24 HOURS
0
0.2
0.4
0.6
0.8
1.0
FRACTION OF LABEL DURATION

Figure 8. Comparison of *in vitro* and *in vivo* performance for ALZET® osmotic minipumps and 24-hour OSMET^{T.M.} drug delivery module (22). *In vivo* performance of subcutaneously implanted ALZET® minipumps was examined in rats on the basis of their urinary excretion of ^{51}Cr-EDTA infused by the minipump; 24-hour urine collections provided the basis for assessment of daily average pumping rates. Individually marked OSMET T.M. modules were orally administered to dogs at 2, 4, 6, 12, 18, and 24 hours prior to sacrifice; recovered modules were assayed for residual drug content, giving an average rate over the time elapsed between feeding and module recovery.
(Copyright 1980 ALZA Corp.)

range of solubilities at desired rates that maintain plasma concentrations of drug within a narrower range than dosage forms specified only by quantity (23); 3) independence of release rate from GI tract pH and motility variations; 4) isolation, from gut contents and from the mucosa, of solid drug not yet delivered; and 5) simplicity of manufacture.

In the OROS® system, solid drug--present alone or with an osmotic driving agent--is surrounded by a semipermeable membrane having one delivery orifice (Figure 9). During pump operation, water from the environment is continuously imbibed across the semipermeable membrane, by osmosis, to produce the fluid drug formulation. The membrane's structure does not allow expansion of the tablet's volume;

thus, fluid must leave the interior of the tablet at the same rate that water enters by osmosis. Fluid--i.e., drug in solution--flows out through the single, small orifice at a constant rate as the OROS® system moves through the gastrointestinal tract, until the last of the solid drug in the core has dissolved. Then the osmotic driving force begins to decline and drug release rate ceases to be constant, but declines in a predictable manner. The membrane is excreted intact.

A typical development process for an OROS® system starts out with the release rate as a given, obtained from pharmacology and pharmacokinetics. Nomograms exist that quantitatively interrelate *in vivo* delivery rate, the solubility properties of the drug, core formulation, and membrane specifications. The desired delivery rate may be achievable by selecting a particular salt of the drug compound or by using straightforward techniques to modify the osmotic pressure of a desirable core formulation or the hydraulic conductivity of the membrane.

The kinetic functioning of the OROS system is so well-defined that there is a simple set of *in vitro* test conditions that are bioanalogous, i.e., predictive of *in vivo* drug release, as described above and elsewhere (10). For each OROS system product it is necessary to validate the predictability of *in vivo* release rate for that particular drug and to determine: the delivery rate needed to attain the desirable concentration in blood; the period of time a single dosage form can deliver drug that will be absorbed; and the

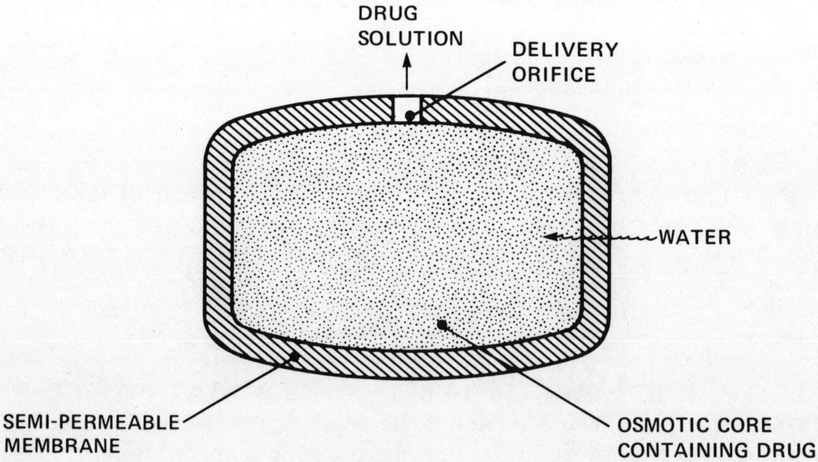
DRUG
SOLUTION
DELIVERY
ORIFICE
WATER
SEMI-PERMEABLE
MEMBRANE
OSMOTIC CORE
CONTAINING DRUG

Figure 9. Elementary osmotic pump cross-section (16). (Copyright 1975 ALZA Corp.)

overall extent and rate of the drug's absorption into the bloodstream vs the extent of its delivery rate from the OROS® system. With the advent of the OSMET$^{T.M.}$ module, such studies can be performed before actually developing a specific OROS system form of the agent.

Studies indicate that absorption times of 12 or 15 hours are achievable with many drugs, despite the common belief that only a few hours are available. That belief may have arisen from studies demonstrating fast absorption of drugs, after ingestion, from dosage forms that delivered most of their drug quickly. The longer absorption times now being demonstrated will offer an opportunity essentially of eliminating three- or four-times-daily dosing; for a substantial class of drugs, once-a-day regimens are within reach.

A number of major drugs in OROS system form are now in development and in various stages of clinical trials. These are collaborative activities between ALZA and other pharmaceutical firms; collaborations with other companies are under discussion.

TRANSDERMAL THERAPEUTIC SYSTEMS

Another technology that we have advanced for systemic treatment is the transdermal therapeutic system (TTS) for release of drug through intact skin into the circulation. The transdermal delivery system represents a major breakthrough. Previously we really had only two approaches to giving medication systemically: tablets, which go back to the Egyptians, and injectable dosage forms, invented about a century ago. The transdermal system is potentially far superior to both.

The transdermal system (TTS) makes it possible to eliminate many variables affecting GI absorption of drugs and to maintain the desired concentration of the drug in the body (18). The system permits drug to go directly through the skin, into the capillaries that run under the skin, and then into the general circulation. This provides the equivalent of a controlled intravenous infusion-- previously the closest approach to rate-controlled drug delivery that we had. The TTS, however, has many advantages over such an infusion, because it does not require use of a needle, or the patient's hospitalization, immobilization, and connection to an I.V. bottle, or a nurse to administer the whole course of treatment.

The transdermal system is contained in a very thin disc that the patient places on intact skin. The unit has multiple layers (Fig. 10) providing the reservoir functions, the controlled-release functions, and the adhesion to the skin that are required to assure continuity of drug transfer into the patient's bloodstream. One site chosen for wearing the transdermal therapeutic system is behind the ear, where skin is particularly permeable.

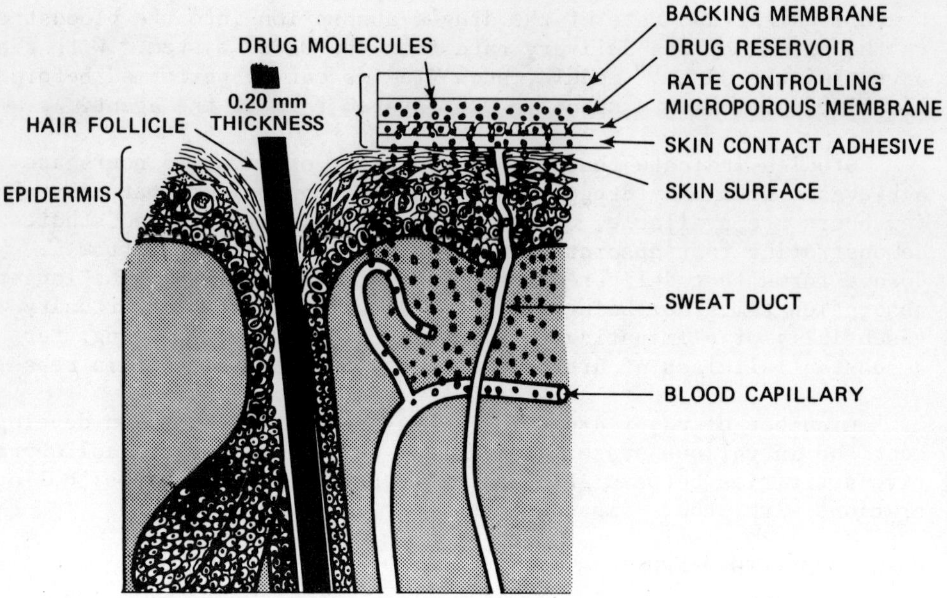
DRUG MOLECULES
0.20 mm THICKNESS
HAIR FOLLICLE
EPIDERMIS
BACKING MEMBRANE
DRUG RESERVOIR
RATE CONTROLLING MICROPOROUS MEMBRANE
SKIN CONTACT ADHESIVE
SKIN SURFACE
SWEAT DUCT
BLOOD CAPILLARY

Figure 10. Schematic representation of a transdermal system in place on the surface of the skin. (Copyright 1975 ALZA Corp.)

The main constraint in applying the transdermal system is the need for very potent drugs, since the skin represents a partial barrier to the passage of many substances. Fortunately, many agents exist in significant areas of medicine that are suitable for transdermal delivery. Transdermal drug delivery could play an important role in treating hypertension, angina (heart pain), peripheral vascular disease, stroke, respiratory ills, and hormone deficiency.

Our first transdermal therapeutic system (TTS) delivers enough scopolamine to prevent motion sickness (24,25) over a 3-day interval. The system reduces this valuable drug's troublesome side effects by controlled delivery of a sufficient amount of it to maintain constant, low plasma concentrations (Fig. 11). We are also studying this system's efficacy in the treatment of vertigo, a condition that produces chronic dizziness. The experimental evidence supporting the product's rate-controlled functioning in vivo is found in ref. 18.

THE MEDICAL POTENTIAL FOR THERAPEUTIC SYSTEMS

Certain areas in medicine appear especially promising for the application of controlled delivery technology. In the Western world,

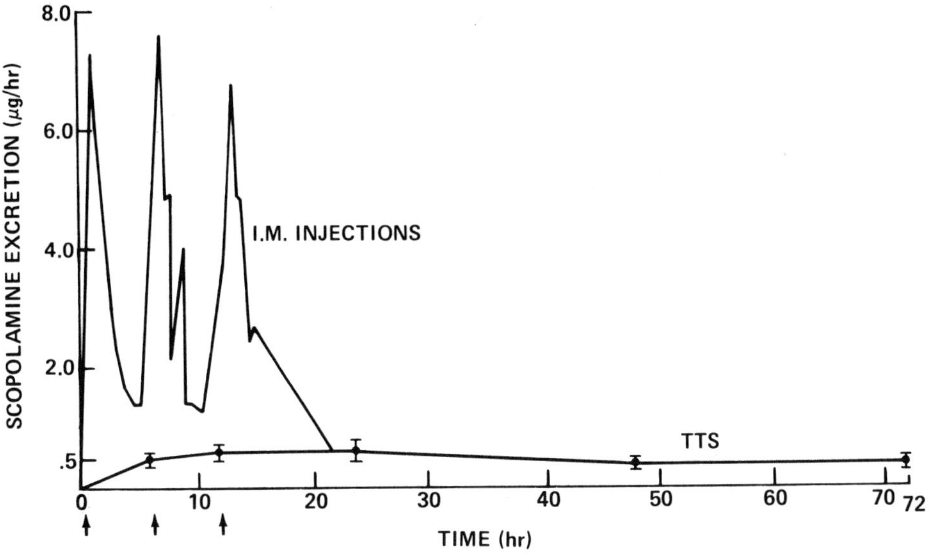
SCOPOLAMINE EXCRETION (μg/hr)
8.0
6.0
4.0
2.0
.5
I.M. INJECTIONS
TTS
0
10
20
30
40
50
60
70
72
TIME (hr)

Figure 11. Comparison of scopolamine excretion from the body when the drug is administered either as multiple intramuscular injections (arrows) or as a single application of a TTS. Excretion levels reflect plasma levels for this drug. (Copyright 1976 ALZA Corp.)

the opportunity exists to transform a great number of existing drugs into significantly superior products by reducing their side-effects. In addition, the long duration of drug release from one application of a therapeutic system will be of great benefit to the individual who has to take medication regularly for a chronic condition such as hypertension or arthritis. Many therapeutic system applications are also likely in pediatrics, in geriatrics, and among the very sick. These three populations share a need to be medicated by a third party, which multiplies the problems of compliance.

Equally attractive are the prospects for controlled-delivery to improve health in the developing countries. There, tropical infections exist--such as malaria and trachoma-- for which no vaccines are available. But drugs are either available or are coming into being that could provide protection with minimal intervention--as do vaccinations--if incorporated into therapeutic systems with long functional lifetimes. Drug therapy could thus enter a new era in its development with the widespread use of controlled-delivery drug systems.

ABSTRACT

Therapeutic systems are rate-controlled dosage forms that administer drugs at specified rates over specified, often prolonged periods of time. These dosage forms, which utilize polymers in their structural and rate-controlling components, represent a new functional use for polymers in medicine. Systems are available or under development to: medicate the eye; administer drugs orally; deliver a contraceptive hormone locally to the uterus; release drug dispersed in bioerodible polymers; and administer potent agents systemically through intact skin. Conventional dosage forms release drug at a rate that is highest initially and thereafter continually declines, producing peaks and troughs in the concentrations of drug in blood and tissues. This pattern can cause an uneven therapeutic effect and tends to trigger side-effects associated with peaks in drug concentration. Therapeutic systems release drug at constant rates selected to maintain drug at low, steady levels that will provide adequate therapy with minimal side-effects. Applications of therapeutic systems are likely to: improve the selectivity, safety, and efficacy of many common drugs; permit the use in human therapy of natural agents having half-lives too short to permit their administration in conventional dosage forms; ease compliance problems among patients with chronic conditions; and provide protection against diseases in tropical areas.

REFERENCES

1. G. Levy, in "Proceedings of the Fifth International Congress of Pharmacology", G. Acheson, ed., S. Karger, Basel, 3, 34 (1973)
2. J. A. Lutterman, A. A. J. Adriaansen & A. van't Larr, Diabetologia, 17, 17 (1979)
3. J. Urquhart, in "Institute of Medicine Conference Proceedings: Pharmaceuticals for Developing Countries", National Academy of Sciences, Washington, D.C., 329 (1979)
4. A. Zaffaroni, in "Abstract of Papers Presented at the 31st International Congress of Pharmaceutical Sciences", Washington, D.D., 19 (1971)
5. F. E. Yates, H. Benson, R. Buckles, J. Urquhart & A. Zaffaroni, in "Advances in Biomedical Engineering", J. H. U. Brown & J. F. Dickson, eds., Academic Press, New York, 5, 1 (1975)
6. A. Zaffaroni, in "Proceedings of the Sixth International Congress of Pharmacology", J. Tuomisto & M. K. Paasonen, eds., Forssan Kirjapaino Oy, Helsinki, 5, 53 (1975)
7. K. Heilmann, "Therapeutic Systems. Pattern-Specified Drug Delivery: Concept and Development", George Thieme, Stuttgart (1978)
8. A. Zaffaroni, in "Proceedings of the International Symposium on Science, Invention, and Social Change", General Electric Co., Schenectady, 73 (1978)
9. J. Urquhart, in "Proceedings of the Sixth International Congress of Pharmacology", J. Tuomisto & M. K. Paasonen, eds., Forssan

Kirjapaino Oy, Helsinki, 5, 62 (1975)
10. J. Urquhart, in "Ophthalmic Drug Delivery Systems", J. E. Robinson, ed., Academy of Pharmaceutical Sciences, Washington, D.C., 105 (1980)
11. B. B. Pharriss, J. Reprod. Med., 20, 155 (1978)
12. G. Trobough, A. M. Guderian, R. E. Erickson, S. A. Tillson, P. Leong, D. A. Swisher & B. B. Pharriss, J. Reprod. Med., 21, 153 (1978)
13. G. Rybo, J. Reprod. Med., 20, 175 (1978)
14. H. S. Brown, G. Meltzer, R. C. Merrill, M. Fisher, C. Ferre & V. A. Place, Arch. Ophthalmol., 94, 1716 (1976)
15. G. Benagiano & H. Gabelnick, J. Steroid Biochem., 11, 449 (1979)
16. F. Theeuwes, J. Pharm. Sci., 64, 1987 (1975)
17. J. E. Shaw & S. K. Chandrasekaran, in "Drug Metabolism Reviews", F. di Carlo, ed., Marcel Dekker, New York, 8, 223 (1978)
18. J. Shaw & J. Urquhart, Trends Pharmacol. Sci., 1, 208 (1980)
19. H. Leeper & H. Benson, Polymer Eng. Sci., 17, 42 (1977)
20. F. Theeuwes & S. I. Yum, Ann. Biomed. Eng., 4, 343 (1976)
21. F. Theeuwes & B. Eckenhoff, in "Sixth International Symposium on Controlled Release of Bioactive Materials" (in press)
22. B. Eckenhoff, in "Controlled and Topical Release Session of the AICHE Meeting", June 10, 1980, Philadelphia, PA
23. F. Theeuwes & W. Bayne, J. Pharm. Sci., 66, 1388 (1977)
24. A. Graybiel, J. Knepton & J. Shaw, Aviat. Space Environ. Med., 47, 1096 (1976)
25. N. Price, L. G. Schmitt & J. E. Shaw, Clin. Ther., 2, 258 (1979)

SECTION IV

DENTAL MATERIALS APPLICATIONS OF POLYMERS

This section is concerned with the use of polymers in dentistry. The fourteen chapters are organized to be a compact survey of dental polymer science. The state of the art of the more important applications is reviewed, original research and newly developing technology are presented, and research trends are predicted. The applications discussed here span the entire field of dental materials - not just dental biomaterials. Biomaterials are generally considered to be those materials designed to reside permanently in the body. By this definition, several important dental polymers are not truly biomaterials, but their technology may be interesting and useful to scientists practicing other disciplines. The chapters are arranged to give overviews of dental polymers (Glenn and Halpern/Karo) followed by specific topics and original research. Subjects included are: particulate filled composites (Abell/Crenshaw/Turner), monomers (Antonucci, Griffith/O'Rear, and Cowperthwaite/Foy/Malloy), denture base polymers (Cornell) polymerization processes (Brauer and Kilian), and ionic polymers (Wilson and Belton/Stupp). The section concludes with three discussions of aspects of structure-property relationships (Draughn, Powers/Fan/Craig, and DeVries/Knutson/Draughn/Reichart/Koblitz).

DENTAL POLYMERS

J. F. Glenn

Milford, DE 19963

INTRODUCTION

Dental materials, that is, all the chemical items used in tooth repair or replacement, covers the broadest range of elements and compounds found in any one chemical industry. The science of dental materials is multidisciplined, requiring its practitioner to be at least knowledgeable himself while maintaining experts for detailed advice on organic, inorganic, physical, bio- and polymer chemistry and related fields of metallurgy, materials science, adhesion, and coatings technology. Not too many years ago the preparation of dental materials was empiricism at its best, with some dentists even compounding their own restorative and prosthetic requirements. Regardless of some opinions to the contrary, the science of dental materials has truly become a "science" utilizing the most sophisticated modern techniques of instrumental analysis, fracture mechanics, and abrasion theory, just to mention a few.

In this discussion, I propose to give only a quick overview of but one segment of dental materials - polymers in dentistry - mentioning the types of dental uses, and leave the important details to be addressed in papers that follow.

DENTAL POLYMERS

Every branch of dentistry depends on natural or synthetic polymeric materials in some form for successful therapy. The most extensively used but not the oldest group of polymers are the acrylic resins - the methyl methacrylates and derivatives. In fact, when one thinks of dental resins, the acrylics almost exclusively come to mind as though dental polymers were equated with the acrylics.

THE ACRYLIC POLYMERS

In 1937, the first use of polymethyl methacrylate appeared as an esthetic denture base formed by injection molding a resin slug into a more or less dry gypsum mold holding porcelain artificial teeth. Not a very satisfactory technique, this was replaced by a predoughed, heat curable mixture of polymethyl methacrylate and methyl methacrylate monomer, called "gel cake".

The handling and stability problems with these gels resulted in replacement by the currently used polymer powder plus liquid monomer added at the time of use. We have come a long way from the carved ivory dentures of George Washington through the era of rubber dentures called "Vulcanite" (there are still a considerable number in use) to the acrylic dentures of the last 40 years. The excellent esthetics and manipulation ease, especially since they followed on the heels of the Vulcanite rubber denture bases has made replacement of the acrylics by other resins a slow and difficult task. Over this 40 year period only marginal improvements have appeared. Notable examples are the addition of crosslinking monomers such as ethylene dimethacrylate to prevent crazing on repair, copolymerization with other acrylates and methacrylates for better durability, internal pigmentation in the polymer beads for shade control, and inclusion of chopped fibers for characterization by vein simulation (Fig. 1).

The improvement of impact resistance in dentures has received attention because unpredictable failures can occur at inconvenient times. Partial grafting of acrylic polymers to butadiene polymers

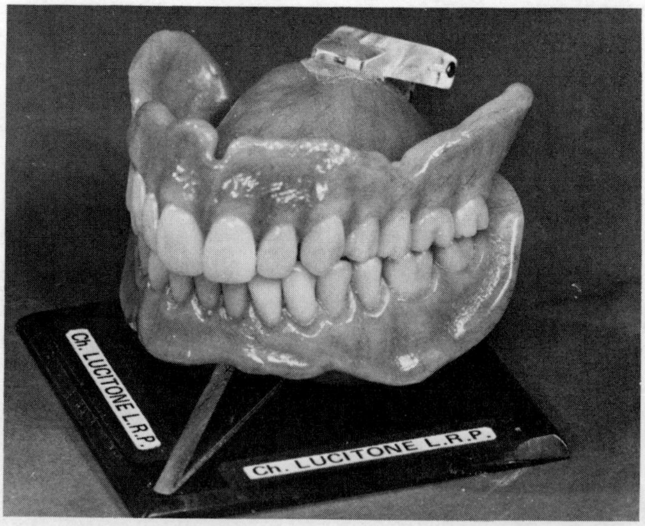
Ch. LUCITONE L.R.P.
Ch. LUCITONE L.R.P.

Figure 1. Acrylic denture characterized for vein similation by fibers.

and the use of vinyl chloride - vinyl acetate copolymers with methyl methacrylate have somewhat improved toughness. However, it is only very recently that more extensive modifications are appearing, such as additions of rubber toughened epoxies and polyurethanes.

Table I compares the important physical properties of the currently used acrylic resin denture bases - the heat cured and self cured bases. These are typically powder/liquid mixtures containing benzoyl peroxide redox systems activated by a tertiary aromatic amine. Among these are the vinyl chloride - vinyl acetate - methyl methacrylate mixtures and the newer pour type denture bases - the latter designed for laboratory convenience and model fit by control of polymerization shrinkage. On this half of the table the important comparative properties are transverse deflection, failure and water sorption.

Shortly after the powder/liquid acrylic denture bases were introduced, they were adapted to customized crown and bridge use, (Fig. 2) first as full crowns and facings to be cemented to natural teeth. Because of poor abrasion resistance of the acrylics in such locations, techniques, have been modified to use metals such as the gold and the cobalt alloys, as the supporting and occluding surfaces with acrylic facings held thereon mechanically. Abrasion of crown

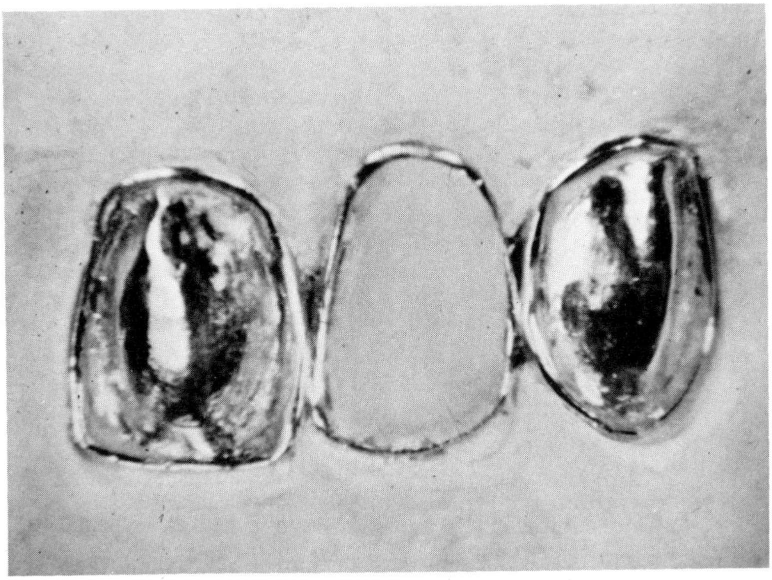

Figure 2. Gold fixed prosthesis with acrylic crown and bridge resin polymerized thereon.

TABLE I

TYPICAL PHYSICAL PROPERTIES OF DENTURE BASE POLYMERS

		Heat Cured Poly (Methyl Meth-Acrylate)	Self Cured Poly (Methyl Methy-Acrylate)	Vinyl-Methyl Meth-Acrylate	Pour Resins
Tensile Strength	MN/m^2	48-62	-	52	-
Compressive Strength	MN/m^2	76	-	70-76	-
Tensile Elongation	%	1-2	1-2	7-10	1-2
Izod Impact	(ft-lb/in)	(3.5)	(2)	(8)	(2-4)
	kg m/cm	0.19	0.11	0.44	0.11-0.22
Transverse Deflection					
3500g	mm	2.0	<1.5 at 2500g (1)	1.9	1.8
5000g	mm	4.0	<4.5 at 4000g (1)	3.9	4.5
failure load	g	6000		7000	5500
KHN Wet	Kg/mm^2	15	12-15	15	15
Coefficient of Thermal Expansion	/°C $x10^{-6}$	81	-	71	-
Water Sorption	mg/cm^2	0.7	<0.8 (1)	0.26	<0.8 (1)
Solubility	mg/cm^2	0.04 (1)	<0.04 (1)	<0.04 (1)	<0.04 (1)

1. ADA Specifications #12 and #13 limits.

and bridge acrylics has been addressed by the inclusion of a few percent of fine (10μm) silica powder with or within the polymer beads. Improvement by this procedure has only been marginal since the surface appearance is adversely affected and the abrasion resistance is not sufficiently increased.

Acrylic resin teeth for denture construction were a natural product for the porcelain tooth manufacturers to produce. Unfortunately they leaped into injection molded teeth which clinically failed by crazing or complete disintegration. Acrylic teeth have been powder/liquid dough molded by a semi-transfer process ever since. In addition to partial crosslinking by ethylene dimethacrylate or allyl methacrylate and excellent shade control, the most recent (within the last 4 years) change to appear out of Germany is the inclusion of up to 50% of microfine silica, i.e., Aerosil OX50 with a particle size of around 40 nm and a surface area of about 50 m^2/g. Only long-term evaluations in clinical use will reveal if this approach is a valid improvement.

With the patenting by Kulzer (1) (Germany) in 1942 of the acrylic self-curing or room temperature polymerizing tertiary amine-benzoyl peroxide redox systems, the resin filling material use of methyl methacrylate (Fig. 3) commenced, first in Germany in the mid 40's and in the U.S., shortly thereafter in 1950. The early acrylic restoratives had serious problems of high thermal expansion (7 or more times that of tooth tissue), low modulus, high polymerization shrinkage, poor wear resistance, and poor color stability. Restorative dentistry is highly important to dental health and since the resins do seem to offer a great potential, the polymeric filling materials area is the most research-active in dental materials by far, and has progressed a long way from the original polymer powder plus benzoyl peroxide - liquid methyl methacrylate plus amine combinations to the present day, rather sophisticated composite resin restoratives (2) composed of two pastes. In the "composites", as they are commonly called, the basic acrylic monomer used as a binder for the mineral filler has been, Fig. 4, an adduct of glycidyl methacrylate with bisphenol-A and called by the acronym "Bis-GMA". Other dimethacrylates - ethylene, triethylene or bisphenol-A dimethacrylate, methacrylate urethanes, (Fig. 5), i.e. reaction product of hydroxypropyl methacrylate with hexamethylene diisocyanate or other diisocyanate (3) are now being incorporated alone or in combination with Bis-GMA. Very little methyl methacrylate is employed in the composite resin restoratives today. The fillers used to paste the binder and control the physical properties of the final restoration, have ranged widely in composition from glass beads to fine (20 mm) silica powders (Fig. 6). The filler component makes up 60-80% of the paste composition. Since both the catalyst and accelerator paste components contain the filler, the final filling in the tooth cavity is equally high in inorganic content. In developing a resin restorative, the interrelationship of binder and filler must be studied, including refractive index, liquid demand, particle packing, surface wetting, particle aspect and thermal expansion.

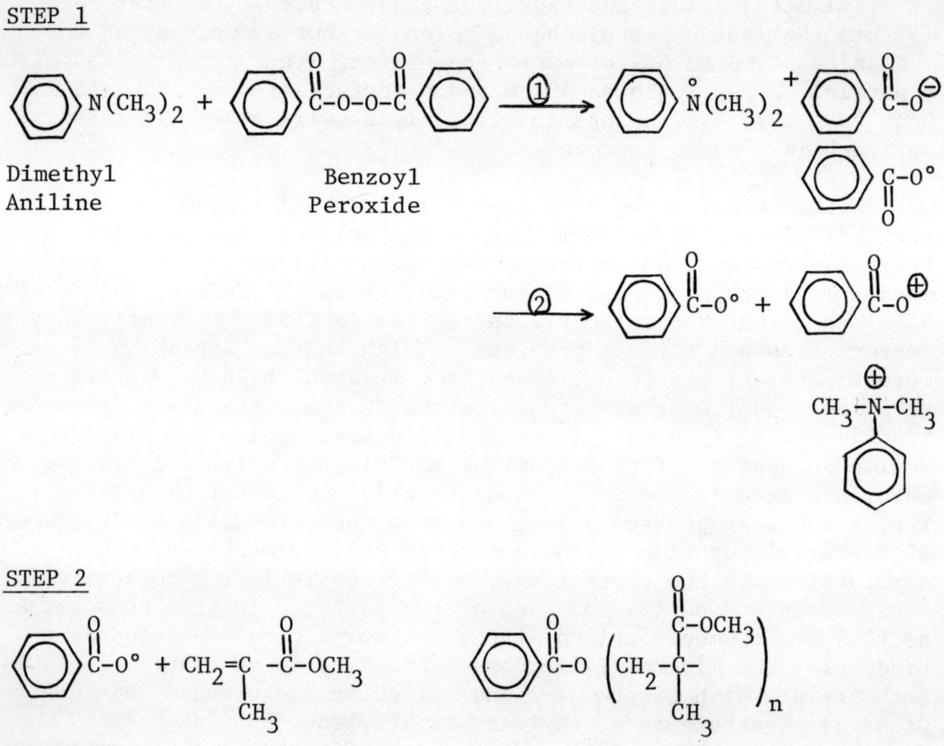
STEP 1
Dimethyl
Aniline
Benzoyl
Peroxide
STEP 2

1. Horner
2. Imoto

Figure 3. Postulated redox activated polymerizations of methyl methacrylate.

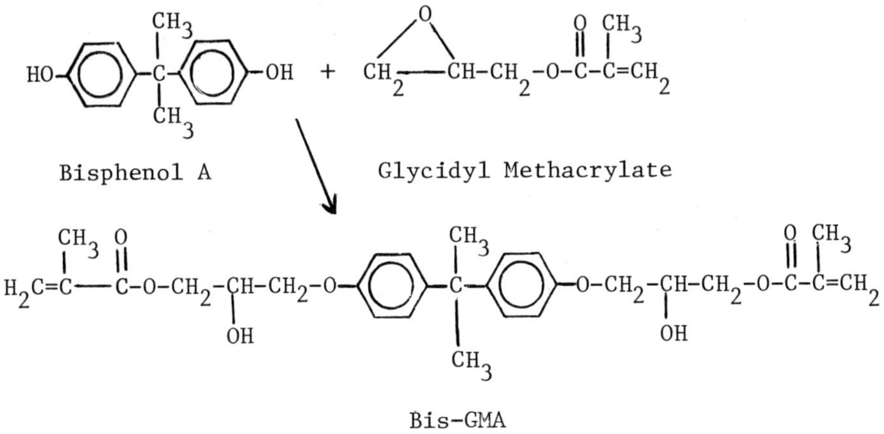
Bisphenol A
Glycidyl Methacrylate
Bis-GMA

Figure 4. Preparation of Bis-GMA (Bowen).

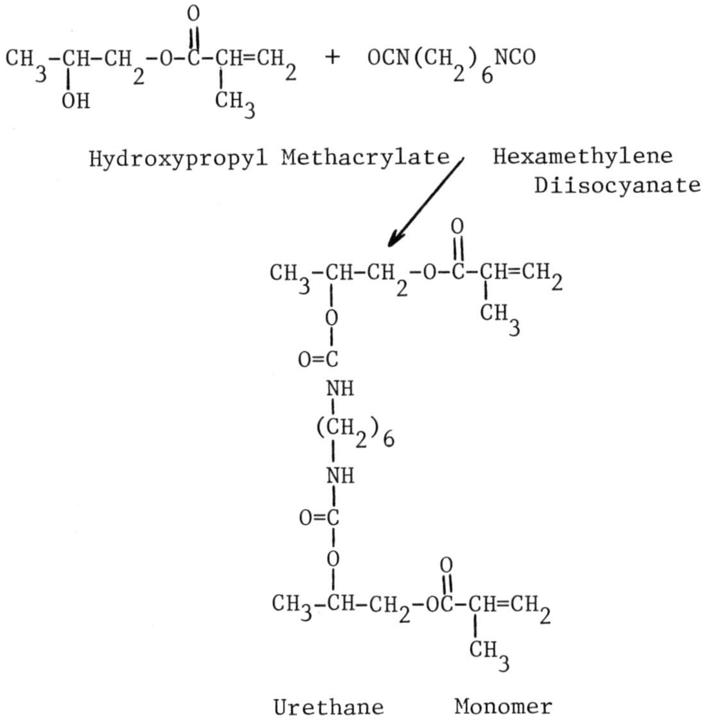
Hydroxypropyl Methacrylate
Hexamethylene Diisocyanate
Urethane Monomer

Figure 5. Urethane dimethacrylate.

E-Glass Fibers
Soda Lime Glass Beads
Calcium Phosphate
Fused Silica
Lithium Aluminosilicate Glass Ceramic
Aluminum Silicates
Barium Boroaluminosilicates
Crystalline Quartz
Calcium Silicate
Pyrogenic Silica

Figure 6. Fillers used in composite restoratives.

The composite resin restoratives have suffered from finishing difficulties (4). That is, polishing results in a rough surface. Shown in Fig. 7 is an SEM photograph of a tooth brushed surface of a composite restorative. It has been demonstrated that this type of surface finish is much more attractive to plaque attachment than the natural tooth surface (5). These problems have lead to the attempt to incorporate microfine silica in the composite resin pastes as has been done in the acrylic teeth. Here the liquid demand of the included microfine particles unfortunately has lowered the capacity to load to only about 40% with a consequent increase in thermal expansion and abrasion loss. While there is currently a rush by manufacturers to market products based on this microfine particles technology, long term clinical observation, as is the case with loaded teeth cited above, will be the judge of efficacy of this approach. Until recently, all composite acrylic restoratives were based on peroxide-amine redox systems using dimethyl p-toluidine as the accelerator. Most self-curing acrylics now use p-tolyl diethanoamine in its place for color stability and curing efficiency improvement. The self-curing acrylic restoratives are now being challenged by cure systems activated by U.V. (6) or visible light (9) exposure, where the photosensitizers are typically benzoin methyl ether or camphoroquinone with a substituted morpholine reducing agent (Fig. 8). Light cured restoratives offer a great manipulation advantage to the dentist, since he can polymerize the filling at will, a very important consideration in a complicated restoration. Problems of lateral cure (8), depth of cure (9), air inhibition, color stability and surface roughness still exist but these are the targets of current intensive development. The composite restoratives are pushing silver amalgam fillings for posterior restoratives and stand a good chance of winning.

Comparative physical property data ranges are summarized in Table 2. The major types of currently used restorative resins are included - the unfilled acrylic restorative, the popular composite resin and the newer microfine silica - filled composite and visible light cured restorative. These are compared with a long established

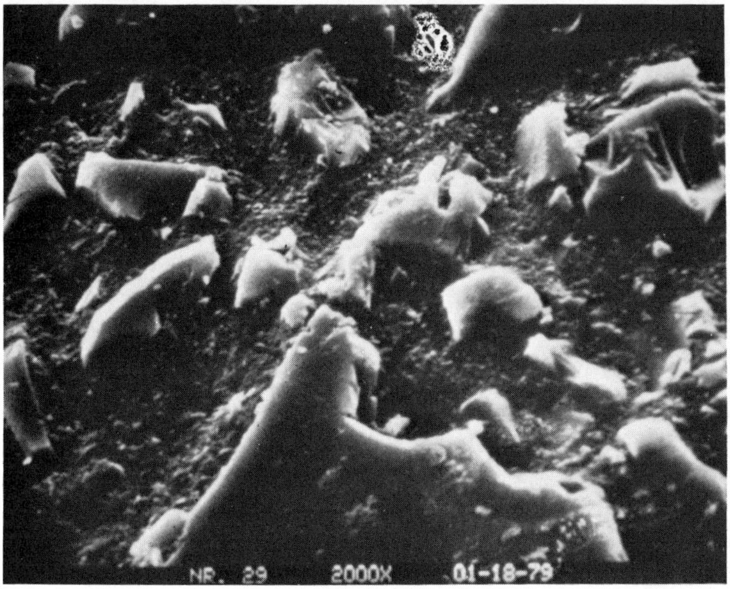
NR. 29 2000X 01-18-79

Figure 7.

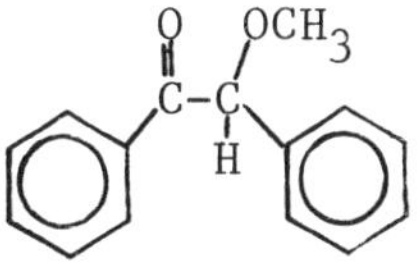

Benzoin Methyl Ether

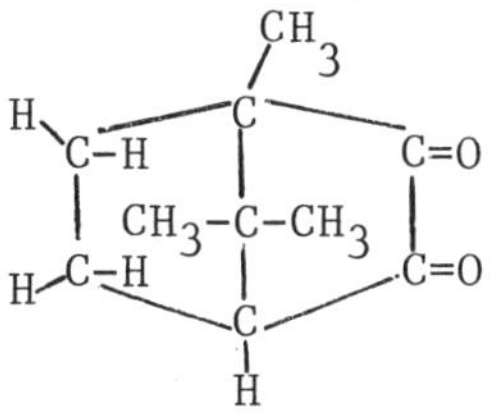

Camphoroquinone

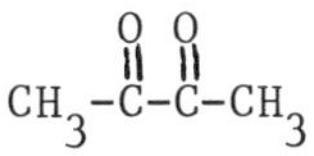

Biacetyl

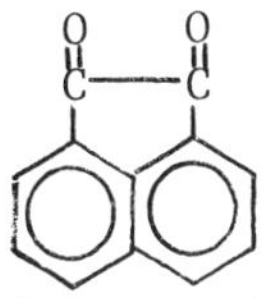

Acenaphthenequinone

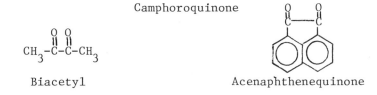

Figure 8. Photoinitiators for polymeric restoratives.

TABLE II

PHYSICAL PROPERTIES OF RESTORATIVE RESINS

	Units	Unfilled	Composite	Micro-Fine Filler	Visible Light Cure	Silicate
Set Time	mins.	4.0	3.25-5.25	3.0	0.3-1	3-5
Shrinkage	%	6-7	13-27	2-5	1-2	1.5
Compressive Strength	MN/m^2	55-72	201-294	138-250	140	158-241
Modulus	GN/m^2	0.26-0.34	2.7-17	3.7-3.9	7.1	-
Tensile Strength	MN/m^2	26-35	10-60	23-36		30-75
Indentation	mm	-	0.055-.057	0.082-	0.052	-
Solubility	mg/cm^2	-	0.65	0.089		
	%	0.17-0.055	0.09-0.5	1-3	0.26	2-7
Thermal Expansion	ppm/°C	80-92	19-55	67	26	7-10
Knoop Hardness	KHN	10-16	15-80	37-52	60-80	60-80

but now little used esthetic anterior filling material - the silicate - the hardenable reaction product of a mix of an alumino silicate glass with aqueous phosphoric acid. Comparison criteria are compressive strength coefficient of thermal expansion and Knoop hardness.

A polymeric filling material which at present is still little used but has major potential when the properties are made more acceptable, is the glass ionomer (10) cement, known by the acronym ASPA - aluminosilicate glass, very similar to the glass of the dental silicate cement and an aqueous polyacrylic acid solution as the setting medium (Fig. 9). Its use currently is practically limited to cervical erosion repair. The more extensive application of this material has been hampered by the lack of esthetic appearance due to opacity and rather slow set.

The acrylic resins also have found application in other, perhaps somewhat more limited-use areas of dentistry. The self-curing methacrylate cements for luting crowns and facings in place were once popular, but have been replaced largely by the easier to use zinc oxide-eugenol cements and the polycarboxylate cements (11). The latter products are composed of a heat condensed zinc oxide powder which, when mixed with a 40% aqueous polyacrylic acid solution, sets by salt formation (Fig. 10). Adhesion to calcium phosphate-containing surfaces such as tooth enamel, by weak chelate bonding, has been demonstrated. Surprisingly enough, these are the

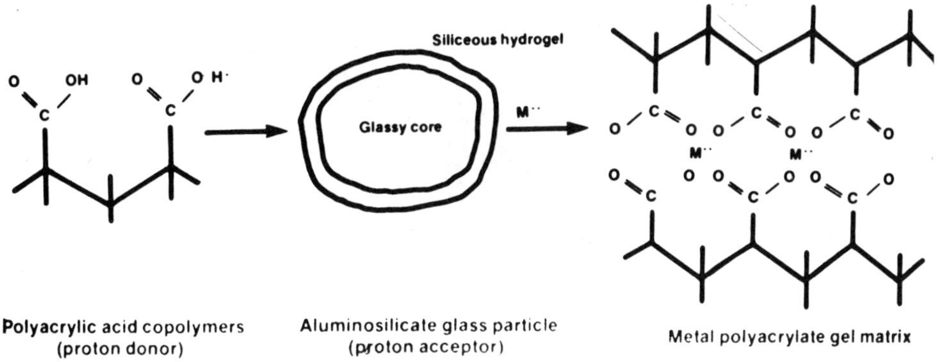
Siliceous hydrogel
Glassy core
Polyacrylic acid copolymers
(proton donor)
Aluminosilicate glass particle
(proton acceptor)
Metal polyacrylate gel matrix

Figure 9. Setting reactions of Chembond.

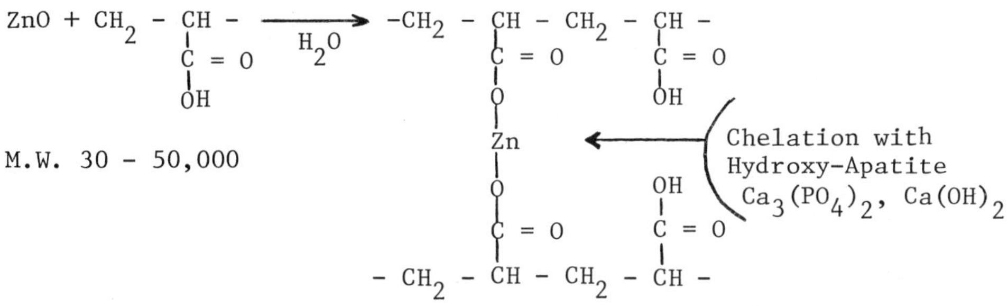
M.W. 30 - 50,000
Chelation with
Hydroxy-Apatite

Figure 10. Zinc polycarboxylate cement - setting and bonding mechanism.

only dental cements with any degree of adhesion to tooth structure.

Pit and fissure sealants (12) used as caries preventive treatment for filling deep fissures in posterior teeth, primarily deciduous molars, are rather popular despite the controversy on the economic efficacy. The sealants now in use are composed of methyl methacrylate, Bis-GMA and/or urethane methacrylate mixtures. They are applied to phosphoric acid-etched occlusal surface in a very fluid state to flow into the surface (Fig. 11), and cured by U.V. or are self curable. They attach to the enamel by mechanical locking into the etched surface rather than by adhesion.

An extension of the use of self or light curing fissure sealant compositions developed for enamel defects, is the application to cements for the attachment of orthodontic brackets to etched enamel (Fig. 12). This use has proven quite satisfactory.

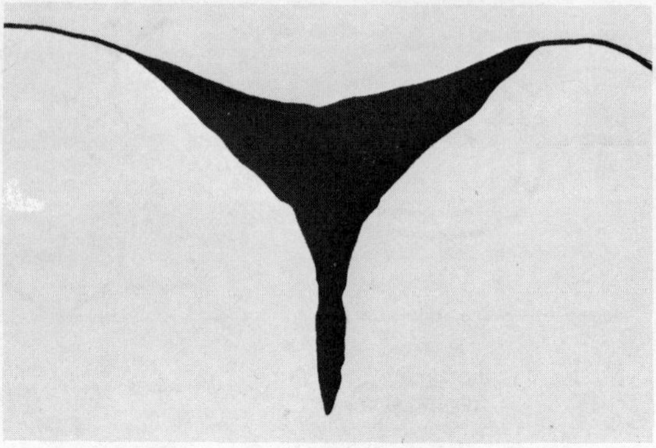

Figure 11. Cured pit and fissure sealant in fissure of molar tooth showing flow of sealout completely into the fissure.

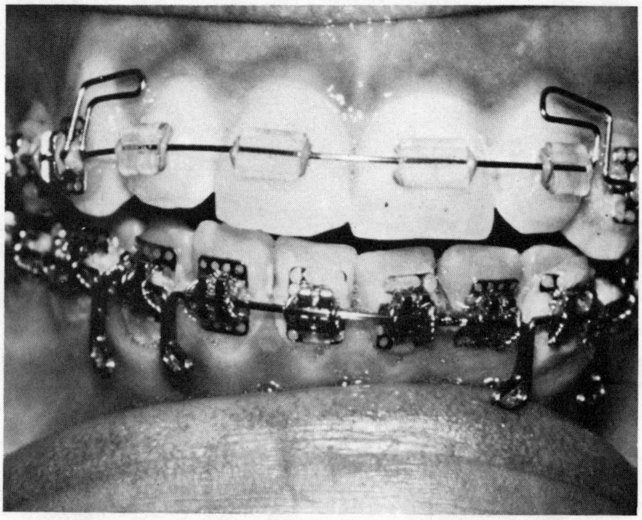

Figure 12. Orthodontic brackets bonded to tooth enamel with UV cured acrylic cement after a phosphoric acid etch.

Preformed acrylic polymer facings (Fig. 13), cemented to fractured, deformed or discolored enamel surfaces by an ultraviolet light curing cement similar to the fissure sealant in composition, are currently in extensive use and seem to fill a cosmetic need.

Self-curing methyl methacrylate polymer-monomer mixtures are employed for denture repair and for relining dentures that have become ill fitting. Many attempts at soft relines have been made by plasticizing methyl methacrylate polymers and attaching to the denture base. None of these has been successful and a real need

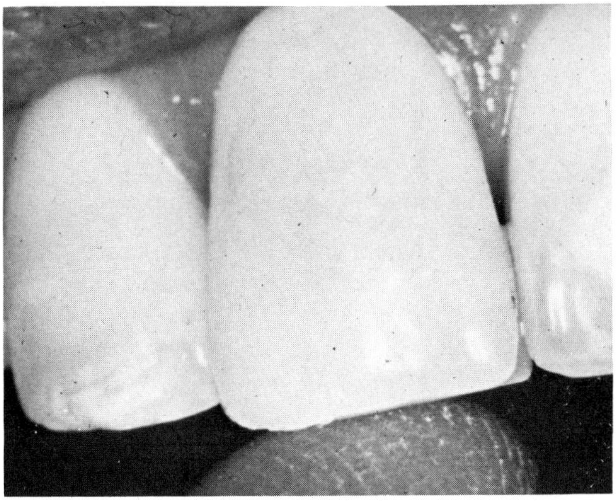

Figure 13. Preformed acrylic facings cemented by self-curing acrylic cement to improve esthetics of deformed, discolored or defective tooth enamel. Central facing is shown in place to cover hyperplasia as evidence in adjacent lateral tooth.

exists for a long term stable material that can be used chairside so that the patient does not give up his denture for lab processing. Soft liners as tissue conditioners for short term use, however, have been quite practical. These are based on ethyl methacrylate polymers, plasticized with an alcohol and/or glycolate ester.

Some effort has been expended trying to adapt the cyanoacrylate tissue adhesives in methyl cyanoacrylate, to dentistry as sealants, coatings and adhesives (14). Thus far there has been no practical application.

OTHER POLYMERS

While it would seem that the acrylic polymers are so widely applied in dentistry to the exclusion of others, there really are many other polymers in constant use, taking advantage of some unique property to fill the dental requirement. Several naturally occurring polymers are typical.

Gutta percha, a poly-trans isoprene, formed into endodontic points, (Fig. 14) is employed almost exclusively for root canal therapy. Cellulose in the form of cotton and paper are excellent absorbants for unwanted moisture and other fluids in the mouth, i.e., points for endodontics and cotton rolls in operative dentistry.

The polygalactans have served as impression materials - the agar-agars (15) as the basis for the heat reversible hydrocolloid

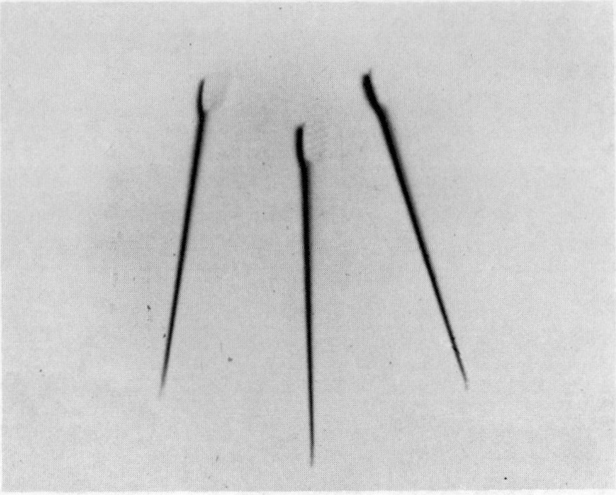

Figure 14. Endontic gutta percha for filling prepared tooth root canal.

for full mouth and single tooth impressions for denture and crown and bridge construction. The algins, also a seaweed product, compounded with calcium sulfate as a water-activated setable gel, (Fig. 15) is finding extensive application in the so-called alginate impressions (16), the workhorse of prosthetic dentistry.

A little known use in dentistry of a common polymer group is in crown forms (Fig. 16) for contouring restoration surfaces in the mouth. Cellulose acetate-nitrate and cellulose triacetate forms are made by dipping an anatomically finished tooth form cast from low fusing alloy, into the resin solution in a volatile solvent. After drying the mold is removed by melting out the alloy form.

BLEND OF:
Potassium alginate
Calcium sulfate
Potassium phosphate

$$\xrightarrow{H_2O} Ca_3(PO_4)_2 + K_2SO_4 \ (1)$$

$$\longrightarrow \text{Calcium alginate} + K_2SO_4 \ (2)$$

(1) Control reaction to allow working time.
(2) Setting reaction to form impression.

Figure 15. Setting mechanism for alginate impression materials.

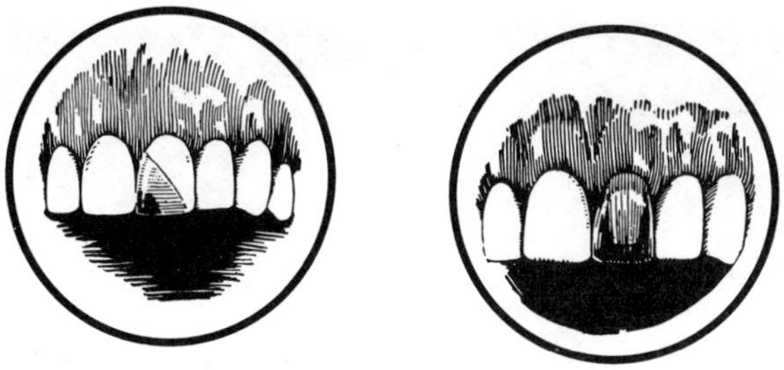

Figure 16. Cellulose acetate crown form used as a matrix for repair of fractured incisal angle fracture.

Possibly the most extensively used "other polymers" are those adapted for crown and bridge impression taking materials (Fig. 17) - the silicones, the polysulfides, and the polyethers. Each group has its advantages and problems and dentists are usually attracted exclusively to one type or other for reasons that are logical to them. All these products are composed (Fig. 18) of two pastes, or a paste and a liquid, where one paste or liquid is the catalyst for setting the mixture fairly rapidly (8-12 minutes). The silicones (17) are typically based on polydimethylsiloxane filled with magnesium silicate, or other mineral powder, and with stannous octoate or dibutyl tin dilaurate. Addition-cured silicones (18) based on vinyl terminated dimethyl siloxane polymers and a precious metal catalyst are the most recent improvements in this area. Even though the silicones have a high thermal change and low tear strength, they are quite convenient to manipulate, therefore popular.

The polysulfides (19) are based on mercaptan end group - polysulfide-ethylene oxide prepolymer, (usually Thiokol LP-2) and are set by crosslinking the -SH groups with lead dioxide, cupric hydroxide, or a hydroperoxide. These materials have better dimensional stability than the silicones and produce reliable impressions but are unpleasant in odor and are more difficult (more messy) to handle. The polyether, the latest addition to the impression group and exclusive to one German producer, (20) is based on an ethylemine-terminated polymeric ether from ethylene oxide (Fig. 19) and sets through crosslinking by cationic polymerization on mixing with an alkyl benzene sulfonate. Of all the elastomeric impression materials, the polyether is the most dimensionally stable. The resultant rubber has a somewhat high durometer value and may be difficult to use in some cases, (especially periodontally involved teeth), but otherwise is very satisfactory.

Over the years as in any field, some polymers have been marketed

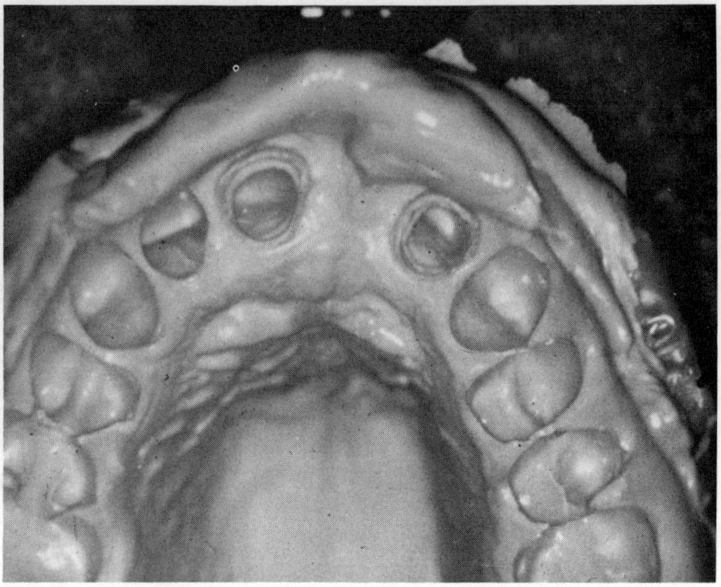

Figure 17. Elastomeric polymer impression for crown and bridge construction technique.

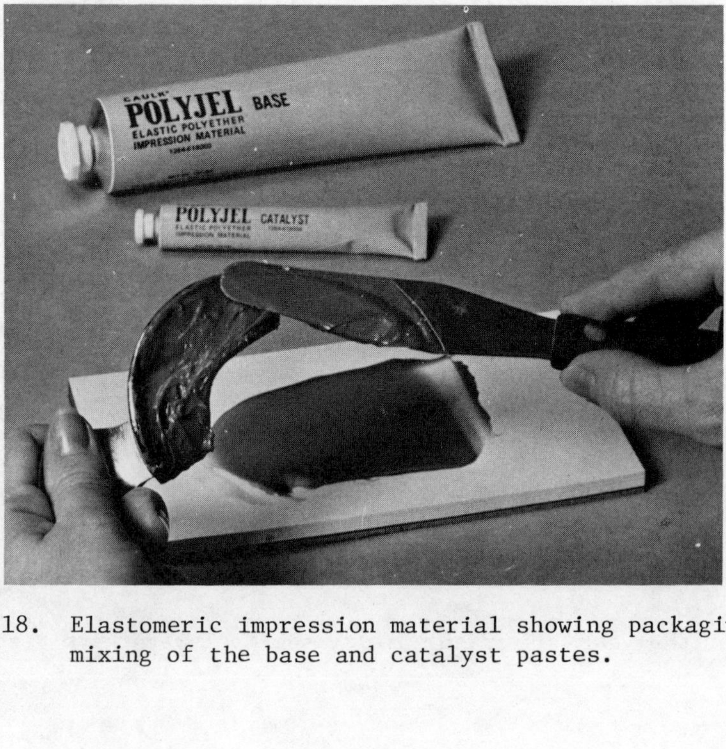
CAULK
POLYJEL BASE
ELASTIC POLYETHER
IMPRESSION MATERIAL
POLYJEL CATALYST

Figure 18. Elastomeric impression material showing packaging and mixing of the base and catalyst pastes.

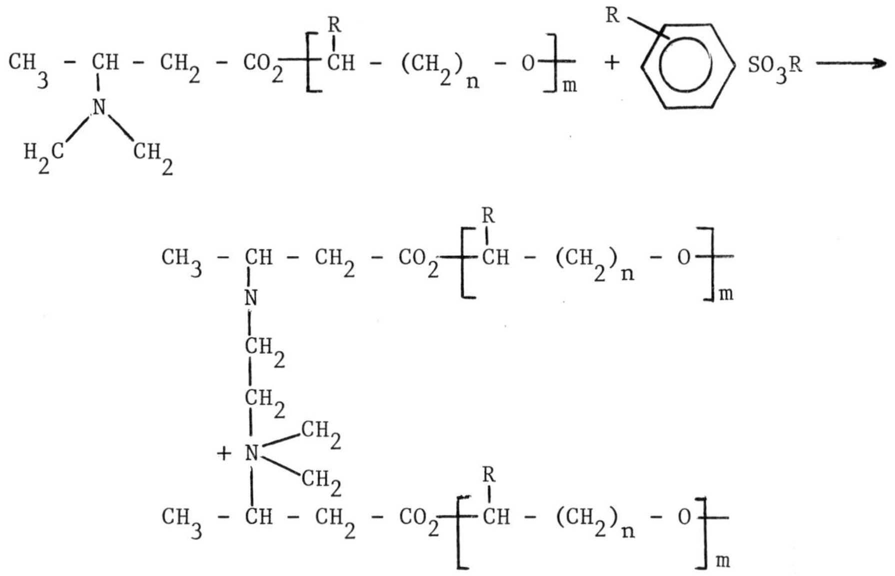

Figure 19. Polyether impression material - setting mechanism.

for dental use but have since been discontinued. Injected polystyrenes, polyamides, and polycarbonate have all been advocated for denture construction. All these have good properties but are extremely difficult to use and do not possess the best esthetics. They therefore have now disappeared from dentistry. The cellulose nitrate denture base proved to be not stable to whiskey and polyvinyl chloride denture bases, hot molded from a preformed blank, split on the median line in clinical function.

Natural rubber, when hard vulcanized with sulfur, was once the material of choice for denture construction, as mentioned earlier. Its difficulties of use and its poor esthetics, and the appearance of the acrylics, has practically removed it from availability.

FUTURE TRENDS

In looking into the future there is always a conflict between what needs to be done and what can or will be done. Restorative dentistry needs an adhesive to tooth tissue, possibly by grafting, for fixing the restoration in the cavity without leakage and extensive undercutting. Prosthetic dentistry needs tougher denture base polymers and abrasion resistant polymeric teeth. However, the restorative material trend will more likely emphasize the development of light cure systems with better photoinitiators and more suitable binders drawn most probably from the more complex polyurethane family. Composite restorative technology will, it seems, because of esthetic appeal, center around primarily microfine filler inclusion and the preparation of specially designed fillers rather than as now adapting existing commercial products. These products then would become serious contenders for posterior restorations replacing silver amalgam. Prosthetics should moderately benefit from the inclusion in denture bases of butadiene-acrylics, rubber toughened fluoropolymers, and again specially tailored urethanes. No great advances are contemplated in redox polymerization systems or dental impression-taking polymers in the near future.

SUMMARY

Polymers have and are making a very important and basic contribution to dental health. Indications are they will not only continue to do so but materially increase in application and replace some of the long time used products of today.

REFERENCES

1. Kulzer G.M.B.H., D.R.P. Application D85578-IV. e/39, July 29, 1943
   French Patent 883679, (March 29, 1943)
   L. Horner and E. Schwenk, Ann., 566, 69 (1950)
   M. Imoto and S. Choe, J. Poly. Sci., 15, 485 (1955)
   F. Knock and J. Glenn, U.S. Patent 2,558,139 (June 26, 1951)
   P. Castan and O. Hagar, Swiss Patent 238,978 (Feb. 1, 1949)
2. R. L. Bowen, U.S. Patent 3,066,122 (1962)
3. J. Foster and R. Walker, British Patent 1,401,805 (July 30, 1975)
   J. Foster, British Patent 1,430,303 (March 31, 1976)
4. S. J. O'Neal and W. B. Eames, Int. Assn. for Dent. Res. Abstract 28 (1973)
   H. C. Lee, J. A. Orlowski, R. W. Glace, P. D. Kidd, and E. Lenabe, Org. Coatings and Plastics Chem., ACS, 34, 403 (1974)
   Hutz, Smiz/RMSU, 85, 253 (1975)
   D. W. Jones, P. A. Jones, and H. J. Wilson, J. Dent., 1, 28, (1972)
   J. M. Powers, L. J. Allen, and R. G. Craig, J. Am. Dent. A., 89, 1118 (1974)
   P. L. Fan, J. M. Powers, and R. G. Craig, J. Dent. Res., 58, Special Issue A, Abstract 665 (1979)
   R. A. Draughn and Alan Harrison, J. Pros. Dent., 40, 220 (1978)
5. R. T. Weitman and W. B. Eames, J. Am. Dent. A., 91, 101 (1975)
6. M. G. Buonocore, J. Am. Dent. A., 80, 324 (1970)
   M. G. Buonocore and J. Davila, J. Am. Dent. A., 86, 1349 (1973)
7. R. T. Lu, Can. Pat. 1,004,200 (1977)
   E. C. Dart and coworkers, Brit. Pat. 1,465,897 (1977)
   E. C. Dart and J. Nemchek, U.S. Pat. 4,071,424 (1978)
   J. B. Cantwell and coworkers, Brit. Pat. 1,498,421 (1978)
   Espe Fabrik Pharmazeutischer Preparate GMBH, Neth. Pat. 7,802,801 (1978)
   T.W.G. Frodsham, U.S. Pat. 4,089,762 (1978)

8. A. K. Abel, K. F. Leinfelder, and D. T. Turner, J. Dent. Res., 58, Special Issue A, Abstract 667 (1979)
9. S. T. Stupp and J. Weertman, J. Dent. Res., 58, Special Issue A, Abstract 949 (1979)
10. A. D. Wilson and B. E. Kent, J. Appl. Chem. Biotech., 21, 313 (1971)
A. D. Wilson and B. E. Kent, Brit. Dent. J., 132, 133 (1972)
11. D. C. Smith, Brit. Dent. J., 125, 381 (1968)
12. A. J. Gwinnett and M. Buonocore, Brit. Dent. J., 119, 77 (1965)
E. I. Cueto and M. Buonocore, J. Am. Dent. A., 75, 121 (1967)
M. Buonocore, J. Am. Dent. A., 80, 324 (1970)
M. Buonocore, J. Am. Dent. A., 82, 1090 (1971)
13. M. Braden, J. Dent. Res., 49, 145 (1970)
14. G. Bourzac, French Pat. 2,116,248 (1972)
15. A. Polly, U.S. Patent 1,672,776 (1928)
16. S. W. Wilding, U.S. Patent 2,249,694 (1941)
17. S. Nitschme and U. Wick, Kunstoffe, 47, 431 (1957)
C. A. Berridge, U. S. Pat. 2,843,555 (1958)
M. Braden and J. C. Elliott, J. Dent. Res., 45, 1016 (1966)
18. B. D. Karstedt, U.S. Pat. 3,715,334 (1973)
3,775,452 (1973)
3,814,730 (1974)
Dow Corning, Brit. Pat. 1,492,616 (1977)
19. E. M. Fetm, J. S. Jorczak, and J. R. Panek, Eng. Chem., 46, 1539 (1954)
S. L. Pearson, Brit. Dent. J., 99, 72 (1955)
M. Braden, J. Dent. Res., 45, 1065 (1966)
E. S. Molnar, U.S. Pat. 3,046,248 (1962)
20. Espe Fabrik Pharmazeutischer Preparate GMBH,
Brit. Pat. 1,044,753 (1966)
French Pat. 1,423,660 (1966)
M. Braden, B. E. Causton, and R. L. Clarke, J. Dent. Res., 51, 889 (1972)

POLYMER DEVELOPMENTS IN ORGANIC DENTAL MATERIALS

B. D. Halpern and W. Karo

Polysciences, Inc.

Warrington, PA 18976

The field of dental materials has made use of the developments in polymer science since the days of Goodyear and his vulcanization process for rubber. Materials such as rubber, celluloid, glyptal, phenol-formaldehyde resins; and, more recently, polystyrene, polyesters, vinyl resins, poly(siloxanes), polysulfides, polycarbonates, polyamides, polyurethanes, polyoxyethylenes, agar, and alginates have been studied and used for various applications.

From the point of view of aesthetics, ease of handling, and availability of raw materials, acrylic and methacrylic acids, their derivatives and their polymers have become most important in dental materials technology.

For convenience, we are reviewing the current status and outlook of polymeric dental materials under the following headings:

1. Impression Materials
2. Acrylic Teeth and Implants
3. Denture Base Materials
4. Restoration and Orthodontic Materials
5. Preventative Dentistry Materials - Pit and Fissure Sealants, Controlled Release Devices

IMPRESSION MATERIALS

To supply properly fitting dental prostheses, exact impressions of the relationship of teeth to oral tissues have to be made. Impression materials range from rigid plaster of paris to semi-rigid gels, such as agar-agar, alginate compositions, and elastomeric types. A certain degree of flexibility is desirable to permit ready removal of castings of undercut or overlapping teeth.

Other properties include the ability to withstand the changes in atmospheric pressure and temperature encountered during air shipment of impressions to centralized dental laboratories, as well as the properties required by ADA Specification No. 19, such as the ability to reproduce a line 25 microns wide.

Current elastomeric impression materials are based on the oxidative polymerization of polysulfides, the condensation of hydroxyl-terminated poly(siloxanes) with alkyl orthosilicates, addition reactions of vinyl poly(siloxane), or the cationic polymerization of ethylenimine-terminated intermediates often referred to as "polyethers".

The recently introduced vinyl poly(siloxane) - based impression material (1) may depend on the addition of poly(siloxane) molecules bearing a few Si-H groups to the double bonds of the vinyl poly-(siloxane). A known addition reaction of this type is catalyzed by salts, such as platinum compounds (2).

In our own laboratory, we have carried out formulation studies on the use of polyurethane chains with pendant or alpha-omega acrylates as fastsetting impression materials using peroxide initiation systems accelerated with aromatic amines. These compositions have a working time between 3 and 5 minutes. They set in less than 10 minutes at room temperature. The temperature rise which is always a problem with polymerization of large masses of acrylate monomers was as little as 5.5°C.

Recent studies (3) have indicated that many different types of impression material undergoes dimensional changes when stored under various conditions of relative humidity. Large dimensional changes are obviously unacceptable, as the end result will be an ill fitting denture. From the standpoint of dimensional changes as a function of relative humidity, the polysulfide - lead dioxide system - being essentially free of hydrophilic groups - was rated somewhat better than any of the other materials.

The polysulfide - lead dioxide system has problems associated with waste disposal of lead-containing materials on the part of the manufacturer and exposure problems of the dentist. Some silicones, while they have good elasticity, suffer from poor tear strength. Dahl (4) considers the polyether impression material which he evaluated as a clinically significant allergen in menopausal females. There clearly is a need for further development of impression materials.

ACRYLIC TEETH AND IMPLANTS

Acrylic teeth have found general acceptance in dental practice. The appearance of natural teeth with all the subtleties embodied in the natural translucent synthetic enamel, such as stains, chips and other common imperfections, and a life-like fluorescence has been simulated with polymer materials. By the appropriate selection of crosslinking agents, the impact resistance and wear characteristics of artificial teeth have been greatly enhanced. High levels of crosslinking agents in the teeth have reduced blanching and crazing of the teeth. The body of the tooth is uncrosslinked or lightly crosslinked so as to permit bonding to acrylic denture bases. The technology of manufacturing acrylic teeth has, by now, been so well developed that it is difficult to visualize any major material changes, particularly since the capital cost for molds is extremely high.

This does not mean that there are no shortcomings to acrylic teeth. In a recent study (5), the wear of acrylic against acrylic teeth was found to be about seven times more than the wear of porcelain teeth. Improvement in the wear resistance of acrylic teeth without loss of the other desirable aesthetic properties would be highly desirable.

Implantation of artificial teeth has received some study. The approach has been to attach a suitable replica of the original tooth to an artificial, porous base. This serves as a root that permits the ingrowth of new living tissue to fix the prosthesis in place. Greenberg and Kamel (6) proposed the use of a porous composite prepared from aluminum oxide particles dispersed in 50 volume % of aqueous poly(acrylic acid). A composite filled with 0.3 microns alumina is said to have a porosity of 38% and compressive strength of 18,000 psi (124 MPa) while a composite filled with 0.05 microns alumina had a porosity of only 15% and a compressive strength of 28,000 psi (192 MPa).

Hodosh and co-workers (7) developed an acrylic system of a suitably pigmented poly(methyl methacrylate), crushed inorganic bone, dinitrosopentamethylenetetramine, a cross-linking monomer mixture, and tributyl phosphate. The dough produced from this composition was polymerized at a high enough temperature for the nitroso blowing agent to render the polymer porous. After cooling, the root area of the tooth was sandblasted to remove the continuous surface and expose an open cellular structure. This root was then implanted. Out of 10 implants (in baboons), four functioned after seven years. The losses had taken place because of changes in pocket depth, gingival recession, or loss of supporting bone. The use of a N-nitrosoamine blowing agent in this work is open to criticism because of the known carcinogenic nature of such compounds. Presumably other blowing agents of less toxic nature might be considered. Grenoble and

Voss (8) have provided a recent review of materials and other aspects of dental implants.

DENTURE BASE MATERIALS

Denture base materials consist primarily of suspension-polymerized methyl methacrylate or copolymers. These are dispersed in a monomer usually containing a cross-linking agent. Suitable dyes, pigments, U.V.-stabilizers, and active free radical initiators are incorporated. The beads may contain residual benzoyl peroxide. To enhance the impact resistance of denture bases, additives similar to those used to improve the impact resistance of PVC may also be added to the polymer phase of the system.

To prepare an acrylic denture base, the general technique is to use dispersions of the polymer powder in acrylic monomer to form a dough or a slurry (the latter in the case of "pour-type" denture base materials) which are conveyed to a mold holding the teeth in proper array for production of the denture. Polymerization may be initiated by thermal decomposition of peroxides or by room temperature cures using amine accelerators in redox type catalyst systems.

From a recent study of the effect of curing cycles (9), it was concluded that the properties of the material varied primarily with the state of polymerization of the matrix monomer, i.e., by the level of residual monomer which may be as high as 2%, rather than the extent to which the polymer beads had dissolved in the monomer. With conventional materials, it has been suggested that the dough in the mold be heated for as much as 7 hours at 70°C followed by 1 hour at 100°C for proper curing.

We understand that new systems are undergoing evaluation which can be cured in a single step at 100°C.

A new denture base resin system which has an unusually high hydrophilic character has been marketed. This material consists, evidently, of a conventional poly(methyl methacrylate) and a monomer composition containing both methyl methacrylate and 2-hydroxyethyl methacrylate.

Both in a dry and in a water-saturated condition, this hydrophilic material has significantly lower mechanical properties, such as Young's modulus, flexural strength, tensile strength, and impact strength, than conventional materials (10). The equilibrium water-uptake of the hydrophilic denture base matrix was estimated to be 6.7% as against about 2% for the conventional PMMA beads. This was thought to give rise to a potential problem of internal strain formation which could lead to inferior mechanical properties.

Radiopacity of denture bases is still an objective that has

not been satisfactorily reached. It is a desirable characteristic because wearers of dentures sometimes inadvertently ingest the denture in part or whole. Incorporation of barium or bismuth compounds usually leads to aesthetically unattractive dentures. Special radiopaque glass has been suggested by the NBS research group (11). Incorporation of highly halogenated compounds, finely dispersed gold, or other atoms with high X-ray absorption have been suggested, but not with much success.

RESTORATION AND ORTHODONTIC MATERIALS

The restoration of carious teeth or teeth with other physical damage which impairs their proper function and aesthetics has led to the development of adhesives, cements, and composite filling materials, which, with minor modification may be used almost interchangibly.

Paffenbarger (12) indicates that dental cements are used as filling materials, luting media, obtundent dressings, impression pastes, and as bases for other restorative materials. He lists six classes of cements. Today, a seventh class, the "glass-ionomer" cements, may have to be added to his classification, since these products straddle his classes C-4 and C-5 compositionally.

The glass-ionomer cements were developed by Wilson and co-workers (13) and introduced in the United States in 1977 (14). Fundamentally, they consist of an aqueous solution of poly(acrylic acid) (or of a copolymer of acrylic acid with other acids) and a leachable powdered glass based on calcium aluminosilicates prepared with a fluoride flux.

In this system, an acid-base reaction is said to form a polysalt gel which binds glass particles into a matrix. Extraction of calcium ions followed by their displacement with aluminum ions is thought to be involved. This cement, like the older zinc oxide-polycarboxylate cements, has some adhesion to enamel and dentin. Marginal leakage is less than that of certain other cements (15). Glass-ionomer cements show no clinical symptoms of discomfort and no histopathological irritation of the pulp (16). Because fluoride ions may be leached from the cement, some anticariogenicity is thought to be observable. Glass-ionomer cements, in the early stages of their cure are quite sensitive to moisture. Absorption of water during this period leads to chalky fillings which erode readily. Use of copolymers of acrylic acid with other monomers need further exploration to improve these cements. Reinforcement with such high modulus fiber as potassium titanate is expected to have an enhancing effect on glass-based cements such as it has had in the case of polycarboxylate cements (17).

In orthodontics, zinc oxide-polycarboxylate cements have been proposed for the retention of orthodontic appliances (18). More recently, it has been observed that poly(acrylic acid) compositions which contain high levels of sulfate ions rapidly developed an extensive growth of gypsum crystals firmly attached to the enamel surface. The use of such a composition is said to enhance the bond strength of orthodontic brackets to enamel surfaces (19). It is anticipated that further development of poly(acrylic acid) copolymers with sulfate end groups will lead to new generations of orthodontic cements. Adhesive attachment of brackets to the teeth are more acceptable aesthetically and less subject to caries attack, if improper dental hygiene is used than conventional methods of attachment.

The composite restoration materials based on the work of R. L. Bowen and co-workers (20) with Bis-GMA and related monomers have had a profound impact on dental practice. In the form of a two paste system - one containing reinforcing fillers, monomer, and peroxide, the other containing monomer with an accelerating amine - composite filling materials have been widely accepted even though there is some clinical evidence that fillings using a powder-liquid system exhibit less marginal leakage (21). A paper on the analysis of several commercial composite materials by Ruyter and Sjovik indicates at least one compositional factor which may be among the causes of the low adhesion of restoration to the tooth cavity (22).

Several commercial composites have been shown to lose weight on exposure to water even though there is a rapid absorption of water (23). The soluble components are believed to be unreacted monomers or low molecular weight polymers. The water sorption may be attributed to the hygroscopicity of Bis-GMA itself, or to the nature of other monomers which are used to reduce the viscosity of Bis-GMA. Recent work by Cowperthwaite and co-workers, (24), has indicated that the various polyethylene glycol - based monomers exhibit higher water sorption than alpha, omega - diol-based methacrylates of roughly comparable molecular weight.

The fillers used in the polymer composites are of importance. Within the last year or two, so called "microfillers" have been introduced into composites. These appear to be fine powders such as "fumed silica" which are embedded in a polymer, said to be based on either the Bis-GMA - or urethane methacrylate-type materials. The microfillers are said to produce a composite restoration which is more readily polished by the dentist than quartz or glass containing materials (25).

The "Bowen-monomer" Bis-GMA may be pictured as the reaction product of bisphenol-A and glycidyl methacrylate. The resultant molecule contains two hydroxyl groups. Recently, analogous monomers have been proposed which do not contain these hydroxyls. Dental restorations made with these monomers are said to be low in water uptake and water solubility. Candidate monomers would be those in which bisphenol-A is bridged to two methacrylate units by C_3H_6 or higher units (26).

In recent years repeated efforts have been made to use UV initiation to cure dental materials *in situ*. However, the energy of UV sources sufficient to accomplish this led to oral tissue damage. Consequently, this approach sees fewer advocates today.

PREVENTATIVE DENTISTRY MATERIALS - PIT AND FISSURE SEALANTS, CONTROLLED RELEASE DEVICES

To reduce dental caries, efforts have been made to develop polymeric coatings for sealing pits, fissures, and other imperfections against further damage. Solutions of Bis-GMA in viscosity-reducing monomers are in use. Initiation by conventional amine-peroxide cures seems preferred to ultraviolet initiated systems. Brauer (27) has attempted to improve commercial pit and fissure sealants by incorporating a photo-sensitive cross-linking agent such as 1-aza-5-acryloxymethyl-3,7-bioxabicyclo[3,3,0]octane ("AADO"). Incorporation of AADO reduced the curing time, reduced the residual monomer at the surface of a sealant, lowered the water solubility of the cured sealant. It may also improve the enamel to sealant bond strength.

The use of tributylboron as an initiator which bonds acrylic monomers to dental tissue has been advocated by Masuhara for some time (28). His "Enamite" - a solution of 2-hydroxy-3-beta-naphthoxypropyl methacrylate, methyl methacrylate, poly(methyl methacrylate), and tri-n-butylboron - is said to show unusually high adhesion to enamel, and, reduced occlusal caries (29). This system has also been used in orthodontics for binding brackets to the frontal face of teeth.

In our own laboratory, we have developed a polymeric system which releases predictable small amounts of fluoride ions (up to 1 milligram per day) at a constant and controllable rate for a period of at least three months. To accomplish this, a Hawley orthodontic appliance was fabricated from poly(methyl methacrylate) beads and sodium fluoride dispersed in a monomer consisting of 2-hydroxyethyl methacrylate and methyl methacrylate. This composite was coated with a rate controlling membrane of poly(methyl methacrylate) with variable amounts of a hydrophilic monomer. The resulting system released fluoride ions over a 100 day period in a linear manner at rates up to approximately 2.0 mg/cm per day (30). The rate was a function of membrane thickness and the degree of hydrophilic monomer contained therein.

In summary, we have tried to outline recent developments in dental materials to illustrate the applications of research ideas in organic and polymer chemistry to one aspect of the health sciences.

There are many areas in which in-depth research will lead to greatly improved materials. For example, in the case of impression materials, better tear strenght, low toxicity, and simplicity of manufacture are sought. The physics and chemistry of hydrocolloids such as agar-agar needs further exploration to control the variations that seemed to be related to the species of agar used to manufacture impression materials. New gel forming compositions are being developed for impression materials. Further studies of the effect of the composition of monomer-polymer systems on impact and wear properties should lead to improvements in acrylic teeth.

The glass-ionomer cement system is considered to be so new that significant improvements may be anticipated. In particular, reduced water sensitivity and improved aesthetics may be expected. Also among the newer concepts which will be exploited further are "microfillers". These may be considered "internally pigmented" reactive polymer systems in which very small particle-size pigments are used. Studies of the relationship of the particle size of the pigment to the organic layer will lead to improvements of the thermal expansion and contraction, refractive index, shrinkage, polish ability, and adhesion of compositions employing "microfillers".

REFERENCES

1. ESPE Dental Products, Lynbrook, N.Y., "Permagum" product bulletins
2. M. E. Nelson, U.S. Patent 3,020,260 (1962)
3. J. W. Bell, E. H. Davies, and J. A. von Fraunhofer, J. Dent., 4, 73 (1976)
4. B. L. Dahl, J. Oral Rehab., 5, 117 (1978)
5. A. Harrison, J. Oral Rehab., 5, 111 (1978)
6. A. R. Greenberg and I. Kamel, J. Biomed. Mater. Res., 10, 77 (1976)
7. L. Gettleman, D. Nathanson, R. L. Myerson, and M. Hodosh, J. Biomed. Mater. Res. Symposium No. 6,243 (1975); M. Hodosh and co-workers, J. Biomed. Mater. Res., 8, 213 (1974); M. Hodosh and co-workers, J. Prosth. Dent., 36, 676 (1976)
8. D. E. Grenoble and R. Voss, Biomat. Med. Dev. Artif. Organs, 4, 133 (1976)
9. R. G. Jagger, J. Oral Rehab., 5, 151 (1978)
10. M. J. Barby and M. Braden, J. Dent. Res., 58, 1581 (1979)
11. H. H. Chandler, R. L. Bowen, and G. C. Paffenbarger, J. Biomed. Mater. Res., 5, 253,335,359 (1971)
12. G. C. Paffenbarger, J. Biomed. Mater. Res. Symp., No. 2 (Part 2), 363 (1972)
13. A. D. Wilson and B. E. Kent, J. Appl. Chem. Biotechnol., 21, 313 (1971); A. D. Wilson, Br. Poly. J., 6, 165 (1974); A. D. Wilson and co-workers, J. Dent. Res., 55, 489, 1023, 1032 (1976); S. Crisp, M. A. Jennings, and A. D. Wilson, J. Oral Rehab., 5, 139 (1978); A. D. Wilson and co-workers, J. Dent. Res., 58, 1065, 1072 (1979) is a selection of references by the originators

of this cement.

14. Council on Dental Materials and Devices, J. Am. Dent. Assoc., 99, 221 (August 1979)
15. E. A. M. Kidd and J. W. McLean, Brit. Dent. J., 147, 39 (1979)
16. M. Yakushiji, T. Kinumatsu, T. Fuchino, and Y. Machide, Bull. Tokyo Dent. Coll., 20, (2), 47 (1979)
17. J. A. Barton, Jr., G. M. Brauer, J. M. Antonucci, and M. J. Raney, J. Dent. Res., 54, 310 (1975)
18. E. Mizraki and D. C. Smith, Brit. Dent. J., 130, 392 (1971); D. C. Smith, J. Canad. Dent. Assoc., 1, 22 (1971)
19. R. Maijer and D. C. Smith, International Association for Dental Research Abstracts, No. 436 (1977)
20. R. L. Bowen, J. Dent. Res., 58, 1493 (1979) a recent review
21. K. Strater, Private communication, October 1979
22. I. E. Ruyter and I. J. Sjovik, Paper given at the International Association for Dental Research Meeting of March, 1978 at Washington, D.C.
23. G. J. Pearson, J. Dent., 7, (1), 64 (1979)
24. G. F. Cowperthwaite, Organic Coatings and Plastics Chemistry, 42, 208 (1980)
25. German Offen. 2,403,211 and 2,405,518 (1975)
26. ESPE Dental Products, Lynbrook, N.Y., "Nimeric and Uvio-Fil" product bulletins
27. G. M. Brauer, J. Dent. Res., 57, 597 (1978)
28. E. Masuhara, Dent. Zahnarzt, Z., 24, (7), 620 (1969)
29. I. Ohmori, K. Kikuchi, E. Masuhara, N. Nakabayashi and S. Tanaka, Bull. Tokyo Med. Dent. Univ., 23, 149 (1976)
30. B. D. Halpern, O. Solomon, L. Kopec, E. Korostoff, and J. L. Ackerman, ACS Symposium Series, No. 33 "Controlled Release Polymeric Formulations", D. R. Paul and F. W. Harris, Editors, American Chemical Society (1976)

LIMITING HARDNESS OF POLYMER/CERAMIC COMPOSITES

A. K. Abell, M. A. Crenshaw, and D. T. Turner

Dental Research Center
University of North Carolina
Chapel Hill, NC 27514

INTRODUCTION

The main deficiency of proprietary composite materials for esthetic restoration of teeth is that they are much less resistant to wear than enamel. Many factors are involved in such a complex service property as some simpler related property. Limiting attention to static mechanical properties, it is to be noted that the proprietary composite materials are not inferior to enamel in respect of compression strength and tensile strength. On the other hand, it is well known that they are much softer, generally having a Knoop hardness number less than 60 Kg/mm^2 whereas values near 380 have been reported for enamel (1,2). An obvious reason for this difference is that enamel contains a much greater ceramic content (95 wt-%) than most proprietary materials (60-80 wt-%). This might be rationalized in relation to wear by supposing that closely packed ceramic particles behave cooperatively and cannot be dislodged as easily from the surface as isolated particles. The objective of the present work is to find how hardness depends on the loading of ceramic particles in a polymeric matrix.

EXPERIMENTAL

Ceramic powders were obtained from a wide variety of sources and included alumina, hydroxyapatite, and lithium aluminum silicate. This latter sample was also used after surface modification with a silane coupling agent. Ceramic powders were also obtained by chloroform extraction of two proprietary tooth restorative composite materials viz. Estilux (Kulzer and Fotofil (Johnson and Johnson). Particle sizes and shapes were noted.

Composites with PMMA (polymethyl methacrylate) as the polymeric matrix were made as follows: Ceramic powders were allowed to sediment in methyl methacrylate monomer in test tubes. The tubes were then exposed to a Cs-137 sourve at a dose rate of 0.8 Mrad/hr; ambient temperature, 35°C. It was noted that, under these conditions, polymerization of the monomer alone was complete after a dose of 3 Mrad. The samples with ceramic were given a dose of 5 Mrad and then heated at 100°C overnight; they were judged to be completely polymerized. After cooling to room temperature, the samples were removed and sectioned with a diamond wheel. The sections were polished with alumina, down to 0.3 μ, and tested with a Kentron Hardness Tester with a Knoop indenter. The load was selected to give an indentation length in the range 0.15-0.30 mm. Ceramic contents were determined by ashing to constant weight.

For purposes of comparison with a densely packed composite material, studies were made of abalone nacre. Specimens of size 0.8 cm x 0.8 cm were cut with a diamond wheel and the outer portion of the shell ground until parallel to the inner surface. The inner surface was then polished to 0.3 μ but leaving a layer of nacre thicker than 0.1 cm. In some experiments the organic matrix, mostly protein, was removed by soaking for two hours in 5% sodium hypochlorite; this was followed by rinsing with distilled water and drying in vacuum. Then, by soaking the specimen in a mixture of methyl methacrylate, ethylene glycol dimethacrylate (5%), and azoisobutyronitrile (0.2%) and by polymerizing at 70°C, the protein could be replaced by crosslinked PMMA. The composite surface was exposed by polishing. For comparison with previous work on other bivalve shell structures (3), hardness measurements were made using a Vickers 136° diamond pyramid indenter under a load of 500 g.

Microstructures were examined by reflected light with a Zeiss Universal Microscope and an ETEC U-1 Scanning Electron Microscope.

RESULTS AND DISCUSSION

In multiphase materials, microhardness measurements are designed either to give discrete hardness numbers for each of the phases or to give a single average value. In the latter case the size of the indentation must be large relative to the microstructural features and this may be effected by increasing the load. Results obtained for two proprietary materials are shown in Fig. 1. For one material there is only a small increase in hardness in the range of loads from 100 to 500 g. For the other, the hardness seems to have leveled off in this same range. However, in the latter case, the limiting mean value of the hardness ranges from 42 to 48. This range is greater than would be acceptable for a single phase material and is, in part, due to a subjective factor in definition of the extremities of the Knoop indentation marks. Although microhardness values have recognized limitations of this kind, they are used because they also serve

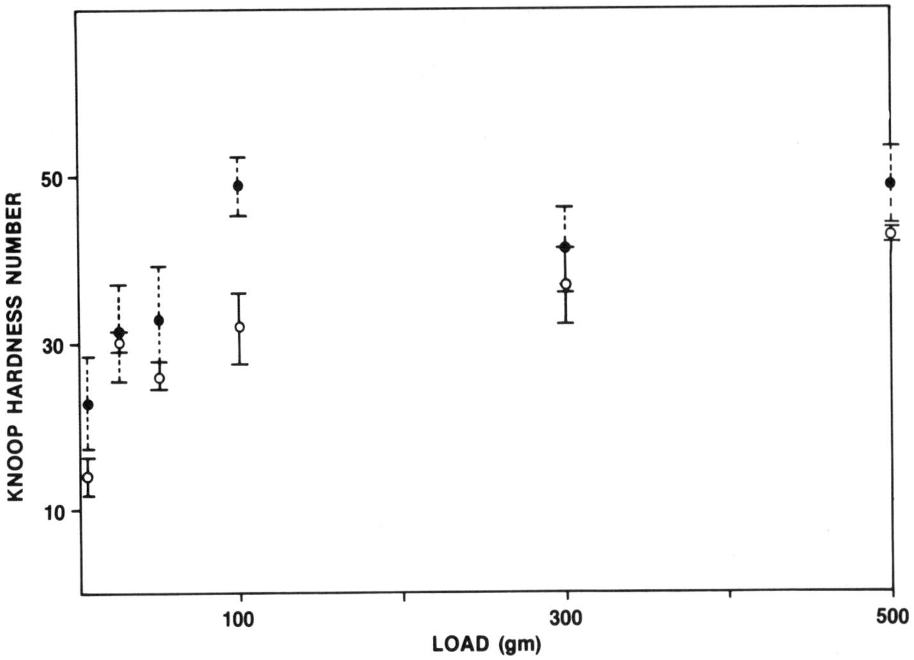
50
30
10
KNOOP HARDNESS NUMBER
100
300
500
LOAD (gm)

Figure 1. Influence of load on Knoop hardness number of composites. Mean values and standard deviations for 12 measurements for Estilux (full line) and Nuva-Fil (broken lines).

to make measurements on composites with high loadings, such as enamel, and even on ceramics themselves. Other tests with more extensive and otherwise defined indentations, such as in the Rockwell test, are invalidated for such extreme cases because of the occurrence of brittle fracture.

Additions of ceramic up to 50 wt-% do not increase the hardness of PMMA greatly above its initial value of 18-20 (Fig. 2). However, above about 60 wt-% the hardness increases markedly reaching a value of 100 for a loading of about 85 wt-%. High loadings of ceramic, and concomitantly high hardness values, were obtained using ceramics recovered from proprietary materials. Presumably they have a suitably graded distribution of particle sizes which favors close packing and also surfaces which have been modified by treatment with a coupling agent. The need to take account of the former factor is well known. However, the role of surface modification in favoring close packing was unexpected. It is best illustrated by reference ot the higher results obtained with a sample of lithium aluminum silicate which had been treated with a silane coupling agent relative to the original, unmodified, sample (Fig. 2). The reason for this effect of surface treatment has not been elucidated.

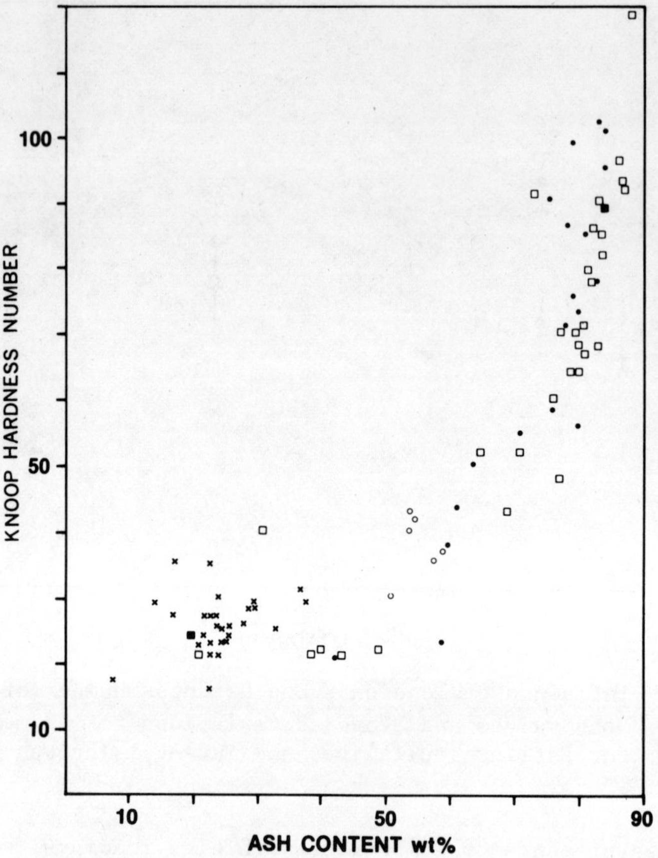
KNOOP HARDNESS NUMBER
100
50
10
ASH CONTENT wt%
10
50
90

Figure 2. Dependence of Knoop hardness number on ash content (wt-% ceramic) in PMMA composites.
Key: □, proprietary fillers: ●, silanated lithium aluminum silicate; O, unsilanated lithium aluminum silicate; X, various unspecified ceramics.

Attempts were made to increase the ceramic content above 85 wt-%. One technique involved packing the ceramic under pressure to make an "aspirin" tablet. This tablet was then degassed and filled under vacuum with methyl methacrylate which was polymerized subsequently by exposure to γ-rays. Disappointingly, the packing achieved in this way was less than by the sedimentation method. Presumably better packing could be achieved by finding conditions in which pressing causes fragmentation of particles or by selection of ceramics with slip mechanisms (4,5). Other approaches under trial include centrifuging and also use of particles with potentially better packing geometry, such as mica.

Concurrently with the above experiments investigations were made of abalone nacre as an example of a close packed composite material with 95 wt-% aragonite in a matrix, mainly of protein. This was chosen rather than tooth enamel, because relatively large test specimens could be obtained and also because of the simple microstructure. The microstructural units are hexagonal plates of aragonite arranged in closely packed stacks which are perpendicular to the surface. The appearance of these stacks, after removal of the matrix, is shown in Fig. 3. This may be compared with the irregular appearance of the microstructure of a proprietary tooth restorative composite material (Fig. 4). It was found that hardness values leveled off for loads in the range 400-1000 g, and subsequently all measurements were made under a load of 500 g for comparison with previous work by Taylor and Layman. The indentation marks were mostly squares with diagonals adequately defined for measurement. Occasionally, however, and erratically, atypical indentations with missing quadrants or in the form of "picture frames" were observed; such abberrants were ignored. Average hardness values varied among shells, and to a lesser extent within a single shell, from 125 to 195. According to Taylor and Layman, the hardness of aragonite is

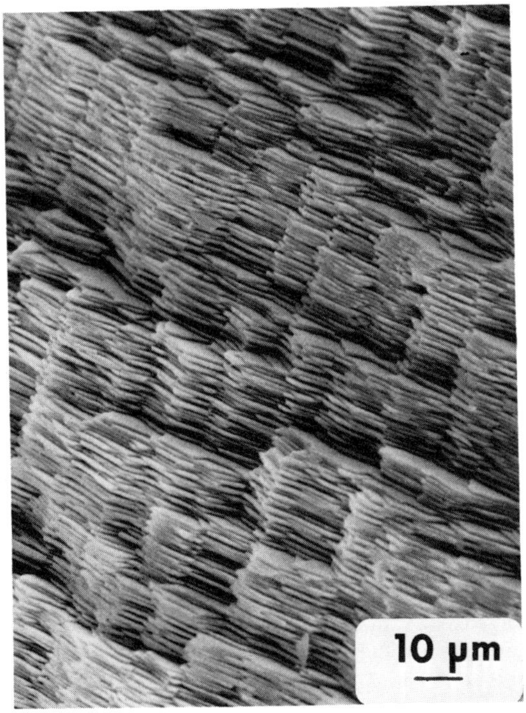
10 μm

Figure 3. Fracture surface of abalone nacre (matrix removed).

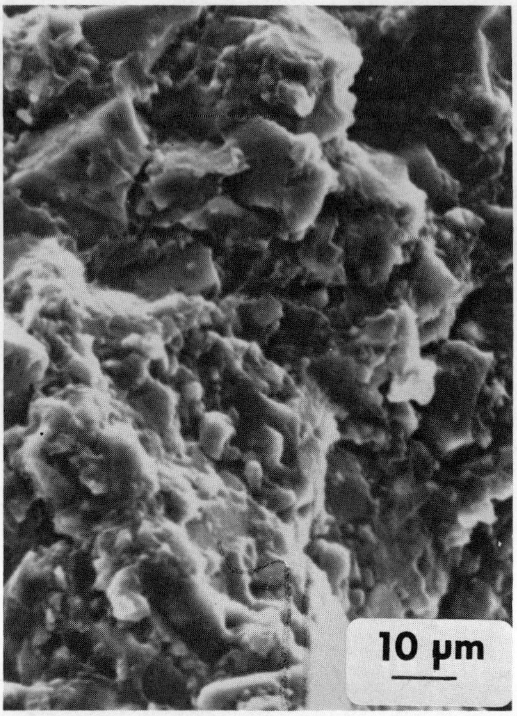
10 μm

Figure 4. Fracture surface of a proprietary material (Nuva-Fil, Caulk).

220. By reference to this value the hardness of the composite material approached but was never greater than that of its ceramic microphase. This observation differs from the report by Taylor and Layman for other different but related bivalve systems viz. "All the shell structures are harder than inorganic calcite or aragonite. This property may indicate the rubber-like nature of the matrix; the indenter stress is probably distributed to other parts of the shell and stored as elastic energy which is released when the indenter is raised."

The possibility raised by the work of Taylor and Layman that a polymeric matrix can raise the hardness of a composite material above that of its ceramic microphase has important potential in the design of such materials. Therefore, despite the unfavorable findings described above for abalone nacre, experiments were made to explore this possibility further. In the experiment in which the matrix of a specimen of nacre was replaced with crosslinked PMMA the hardness was little changed, from 149 $\pm$ 9 to 144 $\pm$ 12. However, this experiment was inconclusive because the replacement of the matrix was confined to a surface layer and results may have been influenced by the unmodified substrate. Further experiments were made to find whether water might affect results through some plasticizer action. To this end a sample was immersed in water for several weeks but, again, the hardness was little changed; from 154 ± 4 to 150 ± 9. Conversely there was not a great change when a specimen was outgassed in a hard vacuum for up to 20 days (Fig. 5A). These results, showing little difference between wet and dry specimens, are in agreement with a statement to the same effect by Taylor and Layman.

Finally, in an extreme attempt to modify the properties of the matrix, specimens were exposed to γ-rays in air. Irradiation of a polymer either results in a continuous decrease in various mechanical properties, for example in tensile strength, or in an initial increase followed by a decrease. The results shown in Fig. 5B are uncertain in respect of initial changes but indicate definitely that the hardness does decrease after higher doses. After the two highest doses some additional observations were made concerning some of the indentation marks. First, colored rings were seen at some distances around the marks suggesting the occurrence of long range effects

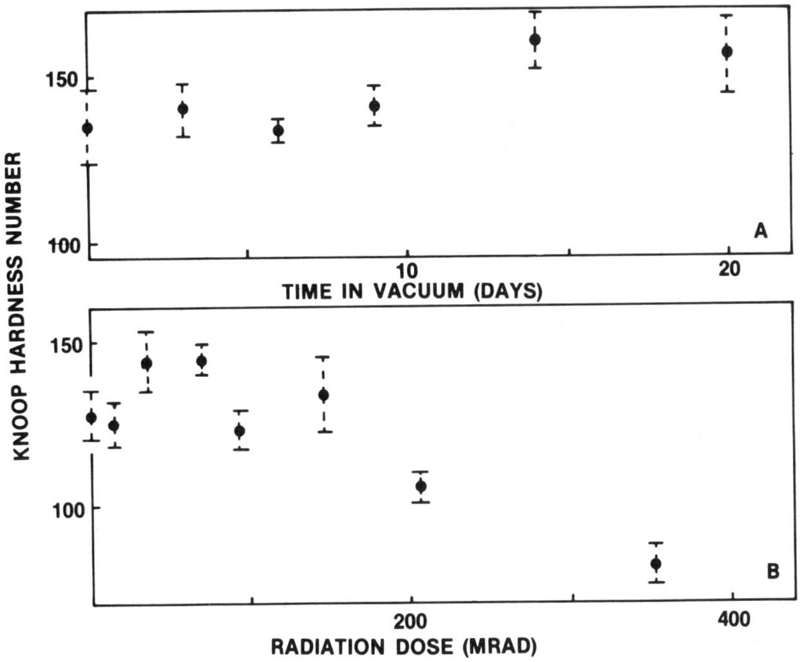
KNOOP HARDNESS NUMBER
150
100
A
10
20
TIME IN VACUUM (DAYS)
150
100
B
200
400
RADIATION DOSE (MRAD)

Figure 5. (A) Influence of storage in vacuum on Knoop hardness number of abalone nacre. (B) Influence of γ-irradiation on Knoop hardness number of abalone nacre.

caused by loading. Second, microcracks were observed mostly originating from the corners of marks suggesting embrittlement of the matrix.

CONCLUDING REMARKS

An empirical relationship has been obtained between hardness and the weight fraction of hard ceramics in a matrix of PMMA. In further, more refined studies it would be preferable to seek a more general relationship which takes account of variables such as the hardness and density of the ceramic. Further studies are also needed to elucidate the role of the polymeric matrix. The present indications are that this is not a sensitive variable, at least in the case of glassy polymers which have reasonable mechanical integrity. For example, similar results were obtained for proprietary materials even though they have matrices which differ considerably from PMMA e.g. copolymers of BIS-GMA. Most of these materials contain less than 60 wt-% ceramic and therefore such similarity is an insensitive test (c.f. Fig. 2). However, recently a proprietary material has been reported to contain 80 wt-% of a strontium glass and to have a Knoop hardness of 82; these data fit the curve for PMMA. Incidentally, it was also reported that this harder material has improved wear resistance (6).

The above remarks apply to rigid, glassy, matrices. In contrast, the matrices in bivalve shell structures are soft and, according to Taylor and Layman, elastic. For such materials Taylor and Layman suggest that the matrix plays an important role in hardness measurements. The present work confirms this suggestion in a negative sense by showing how hardness is decreased by radiation degradation of the matrix. However, it has failed to provide further examples of the remarkable claim that the matrix can endow a composite with a hardness greater than that of its ceramic microphase.

ACKNOWLEDGMENTS

This investigation was supported by NIH research grant number DE-02668 from the National Institute of Dental Research and by NIH grant number RR-05333 from the Division of Research Facilities and Resources. We thank Kerr Sybron Corporation for donating samples of lithium aluminum silicate.

REFERENCES

1. R.G. Craig and F. A. Peyton, J. Dent. Res., 37, 661 (1958)
2. C. L. Davidson, I. S. Hoekstra, and J. Arends, Caries Res., 8, 135 (1974)
3. J. D. Taylor and M. Layman, Palaeontology, 15, 73 (1972)
4. J. S. Hardman and B. A. Lilley, Nature, 228, 353 (1970)

5. J. A. Hersey and J. E. Rees. Nature, Phys. Sci. Ed., 230, 22 March (1971)
6. Product Information Sheet No. 12-79 on "Profile", S. S. White Company, Philadelphia (1979)

NEW MONOMERS FOR USE IN DENTISTRY

Joseph M. Antonucci

National Bureau of Standards

Washington, DC 20234

INTRODUCTION

Dental composites have essentially replaced silicate cements as the preferred esthetic restorative material. Although dental composites exist in a variety of formulations, they all contain the following essential components: 1) a resin system usually comprising one or more monomers, 2) a reinforcing phase consisting of glass, quartz or other filler, 3) a coupling agent such as a vinyl silane for chemically bonding the reinforcing mineral phase with the polymeric matrix, 4) an initiator system for free radical polymerization under ambient conditions, and 5) stabilizers for maximizing the storage stability of the uncured dental materials and the chemical stability of the cured composites. Optionally, such ancillary components as anticariogenic fluoride release agents and coupling agents for promoting adhesion to tooth structure may also be included. Dental sealants have similar compositions except that they are unfilled or only lightly filled materials.

This paper considers only the resin component of such dental materials and will describe several new types of monomers which may have utility in designing alternate monomer systems for such materials.

BACKGROUND

The pioneering work of R. L. Bowen, especially his classic syntheses of the prototypical dental monomer 2,2-bis[p-(2'-hydroxy-3'-methacryloxypropoxy)phenyl] propane, designated BIS-GMA, ushered in the modern era of dental composite and sealant materials (1-6). BIS-GMA may be synthesized from bis-phenol A and glycidyl methacrylate or, alternately, from the diglycidyl ether of bisphenol A and methacrylic acid (1,2) as shown in Fig. 1A. In Fig. 1B are shown the structural formulae of BIS-GMA and, for comparison, methyl methacrylate (MMA), the principal monomer used in typical unfilled acrylic materials.

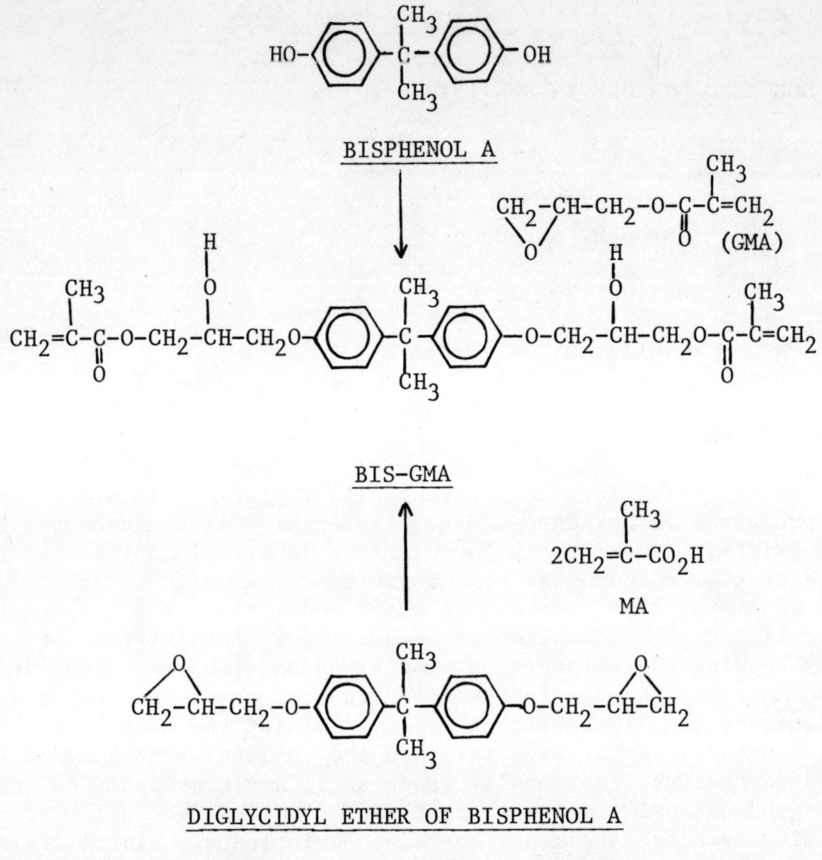
BISPHENOL A
(GMA)
BIS-GMA
MA
DIGLYCIDYL ETHER OF BISPHENOL A

Figure 1A. Synthesis of BIS-GMA.

There are many advantages in using a bulky difunctional monomer such as BIS-GMA, rather than a smaller monofunctional monomer, such as MMA; for example, better marginal adaptability (due to less shrinkage on polymerization) and a probably lower order of toxicity because of reduced volatility and tissue penetration. These desirable properties of BIS-GMA are partially nullified by the high viscosity of the monomer which necessitates the use of diluent comonomers to achieve a workable viscosity for the monomer system. Unfortunately, the dilution of BIS-GMA also results in an increase in polymerization shrinkage. In addition, as obtained commercially, BIS-GMA is not in a high state of putity, as evidenced by batch to batch variations in color, viscosity, acid number, etc. The fact that BIS-GMA itself is

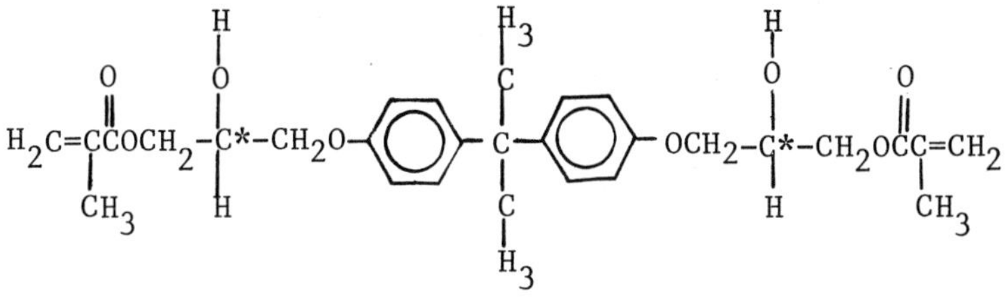

BIS-GMA

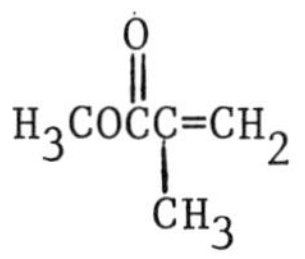

MMA

Figure 1B. Structural formula for BIS-GMA. The asterisks indicate chiral carbon atoms that result in a number of diastereomers. Methyl methacrylate (MMA) is shown for comparison.

a mixture of high molecular weight diastereoisomers makes purification of this monomer by the relatively facile techniques of distillation or crystallization impractical.

There is another type of dimethacrylate that is finding significant use in dental monomers. These are the urethane dimethacrylates, which are synthesized from hydroxyalkyl methacrylates and aliphatic diisocyanates (7,8). The synthesis of this non-aromatic type of monomer is shown in Fig. 2A. The molecular bulk of these monomers is similar to that of BIS-GMA.

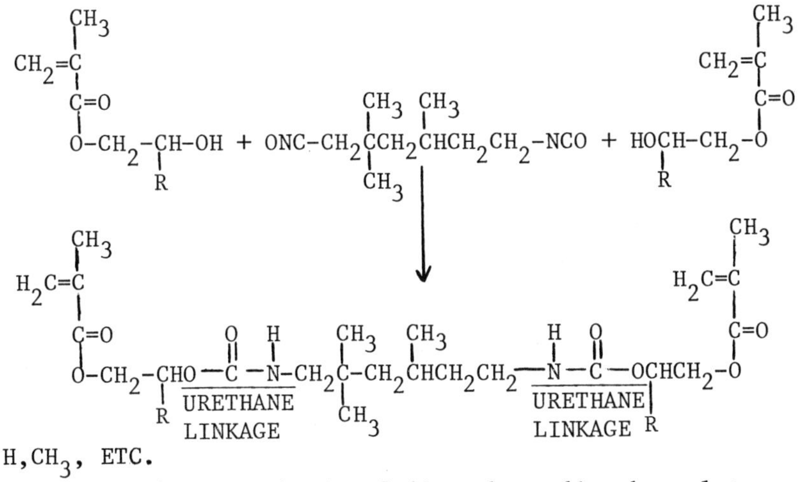
URETHANE
LINKAGE
URETHANE
LINKAGE

R = H,CH_3, ETC.

Figure 2A. Synthesis of diurethane dimethacrylates

Urethane methacrylates with pendant isocyanate groups may also be prepared by suitable modification of reaction conditions from diisocyanates and the appropriate hydroxyalkyl methacrylate. Diisocyanates with NCO groups of different reactivities are especially well suited to the synthesis of isocyanate urethane methacrylates as shown in Fig. 2B for the reaction of tolylene-2,4-diisocyanate (TDl) and 2-hydroxyethyl methacrylate (HEMA) (9,10). Since the isocyanate group of this type of monomer is capable of reacting with functional groups having active hydrogens (e.g. $-NH_2$, -OH, etc.), these monomers have potential as adhesion promoting agents for proteinaceous substrates (e.g. bone, dentin, etc.).

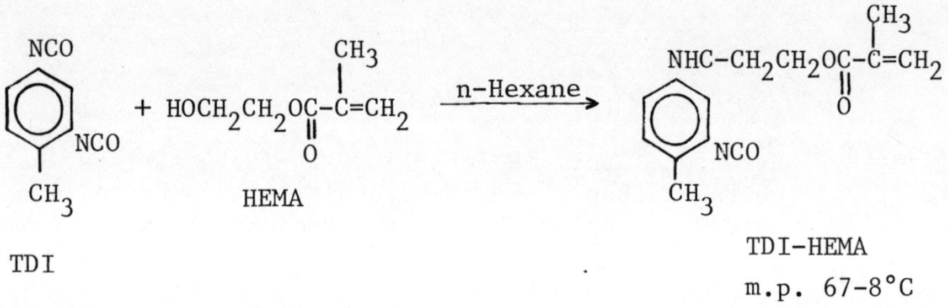
NCO
NCO
CH_3
TDI
+ $HOCH_2CH_2OC(=O)-C(CH_3)=CH_2$
HEMA
n-Hexane
$NHC-CH_2CH_2OC(=O)-C(CH_3)=CH_2$
NCO
CH_3
TDI-HEMA
m.p. 67-8°C

Figure 2B. Typical synthesis of isocyanato urethane methacrylate.

Other substitutes that have been advanced are derivatives of BIS-GMA obtained by selective reaction of the secondary hydroxyl groups, (e.g., esters, urethanes) as shown in Fig. 2C (11,12). Still more hydrophobic variants of BIS-GMA include the dimethacrylate ester of bisphenol A (BIS-MA) and 2,2-bis[p-(2'-methacryloxyethoxy) phenyl] propane (Fig. 2D). These solid monomers, which are relatively easy to purify by recrystallization, should permit the formulation of highly pure monomer systems based on these dimethacrylates (13,14).

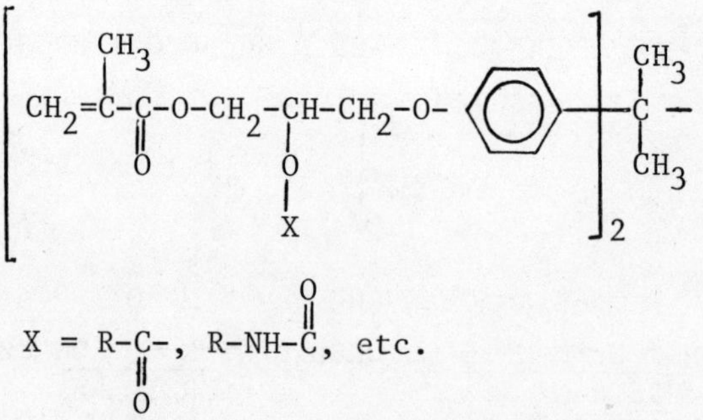
$CH_2=C(CH_3)-C(=O)-O-CH_2-CH(O-X)-CH_2-O-$
$C(CH_3)_2$
2
X = R-C(=O)-, R-NH-C(=O), etc.

Figure 2C. Derivatives of BIS-GMA

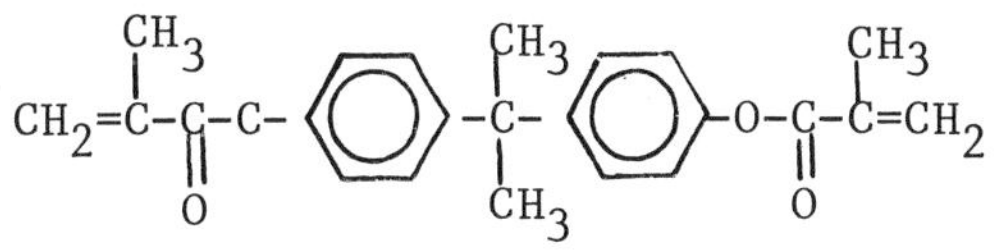

BIS-MA M.P. = 73-74°C

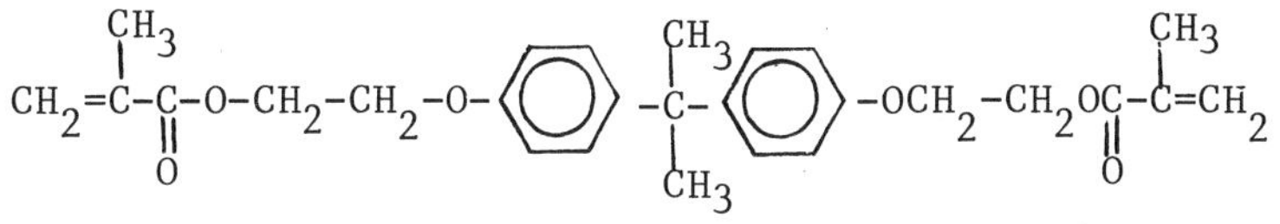

BIS-EMA (ESPE 10) M.P. = 44-45°C

Figure 2D. Non-hydroxylated homologs of BIS-GMA.

EUTECTIC MONOMER SYSTEMS

Aromatic Diester Dimethacrylates

To correct some of the problems associated with the use of BIS-GMA and similar monomers, Bowen devised a unique monomer system based on the use of well characterized crystalline dimethacrylates which on proper admixture form ternary liquid eutectics of workable viscosity (15-17). The structure of this monomer system, based on the bis(methacryloxyethyl) ester of the isomeric benzenedicarboxylic acids, is shown in Fig. 3A (where X = $-\overset{\overset{O}{\|}}{C}-O-$). Despite the excellent promise of this system, a color instability was noted when the usual benzoyl peroxide-tertiary aromatic amine initiator system was used. The color instability is traceable to the formation of a yellow charge transfer complex between the electron poor aromatic diester dimethacrylate and the electron rich amine accelerator (18). By use of alternate activators, e.g., ascorbyl palmitate, this color problem is preventable and excellent composites and sealant materials can be formulated using this monomer system. The synthesis of these aryl diester dimethacrylates is outlined in Fig. 3B.

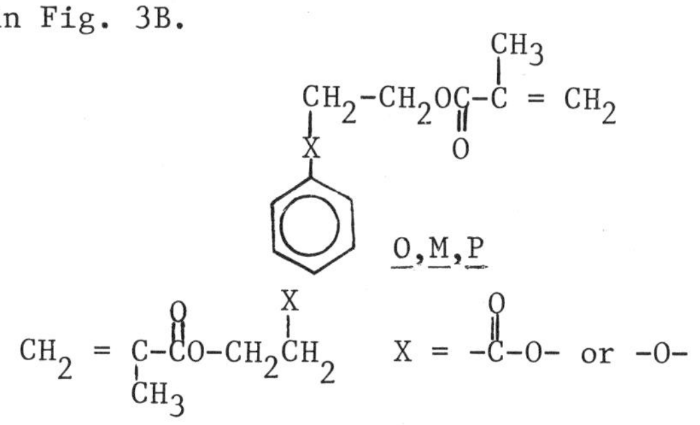
O,M,P
X = -C(=O)-O- or -O-

Figure 3A. Crystalline isomeric monomers capable of forming liquid eutectics.

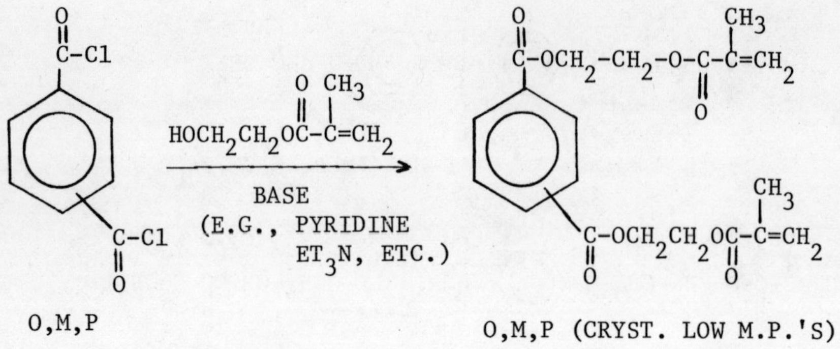

O
C-Cl
C-Cl
O
O CH3
HOCH2CH2OC-C=CH2
BASE
(E.G., PYRIDINE
ET3N, ETC.)
O,M,P
O
CH3
C-OCH2-CH2-O-C-C=CH2
O
CH3
C-OCH2CH2OC-C=CH2
O O
O,M,P (CRYST. LOW M.P.'S)

Figure 3B. Ternary liquid eutectic monomer system synthesis of aryl diester dimethacrylates.

Aromatic Diether and Aromatic Ester-Ether Dimethacrylates

Another approach to circumventing this color problem is to synthesize analogous isomeric crystalline monomers that have phenyl groups of enhanced electron density and, therefore, less affinity for charge transfer complexation with amines. The bis(methacryloxyethyl) derivatives (Fig. 3A) of the isomeric dihydroxybenzenes (X = -O-) (19) and hydroxybenzoic acids (X = -O- and $-\overset{O}{\overset{\|}{C}}-O-$) (20) fulfilled this expectation.

The synthesis of the aryl diether and the aryl ester-ether dimethacrylates are outlined in Figures 3C and 3D, respectively.

Methacryloxyethyl Derivatives of Hydroxybenzaldehydes

The methacryloxyethyl derivatives of the various isomeric hydroxybenzaldehydes also are relatively low melting crystalline monomers that can be formulated to give both ternary and binary liquid mixtures (21). The synthesis of these monomers is shown in Fig. 4. These monomers may have application, if not as the main component of a dental resin binder, at least as a unique type of diluent. Clear, tough polymers that are only partially soluble resulted from their bulk, free radical polymerization. Apparently, the aldehyde functionality enables these monomers to act as chain transfer agents which can delay the onset of gelation, and, perhaps, reduce residual vinyl unsaturation of dental monomer systems which utilize them as diluent comonomers.

In addition, these or similar types of monomers may have utility as adhesion promoting agents for dentin since collagen has a certain natural reactivity towards aldehydes (e.g., maturation of collagenous tissue, tanning of leather).

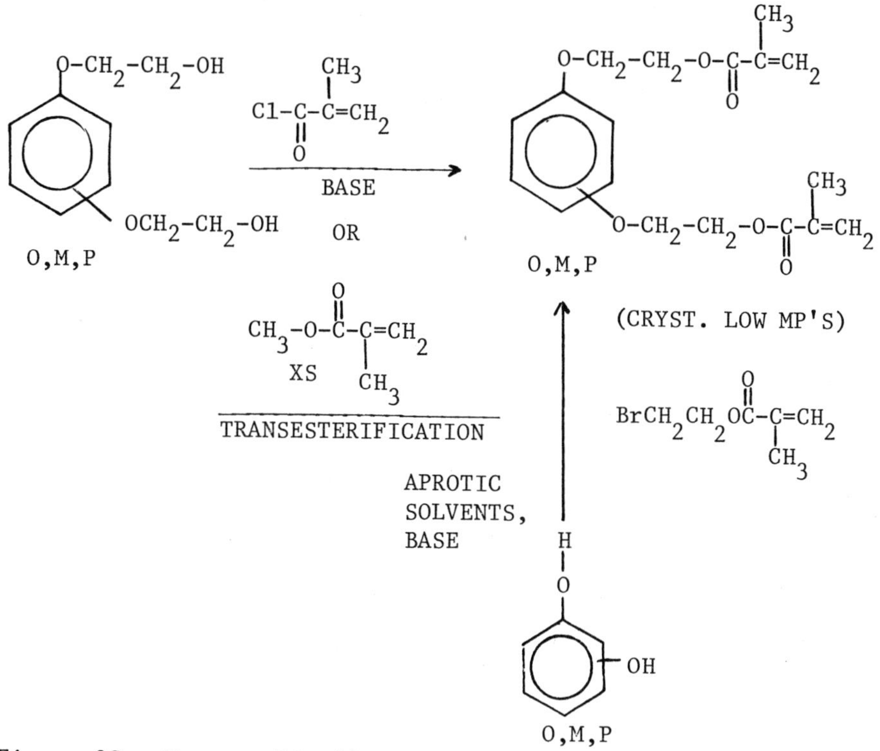
O-CH2-CH2-OH
OCH2-CH2-OH
O,M,P
CH3
Cl-C-C=CH2
O
BASE
OR
O
CH3-O-C-C=CH2
XS
CH3
TRANSESTERIFICATION
CH3
O-CH2-CH2-O-C-C=CH2
O
CH3
O-CH2-CH2-O-C-C=CH2
O
O,M,P
(CRYST. LOW MP'S)
O
BrCH2CH2OC-C=CH2
CH3
APROTIC
SOLVENTS,
BASE
H
O
OH
O,M,P

Figure 3C. Ternary liquid eutectic monomer system synthesis of aryl diether dimethacrylates.

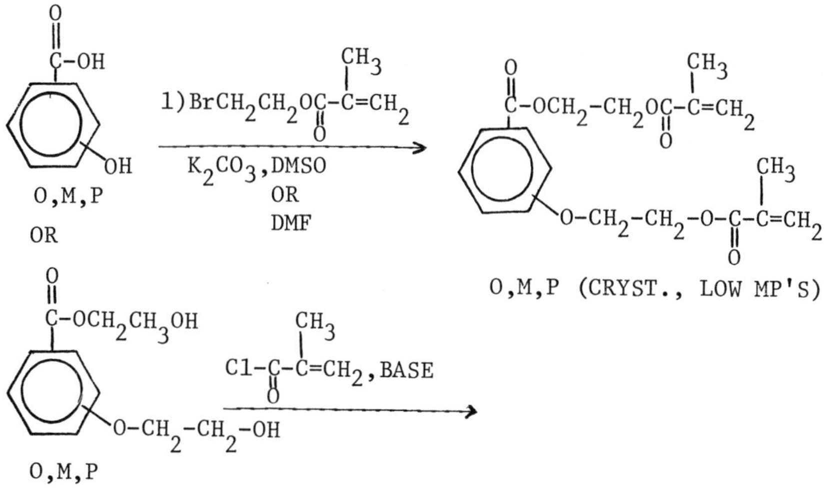
O
C-OH
OH
O,M,P
OR
CH3
1)BrCH2CH2OC-C=CH2
O
K2CO3,DMSO
OR
DMF
O
CH3
C-OCH2-CH2OC-C=CH2
O
CH3
O-CH2-CH2-O-C-C=CH2
O
O,M,P (CRYST., LOW MP'S)
O
C-OCH2CH3OH
O-CH2-CH2-OH
O,M,P
CH3
Cl-C-C=CH2,BASE
O

Figure 3D. Ternary liquid eutectic monomer system synthesis of aryl ester-ether dimethacrylates.

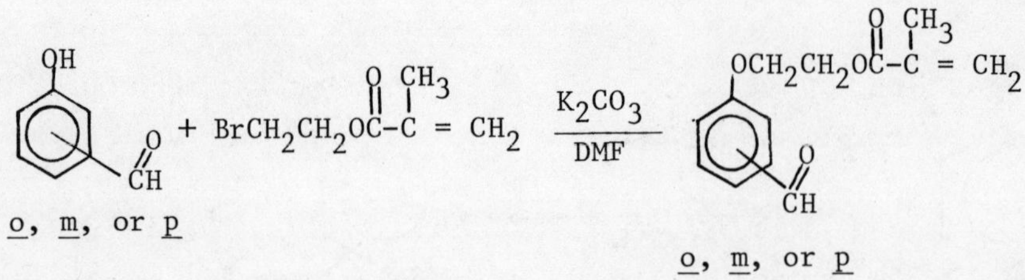
OH
+ BrCH2CH2OC-C = CH2
K2CO3
DMF
OCH2CH2OC-C = CH2
CH
o, m, or p
o, m, or p

Figure 4. Synthesis of isomeric benzaldehyde methacrylates.

Indeed, water insoluble polymers containing pendant benzaldehyde groups have been used for the immobilization of water soluble enzymes (22). Presumably, the coupling reaction involves the formation of a Schiff base as illustrated below:

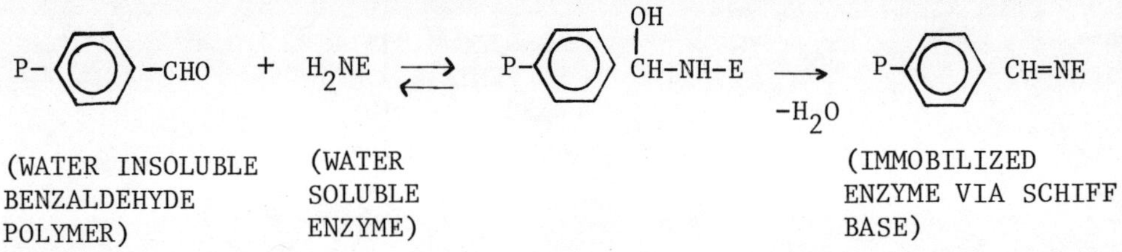
P- -CHO + H2NE
P- CH-NH-E
OH
-H2O
P- CH=NE
(WATER INSOLUBLE BENZALDEHYDE POLYMER)
(WATER SOLUBLE ENZYME)
(IMMOBILIZED ENZYME VIA SCHIFF BASE)

These four isomeric crystalline monomer systems (e.g., a, b and c) can be intermixed to give a variety of other liquid resin binders of high purity. In addition to the practical advantages accruing from the use of monomer systems of high purity, they provide excellent materials for theoretical structure-property studies.

Bicyclic Diacetal Benzaldehyde Dimethacrylates

The synthesis of a new type of monomer, bicyclic diacetal benzaldehyde dimethacrylates (DABDM's), from the para benzaldehyde methacrylate (PBM) and pentaerythritol using p-toluenesulfonic acid (PTSA) and a mixed solvent of dimethyl sulfoxide (DMSO) and toluene, is shown in Fig. 5A (23). This monomer is a white solid, m.p. 104-106°C, which on heating above its melting point polymerizes to a tough, clear, insoluble resin.

Monomers of this type are expected to polymerize via their vinyl groups. However, their acetal groups may also participate in the free radical polymerization, since it is known that cyclic acetals such as shown below undergo a free radical ring opening rearrangement to the corresponding ester (24).

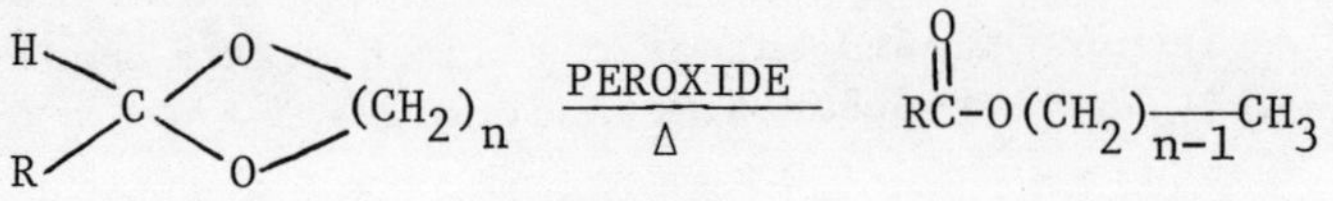
H
R
C
O
O
(CH2)n
PEROXIDE
Δ
RC-O(CH2)n-1CH3

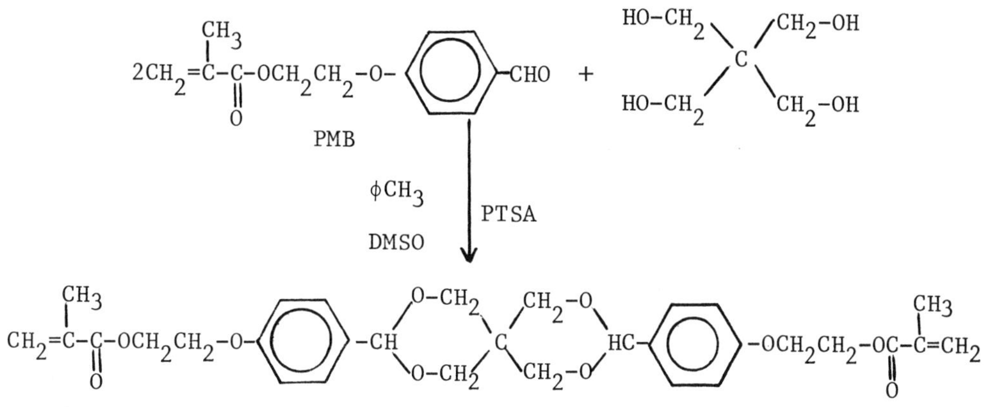
PMB
ϕCH3
DMSO
PTSA

Figure 5A. Synthesis of a bicyclic diacetal benzaldehyde dimethacrylate

While this type of monomer is not of the spiro-o-carbonate type that has been shown to expand on radical ring opening polymerization (Fig. 5B) (25,26), nevertheless, assuming it undergoes just the usual vinyl addition type polymerization, it is still an attractive candidate for use in formulating low shrinking dental resin systems. If on polymerization only free radical rearrangement to the diester form occurs, as shown in Fig. 5C, then polymerization shrinkage should be further reduced.

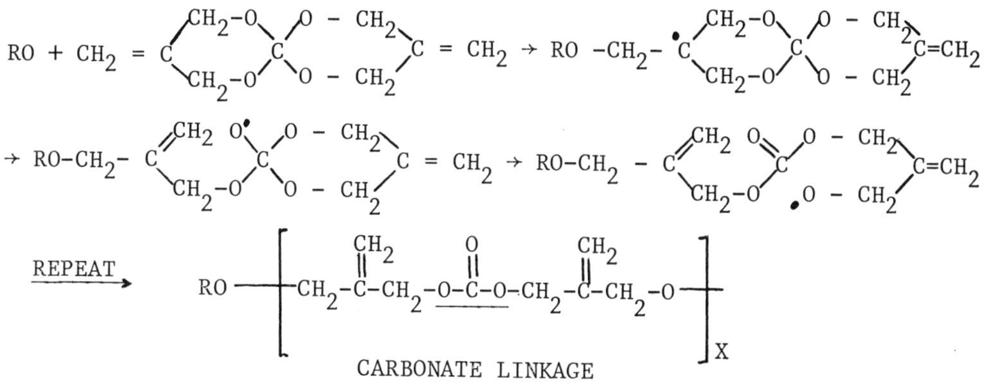
REPEAT
CARBONATE LINKAGE

Figure 5B. Radical ring opening polymerization of spiro ortho carbonate vinyl monomers with expansion in volume (W. J. Bailey et al.).

Diurea Dimethacrylates

Monomers analogous to the previously described diurethane dimeth-

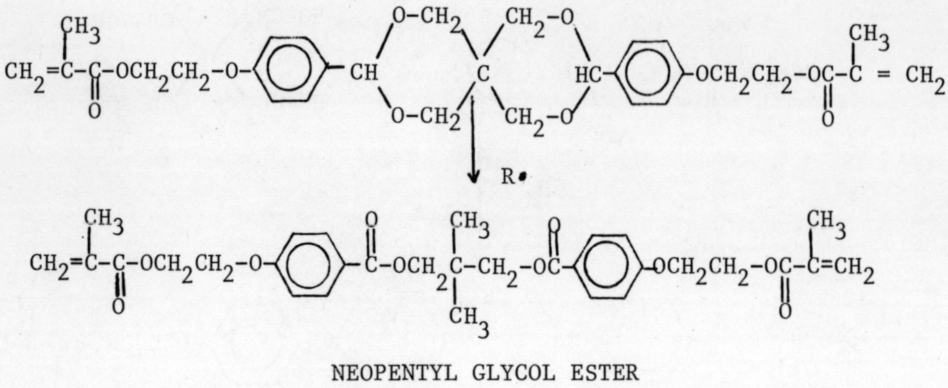
R•
NEOPENTYL GLYCOL ESTER

Figure 5C. Possible mechanism for free radical ring-opening of bicyclic diacetal benzaldehyde dimethacrylates.

acrylates (Fig. 2A) but containing trisubstituted urea groups can be prepared from organic diisocyanates, such as hexamethylene diisocyanate (HDI) and N-t-butyl-2-aminoethyl methacrylate (TBAEM). Due to their bulky nature, and the hindered nature of the urea linkage, it is expected that this monomer would not be as viscous as the corresponding diurethanes and may exhibit even less shrinkage on polymerization (27). The synthesis of this type diurea dimethacrylate is illustrated in Fig. 6.

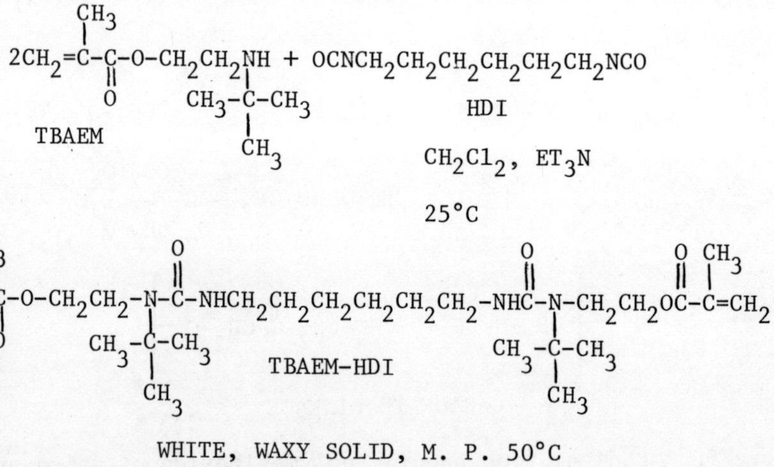
TBAEM
HDI
CH_2Cl_2, ET_3N
25°C
TBAEM-HDI
WHITE, WAXY SOLID, M. P. 50°C

Figure 6. Synthesis of a diurea dimethacrylate.

Polyfluorinated Dimethacrylates

Dental composites that were extremely hydrophobic but mechanically relatively weak were prepared from a resin system consisting of a non-hydroxyl variant of BIS-GMA and the polyfluorinated monomethacrylate, octafluoro-1,1,5-trihydropentyl methacrylate (28). This resin system which contains approximately 38% fluorine by weight yielded quartz composites that had extremely high advancing contact angles (88°) versus water, remarkably low water sorption, and reduced marginal leakage (28-30).

In an effort to formulate monomer systems that will have not only enhanced hydrophobicity but also excellent chemical, physical and mechanical stabilities, a series of thermosetting polyfluorinated dimethacrylates were synthesized as shown in Figs. 7 and 8 (31). These monomers can be formulated into resin systems that will have very low surface energies.

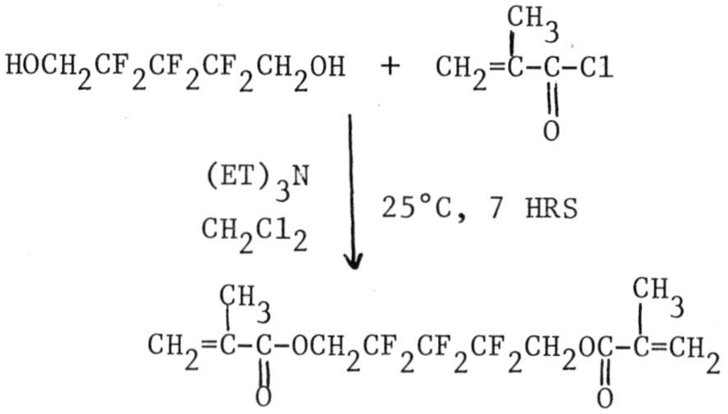
HOCH2CF2CF2CF2CH2OH + CH2=C-C-Cl
CH3
O
(ET)3N
CH2Cl2
25°C, 7 HRS
CH2=C-C-OCH2CF2CF2CF2CH2OC-C=CH2
CH3
O

HFPDMA

MW = 348

VISCOSITY = MOBILE LIQUID ∿ MMA

Figure 7. Synthesis of hexafluoro-1,5-pentanediol dimethacrylate (HFPDMA)

Prepolymer Monomers

The feasibility of using prepolymer monomers as the major component of dental resin systems has recently been demonstrated (31). Highly fluorinated acrylate and methacrylate comb type prepolymer monomers were synthesized from a polyfluorinated oligoether-ol as shown in Figs. 9 and 10. These novel monomers are viscous liquids similar to BIS-GMA but of much larger bulk (MW > 10,000) which should minimize polymerization shrinkage and improve marginal adaptation. They are miscible with a number of other fluorinated and non-fluorinated and non-fluorinated monomers and are easily polymerized by the usual initiator systems.

Composites formulated with monomer system based on the polyfluoroligomethacrylate (PFMA) and high contact angles (80-89°) versus water and excellent esthetic and mechanical properties, e.g., diametral

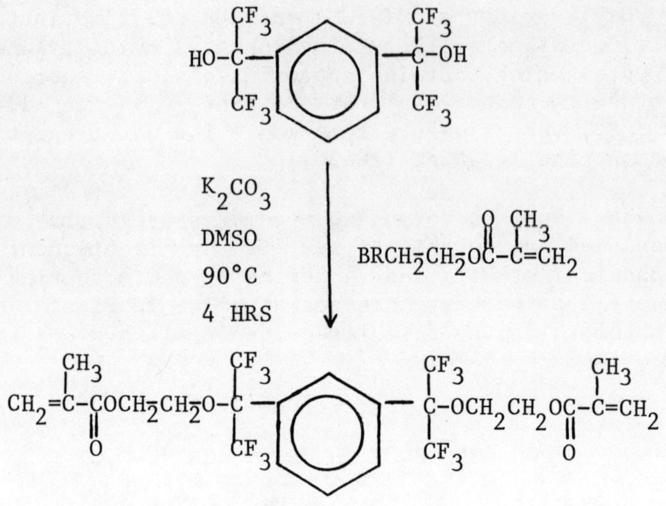

BIS(mMEHFP) Ø

MW = 634

MODERATELY VISCOUS LIQUID (LESS THAN BIS-GMA)

Figure 8. Synthesis of bis-(2-methacryloxyethoxyhexafluoro-2-propyl) benzene [BIS(MEHFP)Ø]

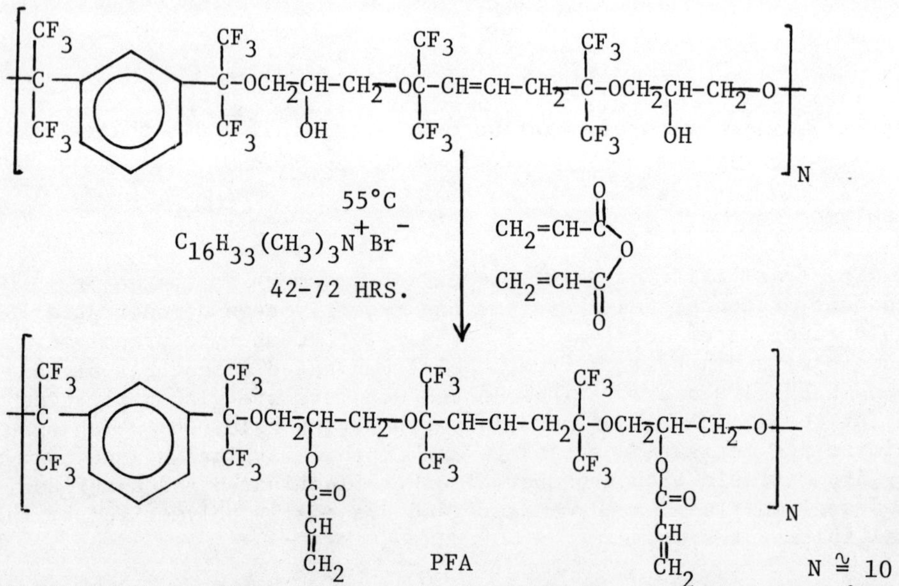

MW OF REPEATING UNIT = 1004; AVG MW = 10,040; VISCOUS LIQUID $\lesssim$ BIS-GMA

Figure 9. Synthesis of polyacrylate (PFA).

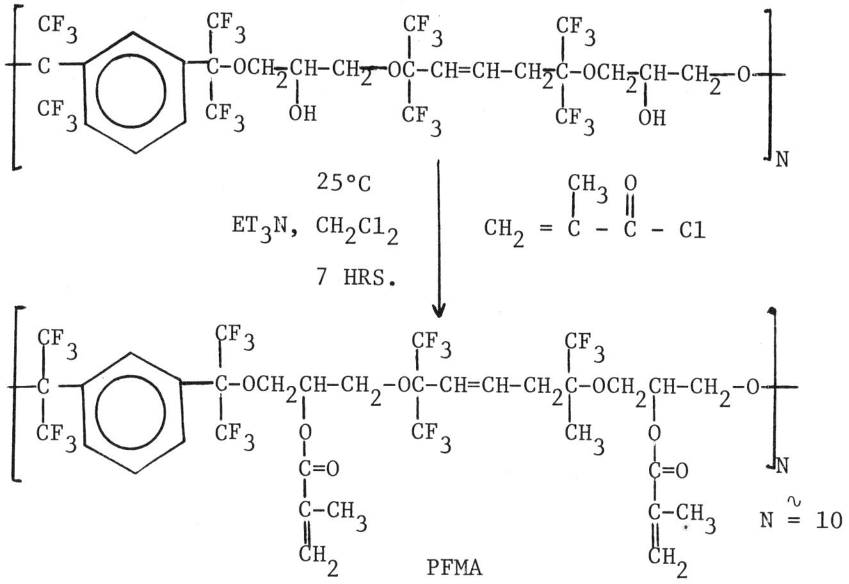
25°C
ET3N, CH2Cl2
7 HRS.
PFMA
N ≈ 10

MW OF REPEATING UNIT = 1032; AVG MW = 10,320

VISCOUS LIQUID $\lesssim$ BIS-GMA

Figure 10. Synthesis of polymethacrylate (PFMA).

tensile strengths > 40 MPa.

In summary, it is hoped that these new monomers will not only be useful in the formulation of improved dental composite restorative and sealant materials, but may also find other medical and industrial applications.

ACKNOWLEDGMENT

This work was supported by the National Institute of Dental Research, Interagency Agreement No. Y01-DE-40015.

SUMMARY

There is a need to further enhance the durability of dental composite restorative and sealant materials. Dimethacrylates such as BIS-GMA (the diadduct of bisphenol A and glycidyl methacrylate) are now widely used to formulate the resin component of these materials. Some of the deficiencies of dental composites and sealants are traceable to impurities and inherent structural imperfections in the monomer systems. Structure-property studies are needed to explore ways of achieving minimal shrinkage on polymerization, reducing water sorption, promoting adhesion, generally optimizing the chemical, physical and mechanical stabilities of the resulting polymers. The synthesis of such an ideal dental monomer system is not an easy task. This paper describes the synthesis of several alternate types of monomers that address some of these problem areas and hopefully have utility as components of dental monomer systems. One novel type monomer system of high purity is derived from isomeric crystalline monomers that can form liquid eutectics. Other types include functional monomers such as benzaldehyde methacrylates, and highly fluorinated thermosetting methacrylates including comb type pre-polymer monomers. The latter monomers are designed to minimize water sorption and polymerization shrinkage and maximize the chemical, physical and mechanical properties of the resin binder.

REFERENCES

1. R. L. Bowen, U. S. Patent 3,066,112 (1962)
2. R. L. Bowen, JADA, 66:57 (1963)
3. R. L. Bowen, ibid, 69, 481 (1964)
4. R. L. Bowen, U. S. Patent 3,179,623 (1965)
5. R. L. Bowen, U. S. Patent 3,194,783 (1965)
6. R. L. Bowen, J. A. Barton, Jr., and A. L. Mullineaux, "Composite Restorative Materials in Dental Materials Research", G. Dickson and J. M. Cassel, Eds., (Proc. 50th Anniversary Symposium, National Bureau of Standards, 1969; NBS Special Publication 354), U. S. Government Printing Office, Washington, DC (1972), p. 73
7. R. J. Foster, and R. J. Walker, Ger. Offen. 2,312,559 (1973)
8. R. J. Foster, Ger. Offen., 2,411,760 (1974)
9. J. M. Antonucci, G. M. Brauer, and D. J. Termini, J. Dent. Res., 59(1), 33 (1980)
10. Z. G. Zemskova, N. G. Matveyevam, and A. A. Berlin, Polymer Sci. USSR, 15, 814 (1973)
11. D. G. Stoffey, and H. L. Lee, U.S. Patent 3,755,420 (1973)
12. D. E. Waller, Ger. Offen., 2,053,901 (1971); U.S. Patent 3,709,866
13. M. Atsuta, N. Nakabayashi, and E. Masuhara, J. Biomed. Mater. Res., 5, 183 (1971)
14. W. Schmitt, and R. Purrman, U. S. Patent 3,810,938 (1974)
15. R. L. Bowen, J. Dent. Res., 49, 810 (1970)
16. J. A. Barton, Jr., C. L. Burns, H. H. Chandler, and R. L. Bowen, ibid, 52, 731 (1973)
17. H. H. Chandler, G. C. Paffenbarger, A. L. Mullineaux, and R. L. Bowen, ibid, 52, 1128 (1973)
18. H. Argentar, and R. L. Bowen, ibid, 54, 588 (1975)
19. R. L. Bowen, and J. M. Antonucci, ibid, 54, 599 (1975)
20. J. M. Antonucci, and R. L. Bowen, ibid, 55, 8 (1976)
21. J. M. Antonucci, ibid, 57, 500 (1978)
22. E. Brown, and R. Joyeau, Polymer, 15, 546 (1974)
23. J. M. Antonucci, unpublished results
24. E. S. Huyser, and Z. Garcia, J. Org. Chem., 27, 2716 (1962)
25. T. Endo, and W. J. Bailey, J. Polymer Sci., Poly. Chem. Ed., 13, 2525 (1975)
26. V. P. Thompson, E. F. Williams, and W. J. Bailey, J. Dent. Res.,

58, 1522 (1979)
27. J. M. Antonucci, unpublished results
28. W. H. Douglas, R. G. Craig, and C.-J. Chen, J. Dent. Res., 58, 1981 (1979)
29. W. H. Douglas, R. G. Craig, and C.-J. Chen, J. Dent. Res., 58, Special Issue A, Abstract No. 1216, p. 395 (1979)
30. R. G. Craig, J. Dent. Res., 58, 1544 (1979)
31. J. M. Antonucci, J. R. Griffith, R. J. Peckoo, and D. J. Termini, J. Dent. Res., 58, Special Issue A, Abstract No. 599, p. 242 (1979)

THE SYNTHESIS OF FLUORINATED ACRYLICS VIA FLUORO TERTIARY ALCOHOLS

James R. Griffith and Jacques G. O'Rear

Naval Research Laboratory
Chemistry Division, Code 6120
Washington, D.C. 20375

INTRODUCTION

Acrylic polymers of linear and of cross-linked varieties are important classes of materials, and the fluorocarbon polymers are likewise important for substantially different reasons. The broad resistance of fluorocarbons to physical and chemical attack suggests that acrylics could be enhanced by the introduction into the molecules of substantial amounts of fluorine, provided such an addition did not compromise the characteristic acrylic properties. Fluorocarbons also possess a range of unusual surface chemical properties which could make for greater versatility in the acrylics if imparted thereto. For example, a fluoroacrylic resin in the liquid, precured state is expected to be of low surface tension and excellent wetting capability for difficult-to-wet fillers, such as powdered Teflon, whereas the cured fluoracrylic can be expected to be relatively non-wetting and non-absorptive of most liquid systems, particularly those that are water based. In order to attempt such an enhancement of acrylic properties, the synthesis of a series of fluorine-bearing acrylics of various functionalities was undertaken.

The problem of introducing substantial quantities of stable fluorocarbon into resins without undue compromise of the strength properties has been previously solved for epoxy systems (1-3). The intermediates for these epoxies were a series of fluoro tertiary alcohols with aromatic nuclei surrounded by perfluorinated aliphatic groups and bearing hydroxyl functionalities derived from hexafluoroacetone. The present synthesis problem entailed the esterification of these intermediates to yield acrylic esters.

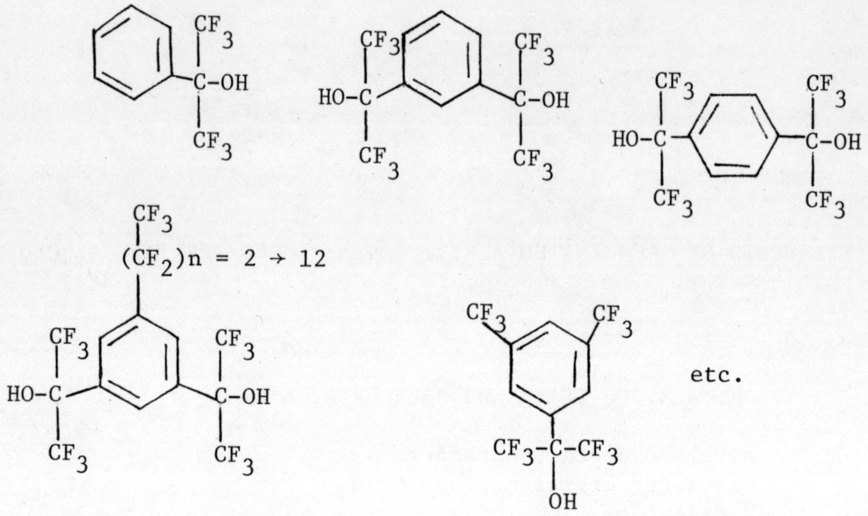
CF3
C-OH
CF3
HO-C
(CF2)n = 2 → 12
CF3-C-CF3
OH
etc.

Attempts to esterify these directly by the use of acrylic acid or acrylic anhydride were not successful. It was found that a convenient, high-yield synthesis could be carried out in fluorocarbon solvent by the reaction of acryloyl chloride and a tertiary amine acid acceptor. Product purification by distillation was generally not satisfactory because of the temperatures required, particularly for the difunctional compounds, but purification by percolation of the fluorocarbon solvent solutions over activated alumina resulted in colorless products of sufficient purity for effective polymerization.

EXPERIMENTAL

Materials

Triethylamine (99% min. by GLC) from Eastman Organic Chemicals was dried over Type 13X molecular sieves. Acryloyl chloride from Aldrich Chemical Company was redistilled before use (bp 73-75°, 99% by GLC). The intermediate fluoro-monol and diols were distilled and/or recrystallized to obtain products exceeding 99.9% purity by GLC.

Synthetic Methods

(2-Hydroxyhexafluoro-2-propyl)benzene acrylate, I. A solution of (2-hydroxyhexafluoro-2-propyl)benzene (95.2 g, 0.39 mole) in Freon 113 (250 ml) was stirred magnetically in a 3-neck flask (1,000 ml) equipped with a dropping funnel, thermometer, dry ice condenser and drying tube (calcium chloride). Triethylamine (39.5 g, 0.39 mole) dissolved in Freon 113 (50 ml) was added dropwise during 10 min. (exotherm: 25 to 40°). At this point an external cooling bath (ice-water) was applied. Then a second dropping funnel charged with acryloyl chloride (35.3 g, 0.39 mole) dissolved in Freon 113 (50 ml) was quickly substituted in the place of the empty dropping funnel. Cautious addition of the latter was completed in 40 min. with stirring at 10 to 20°C. After 2 hrs. of additional stirring, the resulting white slurry was vacuum filtered (350 ml coarse fritted glass filter packed with 10 g of Celite) and washed with Freon 113 (50 ml). Flash evaporation of the straw-colored filtrate left 113.0 g of honey-colored oil analyzing 94.4% I and 5.6% of the starting fluoroalcohol. Methods for decolorization and purification described in the next paragraph led to 83.2 g (71.5% yield) of analytical I (>99.95% purity by GLC) as a colorless viscous liquid; n_D^{25} 1.4336; bp 58-60°/2 mm (4).

<u>1,3-Bis(2-hydroxyhexafluoro-2-propyl)benzene diacrylate, II.</u> This diester was synthesized by a modification of the esterification procedure just described. In the first step 1,3-bis(2-hydroxyhexafluoro-2-propyl)benzene (139.4 g, 0.34 mole) in Freon 113 (350 ml) was reacted with triethylamine (69.0 g, 0.68 mole) in Freon 113 (40 ml). In the second step acryloyl chloride (61.5 g, 0.68 mole) in Freon 113 (40 ml) was added. Vacuum filtration to remove the precipitate, followed by flash evaporation of the filtrate, led to 157.1 g of yellow oil assaying 1.2% of the starting diol, 3.4% of the monoester and 94.6% of II. The yellow oil was dissolved in Freon 113 (1,000 ml), the resulting solution agitated with a mixture of Nuchar (25 g) and Celite (25 g), the latter filtered through packed Celite (30 g) and washed with Freon 113 (500 ml). The clarified filtrate was percolated and washed successively through two chromatographic columns (73 mm OD X 30 cm). Both were fitted with a coarse fritted glass disc, charged with Woelm neutral activated alumina (140 g), packed Celite (40 g) and glass wool (1 g). Flash evaporation of the effluent (finally at 40°/2 mm) left 112.2 g (63.7% yield) of analytical II (100% purity by GLC) as white, almost odorless crystals; mp 65-68°.

<u>1,4-Bis(2-hydroxyhexafluoro-2-propyl)benzene diacrylate, III.</u> Here the comparatively low solubilities of the starting diol and III have required changes in both stoichiometry and chromatography to enhance solubility. Triethylamine (27.0 g, 0.267 mole) in Freon 113 (40 ml) was added to 1,4-bis(2-hydroxyhexafluoro-2-propyl)benzene (41.0 g, 0.10 mole) in Freon 113 (350 ml) to give a homogeneous solution. Acryloyl chloride (19.9 g, 0.22 mole) in Freon 113 (40 ml) was reacted in the usual manner. The precipitate was collected, Freon-washed, dried, and subsequently found to contain 16.1 g of unreacted starting diol. Flash evaporation of the filtrate gave 35.2 g of white crystals assaying 6.5% of the diol, 8.4% of the corresponding monoester and 85.1% of III. Work-up of the residue in 1.3 N NaOH (800 ml), followed by water washes, left 30.0 g of white crystals, which were taken up in ether and the ether extract

water washed and dried. Residue from a small aliquot of the dried extract analyzed 98.0% III and 2.0% monoester. Chromatographing the ether extract over neutral activated alumina, followed by flash evaporation of the effluent, left 25.3 g (48.8% yield) of analytical III (100% by GLC) as white, almost odorless, crystals; mp 113-114°.

<u>1,3-Bis(2-hydroxyhexafluoro-2-propyl)-5-heptafluoropropylbenzene diacrylate, IV.</u> Reaction of 1,3-bis(2-hydroxyhexafluoro-2-propyl)--5-geptafluoropropylbenzene (11.56 g, 0.020 mole) with triethylamine (4.09 g, 0.040 mole), followed by reaction with acryloyl chloride (3.63 g, 0.040 mole), was performed in the customary manner. Filtration left a straw colored filtrate, found by evaporation of an aliquot to contain 0.2% starting diol, 2.1% monoester and 97.7% IV. Customary decolorizing and column chromatography yielded 10.0 g (72.9% yield) of analytical IV (100% by GLC) as white, almost odorless crystals; mp 57-59°; n_D^{60} 1.3684.

<u>1,3-Bis(2-hydroxyhexafluoro-2-propyl)-5-pentadecafluoroheptylbenzene diacrylate, V.</u> Triethylamine (1.60 g, 0.158 mole) was added to 1,3-bis(2-hydroxyheptafluoro-2-propyl)-5-pentadecafluoroheptylbenzene (5.00 g, 0.00643 mole) and then reacted with acryloyl chloride (1.39 g, 0.0154 mole) in the ordinary manner. Work-up of the filtrate gave 5.69 g of cream colored crystals analyzing 0.6% starting diol, 6.0% monoester and 93.4% V. Customary purification steps led to 3.8 g (66.7% yield) of analytical V (>99.95% by GLC) as a clear, viscous oil which crystallized overnight to yield white, almost odorless crystals; mp 35-37°; n_D^{25} 1.3710 (supercooled).

RESULTS AND DISCUSSION

The polymerization of the fluoroacrylics can be effected by typical techniques commonly used for the acrylic class. Free radical initiators, such as 2,2'-azobis (2-methylpropionitrile), produce rapid reactions at 50°C, and ultraviolet light in conjunction with benzoin ethers give polymers at 25°C within short irradiation times. The problem of surface inhibition by oxygen of the air can be avoided by nitrogen blanketing. Since most of the fluoromonomers are solids at room temperature which have strong crystallizing tendencies, it is convenient to dissolve the solids in the liquid monoacrylate I at moderately elevated temperatures and effect the polymerization before crystallization can occur.

Several of the highly fluorinated diacrylate monomers readily wet powdered Teflon, and the cured compositions have a white, semi-translucent appearance which closely approximates that of attractive teeth. These compositions are extremely hydrophobic, and if polished before measurements are made, have static and kinetic friction co-efficients of 0.15 and 0.14, respectively, which are between those of polytetrafluoroethylene and tetrafluoroethylene-hexafluoropropylene copolymer. In addition to possible dental applications, these properties suggest such uses as artificial hip ball and socket prostheses in which the acrylic matrix would provide moldability and structural strength, including cold flow resistance, and the Teflon dispersed phase would provide permanent lubricity to the moving faces. Although not proven in the present studies, it is reasonable to expect that these compositions would be highly impervious to body fluids, including saliva, and resistent to staining.

CONCLUSION

The fluoroacrylic materials appear to be particularly promising for dental and biomedical applications. However, all of the other acrylic uses, such as coating, castings, encapsulants, caulks, etc. based upon linear or three-dimensional polymers can also be realized. The special properties imparted by the presence of fluorocarbon should make these substances uniquely applicable for many purposes.

ACKNOWLEDGEMENT

We are indebted to Robert C. Bowers of the Naval Research Laboratory, Chemistry Division, for the frictional measurements.

REFERENCES

1. J. G. O'Rear, J. R. Griffith and S. A. Reines, Journal of Paint Technology, 43, No. 552, 113 (1970)
2. J. R. Griffith and J. G. O'Rear, Synthesis, 1974, No. 7, 493
3. D. L. Hunston, J. R. Griffith and R. C. Bowers, Ind. Eng. Chem. Prod. Res. Dev., 17, No. 1, 10 (1978)
4. J. R. Roitman and A. G. Pittman, J. Polymer Sci., Polymer Chemistry Edition, 12, No. 7, 1421 (1974)

THE NATURE OF THE CROSSLINKING MATRIX FOUND IN DENTAL COMPOSITE FILLING MATERIALS AND SEALANTS

G. F. Cowperthwaite, J. J. Foy, and M. A. Malloy

Esschem Company Division
Sartomer Industries, Inc.
P. O. Box 56
Essington, PA 19029

INTRODUCTION

Since their introduction, composite resin filling materials have replaced direct filling resins and the silicate cements as the materials of choice for anterior restorations where aesthetics are of the primary importance. These materials have met with wide acceptance in spite of several performance shortcomings. It has been reported that many of the commercial composite resin materials exhibit some leakage at the margins of the restoration (1-4) and gross surface disintegration (5). Many investigators have attempted to trace these problems to some basic properties of the system, such as water sorption and polymerization shrinkage (6). Due to the complex nature of this system, the results of these investigations have not been conclusive.

The composite resin system consists of three major components:

1. Quartz, glass or ceramic filler at about 70% by weight (50% by volume).

2. A high molecular weight monomer included to reduce polymerization shrinkage, usually, 2,2 bis (p-2'-hydroxy-3'-methacryloxy-propoxy) phenylenepropane (Bis-GMA).

3. A low molecular weight diluting monomer to control viscosity.

The choice of this diluting monomer has undergone considerable change since the system was introduced. Originally, methyl methacrylate (MMA) was used, but because of its volatility MMA was replaced by ethylene glycol dimethacrylate (EDMA). EDMA also proved to be too volatile for use. The present system, for the most part, utilizes triethylene glycol dimethacrylate (3EDMA) as diluent (7). Some recent work has proposed the use of other monomer systems to reduce water sorption (8).

The purpose of this work is to study the properties of the individual monomers in order to isolate and understand the variables of this complex system.

EXPERIMENTAL

The materials studied consist of a series of diluting monomers of the structure:

$$CH_2=C(CH_3)-C(=O)-O-(-CH_2CH_2-O-)_N-C(=O)-C(CH_3)=CH_2$$

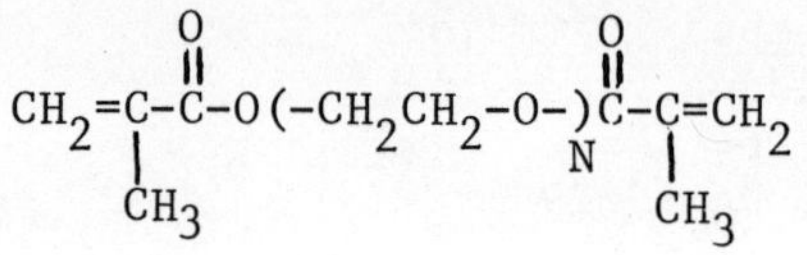

Members of this series include EDMA, N=1; diethylene glycol dimethacrylate, 2 EDMA, N=2; 3 EDMA, N=3; tetraethylene glycol dimethacrylate, 4 EDMA, N=4; and polyethylene glycol dimethacrylate, PEDMA, where N is a mixture of 1 to 6. Also, a series of monomers of the structure:

$$CH_2=C(CH_3)-C(=O)-O(-CH_2CH_2-)_NO-C(=O)-C(CH_3)=CH_2$$

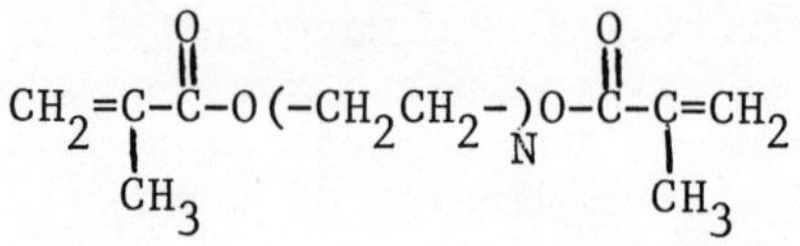

Members of this group include 1,4-butanediol dimethacrylate, BDMA, N=2; 1,6-hexanediol dimethacrylate, HDMA, N=3; and 1,10-decamethylenediol dimethacrylate, DMDMA, N=5. The following individual diluting monomers were also evaluated, neopentyl glycol dimethacrylate, NPGDMA; and trimethylolpropane trimethacrylate, TMPTMA. Two high molecular weight monomers based on the bisphenol A structure were evaluated, Bis-GMA and a related dimethacrylate prepared from ethoxylated bisphenol A (EBPADMA).

The Bis-GMA was used as received from the supplier. The EBPADMA was prepared in our laboratory by the direct esterification using sulfuric acid catalyst. The diluting monomers were obtained from our commercial esterification process and purified by a proprietary column treatment process which reduces residual acidity and color and removes esterification inhibitor. The monomers were characterized as described previously (9) and activated by the addition of 0.25% dimethyl para toluidine (DMT) and 0.5% benzoyl peroxide (BPO). The polymer specimens were prepared in glass cells formed by two 50 millimeter by 50 millimeter OD and 0.5 millimeter wall thickness assembled in a typical sheet casting cell. The activated monomer was introduced into the cell via syringe. The cell was allowed to stand for at least 16 hours at 37.5°C before removal of the polymer sheet. The Bis-GMA sample could not be cast using this method. The monomer was initiated with 1% tertiary butyl hydroperoxide and cured for 16 hours at 100°C. In order to study the effect of this thermal free radical cure versus the redox cure, the sample of EBPADMA was cured using both systems.

The resulting polymer sheets were characterized by determination of the refractive index (10) and density (11). The uncorrected difference between the monomer and polymer specific gravity was taken as the polymerization shrinkage. Using the method of Loshaek and Fox (12) the percentage utilization of unsaturation was calculated from volume contraction based on the assumption of 23.0 cc change in volume per mole of double bonds. The percent of residual double bonds in the sheets was also determined by attenuated total reflectance infrared spectroscopy using a Perkin-Elmer Model 137-D Infracord Spectrophotometer and Buck Scientific Universal Reflectance Unit URU-101.

Water sorption studies were conducted on the cast sheet according to ADA Specification No. 12. The disk thickness was a nominal 0.75 millimeter instead of the required 0.5 millimeter. The test period was extended until the samples reached apparent equilibrium. For most samples the apparent equilibrium was reached at 250-275 days. The samples were then dried to constant weight in a dessicator in order to determine the absolute water sorption. The samples were then submitted for water sorption for a second cycle.

RESULTS

The properties of the polymers evaluated are given in Table 1. The data reported for water sorption is the actual water uptake after the initial extraction and drying and represents a measurement of the actual affinity of the polymer matrix for water.

Table II traces the fate of the double bonds and the efficiency of their utilization in crosslinking the matrix. Data reported is based both on the results of the volume contraction calculations and on the residual double bond concentration measured via infrared. The calculation is based on the assumption that the solubility of the material in water after the long-term extraction represents the residual totally unreacted monomer.

Table III compares the original long-term water extraction study, the solubility determination and the water sorption for the second extraction cycle.

Table I

POLYMER PROPERTIES

Identity	Monomer Molecular Weight	Ref. Index	Spec. Grav. GM./CC.	Polym. Shrinkage %	Max. Water Sorption %
MMA	100	1.460	1.168	24.8	2.00
EDMA	198	1.474	1.220	16.2	4.47
2EDMA	242	1.532	1.228	15.9	3.87
3EDMA	286	1.418	1.220	12.5	5.53
4EDMA	330	1.459	1.225	13.6	7.32
PEDMA	279	1.524	1.229	13.8	7.39
BDMA	226	1.430	1.183	15.9	0.90
HDMA	254	1.435	1.134	14.6	0.83
DMDMA	310	1.469	1.080	12.5	0.36
NPGDMA	240	1.455	1.122	12.2	1.13
TMPTMA	338	1.559	1.170	10.4	0.65
EBPADMA	452	1.589	1.180	5.9	0.44
BISGMA	512	1.531	1.223	6.4	3.55

Table II

FATE OF DOUBLE BONDS

Identity	Polymer Shrinkage	% Reacted Double Bonds	Solubility (Resid. Mon.)	% X-Link
MMA	24.8	98.6	1.3	–
	24.3	95.8	3.5	–
EDMA	16.2	57.1	4.4	18.6
2EDMA	15.9	67.9	6.1	41.9*
3EDMA	12.5	68.2	2.1	38.5
4EDMA	13.6	79.9	1.1	60.8
PEDMA	13.8	68.1	0.9	7.1
BDMA	15.9	63.5	–	–
HDMA	14.6	71.4	1.9	44.7
DMDMA	12.5	78.0	0.5	56.5
NPGDMA	12.2	56.7	2.2	15.6
TMPDMA	10.4	41.7	–	–
EBPADMA	5.9	49.3	–	–
BISGMA	6.4	56.0	0.1**	12.2

* May be high due to incomplete removal of residual monomer in first H_2O extraction.

**May be low due to low diffusion of high MW molecule.

TABLE III

WATER SORPTION

Identity	Water Sorp., %					Solubility	Water Sorp.		
	1 Day	1 Wk.	1 Mo.	8 Mo.	9 Mo.	%	1 Day	1 Wk.	1 Mo.
MMA	1.67	1.60	1.40	0.86	0.76	1.26	1.96	1.95	1.90
	1.45	1.01	0.24	-1.22	-1.50	3.46	2.04	1.96	1.90
EDMA	2.18	1.71	1.23	0.19	0.05	4.35	4.47	4.44	4.31
2EDMA	0.53	0.48	-0.82	-2.11	-2.28	6.10	3.87	2.72	3.54
3EDMA	1.78	3.71	3.51	3.14	3.09	2.11	5.33	5.22	5.18
4EDMA	4.71	6.40	6.20	6.09	6.10	1.03	7.32	7.14	7.07
PEDMA	5.86	6.37	6.25	6.41	6.39	0.87	7.39	7.22	7.19
BDMA	0.90	0.50	-0.13	-1.90	-2.85	-	-	-	-
HDMA	0.94	0.52	0.16	-0.85	-0.97	1.94	0.83	0.71	0.48
DMDMA	0.33	0.23	0.12	-0.17	-0.21	0.54	0.36	0.34	0.32
NPGDMA	0.76	0.57	0.09	-1.00	-1.08	2.16	1.13	1.07	1.00
TMPTMA	0.65	0.44	-0.05	-0.30	-0.53	-	-	-	-
EBPADMA	0.44	0.38	0.25	-0.18	-0.31	-	-	-	-
BISGMA	1.63	2.98	3.19	3.36	3.40	0.14	-	-	-

DISCUSSION

The preliminary work in this study (9) indicated that over the spectrum of polymers evaluated, the polymerization shrinkage was proportional to the molecular weight of the monomer in all cases except for neopentyl glycol dimethacrylate which exhibited much lower shrinkage than would be predicted for a monomer of its molecular weight. This variance from the rest of the series was attributed to the highly branched nature of the NPGDMA. In this study the data indicates that the mechansim for reduced shrinkage of NPGDMA is related to its utilization in crosslinking of the matrix. The extremely low, 15.6%, value obtained from the calculations in Table II was confirmed by ATR infrared examination which indicated a value of 17.5%. Based on this, it appears that the shrinkage in the highly crosslinked network is dependent on both the molecular weight and the efficiency of the utilization of the double bonds in a cross-linking mechanism.

The data in the present study confirms the observations of Loshaek and Fox that the monomers based on the long, straight chain diols, such as tetraethylene glycol dimethacrylate and 1,10-decamethylenediol dimethacrylate, are more efficient crosslinking agents than the short chain ethylene glycol dimethacrylate or highly branched neopentyl glycol dimethacrylate monomers in highly crosslinked networks. Studies of the effect of controlled crosslinking on PMMA indicate that up to 20 weight percent ethylene glycol dimethacrylate, no residual unsaturation can be detected in the resulting polymer via infrared (13). It appears, therefore, that the matrix phase of the composite filling materials and sealants can be optimized for physical properties and utilization of crosslinking agent by selecting the proper monofunctional monomer, such as benzyl methacrylate, and crosslinking the matrix utilizing a highly efficient, long chain dimethacrylate. Studies on this type of system are now underway in our laboratory.

The utilization of crosslinking agent may well also explain the observations of weight loss on long-term water exposure for composite filling materials (8). It has been cited in the literature that the sorption and desorption of water by these systems appear to be a diffusion controlled reaction (14-15). If this is the case, then the migration of unreacted monomer through the system should be dependent on the monomer concentration and molecular size. Our observations of water uptake and solubility presented in Table III indicate that this weight loss may well be due to the leaching of unreacted monomer from the resin matrix. Examination of fully extracted and dried specimens via ATR infrared could detect no residual unsaturation in the samples based on polymethyl methacrylate. For the matrix based on diethylene glycol dimethacrylate with a solubility of 6.10%, the reduction in residual unsaturation indicated a loss of 7.25% monomer, and for the neopentyl glycol dimethacrylate matrix with a weight loss of 2.16%, the infrared analysis indicated a loss of 2.45%. The loss of unreacted monomer on long-term water exposure would also explain the gradual increase in physical properties reported by some observers (16) after long-term water absorption. The residual monomer resin in the matrix would act as a plasticizer and as it diffused through the matrix and was removed into the aqueous phase, the deplasticization of the matrix would exhibit itself by increased compression strength and loss of other mechanical properties

REFERENCES

1. J. Galan, J. Mondelli and J. L. Coradazzi, J. Dent. Res., 55(1), 74 (1975)
2. R. L. Bowen, J. Amer. Dent. Assoc., 74, 439 (1967)
3. E. Asmussen, Acta Odont. Scand., 33, 337 (1975)
4. E. A. Peterson, R. W. Phillips and M. L. Schwartz, J. Amer. Dent. Assoc., 73, 1324 (1966)
5. N. W. Rupp, J. Dent. Res., 58(5) 1551
6. G. Dickson, J. Dent. Res., 58(5), 1535 (1979)
7. E. Asmussen, Acta Odont. Scand., 33, 129 (1975)
8. G. M. Brauer, J. Dent. Res., 57(4), 597 (1978)
9. G. F. Cowperthwaite, M. A. Malloy and J. J. Foy, Int. Assoc. Dent. Res., 1979 Meeting Paper 598
10. ASTM Method D542-50
11. ASTM Method D792-66
12. S. Loshaek and T. G. Fox, JACS, 75 3544 (1953)

13. Z. S. Belokon, A. Ye. Skorobogatova, N. Ya. Gribkova, S. A. Arzhakov, N. F. Bakeyev, P. V. Kozolov and V. A. Kabanov, Vysokomol. Soyed., A18, 1772 (1976)
14. M. Braden, E. E. Causton and R. L. Clarke, J. Dent. Res., 55(5), 730 (1976)
15. D. N. Misra and R. L. Bowen, J. Dent. Res., 56(6), 603 (1977)
16. W. J. O'Brien, Presentation at Conference on Current Status of Composite Restorative Materials and Sealants, Univ. of Michigan, October, 1979

THE DENTAL PLASTICS IN THE FUTURE OF FIXED PROSTHODONTICS

John A. Cornell

INTRODUCTION

The ability to chew has been relished by mankind, and many attempts have been made over the millenia to retain this ability. A gold plate was said to have been fashioned by Petronius in 1565 as a prosthesis for cleft palate. Carved wood, bone, and many other materials have been used in dental prostheses. The oral environment, however, is very stringent. So, it was only the discovery of the rubber vulcanization process by Goodyear in 1839 that led to the ability to form a denture base which is the background of modern denture technology. The development of the dough molding technique in the 1930's and 40's brought the methacrylates to the forefront, which they have not yet relinquished. As with most of the plastic compositions, the ability to determine the testing of properties required for the complex oral environment has aided in the development of new products. The American Dental Association specifications have been designed to identify satisfactory products, but are not always adequate in identifying properties required in new compositions.

Although there has been much improvement in dental prosthetic materials since George Washington's day, many patients and dentists feel present materials can be improved on in various ways. The major improvements possible appear to involve a systems approach to the overall dental delivery system. The physical forces in an ill-fitting or ill-designed appliance quickly exceed even the strongest composite which might be contemplated. Thus, it is equally important to develop the preparation of an appliance so that the materials, the processing, and the use of the appliance have a high prognosis of success even under extreme conditions of use. Where such a tremendous amount of professional time is required to supervise the fitting of an appliance, any saving of time or increased assurance of a successful case should be well worth any additional cost in materials. In point of fact, the economics of dental delivery are placing continuing limits on costs of dental materials for a large number of economic, governmental, and insurance reasons.

PROPERTIES REQUIRED

Table I indicates the broad range and specific requirements deemed desirable for improved or ideal denture base materials. It is obvious that not only are the requirements very limiting to materials of preference, but that specific polymeric materials would often have to be modified to adapt to dental use. Thus, the development might be expected to be at least as difficult as that devoted to any other end use market in present day polymer development.

Table I

DESIRABLE QUALITIES OF IDEAL DENTURE BASE MATERIALS

(1) Physiologic compatability
- Nontoxic
- Noncarcinogenic
- Nonallergenic
- Compatible with physiologic requirements of mucous membranes
- Optimum consistency to maintain or promote tissue health
- Not deleterious to adjacent and underlying tissues
- Conducive to normal salivary flow

(2) Acceptability to patients' senses
- Acceptable to all five senses - sight, sound, smell, taste, and tough
- Able to duplicate and simulate oral tissues as nearly as possible
- Esthetic in color
- Possessing wide selection of color
- Possible for esthetics to be easily modified
- Color stable
- Odorless
- Tasteless
- Possessing instantaneous temperature conductivity
- Light-weight
- Possessing sensation of natural texture

(3) Functional usefulness
- Rigid enough so that teeth penetrate the bolus
- No interference with oral functions of chewing, swallowing, self-cleansing, singing, speech, sneezing, breathing, laughing, coughing, etc.

(4) Hygienic factors
- Sterilizable
- Resistant to stain, calculus, and adherent substances

TABLE I (Continued)

DESIRABLE QUALITIES OF IDEAL DENTURE BASE MATERIALS

- Nonporous to microorganisms
- Nonporous
- Low fluid absorption
- Wettable (low surface tension)
- Easily cleaned

(5) Durability
- Not affected by oral environment -- bacteria, food, medicines, etc.
- Unbreakable (not brittle)
- Not crazing
- Form stability
- Dimensionally stable and statically stable
- Minimal internal strain
- Good bond between different base materials
- Good bond between base and teeth
- Not flammable
- Resistant to weak acids and alkalies
- Resistant to abrasion and wear
- Resistant to strain
- Long lasting

(6) Adaptability to clinical problems
- Adjustable
- Easily polished
- Able to polish junction line
- Easily repaired
- Easily relined
- May need more than one type of material
- May use combinations of materials (soft for tissues, hard for teeth)
- Choice of hardness or softness (various materials for different situations)

(7) Cost factors
- Simple to manipulate
- Simple to process
- Inexpensive equipment for processing
- Average skill required for processing
- No separation medium required
- Easily separated from cast
- Inexpensive
- Moderate cost of fabrication
- Good shelf life
- Predictable properties

(Reference 4).

Table II is essentially a shopping list which can best be implemented by studying the vast field of macromolecular developments, and adapting these concepts to dental materials. Unfortunately few developments can be incorporated unchanged for all the requirements suggested in Table I.

Table II

IDEAS FOR FUTURE RESEARCH IN DENTAL BASE MATERIALS

(1) Chameleon-like
(2) Odor neutralizing (deodorant)
(3) Bacteriostatic
(4) Fungistatic
(5) Self-cleansing
(6) Providing sensation of individual teeth, e.g. anterior teeth not affecting posterior teeth
(7) Able to facilitate patient's ability to perceive textural and thermal qualities
(8) Allowing mucous secretions to pass through
(9) Able to carry therapeutic agents
(10) Adhesive to mucosa
(11) Possessing variable consistency under varying mouth conditions
(12) Selectively resilient compatible with resiliency of the tissues
(13) Resilient with quick recovery, able to recover shape quickly after deforming forces are removed
(14) Compressible on tissue side, but rigid on occlusal side
(15) Shock absorbing
(16) Control or reduce forces which are transmitted through the base to the underlying tissue
(17) Possessing flexibility which can be controlled and varied in processing as desired
(18) Possessing sufficient mass or substance to be easily handled by the patient
(19) Continuously adaptive to tissue changes to maintain close fit
(20) Similar to so-called treatment materials, but long-lasting
(21) Resistant to breakage due to repeated bending forces
(22) Insoluble in oral environment
(23) A method, perhaps photographic, to reproduce the natural color of the patient's gingivae
(24) A timed indicator to indicate to the patient when follow-up consultation is due
(25) Molded directly to tissues of mouth (impressed material becomes the base)
(26) Eliminate the base entirely
(27) Teeth implanted

(Reference 4).

THE NEEDS OF PROSTHODONTIC MATERIAL DEVELOPMENT

The needs which must be considered in any developmental program are quite complex, and have limited the application to prosthodontics of many potential polymeric materials developed for other industries. These requirements include:

(1) Having the capability of being manipulated in the dentist's chair or in the laboratory to produce an appearance similar to oral components;
(2) Having the ability to resist stress in use and yet being capable of being formed;
(3) Requiring the minimum time to manipulate and yet retaining properties within a wide range of time, temperature, environmental conditions, storage conditions, etc.;
(4) The ability to be robust, i.e. to have a low failure rate with minimal final quality control testing;
(5) Being non-toxic at all phases of manipulation and use. Polymers have inherent advantages in being high molecular weight and generally non-ionizable, and hence, lower solubility;
(6) Being capable of functioning for long times in the oral environment when subjected to masticatory and shock forces, temperature, and ultra-violet light, as well as chemical and biological attack from normal oral environment and food solvents such as alcohol, oils, stains, odor-forming compounds, etc.

THE PRESENT LIMITS OF MATERIALS

The theoretical limits of polymers involve the strength of the carbon-carbon covalent bond. Within this limitation, the use of crystalline and amorphous regions improve most physical properties, but the abrasion resistance has been a major limit to the use of plastics in the dental environment. The use of composites has markedly improved modulus problems but has complicated cohesion of the composite and abrasive resistance. Molding and forming properties are also more difficult.

Examples of developments which have been successfully marketed have had the characteristics of:

(1) Studying developments in industrial polymer development such as those of impact polymers -- polystyrene, poly-vinyl chloride and polymethyl methacrylate;
(2) The evaluation of why these materials are not adequate for the dental environment;
(3) The modification of the materials for use in the dental delivery system and the oral environment. (Reference 16)

Other successful developments have been initiated by needs within the dental market such as the demand for a long-working dough for packing a number of denture cases with one mix. Studies of various compositions were made and the possible disadvantages were studied and minimized. The avoidance of premature marketing without adequate clinical evaluation cannot be overstressed. (References 15 and 22)

It is probable that the majority of failures in the use of polymers in prosthetics are due to either design of the prosthesis imparting too much stress; or imperfect manipulation such as incomplete cure, or moisture contamination. Adhesion to tooth structure is also still a major problem, although much improved with acid etch of enamel, and by newer products. Approaching theoretical limits will require prevention of crazing (the effect of solvent on tensile stressed polymers) and the prevention of flow under compression, as well as minimizing residual low molecular weight contaminants.

PROBLEMS IN MATERIAL DEVELOPMENT

There are a number of difficulties in material development. These include:

(1) The difficulty of the oral environment and utilization requirements:
   a. Individual molding without production control;
   b. Extreme moisture and "solvent" biological and u.v. attack conditions with high appearance and design requirements

(2) High per-unit development costs -- small volume. Even the toothbrush is small volume for the mono-filament manufacturer, and the entire dental materials industry is claimed to be smaller than the fortieth largest pharmaceutical company. For example, Table III gives an estimate of total volume.

(3) The Food and Drug Administration requirements for new compositions safety usually require clinical testing. Two clinical studies cost on the order of $300,000 to $500,000, for several years, often more than the total profit on prosthodontic materials and even larger than total sales for at least a ten-year period!

(4) In the United States, industrial research, as differentiated from short-range development, is declining in general, and in the dental industry specifically. Possible reasons are the difficulty of getting and defending broad patents, and the high-risk of not being profitable.

(5) Alternate funding by non-profit and government agencies, such as the American Dental Association, National Institute of Health, or American Dental Trade Association have been somewhat ineffective for long-range dental delivery systems product development. This is at least partially caused by the dental industry, since there is no mechanism for cooperative basic research.

FUTURE OF POLYMERIC MATERIALS

In spite of the difficulties inherent in and compounding prosthodontic materials development, there is an excellent chance that marked changes can be expected in the overall dental delivery system by concepts already in the "pipeline" of materials development.

Table III

ESTIMATE OF U.S. CONSUMPTION OF PLASTICS IN DENTISTRY IN 1978

Appliance	Units made per year (million)	Weight of resin used per unit, g	Total resin used, kg
complete denture	8.0	25	200,000
partial denture	5.8	6	35,000
acrylic teeth	150	0.5	75,000
denture relined or rebased	3	4	12,000
		Total	322,000

These include:

(1) New types of organic-inorganic composites;

(2) Over denture techniques improved by better materials;

(3) Quality control effected by preformed partials (probably requiring redesign of prostheses);

(4) In the long range, basic changes in the dental delivery system, such as early caries detection, and preserving teeth through sealing and coating, bone and periodontal rehabilitation by chemical and physical treatment and caries and periodontal disease prevention stabilizing mouth rehabilitation work. All of these may drastically change the dental delivery system before the year 2000.

REFERENCES

1. American Dental Association, Guide to Dental Materials and Devices, 7th ed., 1974-75, Chicago, Ill., 1974
2. Anon, U. S. Dept. H. E. W., Pub. N.I.H., 77-1198 (1976)
3. Anon, Proofs, (April 13-16, 1977)
4. D. A. Atwood, J. Prosth. Dent. 20, 101 (1968)
5. L. J. Boucher, et al., J. Prosth. Dent., 19, 581 (1968)
6. G. M. Brauer, Am. Dent. Assn. J., 72, 1151 (1966)
7. G. M. Brauer, et al., Am. Dent. Assn. J., 59 270 (1959)
8. A. Breustedt, and A. Tappe, Deut. Stomatol., 20, 295 (1970)
9. H. H. Chandler, R. L. Bowen, and G. C. Paffenbarger, J. Biomed. Mater. Res., 5, 245 (1971)
10. H. H. Chandler, R. L. Bowen, and G. C. Paffenbarger, J. Biomed. Mater. Res., 5, 253 (1971)
11. H. H. Chandler, R. L. Bowen, and G. C. Paffenbarger, J. Biomed. Mater. Res., 5, 335 (1971)
12. H. H. Chandler, R. L. Bowen, and G. C. Paffenbarger, J. Biomed. Mater. Res., 5, 359 (1971)
13. E. C. Combe, and coworkers, Brit. Dent. J., 134, 289 (1973)
14. E. C. Combe, Dent. Pract. Dent. Rec., 22, 51 (1971)
15. J. A. Cornell, U.S. Patent Number 2,947,716

16. J. A. Cornell, U.S. Patent Number 3,427,274
17. R. G. Craig, and F. A. Peyton, eds., Restorative Dental Materials, 5th ed., C. V. Mosby Co., St. Louis, Mo.,(1975)
18. R. G. Craig, "The Effect of Materials, Design and Changes in Dentures on Bone and Supporting Tissues", p. 205-36 (In B. R. Lang, and C. C. Kelsey, eds., International Prosthodontics, Ann Arbor, Univ. Mich., 1973)
19. R. P. Elzay, G. O. Pearson, and E. F. Irish, J. Prosth. Dent., 25, 251 (1971)
20. H. M. Fullmer, J. Dent. Res., 55, D6-D9 (1976)
21. D. Garlen, Cosmetics and Toiletries, 92, 65 (1977)
22. D. J. Gowman, J. Cornell, and C. M. Powers, J. Am. Dent. A., 70, 1200 (1965)
23. B. D. Halpern, and W. Karo, "Dental Applications", Supp. Vol. II, Encyclopedia of Polymer Science and Technology (1977), Wiley
24. R. G. Jagger, and R. Huggett, J. Dent., 3, 15 (1975)
25. R. P. Kusy, and D. T. Turner, J. Dent. Res., 53, 948 (1974)
26. W. B. Love, J. Pros. Dent. 261 (1976)
27. J. F. McCabe, and M. J. Wilson, J. Oral Rehab., 1, 335 (1974)
28. J. F. McCabe, J. Oral Rehab., 2, 199 (1975)
29. G. C. Paffenbarger, J. B. Woelfel, and W. T. Sweeny, Dent. Clin. N. Am., 251 (1965)
30. F. A. Peyton, and D. H. Anthony, J. Prosth. Dent., 13, 269 (1963)
31. F. A. Peyton, Dent. Clin. N. Am., 19(2), 211 (1975)
32. Rustam, Roy, Chapters 7, 13, 14, Guide to Dental Materials and Devices, Eighth Edition (1977), Am. Dent. Assn., Chem. Eng. N., p. 4 (1/10/77)
33. A. Schmidt, p. 77, in K. Eichner, ed., Zahnarztliche Werkstoffe und ihre Verarbeitung, Third Edition, Heidelberg, Alfred Hutteg, (1974)
34. B. R. Lang, and C. C. Kelsey, eds., p. 237, International Prosthodontic Workshop on Complete Denture Occlusion, Ann Arbor, Univ. Mich.,(1973)
35. G. D. Stafford, and W. T. MacCullough, Brit. Dent. J., 131, 22 (1971)
36. D. F. Williams, J. Dent., 3(2), 51 (1975)
37. S. Winkler, ed., Dent. Clin. N. Amer., 19(2), (1975)
38. J. B. Woelfel, G. C. Paffenbarger, and W. T. Sweeny, Am. Dent. Assn. J., 61, 413 (1960)
39. J. B. Woelfel, G. C. Paffenbarger, and W. T. Sweeny, Am. Dent. Assn. J., 62, 643 (1961)
40. J. B. Woelfel, G. C. Paffenbarger, and W. T. Sweeny, Am. Dent. Assn. J., 67, 489 (1963)

INITIATOR-ACCELERATOR SYSTEMS FOR ACRYLIC RESINS AND COMPOSITES

G. M. Brauer

Dental and Medical Materials
Polymer Science and Standards Division
National Bureau of Standards
Washington, DC 20234

Acrylic resins, because of their desirable esthetics, ease of processing, optical clarity that can duplicate in appearance the oral tissues it replaces, satisfactory mechanical properties and excellent biocompatibility, are the materials of choice wherever plastics have found applications in dental practice. The ready acceptability of these materials is the result of the ease with which they can be converted into their final state even under clinical conditions. In practically all dental applications a liquid monomer-solid mixture is cured by a free radical initiated polymerization that is generated by heat, light, an initiator, or a redox initiator-accelerator system adapted to the constraints imposed by the oral environment.

HEAT CURED RESINS

Heat-cured denture base materials were introduced to dentistry in 1937. The liquid portion of the monomer-polymer slurry consists of methyl methacrylate monomer containing 0.006% or less inhibitor, plasticizer and crosslinking agent. The solid contains a modified suspension-polymerized methyl methacrylate and 0.5% to 1% of initiator, usually benzoyl peroxide. This compound has a 10 hour half-life temperature of 72°C (1). Since the polymerization rate is increased with increase in temperature and is directly proportional to the square root of the initiator concentration, the customary denture base curing cycles of 90 min. at 65°C followed by heating to 100°C for 60 min., or heating to 74°C for 9 hrs. will release sufficient free radicals to yield a denture that shows minimum porosity and is fully cured in both the thick portions as well as the thin palate areas. Other peroxides employed for the polymerization of acrylic resins such as diacetyl-di(2,4-dichlorobenzoyl)-or dilauroyl peroxide are less thermally stable than benzoyl peroxide. The first two compounds are sold as a 50% paste dispersion in a phthalate or silicone fluid to improve their storage stability and safety on handling. However, proper precautions should be taken when employing any peroxide. Even benzoyl peroxide dust may explode when subjected to friction or in the presence of contaminants.

Azo compounds such as 2,2'-azo-bis-isobutyronitrile may also be used. These initiators have a high initiating efficiency and their decomposition rates are relatively independent of solvent and concentration.

CHEMICALLY ACTIVATED RESTORATIVE MATERIALS

The thermal decomposition rate of the peroxide or azo initiators developed for curing acrylic resins at elevated temperatures is much too slow to cure acrylic monomers at ambient or mouth temperatures. To cure acrylic composites at room temperature the initiator must be activated to generate enough free radicals so that the polymerization can proceed at a rate that is satisfactory for clinical requirements. An ideal initiator-accelerator system should 1) generate sufficient concentration of free radicals with a minimum of radical wastage, 2) produce enough free radicals so that an adequate working time for the monomer-polymer mix and a required curing time for the specific application is obtained, 3) produce a concentration of free radicals to yield a polymer of a molecular weight distribution which will result in optimum physical properties of the dental restorative, 4) form no undesirable by-products such as esthetically unpleasant or color unstable materials, 5) be tasteless, colorless, non-toxic, non-irritating and completely biocompatible, 6) use constituents that are storage stable for extended time periods under the environmental conditions which may be encountered in transit, in storage, in dental depots or in the dental office, especially in tropical climates, 7) incorporate ingredients that are readily synthesized at a reasonable cost, 8) employ constituents that blend with, dissolve in, or are fully compatible with all components of powders, liquids, or pastes of chemically activated resins or composites.

PEROXIDE-AMINE REDOX SYSTEMS

A great number of redox initiator-accelerator systems have been suggested for dental use. Practically all employ benzoyl peroxide as initiator. Tertiary aromatic amines are most commonly suggested as accelerators. These systems were developed by Schwebel in Germany and patented after World War II (2-4). Two mechanisms for the primary reaction of the aromatic amine with a diacyl peroxide have been proposed: one involving a change-transfer interaction (Fig. 1) (5-8) and the other involving the formation of a quaternary ammonium salt intermediate (Fig. 2) (9,10). Horner concluded that both mechanisms were operating simultaneously (11). Argentar (12) has made a detailed analysis of the reactivity data reported by various investigators for the amine-peroxide system. His conclusions favoring the charge-transfer mechanism are based on two considerations: 1) the correlation of the transition energies of charge-transfer complexes of aromatic amines with the electrophilic substituent σ^+ of Brown and Okamotu (13) of the ring substituents of the amines, and 2) correlation of the reactivity data of the amine-peroxide systems with the same σ^+ values (14). If the quaternary salt formation were involved in the reaction mechanism then σ^- (15) would be the appropriate substituent parameter. A systematic investigation of the effect of amine substituents on the initiation of the free radical polymerization of methyl methacrylate by aliphatic, aromatic and heterocyclic amines, has been made (16). Many amines cause a fast initial reaction, but the polymerization, which is also air-inhibited, does not go to completion. Furthermore, the polymerization rate is often different than one would expect from the relative rates of the amine-accelerated decomposition of benzoyl peroxide in non-polymerizable solvents. Only a small number of tertiary aromatic amines cause rapid polymerization, mainly those with electron donating substituents, particularly in the para position. An increase in the curing rate resulting from the exothermic reaction is also observed. The initiating activity of systems with tertiary aliphatic amines is lower, probably because of the primary combination of the predominant part of $R_1R_2NCHR_3$ radicals with the benzoyloxyl radicals.

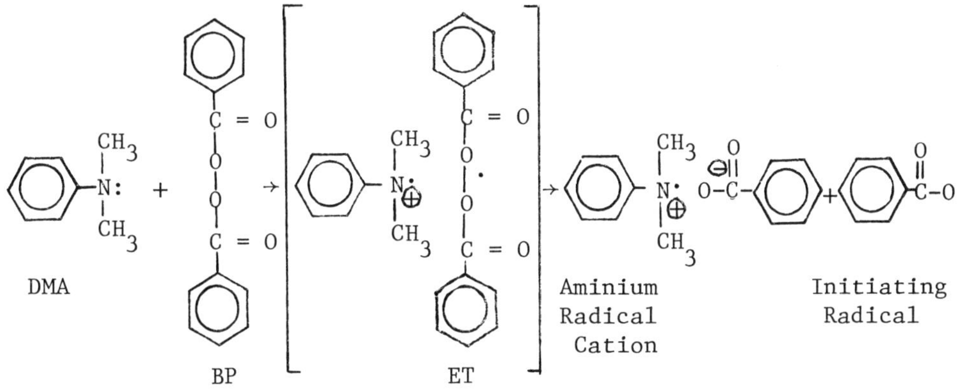
DMA
BP
ET
Aminium
Radical
Cation
Initiating
Radical

Figure 1. Simplified one-electron-transfer (ET) mechanism for the redox reaction of diacetyl peroxides with amine accelerators.

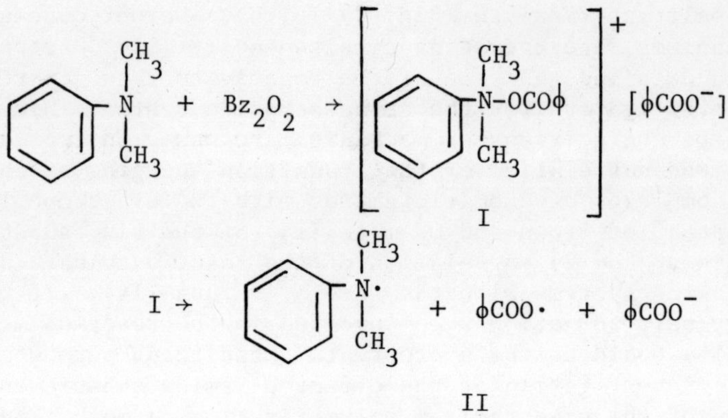
CH3
N
CH3
+ Bz2O2 →
CH3
N-OCOϕ
CH3
[ϕCOO⁻]
+
I
I →
CH3
N·
CH3
+ ϕCOO· + ϕCOO⁻
II

Figure 2. SN_2 mechanism via quaternary ammonium salt intermediate for the redox reaction of diacetyl peroxides with amine accelerators.

Polymers cured in the presence of aromatic amines give products ranging from yellow for dimethyl-p-toluidine to black for N,N-dimethyl-p-phenylenediamine. Substituents that produce a deepening of the shade (bathochromic groups) often increase the rate of polymerization. Rose, Lal and Green (17,18) using temperature rise and peak temperature data, studied the effect of inhibitor, peroxide, initiator, and amine accelerator on the rate of polymerization of poly(methylmethacrylate) slurries. Aromatic peroxides, particularly p-chlorobenzoyl peroxide, appeared most efficient. However, beyond the heat peak the benzoyl peroxide (BP)-dimethyl-p-toluidine (DMPT) system cures faster (19). The smaller the particle size of the powder the more rapid the curing of the resin.

Many of the commercial, chemical activated filling materials and denture resins developed since 1950 employ the BP-DMPT or BP-bis(2-hydroxyl)-p-toluidine (DHEPT) system. The objective of more recent studies has been to develop more reactive color-stable amines with improved biocompatibility and to optimize concentrations of reactants. Amines with increased ring substituents, particularly in the 3,5 or 4 positions, such as N,N-dimethyl-sym-xylidine, decrease the curing time and improve the color stability (20). Accelerating ability is a function of both ring and nitrogen substitution (21). Color stability is a result of secondary condensation reactions between primary amine products of the interaction of BP and tertiary amine; they obviously proceed with participation of the hydrogen•atoms of the benzene ring of the amine (22). Composites containing aromatic amines having a 3,5-dimethylphenyl ring discolor less than those with a 4-methylphenyl ring and much less than those with an unsubstituted phenyl ring (21). N,N-dimethyl-p-tert-butylaniline (21) or N,N-bis(hydroxyalkyl)3,5-di-tert butylanilines (22a) yield hardened composites that have a very light shade and excellent color stability. These compounds have been suggested as accelerators for autopolymerizing pour denture base and reline and repair resins where lighter shades are required to match the tissues (23). Tertiary aromatic amines with large substituents on the nitrogen atom and molecular weight above 400 can be effective accelerators (24) yielding composites with rapid curing times. One recently synthesized high molecular weight tertiary amine is made by the direct addition reaction of the sodium salt of N-phenylglycine, N-methyl-*p*-toluidine and an epoxidized *o*-cresol-formaldehyde novolac (25). It acts as a polymerization accelerator and chelator for calcium and can be incorporated into composites. Such high molecular weight accelerators, because of their low volatility and reduced diffusion rate should be less able to penetrate body tissues and thus be less toxic than the amines presently used.

Tertiary amines with N-methacryloxyethyl groups have been synthesized (26). These amines copolymerize with acrylic monomers. They are incorporated into the resin which should have lower toxicity than polymers cured with non-polymerizable accelerators. Color stability of these resins is similar to those cured with DMPT.

We have synthesized amines which from structure-property considerations, i.e., free energy relationships, are predicted to be highly reactive and compared them with commonly used accelerators for composite restorative materials (27-29). The relationship previously found (12) for co-relating the reactivity of tertiary amines in their reaction with BP as measured by the rate of polymerization of vinyl monomers indicated that amines with aryl (ring) substituents having a σ^+ value (13) close to -0.20 would be the most reactive. The *p*-$CH_2COOC_2H_5$ group has a σ^+ value of -0.16 thus suggesting the potential usefulness of phenylacetic acid or ester derivatives, some of which can be readily synthesized. These amino acids were also selected since they were expected to have promising toxicological characteristics (30-32).

A powder-liquid formulation containing 70% bis(3-methacryloxy-2-hydroxypropyl) bisphenol A (BIS-GMA), 30% triethylene glycol dimethacrylate (TEGDMA), 0.2% butylated hydroxytoluene (BHT) and various amine concentrations in the liquid and a BP-coated silanized spherical silica and BaF_2 containing glass as powder was employed. The overall characteristics of the composites (hardening time, strength and color stability) containing N,N-dimethylaminophenylacetic acid (DMAPAA), its methyl ester (MDMAPAA) or N,N-dimethylaminoglutethimide (DMAG) compare favorably to restorative resins cured with commonly used tertiary amines. Based on hardening times the approximate order of the accelerating ability of the respective amines is: DMAPAA > N,N-dimethyl-*sym*-xylidine > DMPT $\cong$ DMAG < MDMAPAA >> DHEPT. This order of reactivity is dependent on the components (especially monomers) used in the formulations. For the composite studied the maximum tensile and compressive strength of the cured material is obtained over a narrow concentration range of accelerator in the liquid. This range, which is dependent on the type of diluent employed, is approximately the same for all amines.

The biocompatibility of these amines is good. No mutagenic or cytotoxic effects have been observed for pure DMAPAA or DMAG using the Ames test for bacterial mutagenicity and the agar overlay test for cytotoxicity (33).

Our recent objective has been to synthesize homologues and derivatives of DMAPAA(I) and of N,N-dialkylaminophenethanol (II) and to correlate structure and properties of the resulting acids, esters and alcohols. We have prepared the following series of new amines:

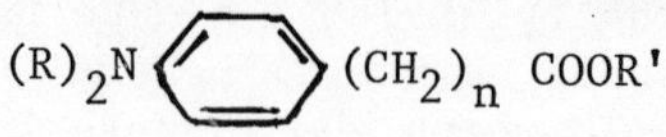
(R)2N
(CH2)n COOR'
(I)

and

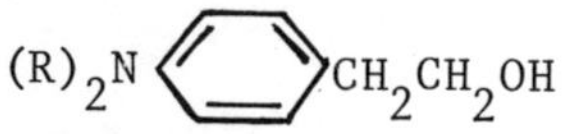
(R)2N
CH2CH2OH
(II)

where $R = CH_3, C_2H_5$
$R' = H, CH_3, C_2H_5$
$n = 1,2$

Compounds were usually synthesized from the primary amines and their structure and purity (99% +) was determined by IR, NMR and GC (Table 1). Composites using a 3 to 1 powder-liquid ratio were prepared from a formulation similar to the one described above. Corning 7724 glass as powder and a resin formulation of 72.4% BIS-GMA, 27.6%, 1,6-hexamethylene glycol dimethacrylate (1,6-HGDMA) was used. The latter monomer replaced the customarily used TEGDMA. Properties of the resulting composites are given in Table 2. All the amines in the concentrations studied gave short setting times (1.25-6 min.), but DEAPAA is the most reactive giving adequate setting times with amine concentrations as low as 3 molal in the liquid. For unfilled resins the rate of curing proceeds most rapidly using a molar peroxide to amine ratio between 1.1 to 1.5 (16,34). A composite prepared from powder coated with 1% BP and a liquid containing 17 mmolal amine has a molar peroxide to amine ratio of 6.5. Thus, for composites a much larger excess of peroxide is required to obtain a minimum setting time. This should be expected since only a small portion of the peroxide is accessible to the amine before the composite is cured. The mechanical properties with 1,6-HGDMA as diluent are not as concentration dependent as was observed with TEGDMA as diluent with optimum properties being obtained in the 15-19 molal amine concentration range in the liquid. Physical properties of the cured composites are excellent (tensile strength 40-55 MPa, compressive strength 241-303 MPa, water sorption 0.5-0.7 mg/cm^2). This tensile strength is considerably higher than the minimum requirement of 34 MPa given in the specification for dental

TABLE I

TERTIARY AROMATIC AMINES SYNTHESIZED

Amine	Abbreviation	Melting or Boiling Point, °C
1. Aminophenethanol		
N,N-dimethyl[a]	DMAPE	MP=52-53
N,N-diethyl[b]	DEAPE	BP=153-154/4 mm
2. Aminophenylacetic Acid		
N,N-dimethyl[c]	DMAPAA	MP=108-112
N,N-diethyl[d]	DEAPAA	BP=180-182/2 mm
3. Aminophenylacetic Acid		
N,N-dimethyl-methyl ester[e]	DMAPAA-ME	BP=107/1 mm
N,N-diethyl-ethyl ester[f]	DEAPAA-EE	BP=135-137/2 mm
4. Aminophenylpropionic Acid		
N,N-dimethyl[g]	DMAPPA	MP=103-106

Method of synthesis: primary amine and [a]CH_3I through quaternary ammonium salt, [b]C_2H_5I in aq. MeOH, [c]CH_3I, [d]acid hydrolysis of ethyl ester, [e]CH_3I in DMSO [f]C_2H_5Br in DMSO, [g]dimethylaminobenzaldehyde + malonic acid. The resulting cinnamic acid was reduced with zinc amalgam.

composite resins (35). If low concentrations are employed in the formulations the cured composites, especially with DEAPAA as accelerator, are nearly colorless. No preceptible change occurs in the color of the specimens containing a UV absorber after exposure for 24 hours to an RS lamp as described in the specification (35). Because of the excellent overall properties, nearly colorless appearance and the potentially improved biocompatibility, compositions using these accelerators should be most useful for formulating improved restoratives.

The storage stability of the components of composites is mainly limited by the poor shelf-life of the BP ingredient (36). Pastes containing BP harden readily when stored at elevated temperatures. Composites made from powder-liquid constituents are more stable. After extended storage, composite mixes show delayed setting and decreased mechanical properties in the cured material. The purity of the BP and the amine selected greatly influences parameters such as reaction rate, color stability and biocompatibility (37). The storage stability of acrylic monomers in the presence of acids is reduced. Thus the amines containing the ethanol or ester

TABLE II

PROPERTIES OF COMPOSITES WITH VARIOUS AMINE ACCELERATORS

Amine Type	mMolal Conc. in Liquid	Setting Time Min.	Tensile Strength MPa	Compressive Strength MPa	H_2O Sorption mg/cm^2
DMPT	15	4.5	45.9(4.2)[a]	276(22)[a]	.60(.05)[a]
"	17	3	43.0(2.6)	246(10)	.58(.00)
"	19	4.5	47.6(3.1)	246(17)	.56(.03)
DMAPE	15	3.5	46.1(0.5)	262(21)	.53(.04)
"	17	3	47.7(2.3)	271(4)	.55(.02)
"	19	3	50.1(2.3)	302(11)	.62(.02)
"[b]	19	2	54.2(1.2)	281(6)	–
"[b]	19	2	52.6(1.8)	263(10)	–
"	21	2.5	39.6(1.7)	286(4)	.67(.05)
DEAPE	15	6	45.6(3.3)	277(9)	.60(.08)
"	17	4	49.1(2.8)	294(9)	.60(.03)
"[b]	17	3	53.4(1.2)	302(13)	.54(.02)
"[h]	17	3	48.6(3.6)	265(8)	.65(.03)
"	19	4.5	48.0(2.7)	279(4)	.54(.01)
DMAPAA	13	2	49.8(3.2)	288(9)	.60(.07)
"	15	2	52.2(3.7)	276(18)	.56(.01)
"	17	2.5	51.6(2.6)	279(11)	.56(.02)
"	19	2	50.2(3.8)	296(7)	.61(.06)
DMAPPA	15	6	41.4(3.4)	241(13)	.55(.06)
"	17	4	43.1(5.1)	263(5)	.55(.01)
"	19	5	45.8(4.2)	258(20)	.58(.02)
DEAPAA	3	4.5	36 (2)		
"[c]	3	1.5	49 (3)		
"	5	2.5	44 (1)	245(7)	.56(.03)
"	7	2.5	45 (1)	275(3)	
"[g]	7	2.5	44 (5)	–	
"	9	2.0	46 (4)	275(10)	
"	11	2.0	46 (4)	284(6)	
"	13	2.0	49 (4)	272(6)	
"	15	1.5	52 (3)	268(9)	
"	17	1.5	54 (4)	303(9)	.58(.02)
"[d]	17	2	52 (2)	280(6)	
"[e]	17	2	50 (3)	294(6)	
"[g]	17	1.25-1.5	55 (2)	–	.57(.02)
"[f]	19	1.5	54 (2)	292(15)	
"	19	1.25-1.5	49 (3)	–	

TABLE II (CONTINUED)

PROPERTIES OF COMPOSITES WITH VARIOUS AMINE ACCELERATORS

Amine Type	mMolal Conc. in Liquid	Setting Time Min.	Tensile Strength MPa	Compressive Strength MPa	H_2O Sorption mg/cm^2
DMAPAA-ME	15	3.5	46.6(2.6)	267(7)	.58(.03)
"	17	3.5	49.8(3.2)	273(6)	.65(.09)
"	19	3	46.0(2.2)	269(10)	.59(.04)
DEAPAA-EE	11	4.5	43.7(5.5)	272(8)	-
"	13	4	51.2(2.8)	278(8)	.54(.02)
"	15	4	48.0(3.9)	276(7)	.54(.01)
"	17	4	46.5(2.9)	283(9)	.55(.01)
"	19	5	47.0(2.6)	276(9)	.59(.04)

[a]Values in parentheses give standard deviation

[b]Prepared with 0.1% BHT

[c]Prepared with no inhibitor

[d]Prepared with 0.3% BHT

[e]Prepared with 0.4% BHT

[f]Prepared with 0.25% BHT

[g]0.4% ultra-violet absorber added

[h]sample prepared with 13.6% 1,10-decamethylene glycol dimethacrylate as solvent.

groups should have better shelf-life than the substituted amino acids.

OTHER REDOX SYSTEMS

The reaction of peroxyesters, hydroperoxides or peroxides, which are more storage stable than BP, with tertiary amines is generally too slow to give a sufficiently rapid cure for acrylic resins. Composite mixes containing *t*-butyl perbenzoate, *t*-butyl hydroperoxides or dicumyl peroxide-tertiary amines do not harden for days. Many other redox systems have been suggested for vinyl polymerizations, only a few have been employed in dental resins. Substitution of *p*-toluene sulfinic acid or sulfinic acid derivatives for tertiary amines yields colorless products (38-42). Most of these compounds have poor shelf-life. They readily oxidize in air to sulfonic acids which do not activate polymerization. Lauroyl peroxide in conjunction with a metal mercaptide (such as the zinc hexadecyl mercaptide) and a trace of copper has been used to cure monomer-polymer slurries containing methacrylic acid (43,44). Acrylic monomers also polymerize rapidly in the presence of N-salt of saccharin with an N,N-dialkylarylamine (45).

Compositions based on polymerizable methacrylate monomers can be polymerized employing a hydroperoxide (t-butyl, cumyl or p-methane) and a thiourea reducing agent (46-49). A possible mechanism of the hydroperoxide-thiourea reaction is given in Fig. 3.

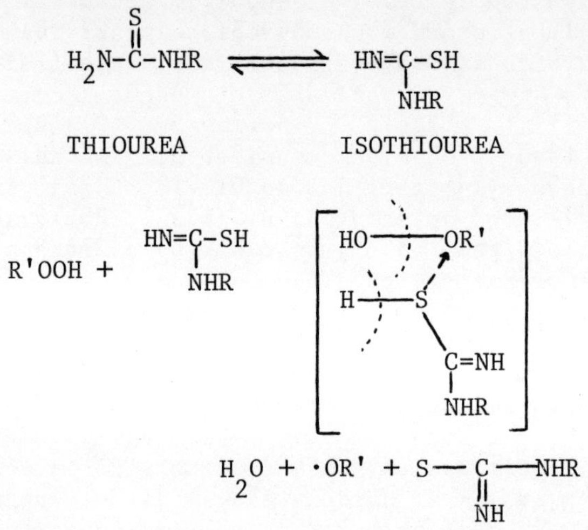

S
H_2N-C-NHR
HN=C-SH
NHR
THIOUREA
ISOTHIOUREA
R'OOH +
HN=C-SH
NHR
HO
OR'
H
S
C=NH
NHR
H_2O + ·OR' + S—C—NHR
NH

Figure 3. Suggested mechanism for the reaction of hydroperoxides with thiourea.

N-substituted compounds such as phenyl-, acetyl- or allyl-thiourea in 0.5 to 1% concentration give composites with excellent color stability. However, the biocompatibility of these compounds has not been established.

Colorless composites with good mechanical properties can be obtained with either t-butyl perbenzoate, cumene- or t-butyl hydroperoxide and ascorbic acid or ascorbyl palmitate systems (50). Mechanisms for the free radical formation are given in Fig. 4. Addition of trace amounts of transition metals in their higher oxidation state (Cu^{+2}, Fe^{+3}) to the perester component further speeds up the polymerization. On admixture with the ascorbic acid derivative the metal cation is reduced to its lower oxidation state which, because it is a potent one electron reductant and will rapidly activate the free radical decomposition of the perester, which it in turn is reoxidized to its higher oxidation state. Means for prevention of oxidation of ascorbic acid or its derivatives on prolonged storage must be developed for these formulations to be suitable for dental application.

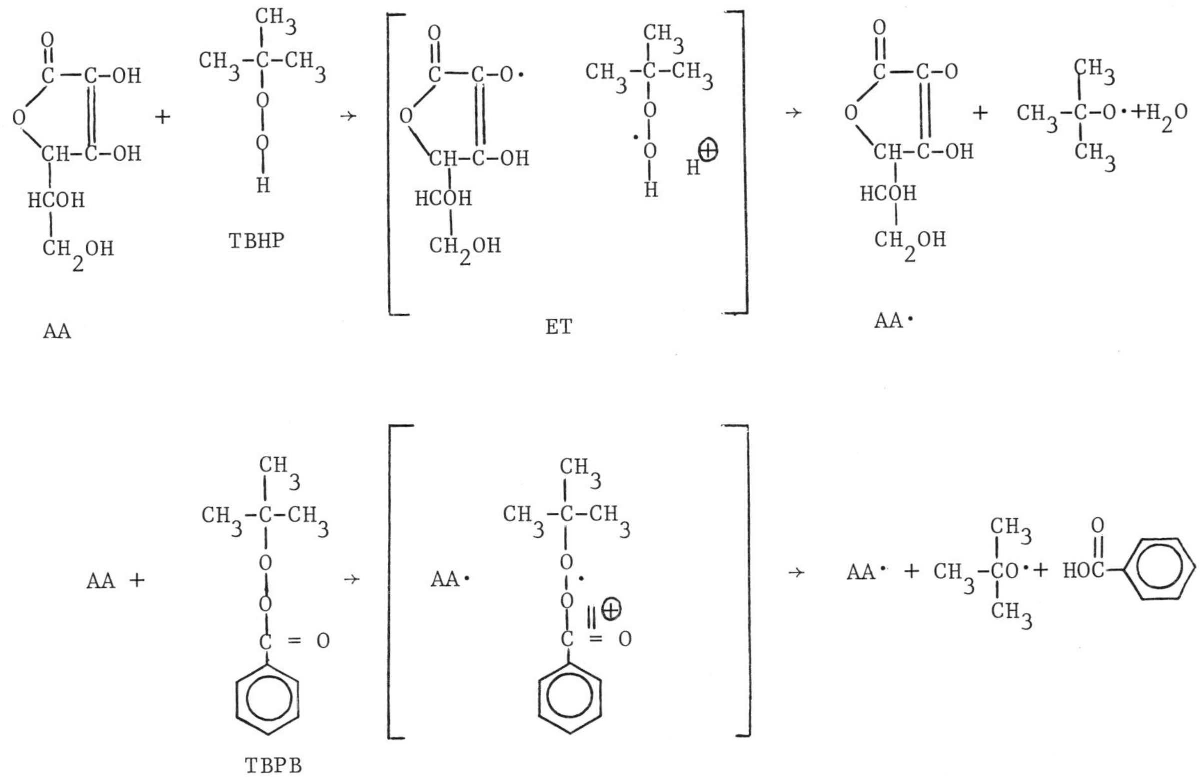
AA
TBHP
ET
AA·
AA +
TBPB
AA·
AA· +

Figure 4. Mechanism for the reaction of peresters and hydroperoxides with ascorbic acid.

Trialkylboranes and their oxidation products initiate polymerization of acrylic monomers as shown in Figure 5 (51-54). Complexes of tributyl borane with amines are more stable in air and can be easily reduced. With this initiator, polymer formation has been postulated to begin at the physiologically moist dentin surface. Grafting of the borane-cured resin to dentin may take place in the presence of moisture and bond strength is retained fairly well after water immersion. Bonding exists only between collagen of the dentin and any retention to enamel is of a mechanical nature.

PHOTOINITIATORS

For specific applications such as composite restoratives, pit and fissure sealants, cavity liners or restoration glazes acrylic formulations containing a photoinitiator can be cured by light (55-59). Usually radiation in the near ultraviolet region around 360 nm is used. The acrylic resin is composed of a dimethacrylate such as BIS-GMA and a diluent. The second component contains the photoinitiator, usually a benzoin alkyl ether or a diketone. BP (55) or phosphite esters (56) may be included to accelerate the hardening of the acrylic resin which should be completed in 30-60 sec. Addition of a photocrosslinking agent reduces curing time and lowers water solubility of UV-cured sealants (57). Irradiation with visible radiation in the 400-500 nm range of a mix comprising a urethane-methacrylate prepolymer, diluent, camphorquinone (initiator), dimethylaminoethyl methacrylate (reducing agent) and silanized glass powder yields a dental restorative with good physical properties (58).

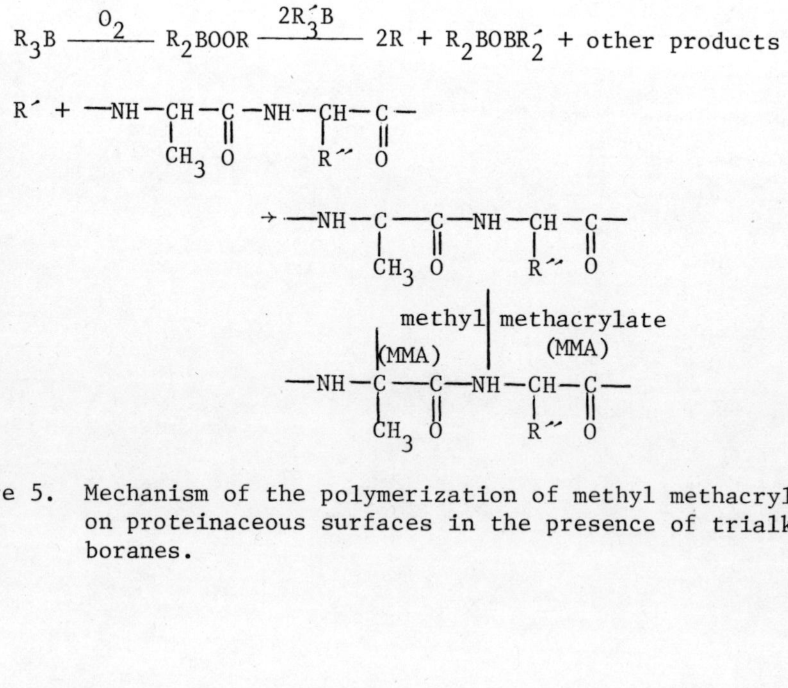
R3B
O2
R2BOOR
2R'3B
2R + R2BOBR'2 + other products
R' + —NH—CH—C—NH—CH—C—
CH3 O R'' O
→ —NH—C—C—NH—CH—C—
CH3 O R'' O
methyl methacrylate
(MMA) (MMA)
—NH—C—C—NH—CH—C—
CH3 O R'' O

Figure 5. Mechanism of the polymerization of methyl methacrylate on proteinaceous surfaces in the presence of trialkylboranes.

ACKNOWLEDGMENT

This work was supported by the National Institute of Dental Research, Interagency Agreement Y001-DE-4001.

SUMMARY

Acrylic resins commonly used in dentistry are cured by a free-radical initiated polymerization. The thermal decomposition of benzoyl peroxide (BP) yields these radicals for the heat cure of denture base materials. To speed up the cure at ambient or mouth temperature, redox initiator-accelerator systems, generally BP-tertiary aromatic amines, are employed. Increasing reactivity of the amine is a function of both ring and nitrogen substitution. Ring substitution, particularly in the 3,5 or 4 positions, minimizes the formation of undesirable colored products. Amines with large substituents on the nitrogen atom can also be effective accelerators. The objective of more recent studies has been to develop more storage stable, yet highly reactive, redox systems, to optimize concentration of reactants, and to employ amines with improved biocompatibility which yield color stable products. Newly synthesized accelerators $(R)_2N\text{-}C_6H_4\text{-}(CH_2)_n$ COOR' and $(R)_2N\text{-}C_6H_4\text{-}CH_2CH_2OH$, where R = CH_3 and C_2H_5; R' = H,CH_3 and C_2H_5 and n = 1 or 2 are very effective accelerators for composites, yielding restoratives with excellent mechanical properties and minimum discoloration. Other redox systems such as BP-sulfinic acids, peroxide-thiourea, hydro-peroxide-ascorbic acid or trialkylborane-oxygen also yield rapid polymerization of acrylic resins. Their instability on prolonged storage or questionable biocompatibility has limited clinical application to date. Coatings or thin layers of restoratives may be cured by ultraviolet or visible radiation using benzoin ethers or diketones as the photoinitiator.

REFERENCES

1. Technical Bulletin 30.30. Evaluation of Organic Peroxides from Half-Life Data, Lucidol Div. Pennwalt Corp., Buffalo, NY
2. Blumenthal. Fiat Report No. 1185, May 27, 1947
3. Deutsche Gold und Silberscheideanstalt vormals Roessler, French Patent 883, 619, March 29, 1943; 895,566, April 3, 1944; Swiss Patent 230,706, January 31, 1944; 237,089, March 31, 1945; Swedish Patent 126,006, July 7, 1949
4. F. E. Knock, and J. F. Glenn, (to L. D. Caulk Co.); U.S. Patent 2,558,139, June 26, 1951
5. L. Horner, and E. Schwenk, Ann., 566, 69 (1950)
6. L. Horner, and C. Betzel, Ann., 579, 175 (1953)
7. L. Horner, and H. Junkermann, Ann., 591, 53 (1955)
8. L. Horner, J. Polym. Sci., 18, 438 (1955)
9. M. Imoto, and S. Choe, J. Polymer Sci., 15, 485 (1955)
10. C. Walling, and N. Indictor, J. Am. Chem. Soc., 80, 5814 (1958)
11. L. Horner, and B. Anders, Chem. Ber., 95, 2470 (1962)

12. H. Argentar, unpublished data, 1976
13. H. C. Brown, and Y. Okamoto, J. Am. Chem. Soc., 80, 4979 (1958)
14. H. Argentar, J. Res. Nat'l. Bur. Stand., 80A, 173 (1976)
15. H. H. Jaffé, Chem. Rev., 53, 191 (1953)
16. G. M. Brauer, R. M. Davenport, and W. C. Hansen, Mod. Plastics, 34, 153 (1956)
17. E. E. Rose, J. Lal, and R. Green, JADA, 56, 375 (1958)
18. J. Lal, and R. Green, J. Polymer Sci., 17, 403 (1955)
19. J. A. Cornell, and C. M. Powers, J. Dent. Res., 38, 606 (1959)
20. R. L. Bowen, and H. Argentar, JADA, 75, 918 (1967)
21. R. L. Bowen, and H. Argentar, J. Dent. Res., 50, 923 (1971)
22. J. Lokaj, and F. Hrabak, Makromol. Chem., 119, 23 (1968)
22a. W. Schmitt, R. Purrmann, and P. Jochum, Ger. Offen. 2,658,538, June 30, 1977; Chem. Abstr., 87, 157,209 (1977)
23. H. Argentar, J. A. Tesk, and E. E. Parry, unpublished data
24. R. L. Bowen, and H. Argentar, J. Dent. Res., 51, 473 (1972)
25. J. M. Antonucci, and R. L. Bowen, J. Dent. Res., 56, 937 (1977)
26. J. Dnebosky, V. Hynkova, and F. Hrabak, J. Dent. Res., 54, 772 (1975); Czech Patent 169,255, May 15, 1977; Chem. Abstr. 88, 190,356 (1978)
27. G. M. Brauer, D. M. Dulik, J. Antonucci, and H. Argentar, Polymer Preprints, 19, (No. 2), 585 (1978)
28. D. M. Dulik, J. Dent. Res., 58, 1308 (1979)
29. G. M. Brauer, D. M. Dulik, J. M. Antonucci, D. J. Termini, and H. Argentar, J. Dent. Res. (in press)
30. The Merck Index, P. G. Stecker, Ed., 8th ed., Rahway, NJ, Merck and Co., Inc., 1968, p. 60.
31. R. W. J. Carney, and G. DeStevens, U.S. Patent 3,657,230, April 18, 1972
32. V. A. Lapkin, Russ. Pharmacol. Tox., 37, 256 (1974)
33. A. H. Jacobsen, and A. H. Pettersen, private communication
34. R. L. Bowen, and H. Argentar, J. Appl. Polymer Sci., 17, 2213 (1973)
35. American Dental Association Specification No. 27 for Direct Filling Resins, JADA, 94, 1191 (1977)
36. G. M. Brauer, N. Petrianyk, and D. J. Termini, J. Dent. Res., 58, 1791 (1979)
37. F. F. Koblitz, T. M. O'Shea, J. F. Glenn, and K. L. DeVries, J. Dent. Res., 56B, No. 791 (1977) (abstract)
38. O. Hagger, Helv. Chim. Acta, 31, 1624 (1948); 34, 1872 (1951)
39. P. Castan, and O. Hagger, U.S. Patent 2,567,803, September 11, 1951; Swiss Patent 255,978, February 1, 1949
40. H. Bredereck, and E. Bäder, Chem. Ber., 87, 129 (1954)
41. H. Brederick, and E. Bäder, German Patent 913,477, June 14, 1954 and German Patent 916,733, August 16, 1954; British Patent 771, 631, April 3, 1957; U.S. Patent 2,846,418, August 5, 1958
42. G. M. Brauer, and F. R. Burns, J. Polymer Sci., 19, 311 (1956)
43. H. J. Stern, L. E. Shadbolt, and W. L. Rawitzer, British Patent 721,641, January 12, 1955
44. L. E. Shadbolt, Ger. Offen., 1,937,871, January 29, 1970

45. J. Lal, U.S. Patent 2,833,753, May 6, 1958
46. S. Takaaki, and M. Yuji, J. Polymer Sci., A-1, 4, 2735 (1966)
47. Sumitomo Chemical Ltd., British Patent 1,177,879, January 14, 1970
48. S. C. Termin, and M. C. Richards, U.S. Patent 3,991,008, November 9, 1976
49. Colgate Palmolive, Neth. Appl., 7609,611, March 2, 1978
50. J. M. Antonucci, C. L. Grams, and D. J. Termini, J. Dent. Res., (in press)
51. K. H. Drause, Kunststoff-Rundschau, 6, 139 (1959)
52. E. Masuhara, Deut. Zahärztl. Z., 24, 620 (1969)
53. C. H. Fischer, M. Strassburg, and G. Knolle, Int. Dent. J., 20, 679 (1970)
54. N. Nakabayashi, E. Masuhara, E. Mochida, and I. Ohmori, J. Biomed. Mat. Res., 12, 149 (1978)
55. D. E. Waller, U.S. Patent 3,709,866, January 9, 1973; German Patent 2,126,419, December 16, 1971
56. W. Schmitt, and R. Purrmann, Ger. Offen., 2,646,416, May 26, 1977
57. G. M. Brauer, J. Dent. Res., 57, 597 (1978)
58. E. C. Dart, J. B. Cantwell, J. R. Traynor, J. N. Jaworzyn, M. E. B. Jones, and I. Thomas, British Patent 1,465,897, March 2, 1977
59. C. E. Dart, J. B. Cantwell, J. R. Traynor, and J. N. Jaworzyn, U.S. Patent 4,089,763, May 16, 1978

THE APPLICATION OF PHOTOCHEMISTRY TO DENTAL MATERIALS

Robert J. Kilian

Johnson & Johnson Dental Products Company
20 Lake Drive
East Windsor, NJ 08520

INTRODUCTION

Within the last 10 years, photochemistry has begun to be used in the field of dental materials for the photocuring of methacrylate monomers. The current applications are the photocuring of 1) composite restoratives for the repair of anterior teeth, 2) liquid bonding agents to help bond these restoratives to the cavity preparations, 3) pit and fissure sealants for the sealing of enamel imperfections on the occlusal surfaces of molars, and 4) orthodontic bracket adhesives for the direct bonding of orthodontic brackets to teeth.

Until recently, only long wave ultraviolet light curing systems (320 - 400 nm) were used. The first of these was a pit and fissure sealant which was reported in 1970 (1). This was followed by a composite restorative in 1973 (2). Shortly thereafter, this material was slightly modified for use as an orthodontic bracket adhesive. In 1978, a composite restorative which utilized a short wave visible light curing system (400 - 500 nm) was reported (3). Since then, a bonding agent has also become available and, in principle, the other applications are also possible.

The main advantage of photocuring over chemical curing for these applications is that the polymerization of the material is not initiated until it is illuminated with the appropriate light. This gives the dentist as much working time as he requires and thus it satisfies the conflicting requirements of long working time and yet short setting time that are difficult to obtain with the chemically cured systems.

The clinical success of both curing systems is dependent upon the clinical technique. Basically, in each system the clinician is being asked to be the polymer chemist and, with respect to technique, the clinician is more familiar with the mixing of two pastes as in the chemically cured systems than he is to the proper illumination of the clinically placed photocured material. Variables such as light intensity, distance of the light from the material, etc. are encountered in this situation. However, appropriate understanding of and education in clinical techniques will lead to proper results.

Three general types of curing lights are currently in use, two for uv and one for visible light. One type of uv source employs a medium or high pressure mercury arc lamp (4) which provides the well-known emission spectrum with principal peaks at 365, 405 and 435 nm (5). A second type of uv source provides a continuum of near-uv phosphor emission similar, in principle, to that obtained from fluorescent lights (6). The visible light sources employ quartz-halogen bulbs of the type that are commonly used in home movie projectors. These bulbs emit a continuum of ultraviolet, visible and infrared light. In all cases, filters are used to reduce or remove light of undesirable wavelengths such as short wave uv (which can be dangerous) and longer wave visible light (which causes glare but does not contribute to the photocuring of the materials).

DISCUSSION

The photochemically cured systems are similar in composition to those which are chemically cured. They employ the same, or similar, monomer systems, i.e. mono- and dimethacrylates; the filler (if used) are similar, i.e. various forms of glass; and they polymerize via free radical mechanisms. The final physical properties of all of these systems, when they are properly cured, are similar and are clinically acceptable for their intended applications.

The principal difference between the two systems is in the way in which the free radical polymerization of the methacrylate monomers is initiated. In one, it is accomplished photochemically (with either ultraviolet or visible light) and in the other it is accomplished chemically with a standard amine-peroxide system.

The main chemical difference between the ultraviolet and the visible light curing systems is the photoinitiator which is employed. The ultraviolet systems use aliphatic ethers of benzoin, with the methyl ether being the typical, although not the exclusive, choice. The visible systems use an alpha-diketone as the photoinitiator. A reducing agent, such as a tertiary aliphatic amine, is also employed as a polymerization accelerator in at least one of the visible systems (7,8).

The reaction mechanisms involved in the ultraviolet light curing of the methacrylate monomers have been discussed in great detail

(9,10) and will not be repeated here. The reaction mechanisms involved in visible light curing are presumed to be similar with the principal difference being in the wavelength of the light which is employed.

Aside from the above-mentioned chemical differences, there is a significant difference between the chemically and the photochemically cured systems in the amount of material which can be polymerized at any one time. With the chemically cured systems there is, in principle, no limit to the bulk of material which may be cured, as long as the amine and the peroxide containing materials are thoroughly mixed. However, with the light cured systems, the amount of material which can be cured at any one time is limited depending upon the intensity of the curing light, the light transmission of the material, and the light absorbency of the surrounding media.

This difference, together with the fact that an auxiliary piece of equipment, the curing light, is needed to polymerize the photocured materials, leads to the observation that the evaluation of photocured dental materials must be done with particular care. It has been found that the results obtained from any evaluation test of a photocured material are very dependent upon the test method and the test parameters which are used. A good example of this is a simple "depth of cure" test which is often used (11). In this test, dental composite is placed into a cylindrical cavity in a mold and illuminated with its respective curing light for 60 seconds. The sample is then removed from the mold, the unpolymerized material is scraped away with a metal instrument, and the length of the remaining material measured with a micrometer. Typical results are shown in Table 1. Within each composite type, each result was significantly different ($p<0.05$) from the others. Every aspect of the test parameters, even the seemingly trivial such as mold diameter and mold length, profoundly affected the results. Consequently, any reported evaluation of photocured dental materials must include an extremely detailed explanation of the tests and the test parameters employed.

The extent of cure of photocured dental materials is dependent upon the intensity of the incident light. This is illustrated by a study of the dependence of Rockwell Hardness on curing light intensity for a visible light cured composite (12). In this study, all measurements of light were made at 468 nm since this was the principal wavelength of cure. The specimens were prepared as follows: A clear 3.2 mm-thick Plexiglas mold having a cylindrical specimen cavity 6.0 mm in diameter and a transmission of 95% at 468 nm was filled with composite. A 1-mm thick glass microscope slide with clear adhesive tape on one side and white reflecting tape on the other side (with a total reflectance of 60%) was placed on the bottom of the mold with the clear adhesive tape in contact with the composite. A 1-mm thick glass microscope slide with clear adhesive tape on one side (with a transmission of 45% at 468 nm) was placed on top of the mold

Table 1

"DEPTH OF CURE" OF PHOTOCURED DENTAL COMPOSITES

Polymerization System	Mold Material	Mold Cavity Size (mm) Dia. Length	"Depth of Cure" (mm) avg	(SD)
Visible Light	PTFE A[a]	3 x 10	8.31	(0.40)
" "	PTFE B[b]	3 x 10	6.49	(0.73)
" "	"	5 x 10	5.55	(0.41)
" "	"	3 x 15	5.01	(0.17)
" "	Steel	3 x 10	2.26	(0.40)
Ultraviolet Light	PTFE A	3 x 10	6.45	(0.20)
" "	PTFE B	3 x 10	4.13	(0.17)
" "	"	5 x 10	3.53	(0.16)
" "	Steel	3 x 10	2.15	(0.14)

[a]PTFE A and B had different translucencies

with the clear adhesive tape in contact with the composite. The slides were clamped to the mold and the tip of the activator light's fiber optic rod was positioned directly over the hole in the mold and 1 mm above it. The specimen was then illuminated for exactly 60 seconds. Then the glass slides were removed, the specimen was stored in water at 37°C for the desired time period (1 hour or 24 hours), and the Rockwell Hardness, F Scale, was determined at the top and the bottom of each specimen. For each experimental condition, ten specimens were measured and the hardness of the top was always determined before the hardness of the bottom.

The absolute spectral irradiance at 468 nm of the curing light was measured with a Gamma Scientific Monochrometer NM-3, Radiometer DR-1, and a specially built integrating sphere to collect and sample the emitted light. The integrating sphere is essential for light measurements made in the "near field". (The "near field" is a distance from the light source of less than 10 source diameters. In this case, the source diameter was 6 mm and the source is designed to be used 1-2 mm away from the material to be cured.) The results obtained are shown in Figure 1.

A number of significant points can be obtained from Figure 1. Firstly, the extent of cure of the composite is clearly dependent upon the light intensity at 468 nm. As the light intensity increased, the Rockwell Hardness increased in a regular fashion. Secondly, at any given time period or light intensity, the tops of the specimens were always harder than the bottoms. This is expected since light attenuation through the specimens results in the tops receiving

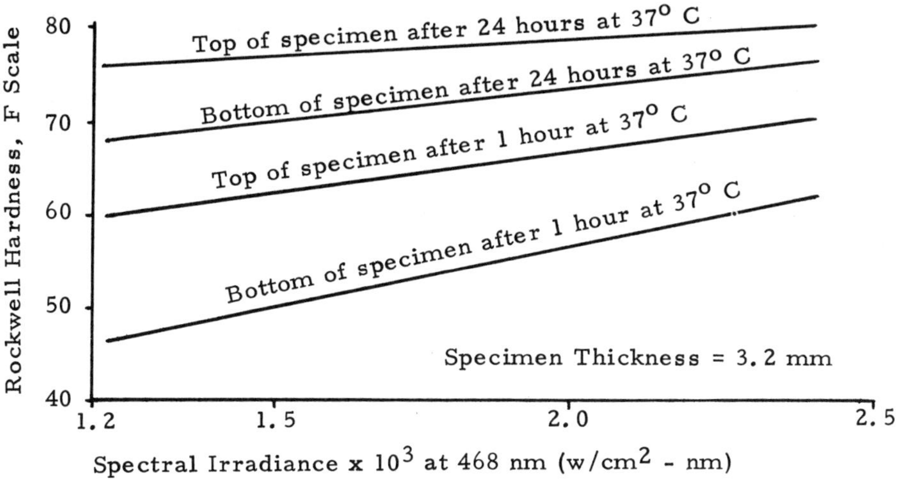

Rockwell Hardness, F Scale
80
70
60
50
40
Top of specimen after 24 hours at 37° C
Bottom of specimen after 24 hours at 37° C
Top of specimen after 1 hour at 37° C
Bottom of specimen after 1 hour at 37° C
Specimen Thickness = 3.2 mm
1.2
1.5
2.0
2.5
Spectral Irradiance x 10³ at 468 nm (w/cm² - nm)

Figure 1. Rockwell hardness, F scale, of a visible light cured dental composite vs curing light intensity at 468 nm.

greater light intensity than the bottoms. Finally, and very importantly, at any given light intensity, the Rockwell Hardness at 24 hours was always higher than the corresponding value at 1 hour. This is particularly significant since it showed that the composite continued to cure after the curing light was extinguished. This is direct evidence that a "dark reaction" is occurring and that two separate chemical reactions take place. The first is the photochemical reaction which initiates the polymerization of the monomers and the second is the thermochemical reaction which continues this polymerization.

CONCLUSIONS

Photochemistry has found a place in the field of dental materials for a variety of applications. The use of photochemically cured systems requires a curing light which is not needed in the amine-peroxide system. However, this is offset by the advantage that the dentist is afforded as much working time as he desires. Although the physical evaluation of dental materials is greatly dependent upon the test methods and test parameters employed, the final physical properties of these materials, when they are properly cured, are clinically acceptable for the intended applications.

A bibliography on the photochemistry of dental materials is included after the references.

REFERENCES

1. M. G. Buonocore, J. Am. Dent. Assn., 80, 324 (1970)
2. M. G. Buonocore and J. Davila, J. Am. Dent. Assn., 86, 1349 (1973)
3. M. A. Bassiouny and A. A. Grant, British Dental J., 145, 327 (1978)
4. K. C. Young, M. Hussey and K. W. Stephen, J. Oral Rehabilitation, 4, 181 (1977)
5. E. U. Condon, (Editor), Handbook of Physics, Second Edition, McGraw Hill, (1967), p. 6-61
6. A. S. Patel, J. Am. Dent. Assn., 94, 1055 (1977)
7. E. C. Dart, et al, U.S. Pat. 4, 110, 184 (1978)
8. E. C. Dart, et al, U.S. Pat. 4, 071, 424 (1978)
9. S. S. Labana, (Editor), "Ultraviolet Light Induced Reactions in Polymers", ACS Symposium Series 25, Americal Chemical Society, Washington, DC (1976)
10. E. W. Mochel, J. L. Crandall and J. H. Peterson, J. Am. Chem. Soc., 77, 494 (1955)
11. R. J. Kilian and T. J. Mullen, J. Dent Res., 59, Special Issue A, Abstract Number 203 (1980)
12. R. J. Kilian, J. Dent. Res., 58, Special Issue A, Abstract Number 603 (1979)

BIBLIOGRAPHY

1. "Harmful Effects of Near-Ultraviolet Radiation Used for Polymerization of a Sealant and a Composite Resin", D. C. Birdsell, P. J. Gannon, and R. B. Webb, J. Am. Dent. Assn., 94, 311 (1977)
2. "Pulpal Response to a New Visible Light-Cured Composite Restorative Material: Fotofil", W. W. Bloch, J. C. Austin, P. E. Cleaton-Jones, H. Wilton-Cox, and L. P. Fatti, J. Oral Pathology, 6, 278 (1977)
3. "Caries Prevention in Pits and Fissures Sealed with an Adhesive Resin Polymerized by Ultraviolet Light: A Two-Year Study of a Single Adhesive Application", M. G. Buonocore, J. Am. Dent. Assn., 82, 1090 (1971)
4. "Bonding of Clear Plastic Orthodontic Brackets Using an Ultraviolet-Sensitive Adhesive", M. E. Cohl, L. J. Green, and J. D. Eick, Am. J. Orthodontics, 62, 400 (1972)
5. "Factors Affecting the Depth of Cure of UV-Polymerized Composites", W. D. Cook, J. of Dent Res., 59, 800 (1980)
6. "Electrical Properties of Visible Light Cured Polymers", R. L. Kolek and J. L. Hammill, J. Soc. Plast. Eng. Tech. Pap, 25, 789 (1979)
7. "Physical and Biological Testing of the Nuva-Lite Apparatus", F. Lampert and R. K. Loew, Zahnaertzliche Welt Rundschau, 83, 696 (1974)
8. "A Comparison of Ultraviolet-Curing and Self-Curing Polymers in Preventive, Restorative and Orthodontic Dentistry", H. L. Lee, J. A. Orlowski, and B. J. Robers, Int. Dent. J., 26, 134 (1976)

9. "Photopolymerized Tooth Restorative Composite Materials, Strength and Microstructure", D. T. Turner, A. K. Abell, and K. F. Leinfelder, J. Dent. Res., 58, Special Issue A, Abstract No. 666 (1979)
10. "Lateral Photopolymerization in Tooth Restorative Composites", A. K. Abell, K. F. Leinfelder and D. T. Turner, J. Dent. Res., 58, Special Issue A, Abstract No. 667 (1979)
11. "Correlation Between Hardness and Abrasion of Photopolymerized Enamel Adhesives", R. J. Reinhardt and J. Vahl, Dtsch Zahnaerztl Z, 32, 625 (1977)
12. "Results Obtained with Two Different Sealants After One Year", W. P. Rock, Brit. Dent. J., 133, 146 (1972)
13. "The Use of Ultra-Violet Radiation in Dentistry", W. P. Rock, Brit. Dent. J., 136, 455 (1974)
14. "Curing Depths of Materials Polymerized by Ultra-Violet Light", N. O. Salako and D. W. Cruickshanks-Boyd, Brit. Dent. J., 146, 375 (1979)
15. "Clinical Comparison of Sealant and Bonding Systems in the Restoration of Fractured Anterior Teeth", Z. Sheykholeslam, M. Oppenheim, and M. I. Houpt, J. Am. Dent. Assn., 95, 1140 (1977)
16. "Restoration of Fractured Incisors with an Ultra-Violet Light-Polymerized Composite Resin", J. J. Watkins and R. J. Andlaw, Brit. Dent. J., 142, 249 (1977)
17. "Fissure Sealants", B. Williams and G. B. Winter, Brit. Dent. J., 141, 15 (1976)
18. "Ultra-Violet Absorption by Two Ultra-Violet Activated Sealants", K. C. Young, C. Main, F. C. Gillespie and K. W. Stephen, J. Oral Rehabilitation, 5, 207 (1978)

IONIC POLYMER GELS IN DENTISTRY

A. D. Wilson

Laboratory of the Government Chemist
Cornwall House, Stamford Street
London, S.E.1., England

INTRODUCTION

An ionic polymer has been defined by Holliday (1) as a polymer which contains both covalent and ionic bonds in its structure. In the form of ionic polymer hydrogels, derived from polyelectrolytes by chemical gelation, they have found application in dentistry. The oldest class is the alginate impression material which is formed by a double decomposition reaction between sodium alginate and the salt of a divalent metal. Other classes are the ionomer and polycarboxylate cements.

Dental cements are a diverse class of material of widely different chemistries and applications (2,3). However, all may be classified as acid-base reaction cements formed by mixing a powder(base) with an acidic liquid. A typical example, the traditional zinc phosphate, is the product of the reaction between a zinc oxide powder and a concentrated solution of phosphoric acid. The cement sets, within minutes, as an amorphous zinc orthophosphate gel is formed.

In 1968 Smith (4) replaced phosphoric acid by poly(acrylic acid) to give the zinc polycarboxylate cements, and in 1971 Wilson and Kent (5) reported a cement formed by the reaction between an ion-leachable glass and an aqueous solution of poly(acrylic acid). They termed this material the glass-ionomer or ASPA cement. (ASPA is the acronym of Alumino Silicate PolyAcrylic Acid). The cement-forming mechanism may be represented as a reaction between two polymers to form a third which acts as a cementing matrix:

Glass + Polyelectrolyte = Polysalt hydrogel + Silica gel
(base) (acid) (matrix) (coating)

In effect, cations are transferred from the glass to the polyelectrolyte causing it to gel as polyanion chains are ionically cross-linked by cations. The system is bland towards living tissue as no monomers are involved in matrix-formation.

LIQUIDS

Composition

The liquids used in glass-ionomer cements are acidic polyelectrolytes and are homo- or copolymers of acrylic acid in concentrated aqueous solutions (6,7). These polyacids are generally prepared by the aqueous polymerization of the unsaturated carboxylic acids using ammonium persulphate as initiator and isopropyl alcohol as the chain transfer agent. Tartaric acid is generally added to the liquids to sharpen the setting of their cement (8). Examples of the liquids used in ionomer cement formulations are given in Table 1.

Liquid Composition and Cement Properties

The properties of glass-ionomer cements are affected by the nature of the parent polyacid, its molecular weight and its concentration in aqueous solution, and also by low molecular weight complexing agents added to the liquid. Strength and resistance to aqueous attack increase with the concentration and the molecular weight of the parent polyacid (9,10). However, the concomitant increases in the viscosity of the liquid make manipulation of the cement mixes more difficult, also working time and setting time are reduced. These factors limit the usable acid concentration to about 50% by weight and the molecular weight of the polyacid to about 10,000. Poly(acrylic acid) has not proved to be the ideal poly(alkenoic acid) in this system because of its tendency to gel after a few weeks when in 50% aqueous solution (11). In this respect a co-polymer of acrylic and itaconic acids is superior to the homopolymer of acrylic acid. A 50% solution of a co-polymer (M = 10,000) has a much lower viscosity than the homopolymer, is indefinitely stable and does not gel (or increase in) viscosity on standing.

Early formulation of ionomer liquids - simple solutions of poly(acrylic acid) - only yielded workable cement pastes which set sharply when mixed with glasses of high fluorine content. This restricted the number of glass formulations which could be used in the system. Later it was found that other chelating and complex forming compounds had a similar effect (6), tartaric acid proved particularly effective and is now incorporated in all glass-ionomer cement formulations (6,8).

Table 1

CHEMICAL COMPOSITION OF LIQUIDS USED IN GLASS-IONOMER CEMENT

	Polyacid			Additives	
Liquid	Type	Mw	% m/m aq.		% m/m aq.
I	Acrylic acid homopolymer	23,000	50.0		
II	Acrylic acid homopolymer	23,000	47.5	Tartaric acid	5.0
III	Acrylic acid homopolymer	23,000	45.2	Tartaric acid Methanol	4.75 5.0
IV	Acrylic acid/ Itaconic acid 2:1 copolymer	10,000	47.5	Tartaric acid	5.0

ION-LEACHABLE GLASSES

Chemical Types

The powders used in glass-ionomer cements are ground ion-leachable aluminosilicate glasses. These glasses vary in appearance and may be transparent, translucent or opaque depending on composition and conditions of preparation. They are prepared by fusing the ingredients together for 1 h and shock cooling the melt by pouring onto a metal tray and then plunging into water. The glass fragments are ground to a fine powder, maximum particle size of between 15 and 50 μm depending on the use to which the material is to be put.

The basic ionomer glass types are:

(I) SiO_2-Al_2O_3-CaO (fusion temperatures 1350-1550°C)

(II) SiO_2-Al_2O_3-CaF_2 (fusion temperatures 1250-1450°C)

(III) Hybrids of (I) and (II)

Practical glasses used in dental compositions are based on type II with the addition of cryolite, Na_3AlF_6, aluminum phosphate, $AlPO_4$, and sometimes aluminum trifluoride, AlF_3. Fluorides act as a flux and serve to reduce the fusion temperature. Some typical compositions of ionomer glasses are given in Table 2.

Table 2

CHEMICAL COMPOSITION % m/m (MOLE RATIOS)

OF SOME IONOMER CEMENT GLASSES

Component	Simple Oxide (3 components)	Simple Fluoride (3 components)	Complex Fluoride (6 components)
SiO_2	35.9 (2.00)	35.0 (1.98)	29.0 (2.99)
Al_2O_3	30.6 (1.00)	30.0 (1.00)	16.5 (1.00)
CaO	33.5 (2.00)	-	-
CaF_2	-	35.0 (1.53)	34.3 (2.71)
Na_3AlF_6	-	-	5.0 (0.15)
AlF_3	-	-	5.3 (0.37)
$AlPO_4$	-	-	9.9 (0.50)

SiO_2-Al_2O_3-CaO glasses. These are clear over a wide range of compositions in the anorthite and gehlanite region of the ternary phase diagram (12). The setting rate of cements derived from these glasses are controlled by their Al_2O_3:SiO_2 ratio and is retarded as the magnitude of this ratio is decreased. When the SiO_2 content of the glass exceeds 60% by weight cement formation does not take place (12).

SiO_2-Al_2O_3-CaF_2 glasses. Clear glasses in this class are formed only over a limited range of compositions, i.e. when the aluminum oxide/calcium fluoride ratio is approximately 1:1 by weight and when the silica content exceeds 35% by weight (12). Again cements cannot be formed from glasses which have a silica content greater than 60% by weight.

Fluorine-containing glasses always form the basis of formulations used in dentistry because a number of advantages are obtained. These are:

(i) The workability and setting characteristics of the cement pastes are improved.
(ii) The strength of the set cements are increased.
(iii) Cements are rendered more translucent.
(iv) Release of fluoride from the cement restoration imparts a cariostatic property to adjacent tooth enamel.

Practical glasses. These glasses are prepared using Na_3AlF_6 as an additional flux which imparts greater translucency to cements prepared from them. The $Al_2O_3:SiO_2$ ratio also controls the setting rate of cements derived from these complex glasses (13).

Acid decomposition. All these aluminosilicate glasses are decomposed by weak acids. Unlike the electrically neutral silica network, the aluminosilicate network carries a negative charge, consequent on the partial replacement of Si^{4+} by Al^{3+}. Thus, this network is vulnerable to attack by protons with release of ions and the formation of silica gel.

SETTING REACTIONS AND MICROSTRUCTURE

The chemical reaction taking place during the setting and hardening of glass-ionomer cements has been elucidated by monitoring the soluble ion content of cement pastes (14,15)(Fig 1) and by recording changes in the infra-red spectra (16). On mixing the powder and liquid into a paste cement-forming reactions take place in a number of overlapping stages:

(1) The aluminosilicate glass is decomposed by hydrogen ions, supplied by the acidic liquid, with liberation of ions and the formation of silicic acid, which subsequently polymerizes to give silica gel.

(2) The liberated ions Al^{3+}, Ca^{2+}, Na^+, F^- and $H_2PO_4^-$ but not silicic acid, migrate and accumulate in the aqueous phase, together with anions generated by ionization of the acid and cations already present.

The progress of the extraction process is revealed by the Na^+/time curve (Fig 2), Na^+ does not precipitate in the reaction, which indicates that 30% of the powder is consumed in the reaction (16).

(3) As the pH of the aqueous phase increases (consequent on the loss of hydrogen ions) the accumulating cations and anions combine and eventually precipitate as salt-like hydrogels. (Fig 1).

The form of the other soluble ion/time curves exhibit maxima, illustrate how the precipitative process is eventually overtaken by the extractive process. The rate of precipitation is different for different cations with calcium being precipitated more rapidly than aluminum. The findings from infra-red spectroscopy indicate that while the polyacrylate of calcium forms within minutes of mixing, that of aluminum only begins to form one hour later. These observations suggest that initial set results from the precipitation of a calcium polyacrylate gel, a contrast with the inessential role played by

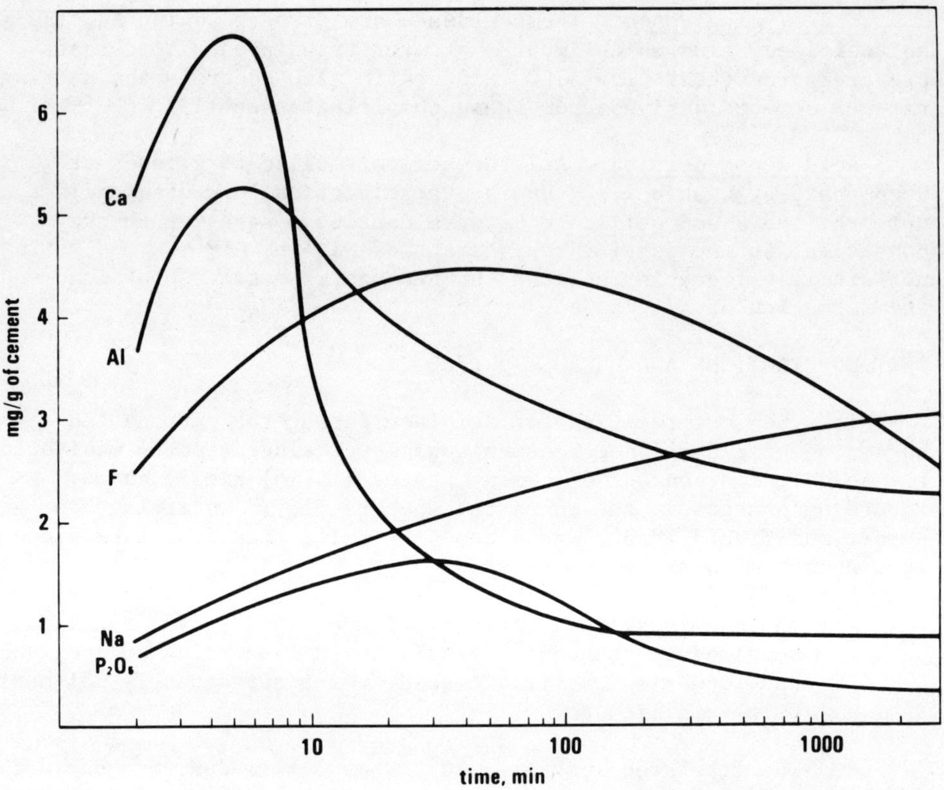
6
5
4
3
2
1
mg/g of cement
Ca
Al
F
Na
P_2O_5
10
100
1000
time, min

Fig 1. Extraction and precipitation of ions in a setting glass-ionomer cement.

calcium in the setting of the dental silicate cement. Only later does aluminum polyacrylate form and contribute to the post-set hardening of theis cement. A schematic representation of the setting reaction is depicted in Fig 2. Ionic reactions continue for several months causing cement strength and modulus to increase. There is also an associated hydration reaction and hardening is associated with an increase in bound water (16).

The set cement contains a filler of partly reacted glass particles which are connected together by a matrix of aluminum and calcium polyacrylates (17). The filler particles consist of a core of unreacted glass sheathed by a layer of siliceous hydrogel (the result of ion-depletion by the acid).

PROPERTIES AND APPLICATIONS

Glass-ionomer cements set within 4-5 minutes (at 37°C) and develop strength rapidly (6), about 170 MPa in compression within

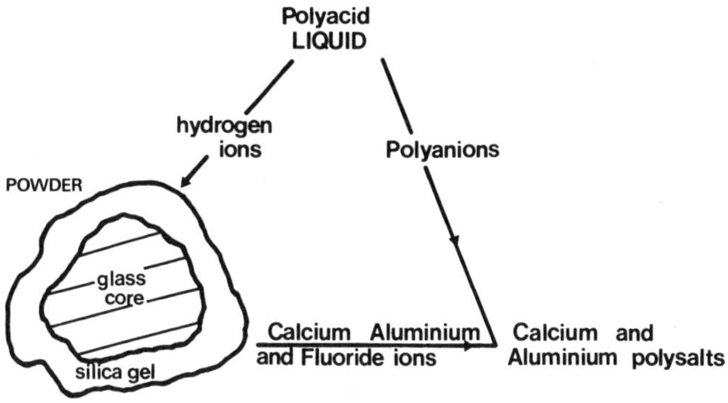
Polyacid
LIQUID
hydrogen
ions
Polyanions
POWDER
glass
core
silica gel
Calcium Aluminium
and Fluoride ions
Calcium and
Aluminium polysalts

Fig 2. Schematic representation of chemical reactions taking place in a setting glass-ionomer cement.

24 hours. Within 7 days a strength in compression of over 200 MPa is attained with a modulus of about 20 GPa. These cements are brittle materials and weak in tension (ca. 12 MPa in 7d). They are translucent (opacity $C_{0.7}$ range between 0.45 and 0.80) and so can be made to match tooth enamel optically. They are adhesive to enamel (bond strength 4 MPa), dentin (bond strength 3 MPa), and base metals. They do not bond to the inert surfaces of noble metals and porcelain. They are bland towards living tissue and adhesive towards enamel and dentin under oral conditions. This combination of properties makes them valuable for use in dentistry. Their adhesive nature minimizes cavity preparation (removal of tooth material to provide mechanical retention is not necessary) and provides a seal against harmful organisms. No cavity liner is needed because they are bland and the fluoride release protects adjacent enamel against caries. They can be used for aesthetic restorations. Their use is contra-indicated in areas of high stress and wear (as in the restoration of posterior teeth).

ACKNOWLEDGMENTS

The author acknowledges the considerable contributions that have been made in this work by his colleagues, Mr. B. E. Kent, Dr. S. Crisp, Dr. H. J. Prosser and Mr. B. G. Lewis and also thanks the Government Chemist for his support over the years.

REFERENCES

1. L. Holliday, "Ionic Polymers", Appl. Sci. Publishers, London, (1975)
2. A. D. Wilson, in "The Scientific Aspects of Dental Materials", Ed. J. A. van Fraunhofer, Butterworths, London,(1975)
3. A. D. Wilson, Chem. Rev., 7, 265, (1978)
4. D. C. Smith, Br. Dent. J., 125, 381 (1968)

5. A. D. Wilson and B. E. Kent, J. Appl. Chem. Biotechnol., 21, 313 (1971)
6. S. Crisp, S. J. Ferner, B. G. Lewis and A. D. Wilson, J. Dent., 3, 123 (1975)
7. S. Crisp and A. D. Wilson, Br. Pat. 1, 484, 454 (1977)
8. A. D. Wilson, S. Crisp and A. J. Ferner, J. Dent. Res., 55, 489 (1976)
9. S. Crisp, B. G. Lewis and A. D. Wilson, J. Dent., 5, 51 (1977)
10. A. D. Wilson, S. Crisp and G. Abel, J. Dent., 5, 117 (1977)
11. S. Crisp, B. G. Lewis and A. D. Wilson, J. Dent., 54, 1173 (1975)
12. A. D. Wilson, S. Crisp, H. J. Prosser, B. G. Lewis and S. A. Merson, (Unpublished Report, Laboratory of the Government Chemist, 1979)
13. B. E. Kent, B. G. Lewis and A. D. Wilson, J. Dent. Res., 58, 1607, (1979)
14. S. Crisp, A. J. Ferner, B. G. Lewis and A. D. Wilson, J. Dent., 3, 125 (1975)
15. J. M. Paddon and A. D. Wilson, J. Dent., 4, 183 (1976)
16. A. D. Wilson, J. M. Paddon and S. Crisp, J. Dent. Res., 58, 1065 (1979)
17. T. I. Barry, D. J. Clinton and A. D. Wilson, J. Dent., Res., 58, 1072 (1979)

ADSORPTION AND IONIC CROSSLINKING OF POLYELECTROLYTES

Daniel Belton and Samuel I. Stupp

Department of Materials Science, Northwestern University, Evanston, Illinois 60201; and Department of Biological Materials, Northwestern University, Chicago, Illinois 60611

INTRODUCTION

Materials formed by polymers containing ionic or ionizable groups in their chemical repeating units (polyelectrolytes) have been of interest in the area of biomaterials over the past two decades. Presently, two largely polyelectrolytic compositions are utilized as dental cements or as restorative materials for hard oral tissues. These two types of dental biomaterials are the zinc polycarboxylate cements (1) and the glass ionomer cements and restoratives (2). Dental impression materials, known as alginates (3), represent an additional example, as they are partly based on an ionizable polysaccharide, alginic acid. Polyelectrolyte-based materials have been tested for other biomedical applications, such as the formulation of cardiovascular elastomers (4), dialysis membranes (5), drug delivery systems (6), or surgical adhesives (7). As a consequence of past and current research activity, it is possible that polyelectrolytes will find application in some of these areas in the near future. It is interesting to consider that biological tissues themselves contain considerable amounts of polyelectrolytic material. This fact alone does not, of course, guarantee that synthetic polyelectrolytes should lead to the successful development of nearly ideal biomaterials from a mechanical, chemical or biocompatibility standpoint. Nonetheless, a strong interest in these materials, and their potential advantages can be justified on the basis of some of their characteristic features, depending on the specific application involved. Some of these characteristic features, and their relevance in the formulation of biomaterials, have been described below.

Generally speaking, the ionic nature of these macromolecules, and, consequently, their hydrophilicity is, in some cases, an important consideration. For example, this property can lead to the development of matrices from these materials which can easily transport ions and organic hydrophilic molecules at normal body temperature. Furthermore, depending on their specific charge bias, they can function as ion-exchange resins that selectively transport certain nutrients, impurities or therapeutic agents. The ability of polyelectrolyte materials to absorb or release controllable amounts of water in aqueous media, depending on the dielectric constant associated with a specific composition, ought to be, as well, as important aspect in biomaterials design. For example, this particular feature can be important in volumetric considerations, for example, the ability of a given material to swell or contract prior to or following implantation. At the same time, variable water content is a versatile characteristic, since it can potentially allow control of mechanical properties in the material involved. From a surface chemistry point of view, polyelectrolytic systems offer the possibility of creating structures which can interact reversibly in aqueous environments with the external hydrophilic surfaces of proteins in native conformation (e.g., plasma proteins with cardiovascular prostheses). This is important since interactions between foreign surface and the hydrophobic interior of native proteinaceous coils can lead to denaturation, and possibly trigger blood clotting reactions (8). On the other hand, synthetic polyelectrolytes can obviously adsorb chemically through ionic or dipolar bonds to hydrophilic tissue surfaces. This property is naturally of great relevance in the development of surgical adhesives for both hard and soft tissue surfaces. Finally, one additional feature which is critical in certain applications is the fact that both covalent and ionic bonds can serve as a source of cohesive strength in a polyelectrolytic solid. This fact is important from the standpoint of generating appropriate mechanical properties and insolubility through ionic and covalent crosslinking in materials for implantation. Since ionic bonds are readily formed, the covalently linked repeating units of a linear polyelectrolyte can be ionically crosslinked *in-situ* in order to generate a strong and insoluble two or three-dimensional matrix. This way, a polyelectrolyte liquid or paste can be surgically placed and shaped before its transformation into a solid mass.

It is, perhaps, fair to say that extensive use of polyelectrolytes as biomaterials has been partly hindered by lack of fundamental and systematic information on these macromolecules pertaining to specific applications as biomaterials. In the present work, the phenomena studied are surface adsorption and ionic crosslinking of poly(alkenoic acid) aqueous solution. These two phenomena are directly relevant to the present use of zinc polycarboxylate cements (ZP) and glass ionomers (GI) in restorative dentistry. Both of these types of dental biomaterials are formulated for clinical use as two-component systems. A liquid, which is an aqueous solution of a poly(alkenoic acid), is used in both products. In the case of

GI materials, the second component is a powder produced from an ion-leachable aluminosilicate glass (9), whereas in ZP cements, the powder is essentially pulverized zinc oxide, containing, in some cases, small amounts of magnesium oxide (10). Both powders are chemically basic, and thus react with the aqueous solution of the polymeric acid. The acid/base reaction that takes place when powder and liquid components are mixed, transforms the paste to a rigid mass within ten to twenty minutes. The mechanistic details of this reaction, as well as the structure/property relations obeyed by the solid product obtained are not well known at this time. Supposedly, the reaction involves the formation of ionic crosslinks between carboxylate groups in the poly(alkenoic acid) and Zn^{+2} and Mg^{+2} ions in ZP cements (17) and Ca^{+2} and Al^{+3} cations in GI products (2).

The unique property of these dental materials is their chemically adhesive quality toward surfaces of hard oral tissues (11,12). This property is crucial in a dental restorative material or cement for two reasons. First of all, a chemical adhesive can potentially seal the margins at tissue/synthetic material interfaces, thus preventing penetration of bacteria which can lead to secondary caries formation. Secondly, an adhesive material does not require the removal of substantial amounts of healthy tissue in order for the clinician to create a cavity possessing a mechanically-retentive shape. The non-adhesive aesthetic restorative materials (having the appearance of natural enamel) used extensively in dentistry at the present time require considerable removal of non-carious tissue during cavity preparation. Also, etching of enamel walls within the cavity with a phosphoric acid solution is necessary in order to induce mechanical interlocking of the paste on the rough, etched surface prior to solidification (13). These materials, known as composites since they are formulated from ceramic fillers and polymerizable organic molecules, present the additional problem of incomplete polymerization at normal oral temperatures (14). Incomplete polymerization is not only detrimental to their mechanical properties and chemical stability, but might also be of concern from a biocompatibility standpoint (e.g., irritation of surrounding tissues as free residual monomer is leached out). The polyelectrolyte-based restoratives, on the other hand, being formulated from macromolecular components, do not present a problem in this respect. In fact, previous studies on their biocompatibility (15,16) have revealed a rather mild response in biological environments. At the present time, clinical use of GI restoratives is quite limited, despite their attractive properties. Their application is restricted to restorations which are not exposed to high mechanical stresses in the oral cavity. Several properties of these materials must be improved before their use is extended to all types of restorations. Mechanical strength should be upgraded, and most importantly, the long-term stability of mechanical properties of the set product in oral fluids must be improved. It is also important to decrease the sensitivity of the setting acid/base interaction to humidity, and to identify the factors that control its kinetics and equilibrium extent of reaction. These two aspects could lead to the development of products which can be more easily manipulated by clinicians. Finally, the extent to which these materials adhere chemically to tissue surfaces must be analyzed and maximized in order to allow formation of tightly sealed interfaces in the oral environment over long periods of time.

Some of the problems of polyelectrolytic dental materials outlined above are being addressed in the present manuscript from a fundamental point of view. The problem of maximizing chemical adhesion is approached through a study of adsorption of poly(acrylic acid) from aqueous solution on tribasic calcium phosphate particles. Poly(acrylic acid) is a close analogue to the polymeric acids used to formulate dental products, which are often copolymers of acrylic acid with small amounts of itaconic or maleic acid. Calcium phosphate, on the other hand, simulates the hydroxyapatite portion of dental tissues. The substrate used for adsorption studies is of interest, since chemical adhesion between these materials and hard oral tissues is believed to be based on ionic interactions between carboxylate groups in the polymeric acids and calcium ions in the apatite lattice. Secondly, the acid/base reaction between powder and liquid components, relevant to clinical manipulation and mechanical properties of the set product, has been studied, using the actual powders and polyelectrolytic solutions of dental products. The experimental tool in the present work has been vibrational infrared spectroscopy. Since the system under investigation naturally involves aqueous media, the spectroscopic analysis has only been possible through the use of the Fourier Transform technique (FTIR). Some details of the experimental methods that led to our results have been described below.

EXPERIMENTAL

Poly(acrylic acid) (obtained from Polysciences, Inc., Warrington, Pennsylvania) of average molecular weight 250,000 was purified by precipitation from p-dioxane solution with benzene. In order to narrow slightly the molecular weight distribution, the purified material was fractionated by precipitation from p-dioxane solution with n-heptane. It is expected (18,19) that material of molecular weight in the very high range of the distribution is eliminated upon addition of small quantities of n-heptane, whereas very low molecular weight material relative to the average value is precipitated when large amounts of the non-solvent are added to p-dioxane solutions of the poly(acrylic acids). Fractionated material was washed and finally stored in doubly-distilled water. Solutions of varying concentrations were prepared by proper dilution, and NaOH was used as a neutralizing base to control the degree of ionization of poly(acrylic acid) (PAA). Neutralized solutions were allowed to achieve equilibrium in sealed glass containers. Adsorption of PAA was induced by adding tribasic calcium phosphate (Fisher Scientific Company) to the aqueous solutions in the amount of 2.18 x 10^{-2} gms/ml of solution. The powder/solution slurry was stirred in the sealed container for 24 hours which was determined to be the necessary time to achieve adsorption equilibrium. The slurry was subsequently separated by centrifugation, and samples of the supernatant liquid, as well as samples of the slurry and original solution, were used for spectroscopic analysis. Experiments involving characterization of acid/base reactions in commercial dental materials were carried out for powder/liquid mixtures in a weight ratio of approximately 1/2. The products analyzed were Fuji Ionomer, type II(G-C Dental Industrial Corporation, Japan), ASPA restorative and Tylok cement (L.D. Caulk Company, Delaware).

FTIR spectroscopic measurements were carried out in a Nicolet model 7199 Fourier Transform Infrared spectrophotometer. In adsorption experiments, the spectrophotometer was coupled to an attenuated total reflectance (ATR), double condensing, variable angle attachment with a liquid cell (Harrick Scientific, Ossining, New York). A germanium, 10 x 5 x 1 mm, 45° reflecting crystal was utilized in these experiments. Kinetic measurements on powder/liquid mixtures were carried out using a Harrick horizontal ATR attachment by placing samples, immediately after mixing, on the surface of a 50 x 20 x 2 mm, germanium reflecting crystal (45°). Digital compensation for water present in the various samples analyzed was accomplished by storing spectra of pure water prior to initiation of each measurement. Upon completion of the experiment, absorbance from water was digitally subtracted from that of samples, yielding the desired difference spectra.

RESULTS AND DISCUSSION

Figure 1A shows the FTIR spectrum of an aqueous solution of poly(acrylic acid) in the range 1810-1340 cm^{-1}. This spectral range includes the stretching vibration from carbonyl groups in the carboxyl side group (1710 cm^{-1}), and that of the carboxylate ion as well (1560 cm^{-1}). As shown in Figure 1B, the absorbance spectrum of pure water in this range essentially obscures these two bands. Figure 1C reveals the result obtained when digital subtraction of B from A is carried out in the FTIR spectrophometer. The bands of interest in the present study are now distinctly revealed in the 1810-1340 cm^{-1} range. The range includes, in addition to the carboxyl and carboxylate absorptions, the CH_2 bending peak used here as an internal band for normalization of data. The subtraction described is, obviously, a key advantage offered by the FTIR in the experimental work described below.

Adsorption Phenomena

Adsorption of PAA in aqueous solution on calcium phosphate

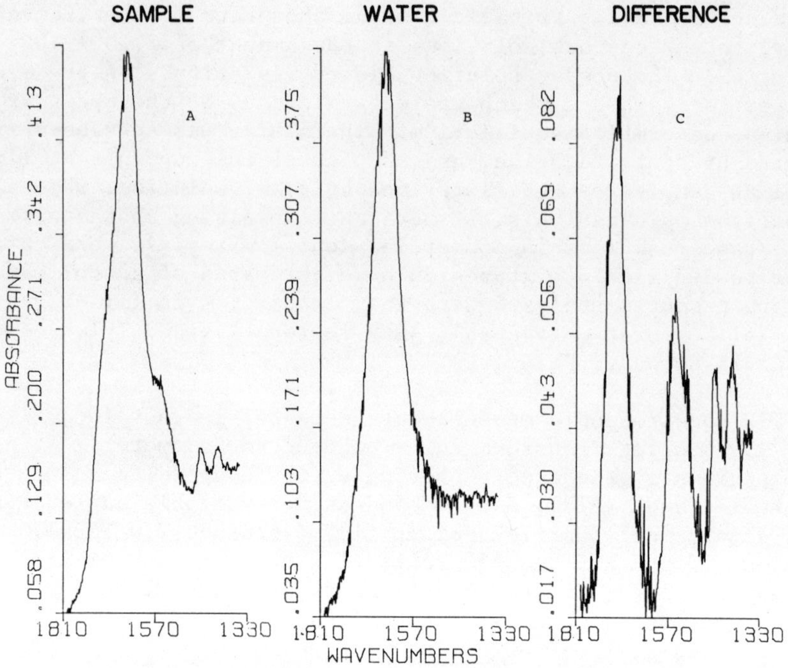
SAMPLE
WATER
DIFFERENCE
A
B
C
ABSORBANCE
.058 .129 .200 .271 .342 .413
.035 .103 .171 .239 .307 .375
.017 .030 .043 .056 .069 .082
1810 1570 1330
1810 1570 1330
1810 1570 1330
WAVENUMBERS

Figure 1. FTIR spectrum of a PAA aqueous solution (A), water (B), and spectrum obtained by subtracting B from A (C).

particles was analyzed spectroscopically as a function of concentration and degree of ionization in the original polyelectrolyte solution. As previously described, ionization was induced by NaOH. Figure 2 reveals qualitatively the relative difference in carboxyl group vs. carboxylate ion absorbance for one specific set of conditions. Spectra were obtained from the following three specimens: A) the original solution, B) supernatant liquid collected once the insoluble calcium phosphate was allowed to settly by centrifugation, and C) a sample of the slurry obtained by mixing the PAA solution with the pulverized ionic substrate. A considerable concentration of carboxylate ions is observed in A which is expected as a consequence of the base-induced ionization in PAA. Both B and C, however, reveal an apparent increase in ionized side group concentration relative to the original solution. The increased level of ionization in the supernatant liquid could be explained on the basis that the finite solubility of the tribasic calcium phosphate leads to further neutralization of the polyelectrolyte. Alternatively, however, this increase could be explained by the presence of microscopic particles of calcium phosphate containing ionically adsorbed PAA on their surface. A higher concentration of carboxylate in the slurry would be expected, of course, on the basis of both PAA

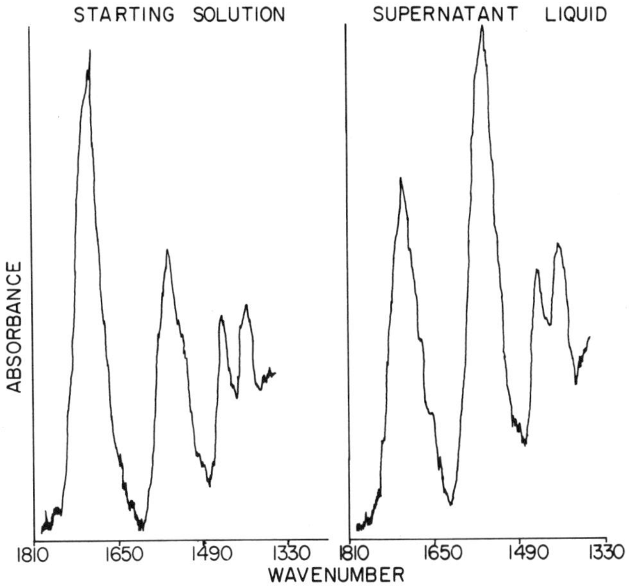
STARTING SOLUTION
SUPERNATANT LIQUID
ABSORBANCE
1810
1650
1490
1330
1810
1650
1490
1330
WAVENUMBER

Figure 2. FTIR spectra of a PAA solution and corresponding supernatant liquid after the solution was exposed to tribasic calcium phosphate for 24 hours.

adsorption on the substrate through the formation of ionic bonds and further neutralization by soluble ions from the substrate.

Quantitative analysis of adsorption trends was carried out from spectroscopic measurements involving original PAA solutions at concentrations of 0.85, 0.62, 0.40 and 0.28 moles/liter and degrees of ionization ranging from 0% to 75%. Two different parameters have been plotted in Figures 3-5 as a function of these molecular variables. The first one (Figure 3), is the amount of adsorbed polyelectrolyte per gram of $CaPO_4$. This parameter was calculated through the extinction coefficient, ε, for the CH_2 bending band (the extinction coefficient was obtained from the slope of integrated absorbance vs. concentration for original PAA solutions). Using ε_{CH_2}, the concentration of PAA in supernatant liquids was calculated, and subtracted from that in original solutions. The parameter plotted in Figures 3 and 4 is, therefore,

$$\text{Amount adsorbed} = \frac{(Co-C_1)Vo}{w_s} \qquad \text{(Equation 1)}$$

where Co = concentration of starting PAA solution, C_1 = concentration of polymer in supernatant liquid, Vo = original volume of solution,

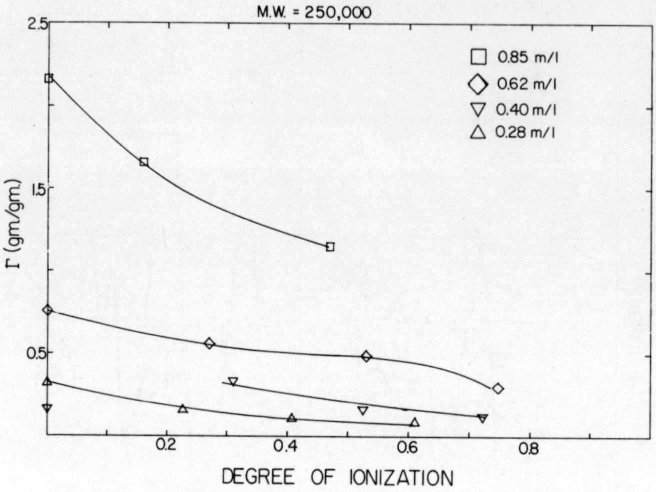

M.W. = 250,000
2.5
1.5
0.5
Γ (gm/gm)
0.2
0.4
0.6
0.8
DEGREE OF IONIZATION
0.85 m/l
0.62 m/l
0.40 m/l
0.28 m/l

Figure 3. Amount of polymer (average molecular weight = 250,000) adsorbed per gram of tribasic calcium phosphate as a function of degree of ionization of the original PAA solution for various concentrations (moles/liter).

and w_s = the weight of calcium phosphate added to the original solution. The second parameter used to characterize adsorption trends was the fraction of repeating units in the adsorbed layer which are ionically bound to the substrate (bound fraction). This fraction is defined as,

$$\text{Bound fraction} = \frac{C_b}{C_a} \qquad \text{(Equation 2)}$$

where C_b = concentration of bound repeating units, and C_a = concentration of adsorbed repeating units (free and bound to the substrate). Term C_a is essentially the concentration differences $Co-C_1$) in Equation (1). Term C_b is approximated from integrated absorbance measurements of the COO^- band in supernatant liquids, and powder/liquid slurries. The COO^- absorbance of the slurries includes that due to free groups in the supernatant liquid and in the adsorbed layer, as well as that of bound groups. Assuming Beer's law holds, the concentration of bound groups should be proportional to the integrated absorbance of carboxylate groups in the slurry. The ratio of this absorbance to C_a has been plotted in Figure 4 as a function of degree of ionization (DI) in original solutions of PAA, suggesting an increase in the number of bound groups with increasing DI values. On the other hand, higher DI values appear to decrease the amount of adsorbed polymer from solution on the ionic substrate.

Increasing degrees of ionization of polyelectrolytes in solution

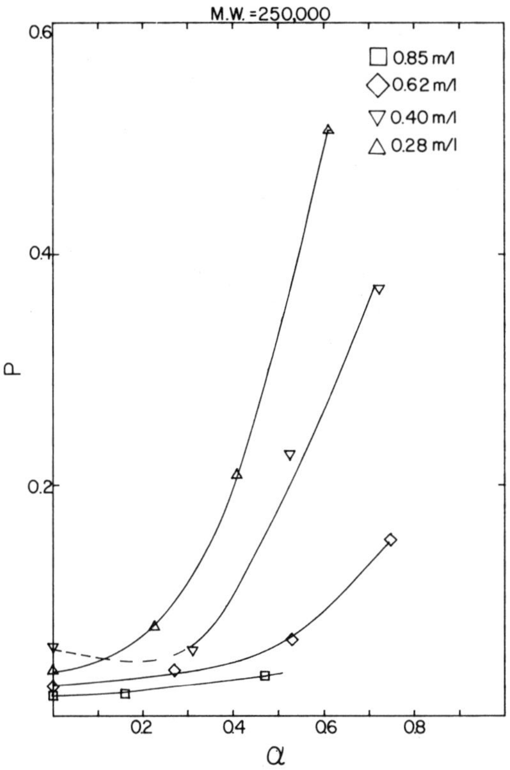
M.W. =250,000
0.6
0.4
0.2
P
0.85 m/l
0.62 m/l
0.40 m/l
0.28 m/l
0.2
0.4
0.6
0.8
α

Figure 4. Apparent sensitivity of P (a parameter proportional to the fraction of carboxylate side groups bound to the substrate's surface) to α (degree of ionization) for PAA solutions of various concentrations (moles/liter).

are known to give rise to higher viscosities due to repulsive interactions between neighboring ionized side groups. These repulsive forces lead to an increase in the radius gyration of molecules in solution, and, therefore, a higher viscosity given the larger hydrodynamic volume per molecule. Higher viscosities are likely to result in lower wetting abilities for the solutions on the ionic substrate. For this reason, one might expect viscosity differences to be partly responsible for the lower amounts of adsorbed polymer observed with increasing DI values. On the other hand, an alternative explanation is the fact that expanded coils would rapidly decrease the internal energy of the substrate after a small amount of polymer has been adsorbed given the larger surface area per molecule.

The fraction of bound groups on the substrate for a given

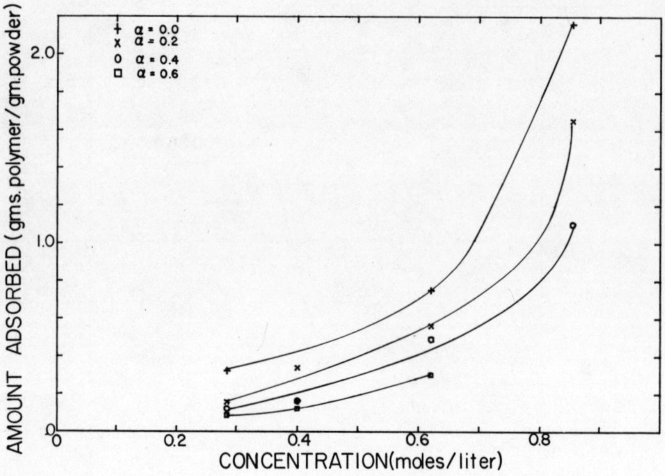
+ α = 0.0
x α = 0.2
o α = 0.4
□ α = 0.6
AMOUNT ADSORBED (gms. polymer/gm. powder)
2.0
1.0
0
0
0.2
0.4
0.6
0.8
CONCENTRATION(moles/liter)

Figure 5. Amount of polymer adsorbed per gram of tribasic calcium phosphate as a function of concentration of the PAA solution.

concentration and DI value is an important parameter, since it should be related to the stability of the absorbed layer. This would be specially significant in applications where adhesive bonds through ionic interactions must persist in aqueous media. The results obtained (Figure 4) can be partly understood in the context that more carboxylate groups are available for ionic adsorption at high degrees of ionization. However, it is also possible that highly ionized macromolecules would decrease substantially their internal energy on adsorption due to stored elastic energy brought about by electrostatically repulsive expansions. Little can be concluded at this point regarding the magnitude of this effect since the parameter plotted from experimental data is not an absolute value, but simply a parameter proportional to the number of bound segments. Furthermore, additional ionization of the polyelectrolyte in solution once brought in contact with the substrate is not all necessarily due to ionic bonding but can also result from partial solubility of the tribasic calcium phosphate.

Ionization Kinetics

The results on ionization kinetics, shown in Figure 6, reveal considerable differences among a zinc polycarboxylate and two glass ionomers in dental use at the present time. These results should not be necessarily interpreted as crosslinking kinetics of the polyelectrolytes by cations such as Zn^{+2}, Ca^{+2}, or Al^{+3} to form the insoluble metal acrylate gel. The plots only represent the rate at which carboxyl groups in the various polyelectrolytes ionize as a consequence of the acid/base reaction. Formation of effective

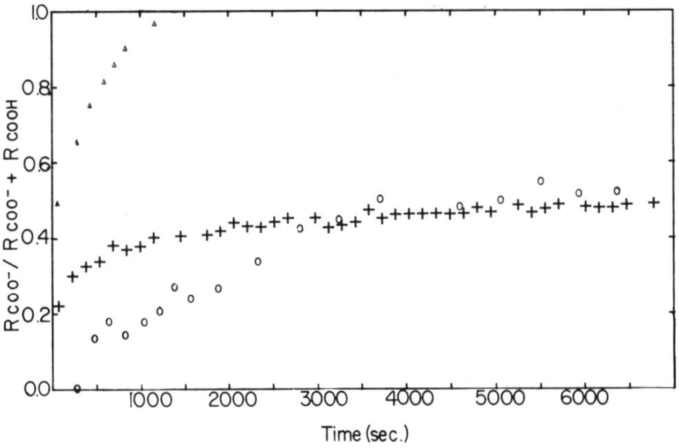

1.0
0.8
0.6
0.4
0.2
0.0
Rcoo⁻ / Rcoo⁻ + R cooH
1000
2000
3000
4000
5000
6000
Time (sec.)

Figure 6. Fraction of ionized carboxyl groups (carboxylate) as a function of time for Tylok polycarboxylate cement (Δ), ASPA (+) and Fuji Ionomer (O). R's represent integrated absorbances normalized with the CH_2 bending band.

crosslinks where the metallic cations are spatially localized on carboxylate groups is a different phenomenon, which might result over longer periods of time as a consequence of ionic diffusion through the polyelectrolyte matrix. In fact, there is no apparent, direct correlation between degree of ionization and modulus of elasticity of the setting material within the short time range where most of the carboxyl groups dissociate into ionized species (20). Nonetheless, it is possible that the ionic crosslink density ultimately reached in a given medium is, in some way, related to the original ionization rate of the polyelectrolyte. This is significant, since the density of ionic crosslinks would, of course, affect mechanical strength and hydrolytic stability of the set material.

An important aspect to consider is the possible relationship between adsorption of the polyelectrolyte on a given substrate and ionic crosslinking in bulk regions of the adhesive. Based on adsorption data, one might suggest that differences in ionization rate could result in varying degrees of adsorption (amount of polymer and bound segments) on calcified tissues. This is inferred on the basis that ionic sites on the tissue surface and cations leached from glasses or oxides compete for carboxylate groups during setting. Also, since the polyelectrolyte's ionization leads to repulsive forces, the adhesive strength and degree of cationic crosslinking should be the result of a subtle balance of electrostatic forces (repulsive and attractive). Rapid ionization, for example, might lead to the adsorption of a thin, strongly bound layer of polymeric material on the tissue surface. However, this layer might not allow a great deal of molecular continuity between the tissue interface and the bulk of the sample (Figure 7). Rapid

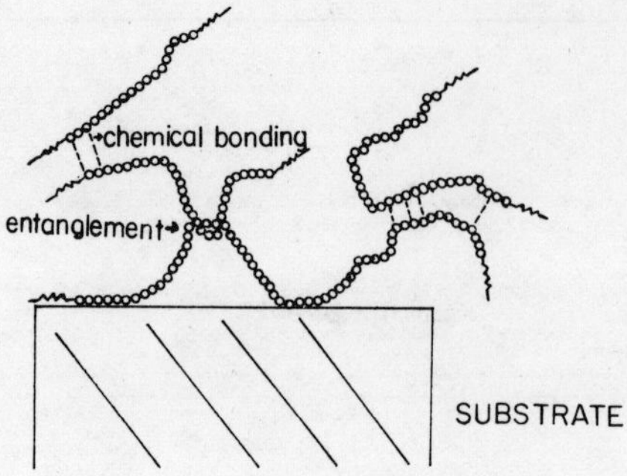
chemical bonding
entanglement
SUBSTRATE

Figure 7. Schematic representation of polymer adsorbed on a substrate and the possible interactions that can preserve molecular continuity from the adsorbed layer to bulk regions of the adhesive material.

ionization rates can also lead to secondary bonding in bulk (among molecules or between side groups and water) to minimize repulsive interactions between ionized groups. This bonded matrix might not be very receptive to crosslinking which requires cations to approach very closely (a few angstroms) the ionized carboxyl group. It is not clear at this point what factors give rise to differences in ionization rate in the three systems studied. These differences could arise from the chemical structure of the polyelectrolyte or the catio-leaching ability of the powders. It is inferred, however, that the amount of elastic energy due to expansion that the polyelectrolyte must store in a given period of time during the acid/base reaction is partly associated with observed rate differences.

CONCLUSIONS

The present work points out the sensitivity of poly(alkenoic acid) adsorption on ionic substrates to molecular variables such as concentration and degree of ionization. For example, increasing degrees of ionization seem to result in reduced amounts of adsorbed polymer. Yet, highly ionized macromolecules appear to adhere more strongly to the substrate's surface. Also, considerable variations in ionization rates of polyelectrolytes through acid/base reactions exist among different dental biomaterials in use at the present time. These variations could result in different degrees of ionic crosslinking which is related to mechanical strength and hydrolytic stability of the set product. Finally, based on model studies on adsorption, further investigations on the role of ionization kinetics on long-term properties are suggested for the purpose of optimizing both adhesion and bulk properties of polyelectrolyte-based biomaterials.

REFERENCES

1. D. C. Smith, Brit. Dent. J., 125, 381 (1968)
2. A. D. Wilson and B. E. Kent, J. Appl. Chem. Biotech., 21, 313 (1971)
3. J. N. Anderson, in "Applied Dental Materials", Blackwell Scientific Publications, Oxford (1976), pp. 218-227
4. H. J. Bixler, L. M. Markley and R. A. Cross, J. Biomed. Res., 2, 145 (1968)
5. A. S. Michaels, et al, US Patent 3,276,598
6. M. J. Lysaght, in "Ionic Polymers", L. Holliday, ed., Applied Science Publishers, Ltd., London (1975), p. 296
7. E. Friedman, Proceedings - Artificial Heart Program Conference, Washington, D. C., June 9-13, 1969
8. S. I. Stupp, J. W. Kauffman and S. H. Carr, J. Biomed. Mater. Res., 11, 237 (1977)
9. A. D. Wilson and S. Crisp, in "Ionic Polymers", L. Holliday, ed., Applied Science Publishers, Ltd., London (1975), Chapter 4
10. B. W. Bertensahw and E. C. Combe, J. Dentistry, 1, 13 (1972)
11. D. C. Smith, J. Can. Dent. Assoc., 37, 22 (1971)
12. D. R. Beech, Brit. Dent. J., 135, 442 (1973)
13. M. G. Buonocore, A. Matsui and A. J. Gwinnett, Archs. Oral Biol., 13, 61 (1968)
14. S. I. Stupp and J. Weertman, J. Dent. Res., 58, Spec. Issue A, 329, Abstract #949 (1979)
15. W. T. Klotzer, L. Tronstad, W. E. Dowden and K. Langeland, Deutsch. Zahnarti. Zeit, 25, 877 (1970)
16. American Dental Association, Guide to Dental Materials and Devices, 6th Ed., pp. 158-167 (1972-1973)
17. D. R. Beech, Arch. Oral Biol., 17, 907 (1972)
18. Newman, Krigbaum, Langier and Flory, J. Polymer Sci., 14, 451 (1954)
19. R. W. Hall, "The Fractionation of High Polymers", in "Techniques of Polymer Characterization", P. W. Allen, ed., Butterworth, London (1959)
20. S. I. Stupp, D. Belton, K. Zak, C. Belting and C. Wilson, J. Dent. Res., 59, Special Issue B, p. 937, Abstract #201,(1980)

EFFECTS OF MICROSTRUCTURE ON COMPRESSIVE FATIGUE OF COMPOSITE RESTORATIVE MATERIALS

R. A. Draughn

Department of Biophysical Dentistry
Medical University of South Carolina
Charleston, SC

INTRODUCTION

Particulate reinforced polymer matrix composites are widely used in dentistry as esthetic restorative materials. Although the normally measured mechanical and physical properties of this class of materials approach the properties of dental amalgam (1) composite restorative materials exhibit limited durability in clinical service. Compared to amalgam restorations, composites undergo loss of material through wear processes and exhibit breakdown at the interfacial region between the restoration and tooth structure.

Long term clinical studies of composite restorations have been reported (2-4). These studies show that when placed in regions subjected to occlusal stress, composites are worn away and require replacement after two to four year of service. Composite restorations generally show little evidence of deterioration for abour one year (5), but rapid wear rates are observed after one year. From the clinical viewpoint, composite occlusal restorations uniformly lose substance over their entire exposed surfaces. There generally are no clinically obvious wear tracks or abrasion marks (6).

A report has been made of a three year study in which occlusal restorations of a commercial material containing quartz particles exhibited excessive loss of anatomic form while experimental materials incorporating strontium or tantalum containing glass filler particles exhibited minimal evidence of wear (7). This study reported that at the end of three years, ninety-one percent of the restorations still present displayed marginal adaptation superior to any amalgam alloy tested in the author's clinic. Substantiated reasons have not been given for the improved durability of the experimental composites.

Statistically valid clinical evaluations of the durability of composite restorative materials have been limited because such studies are expensive, time consuming, and difficult to control. Also, the variation among patients of the factors which affect durability makes interpretation of the results difficult unless there are large differences in the performance of the materials being compared. Thus the need is evident for an understanding of the basic responses of dental composites to conditions of the oral environment.

The dominant processes of wear in the physiological environment have not been established. The oral environment is complex, consisting of fluids with changing composition and pH, cyclic thermal conditions, and cyclic mechanical forces exerted during mastication, tooth brushing and bruxing. It is probable that in such a complex environment, several wear processes are active with the predominant process changing as the conditions of the environment change.

Mechanical fatigue may be a very important consideration in the wear of restorative materials. As two surfaces move relative to one another, cyclic stresses, due to loading and unloading, can initiate and propagate microscopic fatigue cracks in the materials. For long times there may be no macroscopic changes in the material, but when the surface regions have been sufficiently weakened by fatigue, a fragment can be removed from the surface by abrasion or adhesion processes. This forms a flaw in the surface which is susceptible to further damage. Depending upon their nature, the particles produced may cut into adjoining structures and cause abrasive wear. Causative relationships have not been established between fatigue properties and clinical durability of restorations exposed to the cyclic stresses of masticatory function. However, it is apparent that response of restorative materials to cyclic stress can be an important factor in understanding and improving the performance of restorations in the oral environment.

A previous study (8) has reported the compressive fatigue limits of dental composites to be a linear function of compressive strength. In that study, fatigue limit was defined as the stress below which the test specimen would not fail in 5,000 stress cycles. A statistically based test method was used which permitted determination of the mean fatigue limit and its standard deviation. The study showed that when dental composites are subjected to cyclic compressive stresses of magnitude greater than an average of only 64 percent of their compressive strength, failure can be expected to occur in less than 5,000 stress cycles.

The objective of the present study is to measure compressive fatigue properties of commercially available composite restorative materials as a function of stress level and to relate these properties to the microstructures of the materials.

MATERIALS AND METHODS

The seven commercially available composite restorative materials used in this study are listed in Table 1. Photomicrographs of polished crossections of the materials are shown in Figure 1. As is the case with essentially all of the current widely used composite restoratives, the polymer phase in these products consists of approximately 70 percent BIS-GMA and 30 percent low viscosity diluent such as triethylene glycol dimethacrylate (9,10,11). Silane surface treatments are employed to bond the filler particles to the polymer. In this study, filler particle size distribution and volume fraction filler were measured by standard techniques of quantitative metallography. Filler composition was determined by X-ray diffraction. The microstructural data are given in Table 2. The particle size data is presented in terms of the percent particles having a maximum dimension less than the values shown in Table 2. For example, in material A, 99% of the particles are less than 100μm in size, 90% are smaller than 43μm, and 50% are below 11μm.

Table 1

CODE, PRODUCT NAME, AND MANUFACTURER OF COMPOSITE RESTORATIVE MATERIALS

CODE	PRODUCT	MANUFACTURER
A	Adaptic	Johnson & Johnson Dental Products Co. East Windsor, NJ
B	Radiopaque Adaptic	Johnson & Johnson Dental Products Co.
C	Concise	3M Company St. Paul, Minn.
D	Exact	S. S. White Division Penwalt Corp. Philadelphia, PA
G	Prestige	Lee Pharmaceuticals South El Monte, CA
F	Restodent	Lee Pharmaceuticals
E	Simulate	Kerr Manufacturing Co. Division of Sybron Corp. Romulus, Mich.

Cylindrical test specimens, 3.0mm in diameter and 5.7mm. long were prepared by polymerizing the composite resins in a split stainless steel mold. Fifteen minutes after placement in the mold, the specimen ends were ground flat and parallel with 240 grit silicon carbide abrasive paper. Within one-half hour after mixing, the specimens were removed from the mold, placed in 37°C water, and stored for two weeks prior to testing.

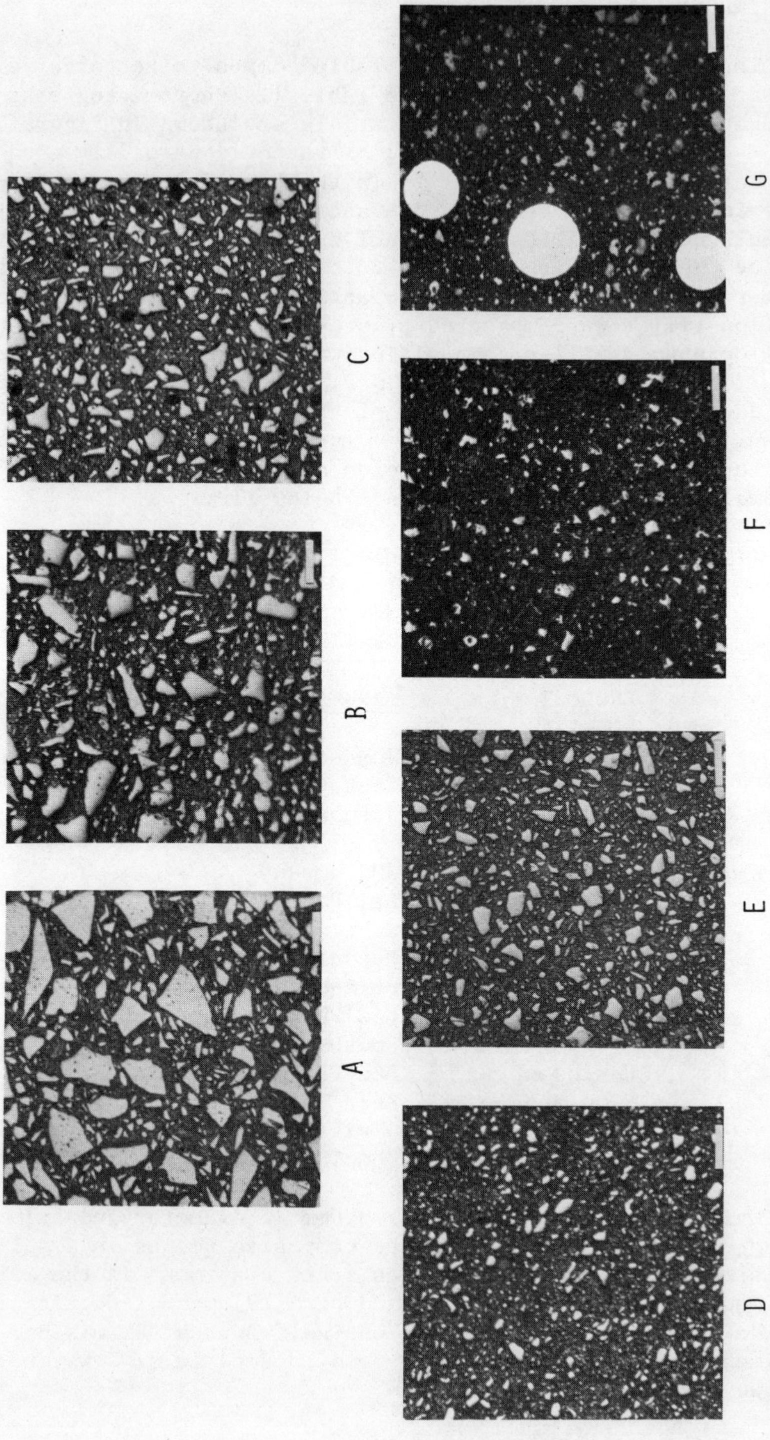
A
B
C
D
E
F
G

Figure 1. Microstructures of Dental Composites. Magnification bar is 50μm long.

Table 2

QUANTITATIVE MICROSTRUCTURAL DATA.
PARTICLE SIZES IN MICRONS.

PRODUCT	FILLER COMPOSITION	V_f	Filler Size S_{99}	S_{90}	S_{50}
A	Quartz	0.55	100	43	11
B	Quartz & Glass	0.55	68	40	12
C	Quartz	0.54	50	25	11
D	Quartz	0.54	29	18	7
E	Silicate & Glass	0.54	42	27	12
F	Quartz	0.45	26	17	4
G	Quartz &	0.43	25	14	4
	Glass Spheres	0.12			80

Testing was done with a hydraulically actuated closed loop testing machine (MTS System 810) operated in the load control mode with a programmed triangular load function having a frequency of 2 Hz. During testing a small static load was placed on the specimen and then various cyclic stresses were applied over that small load. The ratio of the minimum stress to the maximum stress in the tests, the R factor, was 1/10. All testing was done with the specimens submerged in 37° water.

The number of cycles for failure as a function of applied stress level was determined by cycling a minimum of seven specimens of each material at a series of stresses. A minimum of fifteen specimens of each material was tested at 180 MPa in order to provide an adequate data sample for statistical comparisons of the different materials.

RESULTS

Figures 2-5 show the cycles to failure as a function of the maximum applied stress. The date points with horizontal lines drawn through them denote the 5,000 cycle fatigue limits as determined in previous work (8). Fig. 2 indicates the effect of filler particle size for materials containing only quartz particles. Material A, with a small percent of large particles has poor fatigue properties compared to materials C and D which have smaller maximum particle sizes. It is also evident that material C and material D, although they have different maximum particle sizes have very similar fatigue response.

The effect of particle composition is shown in Fig. 3. Material B contains both quartz and silicate glass filler. The fatigue proper

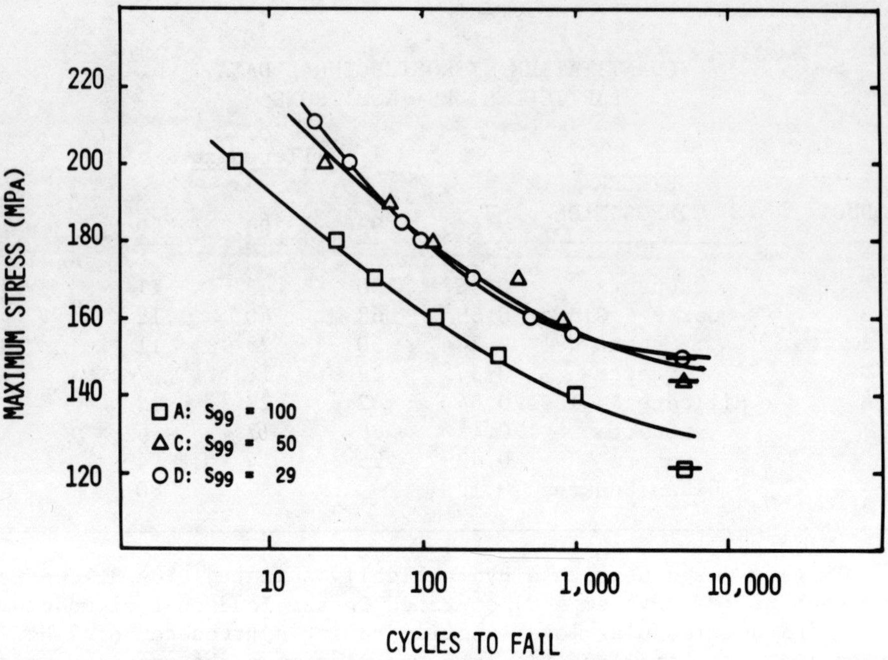
MAXIMUM STRESS (MPa)
220
200
180
160
140
120
10
100
1,000
10,000
CYCLES TO FAIL
□ A: S_{99} = 100
Δ C: S_{99} = 50
○ D: S_{99} = 29

Figure 2. Effect of Quartz Particle Size on Compressive Fatigue.

ties of D which contains only quartz, even though B has a significantly larger maximum particle size. The effect of filler composition is more evident in Fig. 4 which compares a material with quartz filler (D) to a material containing glass and silicate filler (E). Even though the glass/silicate material has a somewhat larger maximum particle size, the fatigue properties are superior to the material containing only quartz particles.

The effect of volume fraction of filler particles is shown in Fig. 5. Materials F and D contain essentially the same particle size distribution, but material F contains only 45 volume percent quartz filler as opposed to the 55 volume percent quartz filler in material D. The 45 percent filler material has superior fatigue properties. Fig. 6 shows the effects of the addition of large (80μm) spherical glass particles to a material containing 45 volume percent of relatively small quartz particles. The presence of the large spheres decreases fatigue properties.

The number of cycles at which failure occurs at a maximum stress of 180 MPa is given in Table 3 along with microstructural information. The standard deviations of the mean cycles to fail are given in parentheses. The cycles to fail data illustrate the high level of

Table 3

CYCLES TO FAIL AT A MAXIMUM CYCLIC COMPRESSIVE STRESS OF 180 MPa. CYCLES JOINED BY VERTICAL BARS ARE NOT SIGNIFICANTLY DEFFERENT. STANDARD DEVIATIONS OF MEAN CYCLES TO FAIL IN PARENTHESES.

	FILLER PROPERTIES				SCHEFFE ANALYSIS	
MATERIAL	V_f	S_{99}	COMPOSITION	CYCLES TO FAIL	ALL DATA	V_f= 0.55
A	0.55	100	Quartz	28 (8)		
D	0.54	29	Quartz	111 (38)		
C	0.54	50	Quartz	134 (50)		
G	0.43	24	Quartz	194 (68)		
	0.12		Glass Spheres			
B	0.55	68	Quartz & Glass	301 (96)		
E	0.55	42	Silicate & Glass	419 (164)		
F	0.45	26	Quartz	1549 (574)		

scatter inherent to fatigue testing. The mean coefficient of variation of the data is 34.8%.

The data at 180 MPa have been statistically analyzed by the Scheffe multiple range test at the 0.05 significance level. The mean cycles to fail which are connected by a single bar in Table 3 are not significantly different according to the statistical analysis. When all the data are considered, only material F, which has the smaller volume fraction of filler, has a significantly different fatigue life.

When the data of material F is omitted and the remaining materials, all of which have essentially 0.55 volume fraction filler, are analyzed an association between particle composition and fatigue life becomes evident. According to this analysis, the materials containing only quartz filler do not have significantly different fatigue lives although particle sizes vary substantially. The materials containing glass and/or silicate fillers have similar fatigue lives which are significantly greater than those of materials containing only quartz filler.

DISCUSSION

The data of Figures 2-6 and Table 3 demonstrate associations between microstructure of dental composites and compressive fatigue properties. The lower fatigue curve of material A in Figure 2, indicates that the presence of a small concentration of comparatively very large particles leads to a decrease in fatigue properties. Such an effect of large particles would be expected if large structural flaws are associated with large, sharp edge particles of the type in material A (12,13). Rigorous statistical analysis of the data of

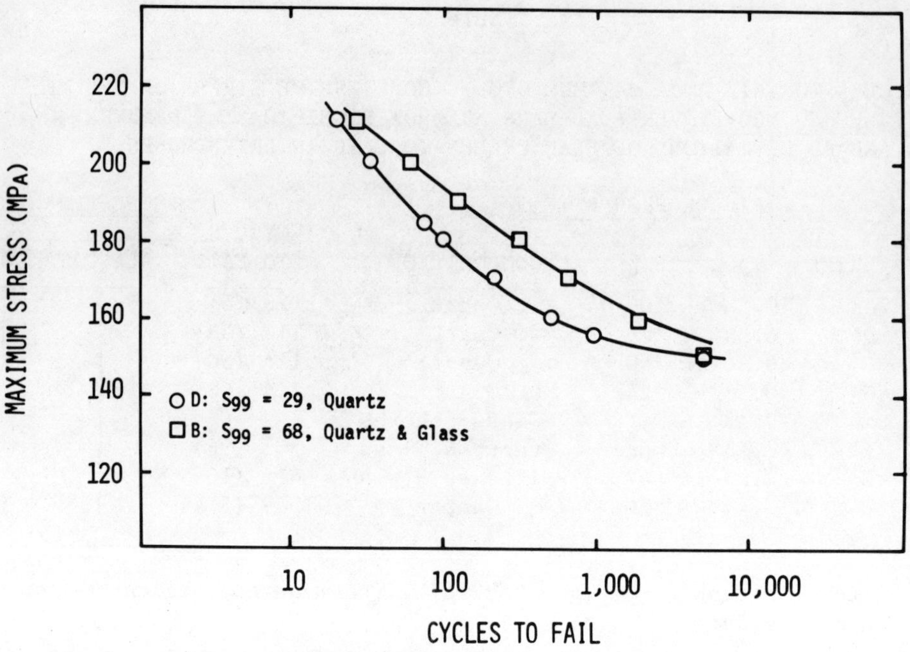
MAXIMUM STRESS (MPa)
220
200
180
160
140
120
O D: S99 = 29, Quartz
□ B: S99 = 68, Quartz & Glass
10
100
1,000
10,000
CYCLES TO FAIL

Figure 3. Effect of Particle Composition on Compressive Fatigue.

Table 3 does not demonstrate a statistically significant difference between material A and materials which do not contain the large particles. It is probable that the high coefficient of variation of the fatigue life data (34.8%) and the relatively small sample size (15) account for the inability to statistically verify the observed effects of very large particles on fatigue life.

Figures 3 and 4 illustrate a dependency of fatigue life on composition of the filler particles. Materials B and E which contain glass and/or silicate particles exhibit longer fatigue lives than materials which contain only quartz particles. This observation is substantiated by the statistical analysis indicated in Table 3. The higher fatigue properties of the glass/silicate filled materials may be due to a lower modulus mismatch between these softer filler particles and the matrix polymer than exists between the hard quartz particles and the polymer. The improved properties might also reflect more efficient and durable bonding between the particles and the polymer.

As shown in Figure 5 and Table 3, there is a statistically significant dependence of fatigue life on volume fraction of filler particles. The material (F) containing 45 volume percent filler has longer fatigue life than any of the materials which contain 55 volume

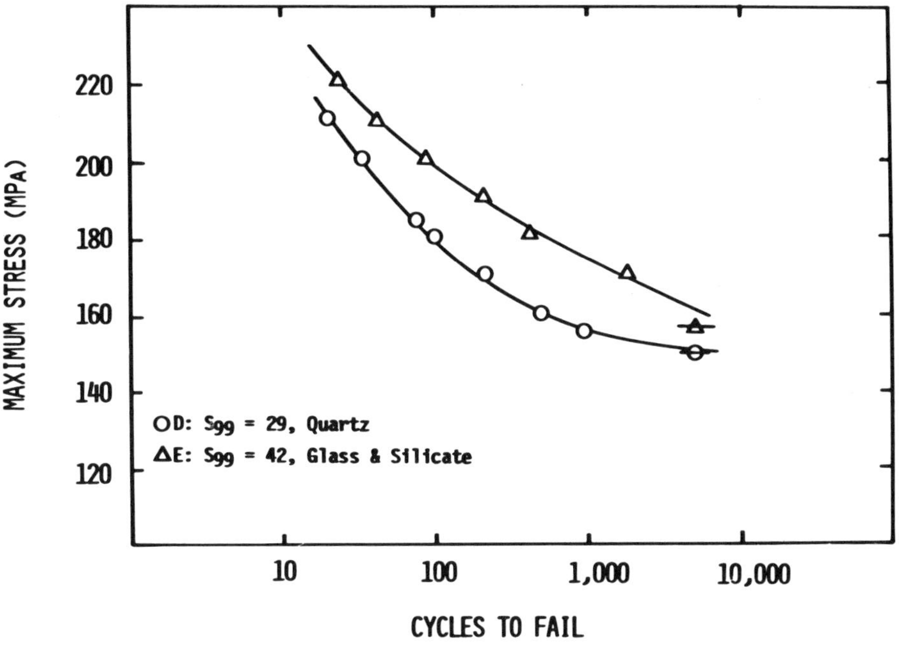
MAXIMUM STRESS (MPa)
220
200
180
160
140
120
OD: S_{99} = 29, Quartz
ΔE: S_{99} = 42, Glass & Silicate
10
100
1,000
10,000
CYCLES TO FAIL

Figure 4. Effect of Particle Composition on Compressive Fatigue.

percent filler. This effect of percent filler particles can be interpreted as evidence that structural flaws which decrease fatigue properties are associated with the particles. Materials with lower volume percent filler have fewer flaws and therefore have better fatigue properties. The lower fatigue life shown in Fig. 6 resulting from inclusion of large spherical particles supports this contention.

SUMMARY AND CONCLUSIONS

The compressive fatigue behavior of commercial composite dental restorative materials has been measured. The seven materials tested have very similar polymer phases but the filler particles differ in size distribution, composition, and volume fraction. Experimental results show a statistically verifiable relationship between volume fraction of filler particles and fatigue life. The material containing 45 volume percent filler particles has statistically longer fatigue life than any of the materials which contain 55 volume percent filler. There is also a statistically verifiable relationship between composition of filler particles and fatigue life. Materials containing glass and silicate particles have significantly longer fatigue lives than materials containing only quartz particles. There appears to be an association between filler particle size and fatigue strength. A small percentage of relatively very large particles yields low

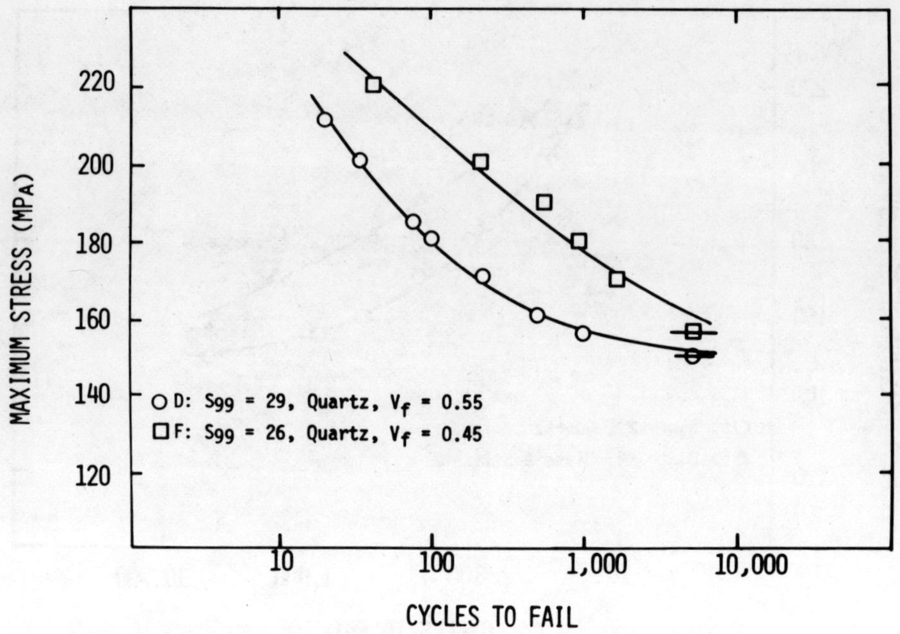
MAXIMUM STRESS (MPa)
220
200
180
160
140
120
10
100
1,000
10,000
CYCLES TO FAIL
○ D: S_{99} = 29, Quartz, V_f = 0.55
□ F: S_{99} = 26, Quartz, V_f = 0.45

Figure 5. Effect of Volume Fraction of Particles on Compressive Fatigue.

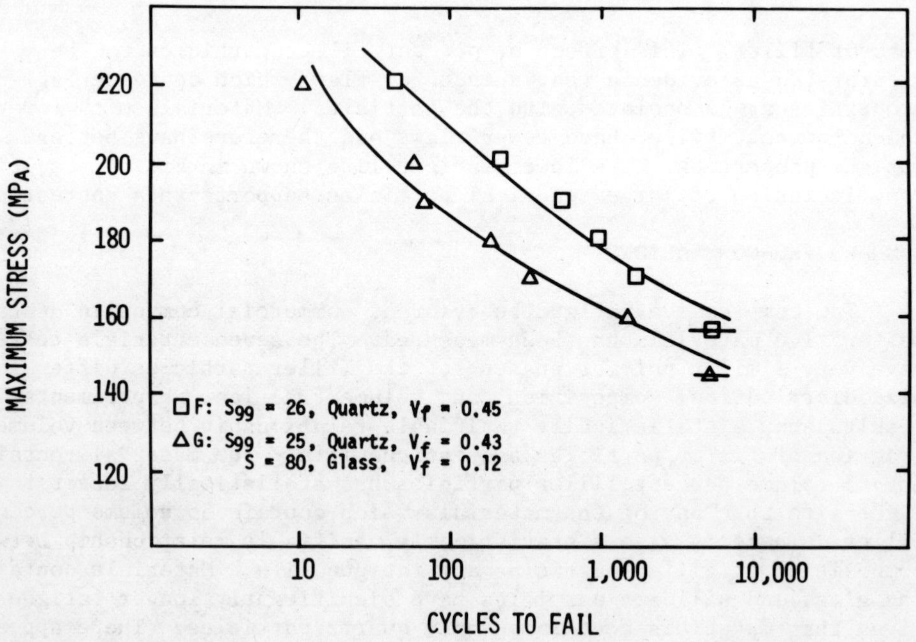
MAXIMUM STRESS (MPa)
220
200
180
160
140
120
10
100
1,000
10,000
CYCLES TO FAIL
□ F: S_{99} = 26, Quartz, V_f = 0.45
△ G: S_{99} = 25, Quartz, V_f = 0.43
S = 80, Glass, V_f = 0.12

Figure 6. Effect of Large Spherical Particles on Compressive Fatigue

fatigue properties.

REFERENCES

1. J. M. Powers, in An Outline of Dental Materials, W. J. O'Brien, and G. Ryge, Editors, W. B. Saunders Co., Philadelphia (1978), 385-413
2. R. W. Phillips, et al., J. Prosth. Dent., 30, 891 (1973)
3. J. P. Moffa, and W. A. Jenkins, Abstract No. 13, J. Dent. Res., Special Issue A, 54, 48 (1975)
4. K. F. Leinfelder, et al., J. Prosth. Dent., 33, 407 (1975)
5. D. B. Nuckles, and W. W. Fingar, J. Amer. Dent. Assoc., 91, 1017 (1975)
6. R. P. Kusy, and K. F. Leinfelder, J. Dent. Res., 56, 544 (1977)
7. J. P. Moffa, and W. A. Jenkins, Abstract No. 206, J. Dent. Res., Special Issue A, 57, 126 (1978)
8. R. A. Draughn, J. Dent. Res., 58, 1093 (1979)
9. G. C. Paffenbarger, and N. W. Rupp, in Kirk-Othmer, Encyclopedia of Chemical Technology, Third Edition, John Wiley and Sons, 7, 550 (1979)
10. B. Causton, in An Outline of Dental Materials, W. J. O'Brien and G. Ryge, Editors, W. B. Saunders Co., Philadelphia, (1978), 92
11. E. Asmussen, Acta Odont. Scand., 33, 129 (1975)
12. L. E. Nielsen, Mechanical Properties of Polymers and Composites, Marcel Dekker, New York, (1974), 411
13. J. Leidner, and R. T. Woodhams, J. Appl. Polym. Sci., 18, 1639 (1974)

WEAR OF DENTAL RESTORATIVE RESINS

J. M. Powers, P. L. Fan, and R. G. Craig

School of Dentistry
University of Michigan
Ann Arbor, MI 48109

INTRODUCTION

The wear of dental composite restorative materials observed clinically in anterior and posterior teeth may be caused by abrasion, sliding, chemical erosion, or some combination of these mechanisms. This paper reviews experimental methodology for the study of wear of dental composite materials in vitro and in vivo.

TWO-BODY ABRASION

Two-body abrasion is a process in which abrasive particles are attached to one (or both) of two rubbing surfaces. Cylindrical specimens (6 mm in diameter and 12 mm in length) were held in a jig under a normal stress of 0.18 MN/m^2. Abrasion was caused by a 600-grit silicon carbide paper (1) or by abrasives such as lithium aluminum silicate glass, quartz, garnet, and alumina (2) attached to the table of a surface grinder. The table moved at a speed of 0.25 cm/sec. Each specimen was abraded for a distance of 10 m with each pass of 25 cm made on a fresh abrasive surface. Wear was determined by measurement of the change in length of the specimen and reported as volume loss per unit length of travel.

Results of two-body abrasion testing on experimental and commercial composites and resins are shown in Table 1 (1,3). These data indicate the effect of silanation, amount of filler, type of filler, and characteristics of the resin on the rate of abrasion. Wear characteristics of various experimental dimethacrylate resins have been studied also as shown in Table 2 (4). Two-body abrasion is a useful method for comparison of differences in formulation of composites.

TABLE 1

RATE OF ABRASION OF EXPERIMENTAL FORMULATIONS AND COMMERCIAL COMPOSITE RESINS AND AN UNFILLED ACRYLIC RESIN (1,3)

Material	Rate of Abrasion, 10^{-4} mm^3/mm of travel		
	Commercial Formulation	Resin with non-silanated filler	Resin without filler
Adaptic (>75 w/o quartz)	3.84 (0.20)*	5.60 (0.20)	18.6 (0.6)
Smile (>75 w/o glass)	7.73 (0.99)	13.8 (1.5)	17.0 (1.1)
Silar (50 w/o 0.2 μm silica)	12.4 (1.0)	--	--
Sevriton (unfilled)	13.3 (1.6)	--	--

*Means of 6 replications with standard deviations in parentheses.

TABLE 2

RATE OF ABRASION OF DIMETHACRYLATE RESINS BY SiC PAPER (4).

Material	Rate of Abrasion, $10^{-4}mm^3$/mm of travel
TEGDMA	22.0 (1.7)*
BisEDMA	17.7 (2.4)
BisGMA + EGDMA (1:1)	15.5 (0.7)
BisEDMA + OFPMA (9:3)	19.1 (1.0)
BisEDMA + OFPMA (3:9)	32.2 (2.8)

*Means of 6 replications with standard deviations in parentheses.

SINGLE-PASS SLIDING

The mechanism of wear during sliding can be studied when a single wear track is examined (3,4,5). The track was formed by causing a diamond hemisphere (360 μm in diameter) to traverse a flat specimen at various normal loads. Parameters such as track width, tangential force and mode of surface failure can be measured.

Wear of dimethacrylate resins used as the matrix phase in dental restorative composites was characterized by single-pass sliding (4).

A 1:1 copolymer of Bisphenol A-bis(2-hydroxy-propyl) methacrylate and ethylene glycol dimethacrylate and a tetraethylene glycol dimethacrylate resin had the lowest values of tangential force and track width in single-pass sliding. Both of these resins showed relatively ductile modes of surface failure (see Table 3) under sliding over the range of normal loads tested. The wear of restorative materials is determined by resistance to penetration as well as mode of deformation during sliding (5).

ACCELERATED AGING

The chemical durability of a composite resin can be tested by accelerated aging in a weathering chamber (6). Disks (36 mm in diameter and 1.3 mm thick were exposed to conditions of accelerated aging for a total of 900 hours in a weathering chamber at 43 C and 90 percent relative humidity. One surface of a sample was subjected continuously to the radiation of a 2500 watt Xenon light source filtered by borosilicate glass and to an intermittent water spray for 18 minutes every two hours. Changes in the exposure surface were studied by surface profile measurements and by SEM.

Conditions of accelerated aging for 900 hours caused erosion of the resin matrices and exposure of filler particles among the products studied. The average surface roughness of four composites increased from 2.25 μm at 0 hours to values between 2.35 and 2.60 μm after aging. This increase in roughness was caused by exposure of filler particles but with little reduction in the initial surface profile. The average surface roughness of four other products decreased from 2.25 μm at 0 hours to values between 1.67 and 2.02 μm after aging. This decrease in average roughness was caused by exposure of filler particles accompanied by a reduction in the surface profile. Accelerated aging holds promise as a model to simulate the erosive wear of composite resins.

CLINICAL STUDIES

The in vivo wear of posterior composite and amalgam restorations can be evaluated clinically in humans using written criteria (7). Twenty-eight paired occlusal restorations were placed in 1st molars and bicuspids using a glass-filled composite (S) and a quartz-filled composite (A). Thirty-four similar restorations were placed using a composite (A) and a dental amalgam (T). Clinical evaluations of margin integrity, anatomic form, and surface texture were made independently by two examiners at 0, 3, 6, 9, 12 and 18 months.

Results for margin integrity are shown in Table 4. Among the 34 composite restorations scored in category B or C of margin integrity, the margin was detected by the explorer in moving from the restoration to adjacent enamel. Among the amalgam restorations, the margin catch was detected in moving from the enamel to the

TABLE 3

RANGES OF NORMAL LOAD OVER WHICH DIFFERENT MODES OF SURFACE FAILURE WERE OBSERVED

Material	Range of normal loads (N)		
	Ductile mode	Brittle mode	Catastrophic mode
TEGDMA	1.0- 6.0	7.0-10.0	--
BisEDMA	1.0- 4.0	5.0- 6.0	7.0-10.0
BisGMA + EGDMA (1:1)	1.0-10.0	--	--
BisEDMA + OFPMA (9:3)	1.0- 3.0	4.0- 7.0	8.0-10.0
BisEDMA + OFPMA (3:9)	1.0- 2.0	3.0	4.0-10.0

TABLE 4

MARGIN INTEGRITY AND SURFACE TEXTURE OF A-S PAIRS AND A-T PAIRS BETWEEN BASELINE (B.L.) AND EIGHTEEN MONTHS

Margin Integrity*

	A						S						A						T					
TIME	A	B	C	D	E	F	A	B	C	D	E	F	A	B	C	D	E	F	A	B	C	D	E	F
B.L.	14						14						17						15	2				
03	8	4					5	7					10	6					6	10				
06	3	7					3	7					9	6					4	9	0	2		
09	3	7					1	9					5	9					3	9	0	2		
12	4	9					2	10	1				3	9	2				0	12	0	2		
18	1	9	2				0	8	4				2	7	4				0	11	0	2		

*For margin integrity A means restorative material is continuous with adjacent tooth structure, B means less than 50% of margin is detectable by sharp explorer, C means more than 50% of margin is detectable by sharp explorer, D means visible evidence of crevice formation into which explorer will penetrate along less than 50% of margin, E means visible evidence of crevice formation into which explorer will penetrate along more than 50% of margin, and F means crevice formation with exposure of underlying dentin.

restoration. The margin integrity rating of B for the composites is indicative of wear, whereas the same rating for the amalgams is indicative of expansion of some areas of the restoration beyond the enamel. The wear of the composites described by the category of margin integrity was not described in most cases by the category of anatomic form. Anatomic form appears to be difficult to evaluate for tooth-colored materials. Interpretation of the margin integrity data for A-S pairs suggests that wear of S may be greater than that of A.

Changes in the category of surface texture were also dramatic. At baseline, a smoother finish was achieved for the amalgams than the composites. Among the A-S pairs, S was always considered smoother than A, even when both A and S were scored as B (rougher than adjacent enamel).

The data indicate that relatively little wear has occurred in vivo in 18 months, but that subtle differences exist among the restorative materials studied.

SUMMARY

An understanding of the wear of composite restorative materials requires in vitro testing that measures chemical as well as mechanical parameters. In vivo clinical testing that uses well-chosen clinical criteria also can be useful in studying wear.

REFERENCES

1. J. M. Powers, L. J. Allen, and R. G. Craig, Am. Dent. A. J., 89, 118 (1974).
2. H. M. Rootare, J. M. Powers, and R. G. Craig, J. Dent. Res., 58, 1097 (1979)
3. P. L. Fan, J. M. Powers, and R. G. Craig, J. Dent. Res., 58, 2116 (1979)
4. J. M. Powers, W. H. Douglas, and R. G. Craig, Wear, 54, 79 (1979)
5. J. M. Powers, J. C. Roberts, and R. G. Craig, J. Dent. Res., 55, 432 (1976)
6. J. M. Powers, and P. L. Fan, J. Dent. Res., 59, 815 (1980)
7. J. B. Dennison, J. M. Powers, and R. G. Craig, J. Dent. Res., 59, (Special Issue A), 318 (1980)

ACKNOWLEDGEMENT

These investigations were supported by Research Grant DE-03416 from the National Institute of Dental Research, National Institutes of Health, Bethesda, MD.

FRICTION AND WEAR OF DENTAL POLYMERIC COMPOSITE RESTORATIVES

K. L. DeVries and Michael Knutson - Department of Mechanical and Industrial Engineering, University of Utah, Salt Lake City, Utah 84112
Robert Draughn - Medical University of South Carolina, Charleston, South Carolina 29403
Jane L. Reichart and Frank F. Koblitz - Dentsply International Inc., York, Pennsylvania 17405

INTRODUCTION AND BACKGROUND

"The objective of restorative dentistry is the reconstruction of missing parts of a tooth or missing teeth. In no other phase of dentistry is it so necessary to have a thorough knowledge of the materials which will be used to restore that missing tooth structure. The study of those reconstructive materials is fascinating when it is placed in its proper perspective--that is, a material that will be used to replace a biological and physiological substance" (1). Gold and amalgam have long been used in making restorations on the human tooth structure. These metals are not perfect replacements for their biological counterparts because they have physiological, mechanical and cosmetic drawbacks.

Extensive research is now being conducted in the field of composite materials. Some dental composites have proven quite successful, and their use is expanding rapidly because of interest from patients and dentists. In vitro laboratory research combined with in vivo clinical studies is providing the needed understanding to design improved dental restorative materials.

Composite materials used in dentistry may be classified for convenience into two groups: those designed for restorative purposes and those designed for use in prostheses. The American Dental Association periodically reports the status (2) and updates the requirements for these materials (3,4).

Prosthetic composites (crown and bridge veneers and denture base resins) (5) are discussed elsewhere in this volume (6,7). It should be noted that these thermosetting polymer particle dispersions are very important biomaterials. An estimated thirty-five million denture wearers in the United States depend on them for effective speech, oral health, mastication, and appearance.

Restorative composites are also very important in dentistry. It is estimated that they are used in more than one hundred and fifty million tooth fillings per year in the United States.

Composite materials may be defined as a combination of materials with distinctive mechanical properties radically different from those of the individual components. According to this definition, materials termed "unfilled composites" (which are polymer bead reinforced), "ionmeric restoratives", etc., would all be considered composite materials. However, the term composite dental restoratives (or dental composites) has come to designate only the class of restoratives comprising oligomeric binders reinforced with inorganic particulates.

Abrasive wear and fracture failure have limited the use of dental composite restoratives (8-21). Mastication forces inducing stresses which approach 2 x 10^8 N/m^2 coupled with exposure to the oral environment can accelerate material failure by either fracture or ablative processes. One must "appreciate that the mouth is ideally designed for the destruction of materials foreign to it" (1). The fracture, wear, and other modes of failure are very complex involving both chemical and mechanical processes (16,17). The combined effects of chemical attack and fracture makes mathematical modeling and analysis of dental composites very difficult. A direct relationship between abrasive wear and compressive strength or hardness has yet to be identified (16,23). Powers et al. have suggested relationships between fracture toughness or fracture resistance with wear (8,17, 24-26). Currently available models of composites are not sufficiently refined to predict the behavior of a wide spectrum of composite materials with much confidence.

Composite materials are designed to take advantage of the combined properties of the individual materials. Often in composites the desired properties are enhanced, and some, in fact, may exceed the qualities of the individual materials. However, there is often a trade off where one property must be sacrificed to enhance another. For example, some tensile strength may be sacrificed to improve abrasion resistance.

The addition of an inorganic filler to a glassy polymer resin reduces curing shrinkage and increases stiffness, compressive strength, and abrasive wear resistance. These desirable effects are often (but not always) accompanied by detrimental effects on tensile strength, fracture toughness, brittleness, etc.

This study has been undertaken to explore the role of the filler-matrix interface with particular emphasis on its effect on friction and wear. The wear of teeth and dental restorative materials has been analyzed by various methods in both the clinic and laboratory (20-42). While this is an ongoing study, the authors felt some of the preliminary results might be interesting to others in the field.

EXPERIMENTAL PROCEDURES AND EQUIPMENT

A device similar to that described in ASTM D2714 was constructed for friction and wear testing in this study. Modifications included replacing the dead weights with a pneumatic loading cylinder and installing load cells and associated electronics to sense the forces.

Composite restorative samples were molded according to American Dental Association Specification Number 27. This configuration was chosen so that samples with the same standard preparatory history could be used for the friction, compressive strength, tensile strength, and fracture toughness tests (17).

The samples were rubbed against a two inch diameter steel disc powered through an auxiliary shaft by a one half horsepower variable speed motor. This is shown schematically in Figure 1. The roughness of the steel disc at the start of each test was kept at approximately a 6 microinch finish. At the completion of each test the steel disc was cleaned with acetone and then polished with 600 grit finishing paper.

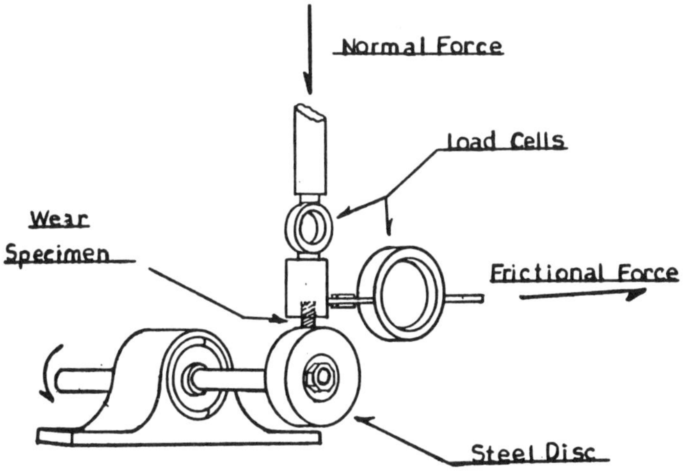
Normal Force
Load Cells
Wear
Specimen
Frictional Force
Steel Disc

Figure 1.

The speed of the disc could be varied from forty-four to three hundred sixty revolutions per minute. Mounted directly above the steel disc was an air cylinder with a two inch diameter bore and four inch stroke. The air cylinder with a regulator was used to apply a constant normal force directly to the polymer sample being tested. The magnitude of the normal force and resulting frictional force was measured by separate load cells. Four active strain gages were attached to each load cell with the output recorded by a strip chart recorder.

The information from the two load cells was used to calculate a coefficient of friction for each sample.

The steel disc was partially submerged in a bath of distilled water. The temperature of this fluid was kept constant by the use of a heat exchanging coil attached to a recirculating pump. Chilled distilled water was used as the cooling medium.

The weight loss due to the effects of wear was determined by weighing the samples prior to and at the completion of each test. The samples were conditioned in room temperature distilled water before each weighing to minimize weighing errors.

In this study, two groups of composite dental materials were tested for wear, coefficient of friction, fracture toughness, and tensile strength. The first group tested, listed in Table I, included eight commerically available materials: six dental composites and two standard control materials, a silver amalgam filling material, and a nylon 6,6. The last two samples were included in the study because of their known history of being resistant to abrasion and wear.

Composite dental restoratives can be classified as conventional (filler particles larger than approximately 0.5 micrometers) and

Table I

COMMERCIAL MATERIAL SPECIFICATIONS

NAME	MANUFACTURER
NYLON 6,6	Lehigh Plastics
EASE ALLOY #75 (Dental Amalgam)	Dentsply International
Adaptic	Johnson & Johnson
ISOPAST	Ivoclar
PROFILE	S.S. White
NUVA-FIL (PA)	Dentsply International
VYTOL	Dentsply International
ADAPTIC RADIOPAQUE	Johnson & Johnson

microfine filled (compositions containing colloidal sized silicas alone or in combination with other fillers). In this study, Isopast was selected as a representative microfine dental composite. The other composite dental restoratives tested were all conventional.

Dental composites can be further characterized as being opaque or transparent to the examining X-rays used by dentists. Profile, Nuva-Fil (PA), Vytol, and Adaptic Radiopaque are all opaque to X-rays. Adaptic and Isopast are transparent to dental X-rays.

For the first series of materials tested, the coefficients of friction, weight loss, and work of wear (the amount of frictional work which is required to remove one gram of material from the sample) were measured and/or calculated. Eight model composite restoratives, listed in Table II, were studied during the second series of tests.

Commercial restorative materials have many desirable properties and have received quite wide acceptance by both patients and dentists. Nevertheless, their properties may not be optimum, and enhancement of their durability characteristics warrants further study. The model composites were studied to assist in the understanding of the mechanisms and processes involved in failure of dental composites, especially the role of interfacial adhesion between filler and matrix.

Glass beads or quartz particles were used as fillers. These particles were given different surface treatments to modify interfacial adhesion. A group of reference samples was formulated by using each filler material with only a simple methanol wash. To study filler-matrix adhesion promoters two different silane coupling

Table II

MODEL COMPOSITE SPECIFICATIONS

FILLER	FILLER TREATMENTS
Glass Beads	Z6032 Silane (Vinyl benzylamine functional silane)
Glass Beads	A174 Silane (methacryloxypropyltrimethoxysilane)
Glass Beads	Dow Corning 200 Silicone Fluid (release agent)
Glass Beads	Methanol Wash (control)
Quartz	Z6032 Silane
Quartz	A174 Silane
Quartz	Dow Corning 200 Silicone Fluid
Quartz	Methanol Wash

agents (A-174, a γ-methacryloxypropyltrimethoxysilane manufactured by Union Carbide) and a vinylbenzylamine functional silane (Z 6032, manufactured by Dow Corning) were used. Each silane was used to treat both glass bead and classified quartz fillers.

To study adhesion reducers, a series of samples were prepared in which the filler particles has been treated with a silicone release agent (Dow 200 fluid, 100 cs). The glass beads used in this study were spherical beads ranging from approximately 0.5 to 5 micrometers in diameter. Figure 2 shows a SEM photograph of the glass beads on a piece of filter paper. The quartz particles were selected to have comparable size ranges. However, caution should be exercised in comparing the adhesion properties of the two fillers, since the quartz particles' surface area and character differed from that of the glass beads. The quartz particles are shown on filter paper in the SEM micrograph as seen in Figure 3.

The wear tests were conducted on the apparatus described above. The normal and tangential (drag) forces were monitored during the entire test to determine the average coefficient of friction for the sample being tested. The normal force was maintained at 135 Newtons

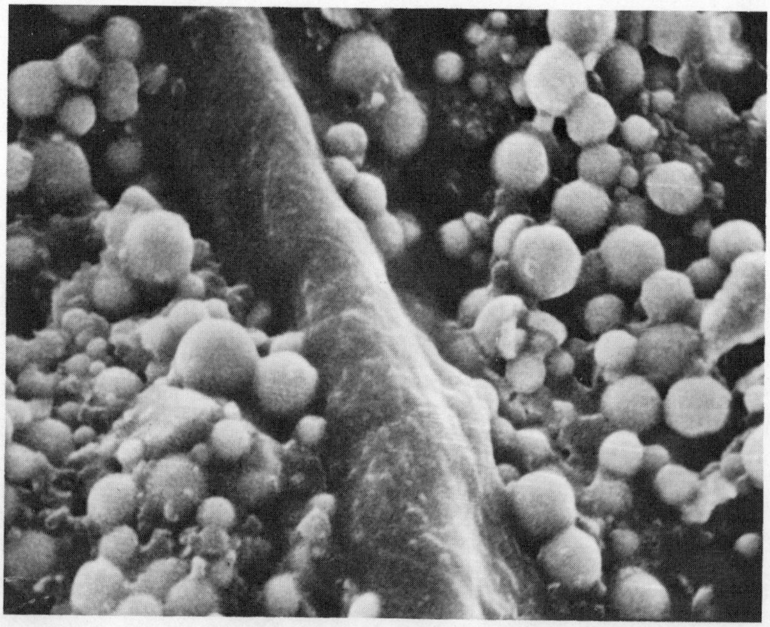

Figure 2.

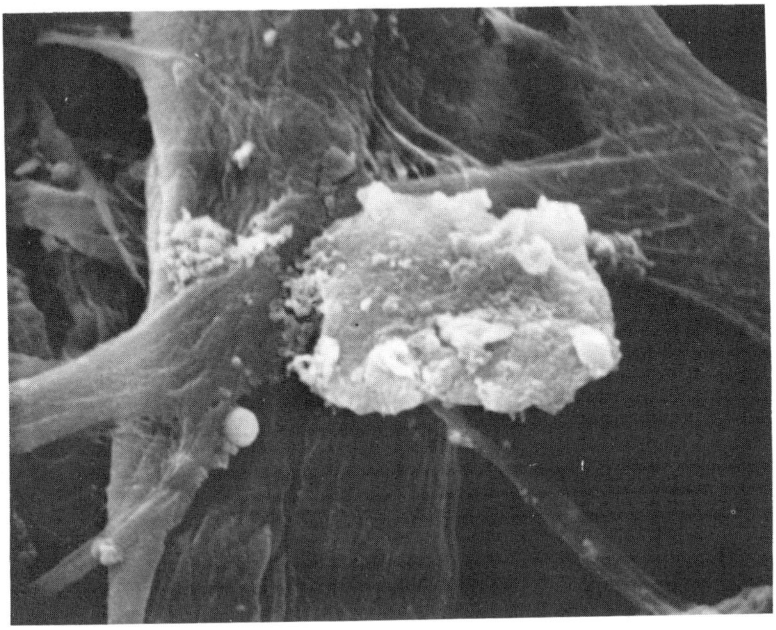

Figure 3.

for an average stress on the 6.4 mm diameter sample of 4.2×10^3 KN/m^2. Weight loss was determined by weighing the sample prior to and at the completion of each wear test.

The diametral tensile and compressive strength tests were performed on the model composites with the aid of an Instron testing machine.

The fracture toughness of the model composites was determined by using the Barker short rod fracture test (29). This technique facilitates accurate determination of K_{IC} values for small diameter specimens.

Subsequent to testing, the wear surfaces and the fracture surfaces of the fracture toughness specimens were examined by Scanning Electron Microscopy, (SEM).

RESULTS AND DISCUSSION

Figures 4 through 8 are SEM photographs of the wear surfaces for the commercial materials. Additional observation of a wear surface was made by transmission electron microscopy (TEM) as shown

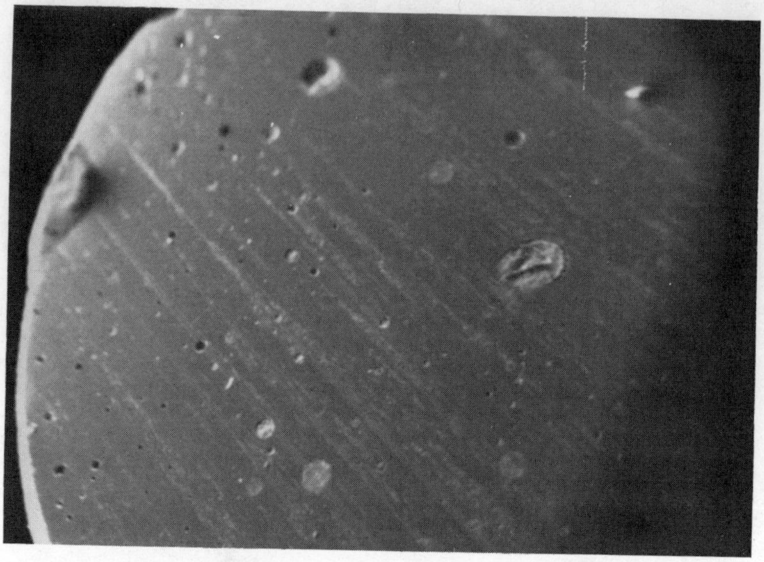

Figure 4. Profile 30x

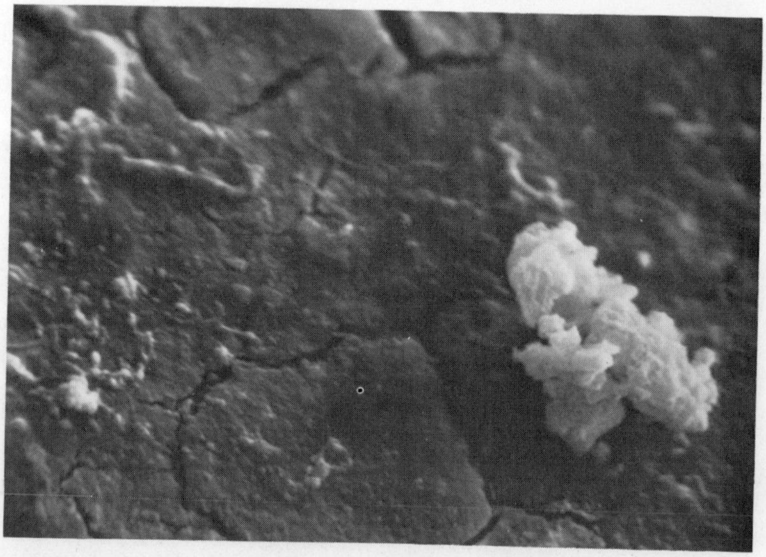

Figure 5. Vytol 2000x

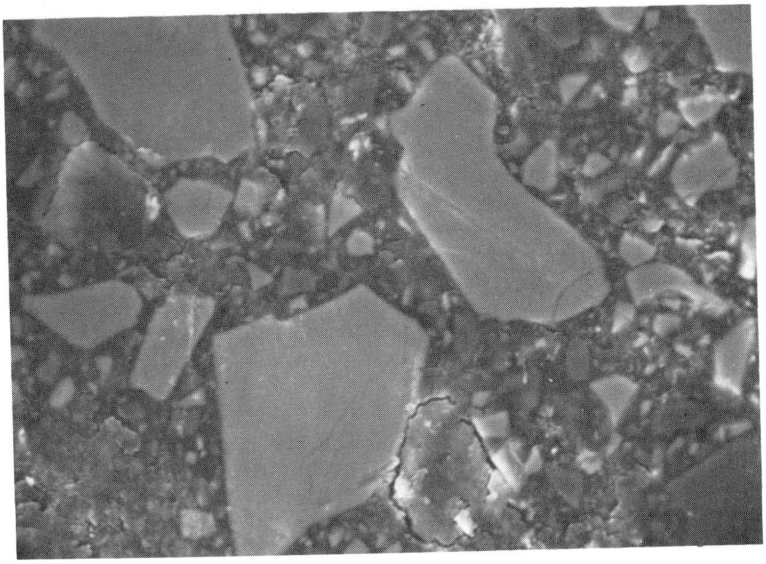

Figure 6. Adaptic Radiopaque 1000x

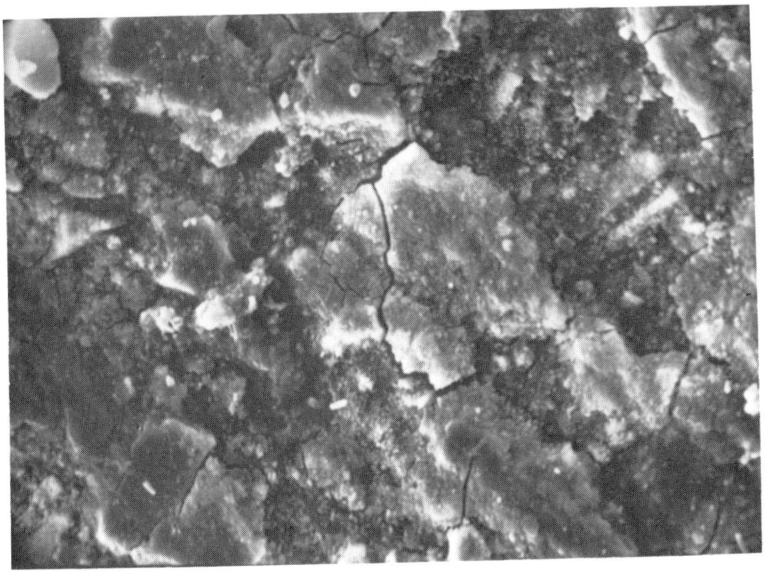

Figure 7. Nuva-Fil 1000x

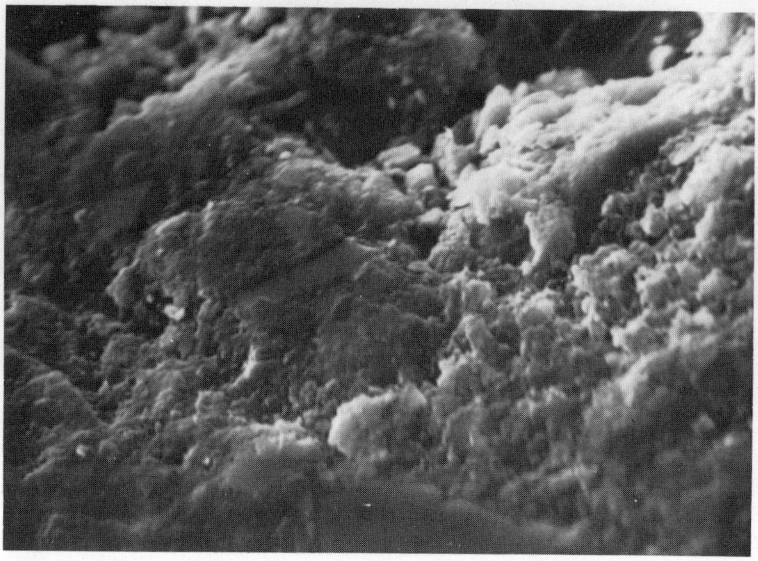

Figure 8. Adaptic 2000x

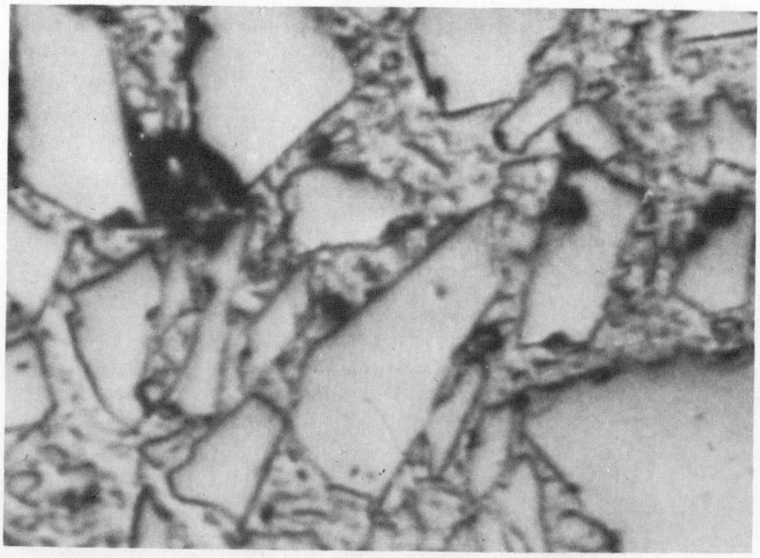

Figure 9. Adaptic 60000x

in Figure 9. This TEM photograph was of a representative commercial restorative, Adaptic, manufactured by Johnson and Johnson. (This TEM was prepared by M. K. Koczak for F. Koblitz.)

The series of photographs of the wear and fracture surfaces of the model composites are shown in Figures 10 through 17.

Table III summarizes the results of the wear experiments on the commercial materials. Table IV summarizes the experimental results for the model composites. The weight loss reported is the gross removal resulting from a wear track of 3000 linear feet. Also shown are the average coefficients of friction, μ; the splitting tensile strength, σ_x; the work of wear, W; and the fracture toughness, K_{IC}.

TABLE III

WEAR PROPERTIES

COMMERCIAL COMPOSITES

MATERIAL	WEIGHT LOSS (GRAMS)	COEFFICIENT OF FRICTION	WORK OF WEAR (N x MTR)
Nylon Lehigh Plastics	0.003	0.3	5.2×10^6
Ease Alloy Dentsply Int.	0.008	0.2	1.5×10^6
Adaptic Johnson & Johnson	0.012	0.6	3.7×10^6
Isopast Ivolclar	0.007	0.3	3.0×10^6
Profile S.S. White	0.004	0.3	3.7×10^6
Nuva-Fil (PA) Dentsply Int.	0.004	0.7	11.0×10^6
Vytol Dentsply Int.	0.007	0.9	7.5×10^6
Adaptic Rdpqe Johnson & Johnson	0.010	0.7	4.5×10^6

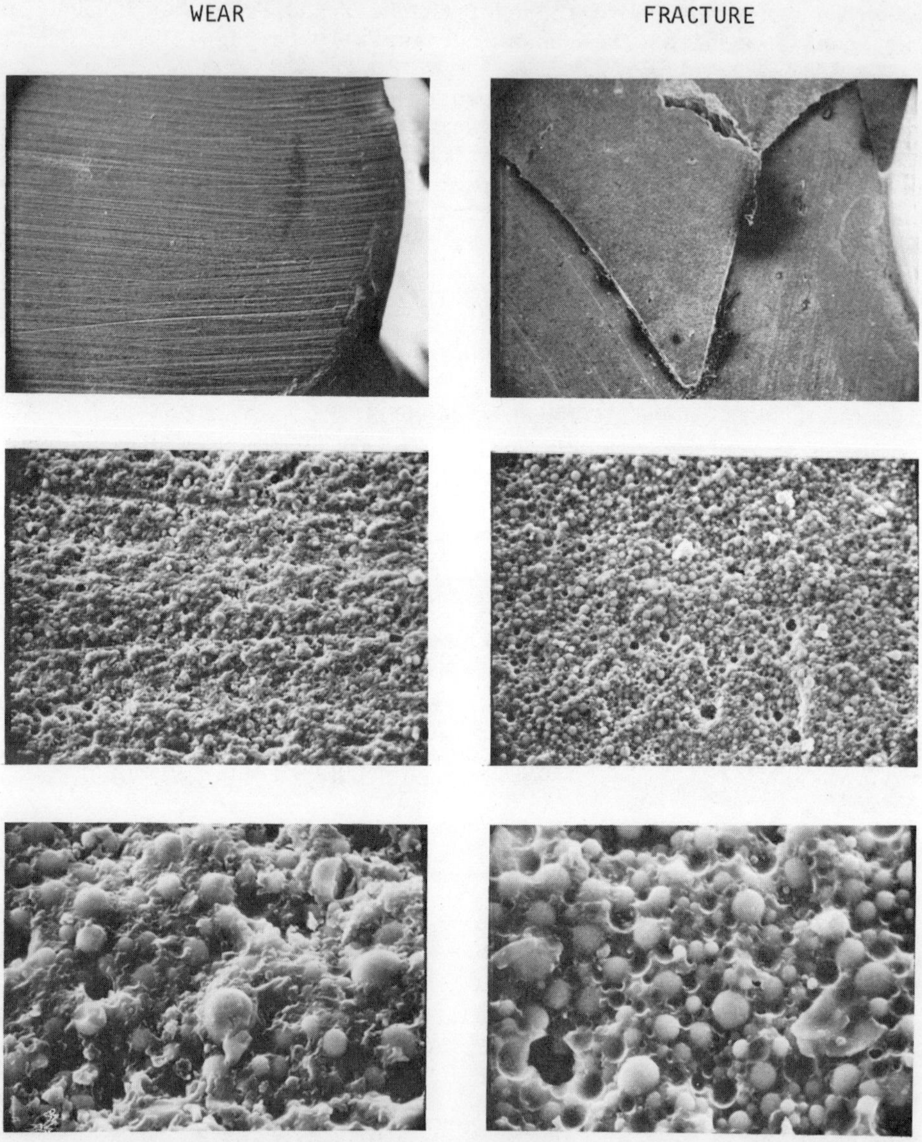
WEAR
FRACTURE

Figure 10. Glass Beads and Z6032 Silane

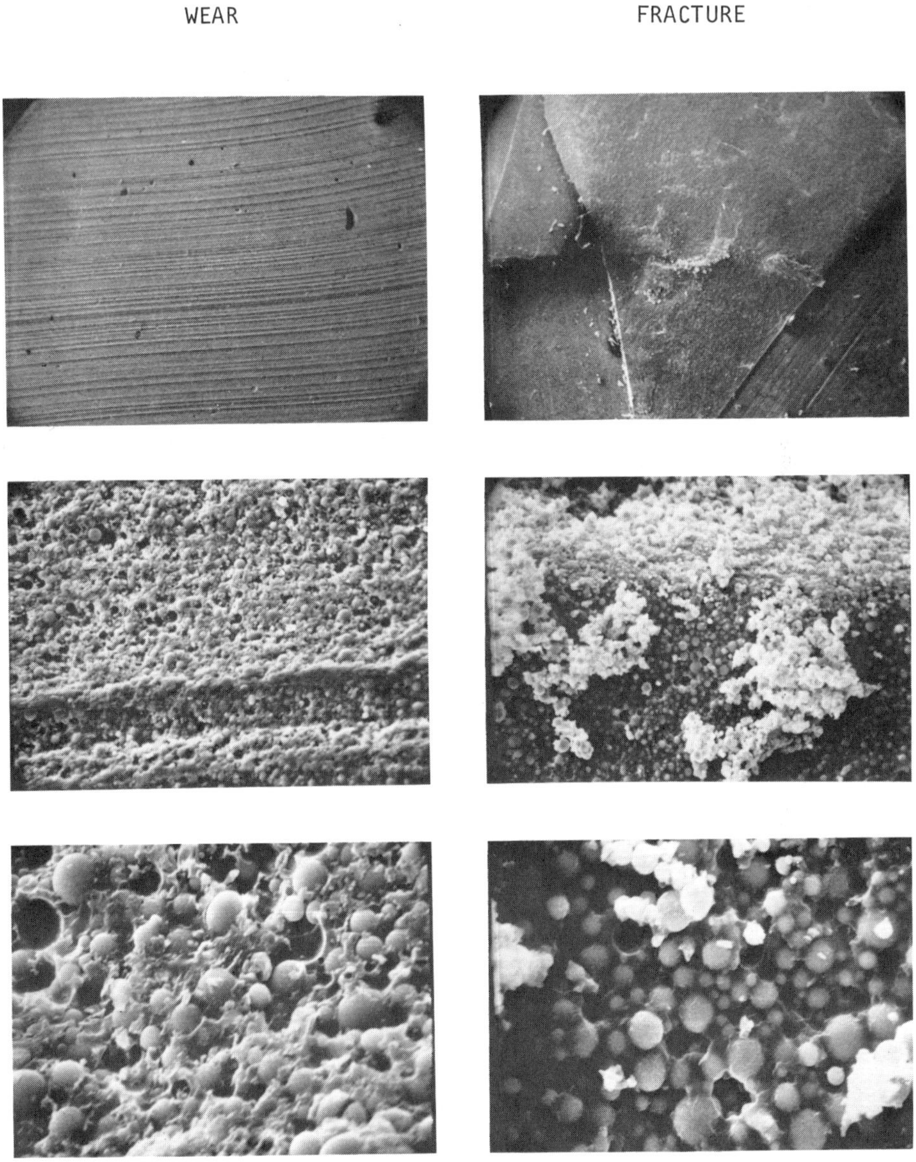
WEAR
FRACTURE

Figure 11. Glass Beads and A174 Silane

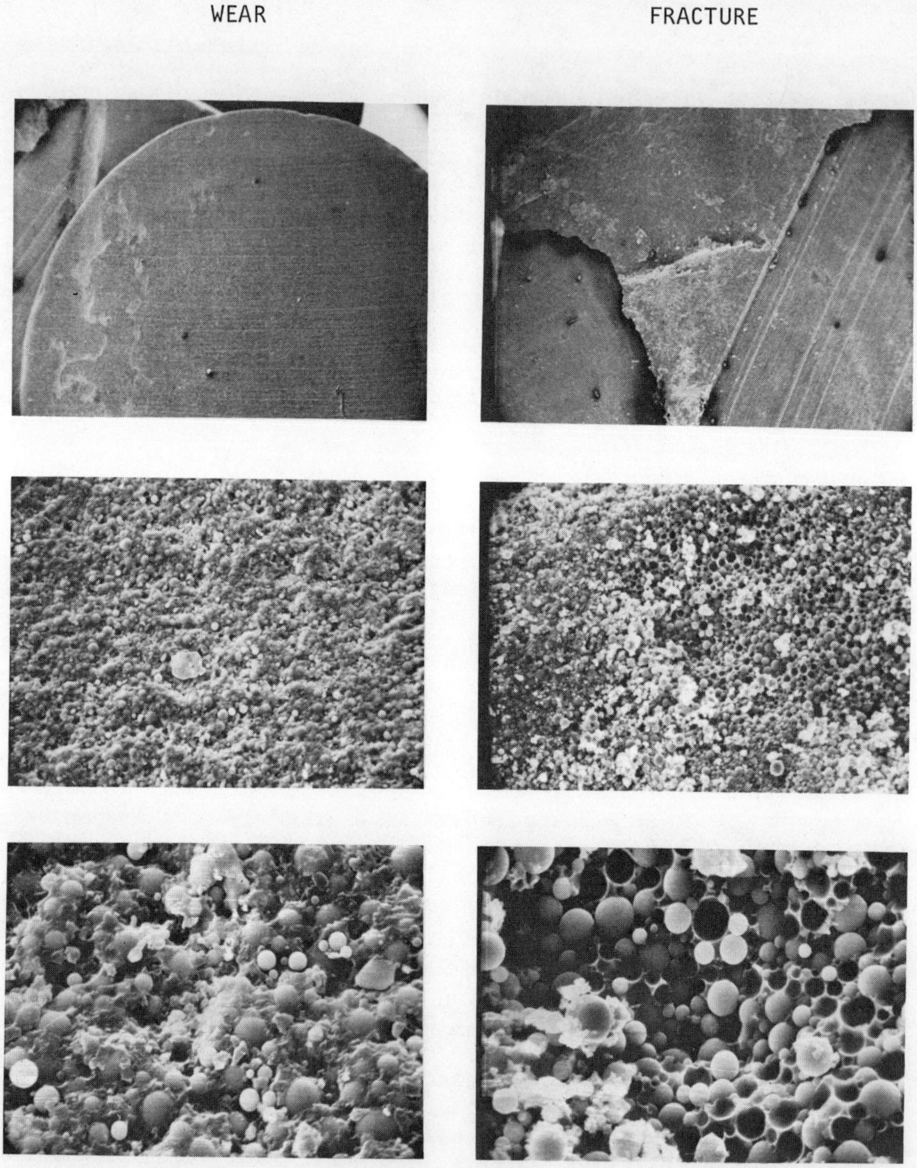
WEAR
FRACTURE

Figure 12. Glass Beads and Dow 200 Silicone

WEAR
FRACTURE

Figure 13. Glass Beads and Methanol Wash

WEAR FRACTURE

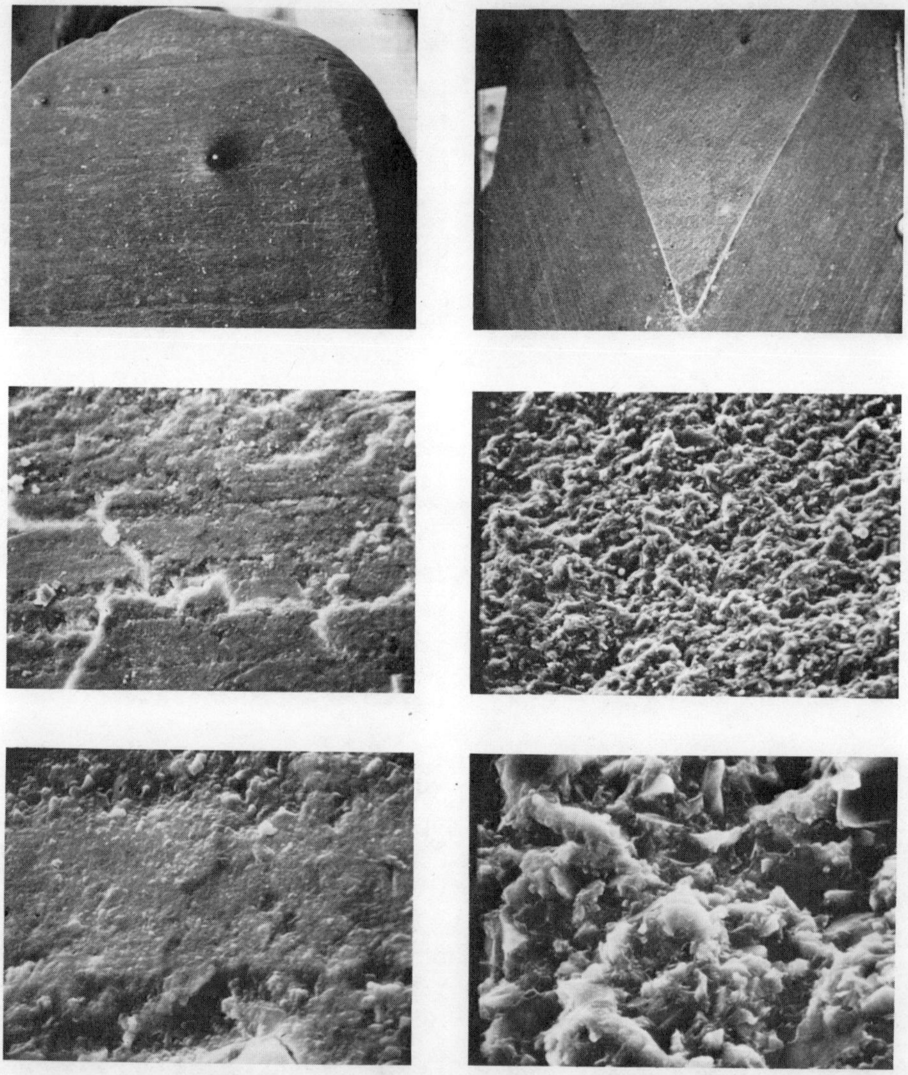

Figure 14. Quartz and Z6032 Silane

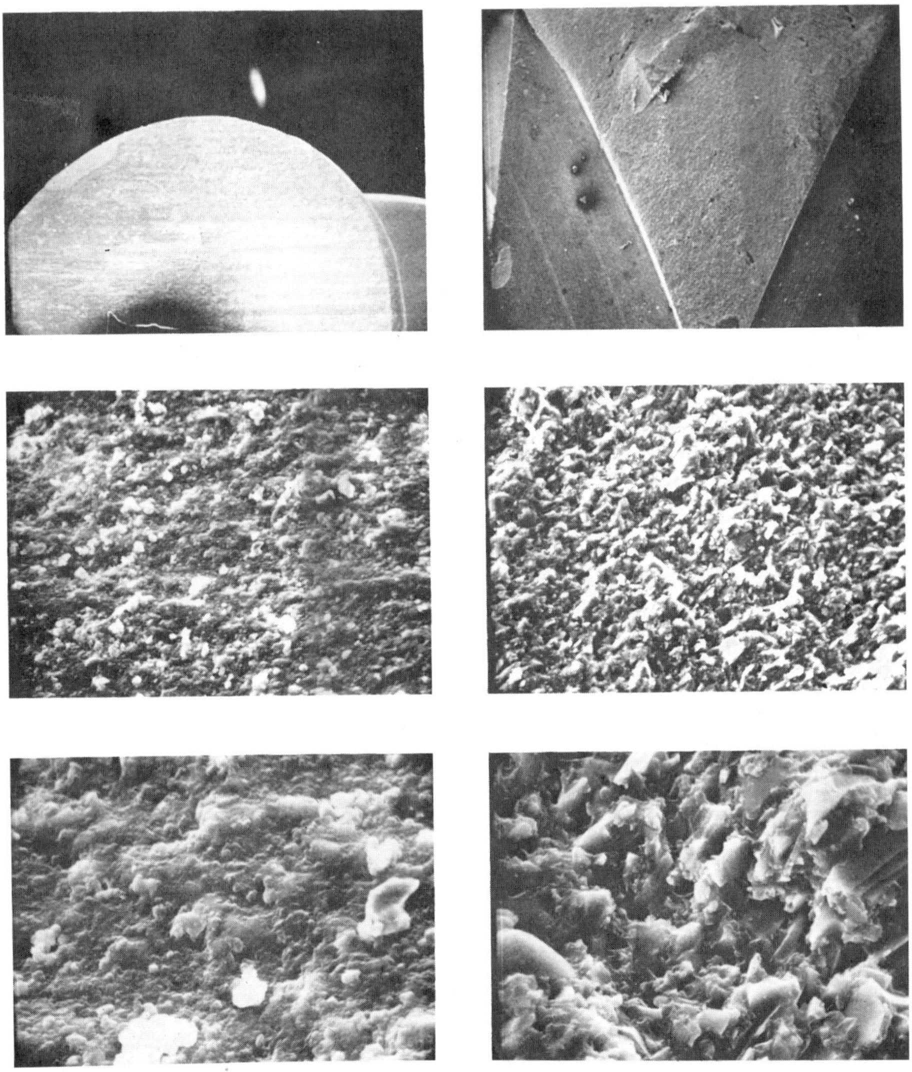
WEAR
FRACTURE

Figure 15. Quartz and A174 Silane

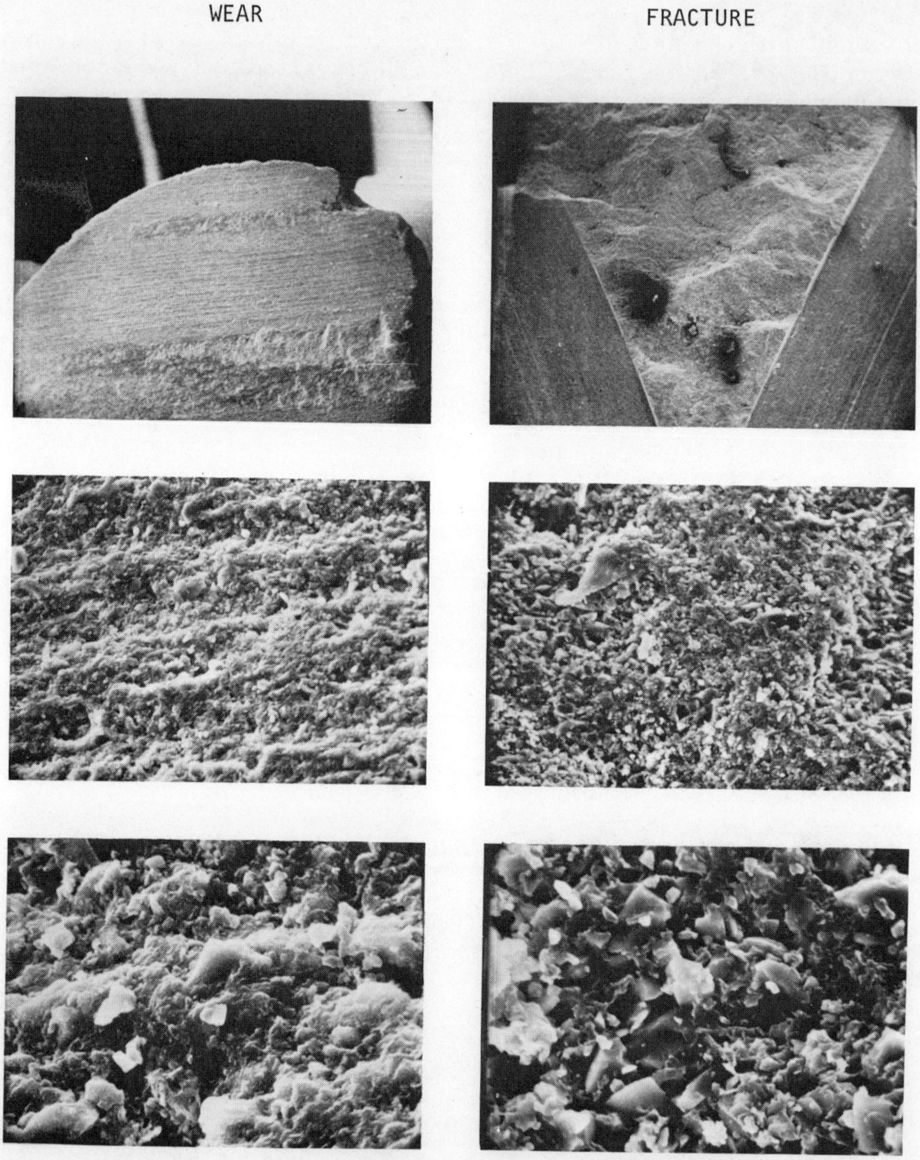
WEAR
FRACTURE

Figure 16. Quartz and Dow 200 Silicone

WEAR
FRACTURE

Figure 17. Quartz and Methanol Wash

TABLE IV

WEAR PROPERTIES

MODEL COMPOSITES

MATERIAL	WEIGHT LOSS (GRAMS)	COEFFICIENT OF FRICTION	WORK OF WEAR (N x MTR)
Glass Beads Z6032 Silane	2.392	0.3	1.1×10^4
Glass Beads A174 Silane	0.199	0.3	1.4×10^5
Glass Beads Release Agent	1.352	0.3	2.0×10^4
Glass Beads Methanol Wash	0.018	0.5	2.3×10^6
Quartz Z6032 Silane	0.014	0.9	4.9×10^6
Quartz A174 Silane	0.003	0.8	2.0×10^6
Quartz Release Agent	0.105	0.7	0.5×10^6
Quartz Methanol Wash	0.015	0.7	3.8×10^6

The range of coefficients of friction (ratio of frictional to normal force) for the commercial materials on this study was 0.2 to 0.9. Nylon, Ease, Isopast, and Profile had relatively low coefficients of friction ranging from 0.2 to 0.4. Adaptic, Adaptic Radiopaque, Vytol, and Nuva-Fil (PA) had coefficients of friction ranging from 0.6 to 0.9.

The SEM micrographs of these later materials revealed comparatively brittle failure in the matrices with evidence of surface fracture and spallation.

Ease and Profile, possibly because of their ease of polishing, required less work to produce a gram of weight loss despite their low coefficients of friction. The frictional work was calculated by multiplying the frictional force times the linear sliding distance.

Nuva-Fil (PA) exhibited low abrasive weight loss and required the most frictional work input to produce the weight loss.

The measured and calculated properties for the model composites are summarized in Tables IV and V. For these composites, the coefficients of friction ranged from 0.3 to 0.9. The quartz filled system exhibited the highest values.

The coefficient of friction did not appear to be highly dependent on the coupling between the matrix and filler.

TABLE V

STRENGTH PROPERTIES

MODEL COMPOSITES

MATERIAL	COMPRESSIVE STRENGTH KN/M^2	TENSILE STRENGTH KN/M^2	FRACTURE TOUGHNESS $KN/M^{3/2}$
Glass Beads Z6032 Silane	140,650	25,400	825.1
Glass Beads A174 Silane	13,340	43,400	1044.9
Glass Beads Release Agent	37,230	1,000	861.6
Glass Beads Methanol Wash	101,350	14,500	691.5
Quartz Z6032 Silane	120,650	41,400	951.7
Quartz A174 Silane	14,350	30,000	1000.5
Quartz Release Agent	28,000	9,000	925.5
Quartz Methanol Wash	113,750	20,000	840.6

The weight loss for these samples varied from about three milligrams to more than one gram. It should be noted, however, that the highest weight loss was for the Z6032 silane treated glass bead filled sample. This sample tended to wear by crumbling and breaking away of relatively large pieces. Similar but much less severe behavior was observed for the A174 treated glass filled sample.

Problems were experienced in the manufacture of the glass filled specimens. Iron and possibly other metallic impurities bound in the glass beads reacted with the matrix resulting in earlier and less well controlled polymerizations than with the quartz filled specimens. Attempts to remove the impurities were unsuccessful. Being cognizant of these observations, although it appears that quartz filled restorative systems were more wear resistant than the glass filled samples, rigorous comparisons and conclusions were not justified.

In the quartz filled systems, increasing the matrix-filler coupling enhanced the wear resistance. Increasing the matrix-filler coupling increased the tensile strength in both glass and quartz filled materials. The fracture toughness, however, was much less dependent on the coupling treatment, although matrix-filler coupling resulted in more brittle materials (reduced fracture toughness). The silicone release agent appeared also to be acting as a plasticizer.

For comparison purposes with the composites, the tensile strength and fracture toughness of PMMA cast rods were typically 65 MN/m^2 and 1 $MN/m^{3/2}$ and maximum values of about 40 $MN/m^{3/2}$ and 1 $MN/m^{3/2}$ for the model composites.

The "conventional composites", Adaptic, Vytol, Nuva-Fil (PA), and Adaptic Radiopaque had relatively high coefficients of friction and exhibited low toughness (brittle failure) in this study.

Resistance to surface destruction by thermal and frictional actions may be among the mechanisms necessary to increase service durability. Frictional heating of the samples during wear tests could have affected the mode of surface failure. We think there is some evidence for this in the micrographs. Softening of the polymer matrices and debonding of fillers from the matrices may be consequences of frictional heating taking place on the wear surface. The temperature of the lubricating medium used in this experiment was kept at a constant temperature of 20°C. There may, however, have been high local heating at the wear interface.

The studies of the model composites indicate that coupling between the matrix and filler plays an important role in the properties of dental restorative materials. This was observed in the SEM micrographs of the wear and fracture surfaces. Those fillers treated with a release agent were rather easily removed from the matrix material during failure. Somewhat tighter bonding was evident in the washed fillers. The silaned samples clearly exhibited the strongest interactions. With these samples there was visible evidence of extensive deformation of the matrix material around the filler particles. This could have been caused by localized softening of the matrices.

The authors believe this constitutes evidence that there may be an optimum amount of coupling between matrix and filler. Too strong filler-matrix interaction may result in brittle, low toughness behavior, while low adhesion between filler and matrix may result in a material which has low tensile strength and poor wear resistance.

To test these postulates, wear studies combined with strength and fracture studies are now in progress. These may give insights into the mechanisms of failure and means of improving service durability of dental composites in the future.

ACKNOWLEDGMENTS

The authors appreciate the assistance of Dr. M. J. Koczak in making the TEM photograph. Major portions of this work were supported by the National Science Foundation Polymer Program (DMR 79-03271-A01). Dentsply International Inc. supported the sample preparation.

REFERENCES

1. M. Shapiro, The Scientific Bases of Dentistry, W. B. Saunders Co., Philadelphia, PA (1966)
2. American Dental Association, 211 East Chicago Ave., Chicago, Ill., 60611, Guide to Dental Materials and Devices, Eighth Ed., (1976)
3. American Dental Association Specification #12 on Denture Base Polymer.
4. American Dental Association Specification #27 on Direct Restorative Materials.
5. F. F. Koblitz, S. D. Steen, and J. F. Glenn, in "Biomedical Applications of Polymers", 97 (H. P. Gregor, Ed., Plenum, 1975)
6. J. F. Glenn, Dental Polymers, (This volume)
7. B. D. Halpern and W. Karo, Polymer Developments in Organic Dental Materials, (This volume)
8. J. M. Powers J. C. Roberts and R. G. Craig, J. Dent. Res., 55, 432 (1976)
9. J. W. Osborne, E. N. Gale and S. W. Ferguson, J. Prosth. Dent., 30, 795 (1973)
10. P. L. Fan, J. M. Powers and R. G. Craig, J. Dent. Res., 58, 2116 (1979)
11. A. Jurecic, J. Dent. Res., 59, Special Issue A, Abstract No. 643 (1980)
12. P. L. Fan and J. M. Powers, J. Dent. Res., 59, Special Issue A, Abstract No. 207 (1980)

13. J. M. Powers and P. L. Fan, J. Dent. Res., 59, 815 (1980)
14. R. W. Phillips, D. R. Avery, R. Mehra, M. L. Swartz and R. J. McCune, J. Prosth. Dent., 30, (1973)
15. K. F. Leinfelder, T. B. Sluder, C. L. Stockwell, W. D. Strickland, and J. T. Wall, J. Prosth. Dent., 33, 407 (1975)
16. R. A. Draughn and A. Harrison, J. Prosth. Dent., 40, 220 (1978)
17. F. F. Koblitz, V. R. Luna, J. F. Glenn, K. L. DeVries and R. A. Draughn, Polymer Eng. and Sci., 19, 607 (1979)
18. S. J. Burns and J. C. Pollett, Invited Lecture, Macromolecular Secretariat, American Chemical Society National Meeting, (April, 1979)
19. R. P. Kusy and K. F. Leinfelder, J. Dent. Res., 56, 544 (1977)
20. R. G. Craig and J. M. Powers, Internal Dent. J., 26, 121 (1976)
21. A. Koran and J. M. Powers, J. Mich. Dent. Assn., 55, 268 (1973)
22. S. Pearlman, The Cutting Edge, HEW Publ. No. NIH 76-670 (1976)
23. A. Harrison and R. A. Draughn, J. Prosth. Dent., 36, 395 (1976)
24. F. F. Koblitz, S. D. Steen, G. P. Anderson and K. L. DeVries, 54th General Session, Int. Assoc. for Dent. Res. (1976)
25. B. E. Causton, J. Dent. Res., 54, 339 (1975)
26. J. C. Roberts, J. M. Powers and R. G. Craig, 54th General Session, Int. Assoc. for Dent. Res. (1976)
27. A. Koran and J. M. Powers, J. Mich. Dent. Assn., 55, 268 (1973)
28. J. C. Roberts, J. M. Powers, and R. G. Craig, J. Mat. Sci., 13, 965 (1978)
29. G. K. Stookey, J. Dent. Res., 57, 36 (1978)
30. J. J. Hefferren, J. Dent. Res., 55, 563 (1976)
31. L. P. Cancro, K. A. Glavan and J. Bianco, J. Dent. Res., 59, Special Issue A, Abstract No. 217 (1980)
32. E. Assmussen, Acta. Odont. Scand., 33, 129 (1975)
33. J. M. Powers, R. W. Phillips and R. D. Norman, J. Dent. Res., 54, 1183 (1975)
34. J. M. Powers, L. G. Allen and R. G. Craig, J. Am. Dent. Assn., 89, 1118 (1974)
35. H. M. Rootare, J. M. Powers, and R. G. Craig, J. Dent. Res., 58, 1097 (1979)
36. J. M. Powers, W. H. Douglas and R. G. Craig, Wear, 54, 79 (1979)
37. J. W. McLean, J. Prosth. Dent., 42, 154 (1979)
38. K. F. Leinfelder and J. B. Sluder, N. C. Dent. J., 35, (Summer, 1978)
39. A. Harrison and T. T. Lewis, J. Biomed. Mater. Res., 9, 341 (1975)
40. E. W. Tillitson, R. G. Craig and F. A. Peyton, J. Dent. Res., 50, 149 (1971)
41. G. Dickson, 56th General Session, Int. Assn. for Dent. Res. (1978)
42. L. M. Barker, J. Eng. Fract. Mech., 9, 361 (1977)

CONTRIBUTORS

INDEX